Symbiotic nitrogen fixation in plants

Contents

Table des matières

1*

IVième partie. Fixation azotée chez les Légumineuses et le milieu

Vième partie. Symbioses de fixation azotée chez les non-Légumineuses

Содержание

Contenido

Parte IV. La fijación de nitrógeno por las leguminosas, en relación con el medio ambiente

Parte V. Simbiosis fijadoras de nitrógeno en plantas no leguminosas

Gardner, Isobel C.	Biology Dept, University of Strathclyde, Glasgow, Scotland
Gibson, A. H.	Division of Plant Industry, CSIRO, Canberra City, ACT, Australia
Graham, P. H.	Centro Internacional de Agricultura Tropical, Apartado Aéreo 67–13, Cali, Colombia
Ham, G. E.	Dept of Soil Science, University of Minnesota, St Paul, Minnesota 55101, USA
†Hamdi, Y. A.	Agricultural Research Centre, Institute of Soil and Water Research, Orman, Giza, Egypt
Hardy, R. W. F.	Central Research Dept, Experimental Station, E.I. du Pont de Memours and Co, Wilmington, Delaware 19898, USA
Havelka, U. D.	Central Research Dept, Experimental Station, E.I. du Pont de Memours and Co, Wilmington, Delaware 19898, USA
‡Hera, C.	Research Institute for Cereals and Industrial Plants (ICCPT), Fundulea, Bucharest, Rumania
Hille, D.	Laboratory of Microbiology, Agricultural University, Wageningen, The Netherlands
Houwers, A.	Laboratory of Microbiology, Agricultural University, Wageningen, The Netherlands
Islam, R.	Soil Microbiology Dept, Rothamsted Experimental Station, Harpenden, Herts, England
Kowalski, M.	Dept of Microbiology, Institute of Microbiology and Biochemistry, M. Curie-Sklodowska University, 20–033 Lublin, Poland
Lambers, R.	Centro Internacional de Mejoramiento de Maiz y Trigo, Mexico
Lawn, R. J.	Dept of Soil Science, University of Minnesota 55101, USA

Lawrie, Ann C.	Dept of Botany, The University, Glasgow, Scotland
Lie, T. A.	Laboratory of Microbiology, Agricultural University, Wageningen, The Netherlands
Masterson, C. L.	Johnstown Castle Research Centre, The Agricultural Institute, Wexford, Ireland
Modi, V. V.	Dept of Microbiology, MS University of Baroda, Baroda 390 002, India
Murphy, P. M.	Johnstown Castle Research Centre, The Agricultural Institute, Wexford, Ireland
Nutman, P. S.	Soil Microbiology Dept, Rothamsted Experimental Station, Harpenden, Herts., England
O'Gara, F.	Dept of Microbiology, University College, Galway, Ireland
Pate, J. S.	Botany Dept, University of Western Australia, Nedlands, Western Australia
Quispel, A.	Research Group on Biological Nitrogen Fixation, Botanical Laboratory, State University of Leiden, Leiden, The Netherlands
Raicheva, L.	N. Pouschkarov Institute of Cell Science, 5, Schosse Bankya, Sofia 24, Bulgaria
Rodriguez-Barrueco, C.	Centro de Edafología y Biología Aplicada, CSIC, Salamanca, Spain
Roughley, R. J.	Horticultural Research Station, Narara, New South Wales, Australia
Shanmugam, K. T.	Dept of Chemistry, University of California, San Diego, La Jolla, California 92037, USA
Silvester, W. B.	Botany Dept, University of Auckland, New Zealand
Sistachs, E.	Instituto de Ciencia Animal, Calle 30, No. 768–1, Nuevo Vedado, Havana, Cuba
Sprent, Janet I.	Dept of Biological Sciences, University of Dundee, Scotland

Strijdom, B. W.	Plant Protection Research Institute, Pretoria, South Africa
Subba Rao, N. S.	Division of Microbiology, Indian Agricultural Research Institute, New Delhi 12, India
Tierney, A. B.	Dept of Microbiology, University College, Galway, Ireland
Truchet, G.	Institut de Cytologie et de Biologie Cellulaire, Marseille, France
Valentine, R. C.	Dept of Chemistry, University of California, San Diego, La Jolla, California 92037, USA
Van Dijk	'Weever's Duin' Biological Station, Costvoorne, The Netherlands
Van Hove, C.	Laboratoire de Physiologie vegetale, Université National, Campus de Kinshasa, Zaire
Vojinović, Z. D.	Institute of Soil Science, Beograd, Topčider, Yugoslavia
Wheeler, C. T.	Dept of Botany, The University, Glasgow, Scotland

Present addresses:

*4142, Hiawatha Drive, Madison, Wisconsin 53711, USA

†State organisation of Land and Land Reclamation, Section of Laboratories, Abu-Ghrib, Baghdad, Iraq

‡Academy of Agricultural and Silvicultural Sciences, Fundulea, Jud Ilfov, Rumania

Preface

> '. . . the leaven
> That spreading in this dull and clodded earth
> Gives it a touch ethereal – a new birth.'

The International Biological Programme has initiated six research programmes on particular aspects of symbiotic nitrogen fixation, and has also stimulated much work in related fields which this volume aims to bring together. The themes of the main programmes concern genetical aspects, culture collections, legume inoculants, field assessment, the environment and fixation in non-leguminous symbioses. These are dealt with here in this order except that the World Catalogue of *Rhizobium* collections, compiled by Professor O. N. Allen and Dr E. Hamatová and edited by Dr F. A. Skinner has already been published (by the IBP Central Office, London, 1973). Mention should also be made of the IBP handbook (no. 15, published by Blackwells Scientific Publications Ltd, Oxford, 1971) Prepared by Professor J. M. Vincent. Most of the chapters are based on papers read at the IBP Nitrogen Fixation Synthesis Meeting held at Edinburgh in September 1973.

A large part of the IBP programme on nitrogen fixation has profited directly from important theoretical and technical developments, especially in microbial genetics, which have occurred in the last decade, largely independently of IBP, and these in their turn have interacted with the IBP programmes and also stimulated much further research. With so much and so varied work in progress the theme co-ordinators and editors thought it unrealistic to restrict this volume to reports of IBP programmes; to have done so would have ignored interesting and important new developments.

For these reasons this volume contains a variety of chapters which when taken collectively portray the current state of knowledge relevant to the different IBP themes. Also reported is a range of other work, some entirely new, that was not even a gleam in the eye of the Scientific Director when SCIBP was conceived. Some of the more controversial aspects of recent work were discussed at an open session at the Edinburgh meeting, a transcript of which forms the volume's final chapter.

Although this approach may lack the virtues of balance, and even of uniform excellence, this volume should nevertheless provide for some years to come valuable source material in wide-ranging fields of research and starting points for future work in symbiotic nitrogen fixation, whether by individuals or in other international programmes. It has also underlined and in some respects quantified for the first time the overriding importance of symbiotic nitrogen fixation in natural and agricultural habitats, and its crucial role in bridging the protein gap.

We are all now acutely aware of the need to conserve energy and although this topic is directly referred to only occasionally in the chapters that follow, it underlies all those concerned with the efficiency of the nitrogen fixing processes, especially in agricultural legumes, and is pertinent to the relationship between fixation and fertiliser use.

Rothamsted, Harpenden,
Herts, 1974

P. S. NUTMAN

PART I

Genetical aspects and taxonomy

1. Recent advances in the genetics of nitrogen fixation

R. A. DIXON & F. C. CANNON

Although biochemical studies of nitrogen fixation have become well-established in the past decade, only in recent years has it been possible to apply molecular genetics to studies involving nitrogen fixation. Because of the importance of symbiotic nitrogen fixation in agriculture, many workers have studied the genetics of the microsymbiont *Rhizobium*, but due to the complexity of the root nodule symbiosis, genetic studies of nitrogen fixation have been restricted to the phenotypic level. The recent development of methods of transfer of the nitrogen fixation genes in free-living bacteria and analysis of mutants defective in nitrogenase has facilitated a molecular genetic approach that will no doubt be extended to symbiotic systems.

This account will be primarily concerned with genes which determine synthesis of the enzyme nitrogenase (*nif* genes) and with associated determinants which allow nitrogen fixation to occur *in vivo*; it does not aim to provide a complete synthesis of data concerning the genetics of nitrogen fixation as more comprehensive reviews of this subject have been published elsewhere (Brill, 1974; Schwinghamer, 1974). In symbiotic systems such as legume root nodules, the interaction between *Rhizobium* and host requires both bacterial and plant gene products. The genetics of the host plant have been discussed by Nutman (1969) and will not be considered here.

Isolation of nif^- mutants

This section will relate specifically to mutants deficient in the enzyme nitrogenase. Undoubtedly, many other mutations will result in a Nif^- phenotype; these will be discussed later. Since rhizobia have not been observed to fix nitrogen in the absence of the host plant, screening methods for isolating nif^- mutants in this system must involve plant tests. Thus, to date it has not been possible to determine whether *Rhizobium* strains with altered symbiotic activity have mutations specifically in *nif* genes; indeed the origin of *nif* genes in this system has not yet been determined (see later section). The first *nif* mutants characterised at the biochemical level were of *Azotobacter vinelandii* (Fischer & Brill, 1969;

Sorger & Trofimenkoff, 1970); later, *nif* mutants of *Klebsiella pneumoniae* (Streicher, Gurney & Valentine, 1971; Dixon & Postgate, 1971) and *Clostridium pasteurianum* (Simon & Brill, 1971) were isolated. The isolation technique for such mutants involves mutagenesis, followed generally by a penicillin enrichment technique. The mutants can be selected by their small colony size compared with wild-type on ammonium-free medium.

Most *nif* mutants isolated so far lack functional nitrogenase activity both *in vivo* and *in vitro*. Shah *et al.* (1973) have further characterised a series of *nif* mutants of *A. vinelandii* by examining biochemical cross-reactions using purified nitrogenase components from wild-type, and serological cross-reactions using antisera to these components. Mutants lacking the activity of nitrogenase Component I, Component II or both proteins were detected, although in some of the mutants the inactive proteins could be identified as immunologically cross-reacting material. Electron paramagnetic resonance spectroscopy (EPR) studies were also used to distinguish mutants; those retaining active Component I showed a corresponding EPR signal at $g = 3.65$.

Control of *nif* genes

The inhibition of nitrogenase synthesis by ammonia and by other fixed nitrogen sources has been reported for most nitrogen-fixing organisms. After exhaustion of ammonia, batch cultures of nitrogen-fixing organisms show a lag period prior to nitrogenase synthesis (Goerz & Pengra, 1961; Mahl & Wilson, 1968; Witz, Detroy & Wilson, 1967; Strandberg & Wilson, 1968). In sulphate-limited chemostat cultures of *Azotobacter chroococcum* this lag is absent (Drozd, Tubb & Postgate, 1972) and in *K. pneumoniae* is shortened considerably by the addition of amino acids (Yoch & Pengra, 1966) or by growth in sulphate-limited continuous culture (Tubb & Postgate, 1973). The length of lag in *K. pneumoniae* is thus dependent on nutritional status and the extensive lags apparent with nitrogen-limited cultures may reflect a depletion of the amino acid pool observed under these conditions (Yoch & Pengra, 1966).

Synthesis of nitrogenase under an inert gas phase has been shown by several workers, and the detection of *K. pneumoniae* nitrogenase in a vacuum after exhaustion of limiting ammonia has led Parejko & Wilson (1970) to conclude tentatively that nitrogenase synthesis is a derepression, not an induction phenomenon. Shah, Davis & Brill (1972) showed that both Component I and Component II of nitrogenase in *A. vinelandii* were co-ordinately synthesised during derepression and co-ordinately lost during repression of the enzyme; thus the structural *nif* genes are apparently co-ordinately regulated in this organism. Gordon & Brill (1972) isolated revertants from a *nif*$^-$ mutant of *A. vinelandii* which lacked both Component I and Component II activities; several revertants produced nitrogenase constitutively and were insensitive to ammonia repression. These observations suggest that the structural genes for nitrogenase are controlled by a common regulatory gene.

The locus of ammonium repression has been demonstrated by Tubb & Postgate (1973). Samples taken from fully repressed, sulphate-limited continuous cultures were allowed to derepress and were treated with rifampicin (an inhibitor of initiation of mRNA synthesis), chloramphenicol (an inhibitor of protein synthesis) or ammonia. Under these conditions it was possible to separate translation from transcription during derepression (Fig. 1.1). Chloramphenicol stopped nitrogenase synthesis almost immediately, whereas rifampicin allowed enzyme synthesis to continue for about 40 minutes as a result of translation of preformed mRNA. NH_4^+ gave a similar effect to rifampicin although it actually stimulated translation as well as exerting repression. Since NH_4^+ paralleled the effect of rifampicin, it follows that ammonia represses nitrogenase synthesis at the transcriptional level. By plotting successive increments of nitrogenase activity during linear derepression in the presence of rifampicin, the coding capacity of preformed nitrogenase mRNA was found to be 4.5 minutes in the presence or absence of NH_4^+.

Other fixed nitrogen sources such as nitrate, asparagine and urea (Wilson, Hull & Burris, 1943) and glutamine (Parejko & Wilson, 1970) may inhibit nitrogenase synthesis under certain conditions. These compounds are probably metabolised to NH_4^+ which then effects repression. Sorger (1969) isolated a chlorate-resistant mutant of *A. vinelandii* which lacked nitrate reductase and showed that nitrate did not significantly repress nitrogenase synthesis in this mutant. Tubb & Postgate (1973) observed that L-aspartate repressed the synthesis of nitrogenase in sulphate- or carbon-limited continuous cultures, but not in nitrogen-limited cultures. They proposed that in sulphate- or carbon-limited conditions L-aspartate would be rapidly de-aminated to repressive NH_4^+, whereas in nitrogen-limited conditions accumulation of NH_4^+ would not occur. Gratuitous co-repressors of the *nif* genes were sought by Sorger (1968). He found that methylamine and 2-methylalanine repressed nitrogenase synthesis in *A. vinelandii* and claimed that a mutant resistant to 2-methylalanine was partially constitutive for nitrogenase. However, these results were obtained without the use of the acetylene reduction technique, and were later shown to be an artefact due to non-specific growth inhibition on media containing certain carbon sources. (St John & Brill, 1972; Drozd, Tubb & Postgate, 1972).

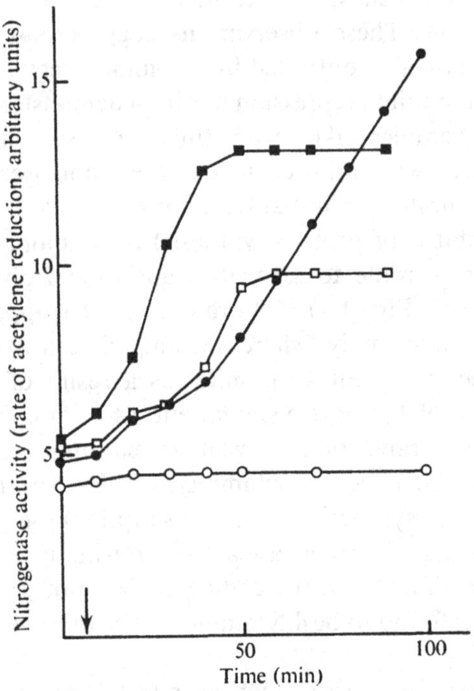

Fig. 1.1. Effect of inhibitors on the derepression of nitrogenase in *Klebsiella pneumoniae* M5a1. Samples (2 ml) from a sulphate-limited chemostat culture grown with a fully repressive concentration of ammonium succinate were transferred to flasks (25 ml) which had been flushed with argon and contained acetylene ($P_{C_2H_2}$ = 0.06 atm) and excess (0.5 mM) sodium sulphate. Under these conditions excess ammonia was exhausted and at time zero the samples were in an early stage of derepression. After 5 min (arrowed) rifampicin (200 μg/ml) □, ammonium succinate (200 μg/ml) ■, or chloramphenicol ○, were added and further derepression was compared with a control, ●, to which no addition was made. (From Tubb & Postgate, 1973.)

Phenotypic expression of the *nif* genes

In all nitrogen-fixing systems and in symbiotic associations in particular, many other genetic determinants, most of which are probably non-specific, are required for expression of the *nif* genes *in vivo*. The general metabolism of nitrogen-fixing organisms must be capable of supplying the additional ATP required and providing a suitable method of donating electrons to nitrogenase. In addition hydrogen ions must be excluded from the nitrogenase and the enzyme must be protected from oxygen damage (see Postgate, 1974).

Free-living bacteria

Although non-nitrogen-fixing mutants have been isolated in several laboratories, none defective specifically in the electron transport pathway to nitrogenase has yet been described. Some of the temperature-sensitive non-nitrogen-fixing mutants of *A. vinelandii* isolated by Benemann, Sheu & Valentine (1971) may have lesions in this pathway, although these mutants have not yet been characterised at the biochemical level. Since nitrogen-fixing organisms have a specific growth requirement for iron and molybdenum, genes involved in the transport of these metals, analogous perhaps to the *ent* (Cox *et al.*, 1970) and *chlD* (Glaser & De Moss, 1971) mutations in *Escherichia coli*, may be important in determining the *Nif*$^{+}$ phenotype, although the effect of such mutations has yet to be ascertained in nitrogen-fixing organisms.

Two pathways of NH_4^+ assimilation to glutamate are now known, one involving the NADP-specific enzyme, glutamate dehydrogenase (GDH), the other glutamine synthetase (GS) and glutamine-2-oxoglutarate amino-transferase (GOGAT or glutamate synthetase). The levels of GDH and GOGAT in bacteria are dependent on the nutritional status of the population, but in general, bacteria grown in NH_4^+-limited conditions show a significant decrease in GDH activity (Meers, Tempest & Brown, 1970). Nagatani, Shimizu & Valentine (1971) have concluded that the GS/GOGAT pathway performs a major role in assimilation during nitrogen fixation. R. S. Tubb (unpublished) has also observed that this pathway predominates in nitrogen-fixing chemostat cultures of *K. pneumoniae*, whether sulphate- or N_2-limited, since GOGAT levels are high in these conditions whereas GDH activities are extremely low. Nagatani *et al.* (1971) have isolated GOGAT deficient mutants (*asm*$^{-}$) of *K. pneumoniae* which cannot utilise N_2 or a variety of compounds as sole nitrogen sources. These mutants, however, synthesise nitrogenase when grown in the presence of aspartate (Streicher, Gurney & Valentine, 1972), a nitrogen source which is presumably assimilated by a different route. Under certain conditions the GS/GOGAT pathway is thus of great importance in determining expression of *nif* genes.

The lag periods observed prior to derepression of nitrogenase in several organisms have been attributed to depletion of the amino acid pool in nitrogen-limited conditions (see previous section). When the *nif* genes from *K. pneumoniae* are derepressed under an argon gas phase, only a limited amount of nitrogenase is synthesised (Parejko & Wilson, 1970), but the final enzyme activity is increased when casamino acids are added prior to derepression (Tubb & Postgate, 1973). The minimal level of nitrogenase required for N_2-dependent growth in *K. pneumoniae* can be synthesised from endogenous nitrogen sources even though protein synthesis in these circumstances is subjected to the general 'step-down' imposed upon metabolism under nitrogen-starved conditions. However, when the *nif* genes were transferred from *K. pneumoniae* to certain prototrophic strains of *E. coli* (see later), the resulting *Nif* hybrids grew poorly on NH_4^+-free medium and failed to synthesise sufficient nitrogenase for growth in the absence of exogenous combined nitrogen. When such hybrids were grown on a medium containing NH_4^+ in the chemostat and were derepressed by changing to NH_4^+-free growth medium, nitrogenase-synthesising clones were selected out of the population (R. S. Tubb, personal communication). The problem of nitrogenase synthesis under nitrogen-starved conditions must thus be borne in mind when genetic crosses involving *nif* are employed.

Root nodule symbiosis

In legume symbiotic systems, the expression of nitrogen fixation appears to be subject to far more genetic variation than in free-living organisms. Changes in symbiotic ability are most common in *Rhizobium* and are often characterised in relation to effectiveness (*Eff*), the ability to express nitrogen fixation in the nodule, and infectiveness (*Inf*), the ability of the bacterium to infect host plants. Resistance to various amino acids, metal ions, phenolic compounds, bacteriophage and antibiotics can result in loss of effectiveness and in some cases infectiveness (for a more detailed account see Schwinghamer, 1974). The antibiotics viomycin and neomycin appear to be very closely associated with loss of effectiveness in *Rhizobium* and provide a method for isolating ineffective strains without the use of mutagens (Schwinghamer, 1964, 1967). There is evidence that viomycin sensitivity is a plasmid-linked character in *Rhizobium trifolii* (Cannon, 1972).

Mutations to auxotrophy often cause loss of symbiotic ability in *Rhizobium*; such mutants are perhaps more advantageous for genetic study. Specific lesions can be distinguished from multiple mutations by selecting prototrophic revertants or recombinants and examining these for restoration of symbiotic ability. Auxotrophic mutations which give rise to ineffectiveness in *Rhizobium* appear to fall into three groups:

(1) Mutations which are apparently not directly associated with symbiotic ability. Restoration of the prototrophic phenotype does not result in restoration of effectiveness; e.g. many auxotrophic mutants (Lorkiewicz *et al.*, 1971), lysine-requiring mutants (lys^-) (Kowalski, 1970*b*). Such mutants are of little value in genetic studies of symbiotic characters.

(2) Mutations which are apparently associated with symbiotic ability at the phenotypic level. Effectiveness is restored both by reversion of the mutation to prototrophy and also by adding the specific growth requirement to the medium in plant nodulation tests; e.g. mutants requiring riboflavin (Schwinghamer, 1969) and leucine (Kowalski & Dénarié, 1972).

(3) Mutations which are closely associated with the symbiosis. Effectiveness is not restored by the addition of the growth requirement to the plant medium but is restored by reversion to prototrophy; e.g. purine and pyrimidine requiring mutants (Dénarié, 1969; Scherrer & Dénarié, 1971).

Some auxotrophic mutations have no effect on symbiotic properties (J. E. Beringer, personal communication) whereas mutations for glycine requirement appear to increase effectiveness (Dénarié, 1969; Scherrer & Dénarié, 1971). There is some evidence that restoration of effectiveness by back-mutation to prototrophy can give rise to strains which are more effective than wild-type (see Schwinghamer, 1969). A model for improving inoculant strains by this method has been devised (Bergersen *et al.*, 1971).

Gene transfer in nitrogen-fixing bacteria

Although there have been many reports of genetic transfer in *Rhizobium*, there have been few systematic studies of characters determining symbiotic ability. In some cases, unequivocal evidence for genetic recombination has not been provided since donor or recipient strains carrying appropriate genetic markers have not been employed.

Both intraspecific and interspecific transformation have been described in *Rhizobium* and the conditions for obtaining competent cultures were first determined by Balassa (1963) and Zelazna (1964*a*, *b*). Competence in *Rhizobium* depends upon nutritional status (Raina & Modi, 1969, 1971;

Zelazna-Kowalska & Lorkiewicz, 1971; Doctor & Modi, Chapter 6) but in general, cultures become competent during the early log phase of growth, although *R. japonicum* gains competence in the late log phase (Mareckova, 1969; Raina & Modi, 1972). The most commonly used markers have been auxotrophy and streptomycin resistance. However, many studies on the transformation of symbiotic characters in *Rhizobium* remain unconvincing without the appropriate use of reliable markers (see Schwinghamer, 1974).

Although lysogeny is widespread in rhizobia, transduction has so far only been demonstrated in *R. meliloti*. Kowalski (1970*a*) screened twenty-one temperate phages and found that twelve were able to transduce streptomycin resistance at a frequency of 10^{-5} to 10^{-7}. A *Lys*$^-$ ineffective mutant was also transduced to prototrophy but effectiveness was not restored (Kowalski, 1970*b*). Recently, Kowalski & Dénarié (1972) have shown that effectiveness is correlated with leucine dependence since *Leu*$^-$ mutants, when transduced to prototrophy, regain effectiveness. Specialised transduction of *cys* has also been demonstrated in *R. meliloti* (Sik & Orosz, 1971).

There have been few reports of conjugation in *Rhizobium*. Higashi (1967) reported the conjugal transfer of infectiveness. Cultures of *R. trifolii* and *R. phaseolii* were mixed and the *R. trifolii* donor was counter-selected with specific phage. The surviving bacteria were capable of infecting clover plants and had physiological and immunological characteristics of *R. phaseolii*, although only one nodule isolate was studied, and it is possible the counter-selection procedure did not completely eliminate the donor. Heumann and his co-workers (Heumann 1968; Heumann, Pühler & Wagner, 1971) have studied conjugation extensively in a strain designated *R. lupinii*, although these workers have not yet reported the symbiotic characteristics of these bacteria. Lorkiewicz *et al.* (1971) crossed several auxotrophic mutants of *R. trifolii* and in some cases obtained evidence for conjugation in this organism. Bose & Venkataraman (1969) reported conjugation in *R. leguminosarum* using drug resistance markers; however, the frequency of recombination was very low and could be accounted for by spontaneous mutation. Beringer (1972*a*) obtained no evidence for conjugation in this species; presumptive 'recombinants' which appeared at low frequency, in fact resulted from cross-feeding on the selective plates. Attempts have been made to introduce various sex factors into *Rhizobium* in order to promote conjugation in the genus. Derepressed I- or F-like factors or the *Pseudomonas* sex factor FP2 are not transferrable by conjugation to most

Rhizobium strains, although P-like plasmids such as the R factor RP4 can be transferred to these bacteria (Datta & Hedges, 1972; Beringer, 1972*b*). RP4 can be transferred in *R. leguminosarum* at a frequency as high as 10^{-1} per donor cell but transfer of the *Rhizobium* chromosome mediated by this R factor has not yet been demonstrated (Beringer, 1972*b*).

Methods of gene transfer have only recently been developed in free-living nitrogen-fixing bacteria but this has greatly facilitated the genetic analysis of the *nif* genes. Streicher, Gurney & Valentine (1971) first reported generalised transduction in *Klebsiella pneumoniae* with phage P1 and by performing a series of two-factor crosses, showed that many *nif* mutations are clustered close to the *his* operon in this organism. The frequency of co-transduction between *his* and most *nif* mutations was between 30 and 80% (Streicher, Gurney & Valentine, 1972). A more detailed account of this work is described elsewhere in this volume (see Shanmugan & Valentine, Chapter 2).

A conjugational system was developed in *Klebsiella* by introducing the derepressed R factor R144*drd*3 into *K. pneumoniae* M5a1 (Dixon & Postgate, 1971). This R factor promoted the transfer of *Klebsiella* chromosomal markers at similar frequencies (approximately 10^{-5} per donor cell) to those obtained with this sex factor in *E. coli* K12. In addition the *nif* genes were transferred from wild-type to various auxotrophic recipients of strain M5a1; an analysis of *Nif*$^{+}$ recombinants indicated that several *nif* genes were linked to *his* in agreement with Streicher *et al.* (1971). In order to obtain a more useful donor strain, *K. pneumoniae* M5a1, carrying R144*drd*3, was irradiated with ultra-violet light and clones which transferred *his* at high frequency were selected. One such clone, designated HF3(R144*drd*3) gave polarised transfer of chromosomal markers from a fixed origin, with the order O, *his*, *nif*... *lys*... *leu*... *trp* and in interrupted mating experiments the kinetics of chromosome transfer were similar to those of an Hfr strain (Dixon, Cannon & Postgate, unpublished data). However, HF3(R144drd3) did not precisely resemble an Hfr strain or an F′ donor, since the R factor remained autonomous and did not carry either *his* or *nif*. Evidently in HF3, the R factor has acquired a specific affinity for the *his* region of the M5a1 chromosome similar to the affinity of the R factor R1 for the *trp* region in *E. coli* K12 (Pearce & Meynell, 1968).

Using HF3(R144*drd*3) as donor, intergeneric transfer of *nif* from *K. pneumoniae* to *E. coli* (an organism which does not naturally fix nitrogen) has been achieved (Dixon & Postgate, 1972). In these matings

E. coli strain C was chosen as a recipient, since this strain is naturally non-restricting and non-modifying with respect to the known host specificity types (see Arber & Lynn, 1969). Since direct selection of *nif* involves problems of phenotypic expression, *his* was the selected marker.

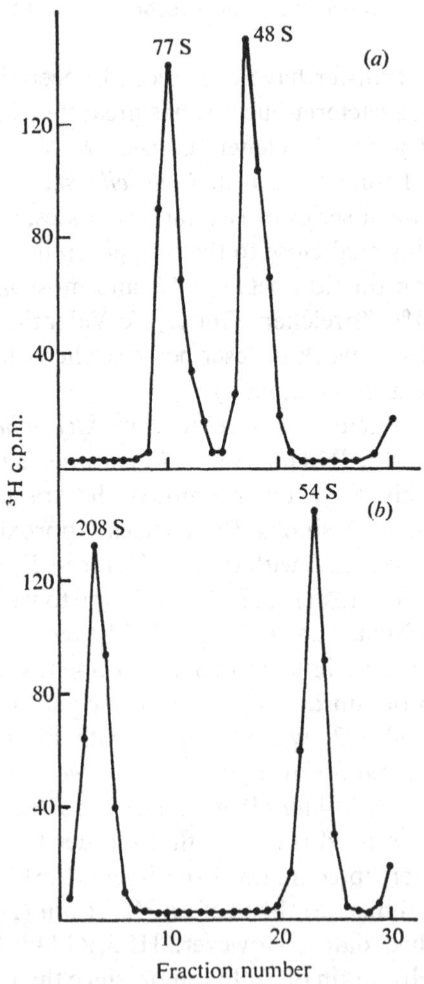

Fig. 1.2. Profiles of ³H-labelled DNA from *E. coli* hybrid C–M7 analysed by velocity centrifugation in (*a*) neutral and (*b*) alkaline sucrose gradients. Supercoiled DNA was purified by dye-buoyant density centrifugation, dialysed overnight and then layered on sucrose gradients with Col E1 DNA as a sedimentation marker. Centrifugation was carried out for (*a*) 45 min and (*b*) 30 min at 120 000 *g* at 20 °C. After centrifugation fractions were collected on Whatman No. 3 filter disks, precipitated with cold trichloroacetic acid and assayed for ³H radioactivity. A similar profile was obtained with *E. coli* C–603 (R144*drd*3). The species of 77 S in neutral and 208 S in alkaline sucrose corresponds to the covalently closed circular (CCC) form of R144*drd*3. The other species present (48 S in neutral and 54 S in alkaline gradients), is the open circular form of this R-factor.

A high proportion of the *E. coli his*$^+$ hybrids obtained were also *Nif*$^+$, confirming the linkage of *his* and *nif* in *K. pneumoniae*. The *nif*$^+$ hybrids were positively identified as *E. coli* both biochemically and serologically. The *nif* genes remained susceptible to NH_4^+ repression in *E. coli* indicating that both the structural and regulatory *nif* genes had been transferred, thus supporting the concept of a *nif* operon. *E. coli Nif*$^+$ hybrids had varied stabilities. One hybrid designated C–M8 produced *Nif*$^-$ segregants at very high frequency and could only be maintained by positive selection. Two other hybrids, C–M9 and C–L4, were more stable, whereas a further hybrid, C–M7, was extremely stable and rarely gave rise to *Nif*$^-$ segregants.

Research in our laboratory has been partly directed towards establishing the physical state of the *Klebsiella* DNA in the *Nif*$^+$ hybrids. Extrachromosomal DNA was extracted from the hybrids using the dye-buoyant density procedure (Bazaral & Helinski, 1968) and was further analysed on sucrose density gradients. In the stable hybrid C–M7, the only supercoiled DNA detected had a molecular weight of 69×10^6 daltons and was identified as R144drd3, the R factor used to promote *nif* transfer (Fig. 1.2). The elimination of R144drd3 by superinfection with a plasmid of the same compatibility group did not result in the loss of the *His*$^+$ *Nif*$^+$ phenotype, indicating that the *Klebsiella* genes were not associated with R144 in hybrid C–M7 (Cannon *et al.*, 1974*a*). The situation was more complex in the less stable hybrids C–M9 and C–L4; these hybrids contain newly derived plasmids in addition to R144drd3. In C–M9, both *his* and *nif* were probably borne on a plasmid of molecular weight 9.5×10^6 daltons since this plasmid was absent in a *His*$^-$ *Nif*$^-$ segregant of the hybrid; another plasmid (molecular weight 118×10^6 daltons) of unknown function was also present in C–M9 (Figs. 1.3 and 1.4). Four plasmids were detected in hybrid C–L4, one identifiable as R144drd3; we have so far been unable to assign positively a function to the other three plasmids (Cannon *et al.*, 1974*b*). Further evidence for the plasmid location of *nif* and *his* in the less stable hybrids has come from mating experiments. In C–M9 and C–L4 these determinants are mobilised by R factors at high frequency and can be readily transferred to other strains of *E. coli*. In contrast, *his* and *nif* are mobilised at very low frequency in comparable matings with C–M7, suggesting that the *Klebsiella* genes are possibly integrated into the chromosome of this hybrid.

The *Nif*$^+$ hybrids described above are resistant to the bacteriophages P1, ϕX174, λvir, and λh80, whereas the original *E. coli* strain used as a recipient is sensitive to these phages; phage resistance is probably due to the absence of receptor sites since C–M7 does not adsorb ϕX174. Since

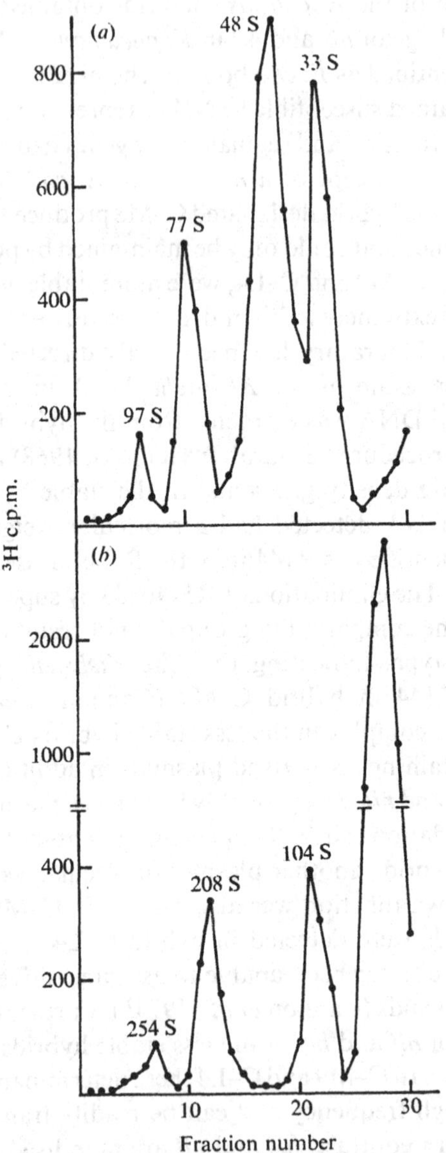

Fig. 1.3. Characterisation of ³H-labelled plasmid DNA from *E. coli Nif*⁺ hybrid C–M9 by velocity centrifugation in (*a*) neutral and (*b*) alkaline sucrose density gradients. The procedure used was identical to that described in the legend to Fig. 1.2. The CCC species of 7 S in neutral and 208 S in alkaline sucrose corresponds to R144*drd*3 (mol. w. 69×10^6 daltons). Two further CCC species are present; a large plasmid of 97 S in neutral and 245 S in alkaline sucrose (mol. wt. 118×10^6 daltons) and a small plasmid of 33 S in neutral and 104 S in alkaline sucrose (mol. wt. 9.5×10^6 daltons).

ϕX174 is a phage specific for rough mutants, it is probable that the *Klebsiella rfb* genes which map close to *his* in *K. pneumoniae* have also been transferred to *E. coli*. S. Primrose (personal communication) has isolated a phage, termed EC1, which is specific for our *E. coli* hybrids. This phage has been extremely useful in selecting segregants of hybrid C–M7; mutants resistant to the phage have deletions in the *rfb–his* segment

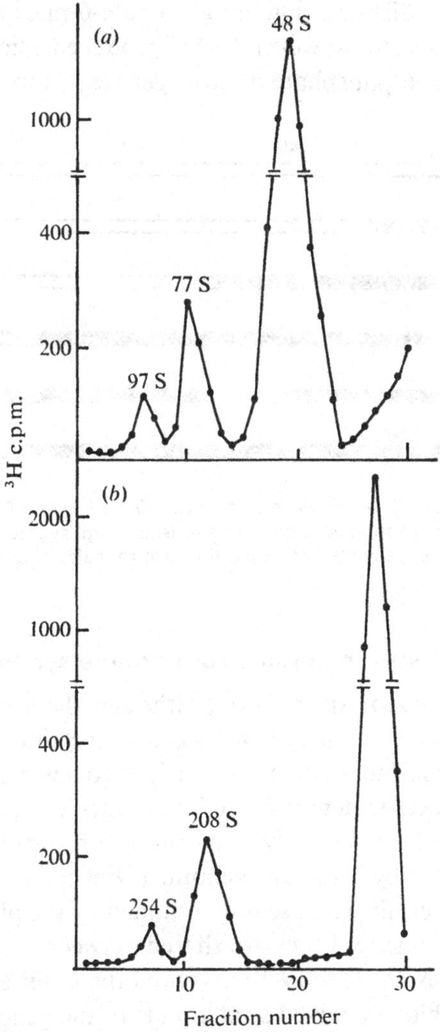

Fig. 1.4. Characterisation of plasmid DNA from *E. coli His⁻ Nif⁻* segregant C–M91 by velocity centrifugation in (*a*) neutral and (*b*) alkaline sucrose density gradients. The procedure used was as described in the legend to Fig. 1.2. Note that the profiles are the same as for C–M9 except that the small plasmid of molecular weight 9.5×10^6 daltons is absent.

derived from *Klebsiella.* A deletion analysis indicates that the gene order in this region is probably *rfb–nif–gnd–his* (Fig. 1.5). Although the original recipient *E. coli* strain possessed gluconate-6-phosphate dehydrogenase (*Gnd*$^+$), we obtained segregants which were *His*$^-$*Gnd*$^-$ *Nif*$^-$*Rfb*$^-$, strongly suggesting that the *Klebsiella nif* region was integrated into the chromosome of hybrid C–M7 (Cannon *et al.*, 1974*a*). Duplication of the *gnd* gene was obvious in hybrid C–M9 since three electrophoretically distinguishable gluconate-6-phosphate dehydrogenases could be detected; however, C–M7 produced a single heteromeric species of gluconate-6-phosphate dehydrogenase (Cannon *et al.*, 1974*b*).

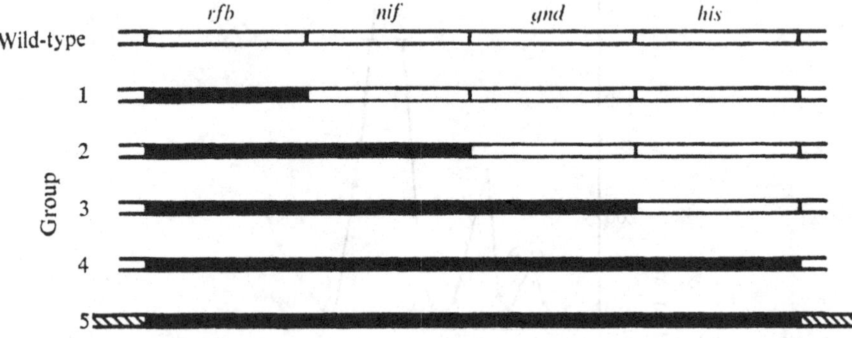

Fig. 1.5. Deletion analysis of the *rfb–his* region in *E. coli Nif*$^+$ hybrid C–M7. The deletions were obtained by selecting spontaneous mutants resistant to phage EC1. The extent of the deletions is shown in black; the markers are not drawn to scale. The hatched regions may or may not be deleted.

The origin of *nif* genes in legume root nodule symbiosis

This is still a subject for speculation, although there have been some recent developments which may help to clarify the situation. The establishment of the symbiotic system obviously involves plant genes and it has been unequivocally demonstrated that the leghaemoglobin apoprotein which is physiologically important for nitrogen fixation in nodules, is coded for by the plant genome (Dilworth, 1969). However, it is not clear whether nitrogenase is determined by the plant or bacterial genes, or whether some *nif* genes are shared between the bacterium and the host. Dilworth & Parker (1969) favoured the latter explanation and proposed that symbiosis evolved so that part of the genetic information for nitrogenase was 'banked' with the plant and that some 'nif' genes were retained by the bacterium (using the assumption that *Rhizobium* was once a free-living nitrogen-fixer; Parker, 1968). After infection of the host plant by *Rhizobium*, the information for nitrogenase components coded by the plant could be transferred to the bacteroids either as nuclear or extrachromosomal DNA, or possibly as plant mRNA. Rake (1972) has found no DNA homology between *Glycine max* and its symbiotic bacterium, *R. japonicum*. According to Rake the technique was only sensitive to within 0.6 % of the *Rhizobium* genome so that a homology of about 800 genes with the host genome could have remained undetected; this would be adequate to code for all the nitrogenase proteins. Rake concluded that although DNA exchange could occur between host and symbiont, large segments of bacterial DNA were not stably integrated into the plant nucleus. In contrast, Quille, Quetier & Huguet (1968) have observed a satellite band of density 1.722 g cm^{-3} in caesium chloride density gradients of DNA extracted from the nuclei of roots and stems of *Glycine hispida*. This density corresponds precisely to the buoyant density of *Rhizobium* DNA (Schildkraut, Marmur & Doty, 1962). It is also conceivable that *Rhizobium* DNA could be maintained in the legume as closed circular DNA, analogous to plant mitochondrial DNA (Avers *et al.*, 1968) or chloroplast DNA (Manning *et al.*, 1971).

It is therefore feasible that genetic exchange occurs between host and bacterium in the legume symbiosis, but we believe that a sharing of *nif* genes between plant and *Rhizobium* is an unlikely possibility. In the free-living organism *K. pneumoniae*, both structural and regulatory *nif* genes are clustered together forming a *nif* operon which is readily transferred as a complete nitrogenase system in intergeneric matings. The division of *nif* genes between *Rhizobium* and its plant host would be a most complex evolutionary achievement; such separated *nif* determinants would require sophisticated methods of regulation in order to obtain co-ordinate derepression of nitrogenase proteins in the bacteroids. Dilworth & Parker (1969) ruled out the possibility that all the *nif* genes could be carried by the plant since such plants would have evolved the ability to fix nitrogen in the absence of symbiosis. Another possibility, that the *nif* genes are present in *Rhizobium* but are completely repressed except in the nodule, was considered unlikely since constitutive *nif* mutants would be expected to occur. Secondly, attempts to obtain nitrogen-fixing *Rhizobium* cultures using controlled cultural conditions (see Bergersen, 1971) or by mutation (Phillips, Howard & Evans, 1973) have been unsuccessful, although in anaerobic conditions with nitrate as terminal electron acceptor, *R. japonicum* produces cytochrome patterns similar to those in bacteroids (Daniel & Appleby, 1972) and factors capable of transferring electrons to bacteroid nitrogenase (Phillips,

Daniel, Appleby & Evans, 1973). Nevertheless, in view of the stringent requirements for the phenotypic expression of *nif* genes (see previous section), we believe that *Rhizobium* could possess genes for nitrogenase synthesis but that expression of *nif* requires interaction with plant-specific gene products. It is perhaps significant that the DNA of bacteroids decreases by 40–70 % during nodule development (Dilworth & Williams, 1967; F. C. Cannon, unpublished data). The decline in DNA content has been attributed to the marked change in growth rate; however, preliminary work in our laboratory indicates that the physical state of the bacteroid DNA differs from that of cultured *Rhizobium* cells.

Phillips, Howard & Evans (1973) have performed the arduous task of comparing the amino acid compositions of Mo–Fe proteins of nitrogenase purified from different legume–*Rhizobium* associations and have come to the tentative conclusion that *Rhizobium* carries the genetic information for the synthesis of the Mo–Fe protein. It is therefore possible that *Rhizobium* carries a complete *nif* system; further genetic experiments could help to resolve this problem.

Conclusions and future prospects

Recent work has established that the *nif* genes are amenable to genetic analysis and manipulation. The *nif* genes in *K. pneumoniae* most probably form an operon, repressed by NH_4^+ at the transcriptional level. The biochemical similarity of nitrogenases purified from several nitrogen-fixing bacteria suggests that the *nif* determinants of diverse organisms may be closely related. However, phenotypic expression of *nif* appears to be very complex and may present a problem in future attempts involving genetic transfer of nitrogen fixation; in aerobic organisms the problem of excluding oxygen from nitrogenase (see Postgate, 1974) may prove a difficult barrier to circumvent in practice.

Evidently a more detailed mapping of the *nif* genes is required and in particular it will be necessary to isolate more regulatory mutants if the control of the *nif* system is to be fully understood. The recent advances in molecular genetics could be fully applied to the *nif* genes; the isolation of a specialised transducing phage carrying *nif* would provide a useful method of purifying *nif* DNA. The demonstration of legume–*Rhizobium* symbiosis in tissue culture (Holsten *et al.*, 1971; Child, La Rue & Haskins, 1973) has been a major development in techniques for studying the legume root nodule symbiosis. Genetic techniques could perhaps be employed to extend the host range of the symbiotic system. The use of protoplast fusion to obtain somatic cell hybrids (Carlson, Smith & Dearing, 1972) could help to achieve this goal. Recently, protoplasts have been isolated from legume root nodules (Davey, Cocking & Bush, 1973); such protoplasts will undoubtedly provide suitable material for future genetic studies.

The importance of biological nitrogen fixation in terms of the world protein supply has been strongly emphasised (Hardy *et al.*, 1971; Nutman, 1971) and no doubt the ultimate goal in genetics studies of nitrogen fixation will concern the transfer of *nif* genes to higher plants. Bacterial DNA can be maintained and transcribed in plant cells (Stroun & Anker, 1971) and some bacterial genes can be expressed in whole plants for several generations (Ledoux, Huart & Jacobs, 1971; Ledoux *et al.*, 1972). Specialised transducing phages carrying the *gal* and *lac* operons from *E. coli* can be used as vectors in the transfer of bacterial genes (Doy, Gressholf & Rolfe, 1973; Johnson, Griersen & Smith, 1973). If the *nif* genes can be attached to such a phage genome, a similar transfection of plant cells could be attempted.

We wish to thank S. B. Primrose and R. S. Tubb for their valuable contributions to this work and J. R. Postgate and F. J. Bergersen for useful discussions.

References

Arber, W. & Lynn, S. (1969). DNA modification and restriction. *Annual Review of Biochemistry*, **38**, 467–500.

Avers, C. J., Billheimer, F. E., Hoffman, H. P. & Pauli, R. M. (1968). Circularity of yeast mitochondrial DNA. *Proceedings of the National Academy of Sciences, USA*, **61**, 91–7.

Balassa, G. (1963). Genetic recombination in *Rhizobium:* a review of the work of R. Balassa. *Bacteriological Reviews*, **27**, 228–41.

Bazaral, M. & Helinski, D. R. (1968). Circular forms of colicinogenic factors E1, E2 and E3 from *Escherichia coli. Journal of Molecular Biology*, **36**, 185–94.

Benemann, J. R., Sheu, C. W. & Valentine, R. C. (1971). Temperature sensitive nitrogen fixation mutants of *Azotobacter vinelandii. Archiv für Mikrobiologie*, **79**, 49–58.

Bergersen, F. J. (1971). Biochemistry of symbiotic nitrogen fixation in legumes. *Annual Review of Plant Physiology*, **22**, 121–140.

Bergersen, F. J., Brockwell, J. Gibson, A. H. & Schwinghamer, E. A. (1971). Studies of natural populations and mutants of *Rhizobium* in the improvement of legume inoculants. In: *Biological nitrogen fixation in natural and agricultural habitats, Plant and Soil* special volume (ed. T. A. Lie & E. G. Mulder, pp. 3–16. M. Nijhoff, the Hague.

Beringer, J. E. (1972*a*). Genetic studies with *Rhizobium leguminosarum*. John Innes Institute sixty-third Annual Report. JII, Norwich.

Beringer, J. E. (1972*b*). R. factor transfer experiments in *Rhizobium leguminosarum*. John Innes Institute sixty-third Annual Report. JII, Norwich.

Bose, P. D. & Venkataraman, G. S. (1969). Recombination in *Rhizobium leguminosarum*. *Experientia*, **25**, 772.

Brill, W. J. (1974). Genetics of nitrogen fixing organisms. In: *The biology of nitrogen fixation* (ed. A. Quispel), pp. 639–60. North-Holland, Amsterdam.

Cannon, F. C. (1972). A study of effectiveness and viomycin resistance in relation to the genome structure of *Rhizobium*. PhD Thesis, University College, Galway, Ireland.

Cannon, F. C., Dixon, R. A., Postgate, J. R. & Primrose, S. B. (1974*a*). Chromosomal integration of *Klebsiella* nitrogen fixation genes in *Escherichia coli*. *Journal of General Microbiology*, **80**, 227–39.

Cannon, F. C., Dixon, R. A., Postgate, J. R. and Primrose, S. B. (1974*b*). Plasmids formed in nitrogen-fixing *Escherichia coli–K. pneumoniae* hybrids. *Journal of General Microbiology*, **80**, 241–51.

Carlson, P., Smith, M. & Dearing, R. D. (1972). Parasexual interspecific plant hybridisation. *Proceedings of the National Academy of Sciences, USA*, **69**, 2292–4.

Child, J. T., La Rue, T. A. & Haskins, P. H. (1973). Facile establishment of nitrogenase in tissue culture. *Plant Physiology*, **51**, 34P.

Cox, G. B., Gibson, F., Luke, R. K. J., Newton, N. A., O'Brien, I. G. & Rosenberg, H. (1970). Mutations affecting iron transport in *Escherichia coli*. *Journal of Bacteriology*, **104**, 219–26.

Daniel, R. M. & Appleby, C. A. (1972). Anaerobic-nitrate, symbiotic and aerobic growth of *Rhizobium japonicum*; effects of cytochromes P-450, other haemoproteins, nitrate and nitrite reductases. *Biochimica et Biophysica Acta*, **275**, 347–54.

Datta, N. & Hedges, R. W. (1972). Host ranges of R factors. *Journal of General Microbiology*, **70**, 453–60.

Davey, M. R., Cocking, E. C. & Bush, E. (1973). Isolation of legume root nodule protoplasts. *Nature, London*, **244**, 460–1.

Dénarié, J. (1969). Une mutation provoquant l'auxotrophie pour l'adenine et la perte du pouvoir fixateur d'azote chéz *Rhizobium meliloti*. *Comptes Rendus de l'Academie des Sciences*, Ser. D, **266**, 836–8.

Dilworth, M. J. (1969). The plant as the genetic determinant of leghaemoglobin production in legume root nodules. *Biochimica et Biophysica Acta*, **184**, 432–41.

Dilworth, M. J. & Parker, C. A. (1969). Development of the nitrogen fixing system in legumes. *Journal of Theoretical Biology*, **25**, 208–18.

Dilworth, M. J. & Williams, D. C. (1967). Nucleic acid changes in bacteroids of *Rhizobium lupini* during nodule development. *Journal of General Microbiology*, **48**, 31–6.

Dixon, R. A. & Postgate, J. R. (1971). Transfer of nitrogen fixation genes by conjugation in *Klebsiella pneumoniae*. *Nature, London*, **234**, 47–8.

Dixon, R. A. & Postgate, J. R. (1972). Genetic transfer of nitrogen fixation from *Klebsiella pneumoniae* to *Escherichia coli*. *Nature, London*, **237**, 102–3.

Doy, C. H., Gresshoff, P. M. & Rolfe, B. G. (1973). Transgenesis of bacterial genes from *Escherichia coli* to cultures of *Lycopersicon esculentum* and haploid *Arabidopsis thaliana* plant cells. In: *Biochemistry of gene expression in higher organisms* (ed. Pollack & Lee). New Zealand Book Co., Sydney. (In press.)

Drozd, J. W., Tubb, R. S. & Postgate, J. R. (1972). A chemostat study of the effect of fixed nitrogen sources on nitrogen fixation, membranes and free amino acids in *Azotobacter chroococcum*. *Journal of General Microbiology*, **73**, 221–32.

Fischer, R. J. & Brill, W. J. (1969). Mutants of *Azotobacter vinelandii* unable to fix nitrogen. *Biochimica et Biophysica Acta*, **184**, 99–105.

Glaser, J. H. & De Moss, J. A. (1971). Phenotypic restoration by molybdate of nitrate reductase activity in *chlD* mutants of *Escherichia coli*. *Journal of Bacteriology*, **108**, 854–60.

Goerz, R. D. & Pengra, R. M. (1961). Physiology of nitrogen fixation by a species of *Achromobacter*. *Journal of Bacteriology*, **81**, 568–72.

Gordon, J. K. & Brill, W. J. (1972). Mutants that produce nitrogenase in the presence of ammonia. *Proceedings of the National Academy of Sciences, USA*, **69**, 3501–3.

Hardy, R. W. F., Burns, R. C., Herbert, R. R., Holsten, R. D. & Jackson, E. K. (1971). Biological nitrogen fixation: a key to world protein. In: *Biological nitrogen fixation in natural and agricultural habitats, Plant and Soil*, special volume (ed. T. A. Lie & E. G. Mulder), pp. 561–90. M. Nijhoff, The Hague.

Heumann, W. (1968). Conjugation in star-forming *Rhizobium lupini*. *Molecular and General Genetics*, **102**, 132–44.

Heumann, W., Pühler, A. & Wagner, E. (1971). The two transfer regions of the *Rhizobium lupini* conjugation. *Molecular and General Genetics*, **113**, 308–15.

Higashi, S. (1967). Transfer of clover infectivity of *Rhizobium trifolii* to *Rhizobium phaseoli* as mediated by an episomic factor. *Journal of General and Applied Microbiology*, **13**, 391–403.

Holsten, R. D., Burns, R. C., Hardy, R. W. F. & Herbert, R. R. (1971). Establishment of symbiosis between *Rhizobium* and plant cells *in vitro*. *Nature, London*, **232**, 173–5.

Johnson, C. B., Grierson, D. & Smith, H. (1973). Expression of λp*lac*5 DNA in cultured cells of a higher plant. *Nature New Biology*, **244**, 105–6.

Kowalski, M. (1970*a*). Transducing phages of *Rhizobium meliloti*. *Acta Microbiologica Polonica*, Ser. A, **2**, 109–14.

Kowalski, M. (1970*b*). Genetic analysis by transduction of *Rhizobium meliloti* mutants with changed symbiotic activity. *Acta Microbiologica Polonica*, Ser. A, **2**, 115–22.

Kowalski, M. & Dénarié, J. (1972). Transduction d'un géne contrôlant l'expression de la fixation de l'azote chez *Rhizobium meliloti*. *Comptes Rendus de l'Academie des Sciences*, Ser. D, **275**, 141–4.

Ledoux, L., Brown, J., Charles, P., Huart, R., Jacobs, M., Remy, J. & Watters, C. (1972). Fate of exogenous DNA in mammals and plants. In: *Workshop on mechanisms and prospects of genetic exchange, Advances in the Biosciences* 8 (ed. G. Raspé), pp. 347–67. Pergamon Press, Oxford.

Ledoux, L., Huart, R. & Jacobs, M. (1971). Fate of exogenous DNA in *Arabidopsis thaliana.* In: *Informative molecules in biological systems* (ed. L. Ledoux), pp. 159–70. North-Holland, Amsterdam.

Lorkiewicz, Z., Zurkowski, W., Kowalczuk, E. & Gorska-Melke, A. (1971). Mutagenesis and conjugation in *Rhizobium trifolii. Acta Microbiologica Polonica*, Ser. A, **3,** 101–7.

Mahl, M. C. & Wilson, P. W. (1968). Nitrogen fixation by cell free extracts of *Klebsiella pneumoniae. Canadian Journal of Microbiology*, **14,** 33–88.

Manning, J. E., Wolstenholme, D. R., Ryan, R. S., Hunter, J. A. & Richards, O. C. (1971). Circular chloroplast DNA from *Euglena gracilis. Proceedings of the National Academy of Sciences, USA*, **68,** 1169–73.

Mareckova, H. (1969). Transformation in *Rhizobium japonicum. Archiv. für Mikrobiologie*, **68,** 113–15.

Meers, J. L., Tempest, D. W. & Brown, C. M. (1970). 'Glutamine (amide): 2 oxoglutarate amino transferase oxido-reductase (NADP)', an enzyme involved in the synthesis of glutamate by some bacteria. *Journal of General Microbiology*, **64,** 187–94.

Nagatani, H., Shimizu, M. & Valentine, R. C. (1971). The mechanism of ammonia assimilation in nitrogen-fixing bacteria. *Archiv für Mikrobiologie*, **79,** 164–75.

Nutman, P. S. (1969). Genetics of symbiosis and nitrogen fixation in legumes. *Proceedings of the Royal Society of London*, Ser. B, **172,** 417–37.

Nutman, P. S. (1971). Perspectives in biological nitrogen fixation. *Science Progress, Oxford*, **59,** 55–74.

Parejko, R. A. & Wilson, P. W. (1970). Regulation of nitrogenase synthesis by *Klebsiella pneumoniae. Canadian Journal of Microbiology*, **16,** 681–5.

Parker, C. A. (1968). On the evolution of symbiosis in legumes. In: *Festskrift til Lauritz Jensen*, pp. 107–16. Gadgaard Nielsens Boktrykkeri, Lemvig, Denmark.

Pearce, L. E. & Meynell, E. (1968). Specific chromosomal affinity of a resistance factor. *Journal of General Microbiology*, **50,** 159–72.

Phillips, D. A., Daniel, R. M., Appleby, C. A. & Evans, H. J. (1973). Isolation from *Rhizobium* of factors which transfer electrons to soybean nitrogenase. Plant Physiology, **51,** 136.

Phillips, D. A., Howard, R. L. & Evans, H. J. (1973). Studies on the genetic control of a nitrogenase component in leguminous root nodules. *Physiologica Planatarum*, **28,** 248–53.

Postgate, J. R. (1974). Prerequisites of biological nitrogen fixation in free-living organisms. In: *The biology of nitrogen fixation* (ed. A. Quispel), pp. 663–86. North Holland, Amsterdam.

Quille, E., Quetier, F. & Huguet, T. (1968). Etudes des acides désoxyribonucléiques des vegetaux. Formation d'un ADN nucleaire rich en G + C lors de la blessure de certaines plantes superieures. *Comptes Rendus de l'Academie des Sciences*, Ser. D, **266,** 836–8.

Rake, A. V. (1972). Lack of DNA homology between the legume *Glycine max* and its symbiotic *Rhizobium* bacteria. *Genetics*, **71,** 19–24.

Raina, J. L. & Modi, V. V. (1969). Genetic transformation in *Rhizobium. Journal of General Microbiology*, **57,** 125–30.

Raina, J. L. & Modi, V. V. (1971). Further studies on genetic transformation in *Rhizobium. Journal of General Microbiology*, **65,** 161–5.

Raina, J. L. & Modi, V. V. (1972). Deoxyribonucleate binding and transformation in *Rhizobium japonicum. Journal of Bacteriology.* **111,** 356–60.

Scherrer, A. & Dénarié, J. (1971). Symbiotic properties of some auxotrophic mutants of *Rhizobium meliloti* and of their prototrophic revertants. In: *Biological nitrogen fixation in natural and agricultural habitats* (ed. T. S. Lie & E. G. Mulder), pp. 39–45. M. Nijhoff, The Hague.

Schildkraut, C. L., Marmur, J., & Doty, P. (1962). Determination of the base composition of deoxyriboneucleic acid from its buoyant density in CsCl. *Journal of Molecular Biology*, **4,** 430–43.

Schwinghamer, E. A. (1964). Association between antibiotic resistance and ineffectiveness in mutant strains of *Rhizobium* spp. *Canadian Journal of Microbiology*, **10,** 221–33.

Schwinghamer, E. A. (1967). Effectiveness of *Rhizobium* as modified by mutation for resistance in antibiotics. *Antonie van Leeuwenhoek Journal of Microbiology and Serology*, **33,** 121–36.

Schwinghamer, E. A. (1969). Mutation to auxotrophy and prototrophy as related to symbiotic effectiveness in *Rhizobium leguminosarum* and *R. trifolii. Canadian Journal of Microbiology*, **15,** 611–22.

Schwinghamer, E. A. (1974). Genetic aspects of nodulation and nitrogen fixation by legumes: the microsymbiont. In: *Dinitrogen fixation* (ed. G. W. Hardy). Wiley & Co., New York. (In press.)

Shah, V. K., Davis, L. C. & Brill, W. J. (1972). Nitrogenase. I. Repression and derepression of the iron–molybdenum and iron proteins in *Azotobacter vinelandii. Biochimica et Biophysica Acta*, **256,** 498–511.

Shah, V. K., Davis, L. C., Gordon, J. K., Orme-Johnson, W. H. & Brill, W. J. (1973). Nitrogenase. III. Nitrogenaseless mutants of *Azotobacter vinelandii*: activities, cross-reactions, and EPR spectra. *Biochimica et Biophysica Acta*, **292,** 246–55.

Sik, T. & Orosz, L. (1971). Chemistry and genetics of *Rhizobium meliloti* phage 16–3. In *Biological nitrogen fixation in natural and agricultural habitats* (ed. T. S. Lie & E. G. Mulder), pp. 57–62. M. Nijhoff, The Hague.

Simon, M. A. & Brill, W. J. (1971). Mutant of *Clostridium pasteurianum* that does not fix nitrogen. *Journal of Bacteriology*, **105,** 65–9.

Sorger, G. J. (1968). Regulation of nitrogen fixation in *Azotobacter vinelandii* OP and in an apparently partially constitutive mutant. *Journal of Bacteriology*, **95,** 1721–6.

Sorger, G. J. (1969). Regulation of nitrogen fixation in *Azotobacter vinelandii* OP: the role of nitrate reductase. *Journal of Bacteriology*, **98,** 56–61.

Sorger, G. J. & Trofimenkoff, D. (1970). Nitrogenaseless mutants of *Azotobacter vinelandii. Proceedings of the National Academy of Sciences, USA*, **65,** 74–80.

St John, R. T. & Brill, W. J. (1972). Inhibitory effect of methylalanine on glucose grown *Azotobacter vinelandii. Biochimica et Biophysica Acta*, **261,** 63–9.

Strandberg, G. W. & Wilson, P. W. (1968). Formation of the nitrogen-fixing enzyme system in *Azotobacter vinelandii. Canadian Journal of Microbiology*, **14,** 25–31.

Streicher, S. L., Gurney, E. G. & Valentine, R. C. (1971). Transduction of the nitrogen-fixing genes in *Klebsiella pneumoniae. Proceedings of the National academy of Sciences, USA*, **68,** 1174–7.

Streicher, S. L., Gurney, E. G. & Valentine, R. C. (1972). The nitrogen-fixation genes. *Nature, London*, **239,** 495–9.

Stroun, M. & Anker, P. (1971). Bacterial nucleic acid synthesis in plants following bacterial contact. *Molecular and General Genetics*, **113,** 92–8.

Tubb, R. S. & Postgate, J. R. (1973). Control of nitrogenase synthesis in *Klebsiella pneumoniae. Journal of General Microbiology*, **79,** 103–17.

Wilson, P. W., Hull, J. F. & Burris, R. M. (1943). Competition between free and combined nitrogen in the nutrition of *Azotobacter. Proceedings of the National Academy of Sciences, USA*, **29,** 289–94.

Witz, D. F., Detroy, R. W. & Wilson, P. W. (1967). N_2 fixation by growing cells and cell-free extracts of *Bacilliaceae. Archiv für Mikrobiologie*, **55,** 369–81.

Yoch, D. C. & Pengra, R. M. (1966). Effect of amino acids on the nitrogenase system of *Klebsiella pneumoniae. Journal of Bacteriology*, **92,** 618–22.

Zelazna, I. (1964*a*). Transformation in *Rhizobium trifolii*. I. The influence of some factors on the transformation. *Acta Microbiologica Polonica*, **13,** 283–90.

Zelazna, I. (1964*b*). Transformation in *Rhizobium trifolii*. I. Development of competence. *Acta Microbiologica Polonica*, **13,** 291–8.

Zelazna-Kowalska, I. & Lorkiewicz, Z. (1971). Conditions for genetical transformation in *Rhizobium meliloti. Acta Microbiologica Polonica*, Ser. A, **3,** 21–8.

2. Genetic analysis of nitrogen fixation in *Klebsiella pneumoniae*

K. T. SHANMUGAM & R. C. VALENTINE

Many genes are required for nitrogen fixation including those for nitrogenase and its regulation as well as genes specifying ancillary enzymes which support the nitrogen fixation process (Streicher, Gurney & Valentine, 1972*a*). Recently a cluster of *nif* (nitrogen fixation) genes which probably comprises the major *nif* operon has been discovered near *his* on the genetic linkage map of *Klebsiella* (Streicher, Gurney & Valentine, 1971). There is much indirect evidence that genes in this cluster code for nitrogenase. Most *nif* lesions map in this cluster (Streicher *et al.*, 1972*a*); most of these strains are devoid of nitrogenase activity although some mutants are still capable of synthesizing one or more of the three different subunit proteins comprising the nitrogenase complex (Streicher *et al.*, 1972*b*). An elegant proof of the importance of this cluster of *nif* genes was recently provided by Dixon & Postgate (1972) who constructed nitrogen-fixing hybrids of *Escherichia coli* by transferring this segment of DNA from *Klebsiella* to *E. coli*. We have found also that long deletions which extend through this region abolish nitrogenase activity (Streicher *et al.*, 1972*a*).

Because of the essential nature of this cluster of genes for nitrogenase activity we are undertaking a fine-structure genetic analysis of the region in *Klebsiella* to determine the precise location of the structural genes of nitrogenase and the regulatory elements controlling these genes. In addition, *nif*$^-$ loci which do not map in this cluster are known (Streicher *et al.*, 1972*a*) and will continue to be studied with the aim of determining their contribution to the nitrogen fixation process.

The mapping of essential *nif* functions depends largely on the utility of several basic genetic techniques such as transduction and conjugation and for this reason we have chosen to discuss the current status of nitrogen fixation genetics in the framework of these crucial methods.

Transduction

The discovery that coliphage P1 mediated generalized transduction of the *nif* genes (Streicher *et al.*, 1971) provided the cornerstone for genetic studies. Transductional analysis has yielded several important facts, the most important of which was the cotransductional mapping of an essential cluster of *nif* genes near the *his* operon on the *Klebsiella* linkage map. Most chemically induced *nif*$^-$ mutations were found to be cotransduced with *his*D at a frequency of 30–80 % (Streicher *et al.*, 1972*a*). It is interesting to note that this tract of DNA, calculated from the cotransduction frequencies, would be more than sufficient to code for nitrogenase and ancillary *nif* genes.

In addition to the majority of *nif*$^-$ lesions which are located near *his*, occasional *Nif*$^-$ phenotypes are found which do not map in this region. P1-mediated genetic crosses (two-factor crosses) among these *nif*$^-$ strains as well as *nif*$^-$ strains which map near the *his* operon have revealed that several different loci giving the *Nif*$^-$ phenotype are present on the chromosome. So far five different genetic classes of *Nif*$^-$ mutants including the cluster near *his* are recognized, although the map location of most of these types is not yet known (Streicher & Valentine, 1973*a*). It should also be emphasized that with the exception of the *Nif*$^-$ pleiotropic *Asm*$^-$ mutants (Nagatani, Shimizu & Valentine, 1971) and the *Nif*$^-$ pleiotropic *His*D$^-$ mutants (Streicher *et al.*, 1972*a*) only a few members of the remaining phenotypes have been isolated for study. The biochemical functions of the pleiotropic *Asm*$^-$ and *His*D$^-$ mutants are of considerable interest since they may represent various regulatory as well as ancillary enzymes and proteins essential for nitrogen fixation.

P1 has also been used for ordering several *nif* loci within the *nif* cluster by three-factor crosses (Streicher *et al.*, 1972*a*).

The construction of P1-sensitive strains of *Klebsiella* from strains resistant to this phage (Streicher *et al.*, 1972*a*) increases the utility of P1 for genetic studies with this group of organisms. There remains the possibility of creating P1-sensitive strains of other types of nitrogen-fixing bacteria, and this might allow studies of the comparative biochemistry and genetics of nitrogen fixers from various ecological groups, including the root nodule bacteria. There is considerable interest in the construction of specialized transducing phages carrying the *nif* segment of DNA, phages which would be of great use as sources of 'pure' *nif* template DNA for molecular biologists and biochemists interested in regulation and synthesis of nitrogenase in cell-free systems. These phages would also be of great interest for geneticists. Lambdoid phage 424, but not λ, was found to infect *K. pneumoniae* M5A1 (Streicher *et al.*, 1972*a*). Also the availability of nitrogen-fixing strains of *E. coli* may now permit the construction of λ*nif* from this source.

Conjugation

The major applications of this method have been the generalized mapping of the *Klebsiella* chromosome (Matsumoto & Tazaki, 1970, 1971), the transfer of *nif* (Dixon & Postgate, 1971), and the construction of *Klebsiella* × *E. coli* hybrids (Dixon & Postgate, 1972), interesting new strains capable of nitrogen fixation. The repression of nitrogenase synthesis by NH_4^+ in these hybrids suggests that regulator genes are associated with the *nif* cluster. This is discussed by Dixon & Cannon in Chapter 1.

Deletion analysis

This method, newest to be adapted for studies of *nif* genetics in our laboratory, is considered here because of its considerable potential application to this field; it will be published in detail elsewhere.

Use of phage for isolation in nif *deletions*

We have observed that phage-resistant loci for several virulent *Klebsiella* phages are located on the genetic linkage map of this organism near *nif*. Mapping of the resistance markers places them in the vicinity of the *rfb* (rough B) region, a set of genes coding for enzymes involved in synthesis of polysaccharides of the cell wall. The biochemical basis of the resistance markers is not understood but we presume that alterations of phage receptor sites on the cell surface are responsible.

The location of phage resistance loci near *nif* provides a powerful tool for selecting deletions extending into the *nif* region of DNA. It should be recalled that a high percentage of phage resistance mutations are deletions of various lengths including, in many instances, not only phage resistance loci but adjacent genes as well (Spudich, Horn & Yanofsky, 1970). The maximum extent of deletions of different regions of the chromosome is probably determined by a number of poorly understood factors including the presence of 'lethal genes' in the region. Using the eductant procedure of Sunshine & Kelly (1971) we have observed a strikingly large deletion of the *nif* region of *Klebsiella*, a deletion which presumably shortens the bacterial genome by several percent of its length. *Klebsiella* strains harboring these large deletions are isolated from phage P2 lysogens; such strains which have lost their nitrogen fixing capacity require only histidine for growth on a glucose–ammonium salts medium.

Lethal genes, therefore, are not present on this segment of DNA, and this permits a free hand for construction of deletions.

Strains carrying deletions in the *nif* region were isolated as phage-resistant clones as follows. A fresh culture of M5A1 was inoculated into 10 ml of L-broth containing 0.3 % sucrose in a 16 × 150 mm screw-cap tube and incubated at 37 °C without shaking for about 3–5 days. For enriching deletion mutants resistant to *Klebsiella* phage 3 (K3) 0.1 ml of this culture was inoculated into 9 ml of L-broth in a 125 ml flask and grown at 37 °C with shaking to a cell density of about 10^7 to 5×10^7 cells/ml; this culture was infected with 0.1 ml of a phage stock of K3 (1.8×10^9 plaque forming units/ml) and incubation continued until cell lysis was complete. The lysed cell suspension was next centrifuged at 10 000 *g* for 5 min and the cell pellet collected and washed once in 10 ml of sucrose-minimal medium and resuspended in the same medium. The cell suspension was incubated for 30 min at 37 °C with shaking. Penicillin G (5 mg/ml, 1635 units/mg) was added and the incubation continued for 4 hr to enrich for histidine auxotrophs. The cells were washed three times in sucrose-minimal medium, resuspended in 2 ml of L-broth, and samples (0.1 ml) of cells escaping penicillin lysis inoculated into 0.5 ml of L-broth in a 16 × 100 mm tube and incubated at 37 °C until maximum density was reached. This culture was diluted in saline so as to give single colony formation, spread onto L-broth plates and incubated at 37 °C. Colonies were transferred by replica plating onto sucrose-minimal plates with/without histidine to score for their histidine requirement, and histidine-requiring clones (putative *nif* deletions) purified by restreaking.

A second double auxotroph (*His*$^-$, *Nif*$^-$) method which we initially devised for selection of deletions of *nif* has in fact turned out not to yield *nif* deletions but rather a class of pleiotropic, revertable, point mutations affecting growth on nitrogen-free plates (E. Gurney & R. C. Valentine, unpublished data). We mention this technique briefly because it might benefit others making similar attempts. A penicillin enrichment for double *His*$^-$, *Nif*$^-$ auxotrophs should yield some strains carrying deletions of both regions because two independent point mutations should happen only rarely compared to a deletion. This experiment did produce many pleiotropic *His*$^-$, *Nif*$^-$ auxotrophs but further analysis showed these to be the result of a revertable (single?) point mutation affecting both pathways (histidine requirement and inability to grow on N_2-free plates). Most of these lesions were found to map in the *his*D gene of the *his* operon, blocking synthesis of histidinol dehydrogenase. In a still unknown fashion this lesion blocks colony formation on agar plates with N_2 as nitrogen source but does not block induction of nitrogenase in liquid medium. This might be caused by a regulatory defect of nitrogen metabolism in these strains.

Biochemical analysis

The segment of DNA near *nif* codes for a variety of enzymes which may be assayed in individual deletion strains yielding a 'biochemical map' of the particular deletion. For instance, if a strain is missing all enzymes (gene products) of this region from *gnd* to *nif* we assume that this entire tract of DNA is missing. Lack of these gene products can also be observed as physiological differences among these deletions and the wild-type. Growth studies showed that deletions such as $\Delta 4$ and $\Delta 30$ extend into the histidine operon as shown by the requirement of histidine for growth. The termination of these deletions ($\Delta 4$ and $\Delta 30$) within the *his* operon is shown by their ability to utilize histidinol (*his*D). These two deletions, however, retain their ability to grow with N_2 as nitrogen source, while, for example $\Delta 16$ and $\Delta 17$ do not utilize N_2. Deletions 16 and 17 also have an absolute requirement for histidine indicating that the genes coding for both nitrogenase and histidine biosynthesis are deleted. Deletion 17 has also lost ability to utilize shikimic acid (an aromatic hydrocarbon) as sole source of carbon and energy. We have observed too that $\Delta 17$ requires an unidentified growth factor(s) which can be satisfied by 40 μg/ml of yeast extract. The major conclusion from growth experiments is that spontaneous phage-resistant mutants are often deletions, many of which extend into or through the *nif* segment of DNA. Table 2.1 summarizes the enzymatic analysis of these deletion strains.

Gluconate-6-phosphate dehydrogenase (*gnd*) catalyzes the $NADP^+$-linked oxidation of gluconate-6-phosphate yielding ribulose-5-phosphate, carbon dioxide and NADPH. Deletions lacking this enzyme are still able to metabolize gluconic acid as an energy source via an alternate pathway. The enzyme is readily measured by following the increase in pyridine nucleotide absorption at 340 nm or by coupling NADPH to a thiazolyl blue dye system (Peyru & Fraenkel, 1968). All of the deletions discussed here are missing this enzyme (*gnd* product) as shown in Table 2.1.

Imidazolyl acetol phosphate: L-*glutamate aminotransferase* (*his*C) catalyzes a reversible transamination step in histidine biosynthesis and is assayed spectrally by the increase in absorption at 280 nm of a reaction product (Martin *et al.*, 1971). On the *Salmonella* linkage map this enzyme is located near the *his* operator. All deletions except Δ30 are missing *his* C gene activity.

L-histidinol dehydrogenase (*his*D) catalyzes the NAD-linked dehydrogenation of histidinol, the terminal reaction of histidine biosynthesis (Martin *et al.*, 1971). NADH production is followed spectroscopically. The gene for this enzyme is located near the operator end of the *his* operon between the *his*C gene and operator. Cells with this gene intact are able to substitute histidinol for histidine by oxidizing it and thus fulfilling their nutritional requirement for this amino acid. Note that two of the different classes of deletions retain the *his*D gene (Table 2.1).

Table 2.1. *Properties of deletion mutants*

	Enzyme activity (nmoles/min/mg protein)				
Deletion	Gluconate-6-phosphate dehydrogenase (*gnd*)	Amino-transferase* (*his*C)	Histidinol dehydrogenase (*his*D)	Nitrogenase† (*nif*)	Shikimate‡ permease (*shi*A)
Wild-type	52.0	18.8	41.0	2.9	+
Δ30	0.0	4.4	8.1	3.3	+
Δ4	0.0	<0.5	8.3	3.4	+
Δ16	0.0	0.0	0.0	0.0	+
Δ17	0.0	0.0	0.0	0.0	−
Educant	0.0	0.0	0.0	0.0	+

* Imidazolyl acetol phosphate: L-glutamate aminotransferase activity.
† μmoles ethylene produced/h/mg protein.
‡ See text for details.

Nitrogenase (*nif*) is measured in whole cells as the ability to catalyze the reduction of acetylene to ethylene. It is important to note from Table 2.1 that deletions which end in the histidine operon (Δ4 and Δ30) still retain the *nif* gene cluster (and thus the capacity to reduce acetylene). Nitrogenase activity is absent in the other deletions.

Shikimate permease (*shi* A). The shikimate utilization (*shu*) genetic system, which includes the permease (*shi* A), is of particular interest because it represents a new selective locus near *nif*, a region of the chromosome that is poorly marked genetically. We have observed that *K. pneumoniae* utilizes shikimic acid as a carbon source, presumably catabolizing it through the muconate, adipate pathway observed in *Moraxella* (Cánovas, Wheelis & Stanier, 1968). It is of interest that *E. coli* which does not catabolize shikimate as an energy source nevertheless synthesizes a permease for shikimic acid (Pittard & Wallace, 1966).

This permease (*shi*A) which maps near *his* is utilized by *E. coli* auxotrophs, blocked before the shikimic acid level, to transport shikimic acid for synthesis of aromatic amino acids. It seems probable, in *Klebsiella*, that deletions which extend into the shikimic acid permease locus would produce the *Shu*$^-$ phenotype. [^{14}C] shikimate uptake experiments are in progress to confirm the absence of shikimate permease in Δ17. Genetic experiments discussed in the next section show that Δ17 lacks only the permease.

In summarizing the biochemical information, all deletion strains examined are missing gluconate-6-phosphate dehydrogenase (*gnd*) activity suggesting that all deletions extend beyond this point. Deletions 30 and 4 produce histidinol dehydrogenase, the terminal enzyme of the histidine biosynthetic pathway, which converts histidinol to histidine. It should be recalled that histidinol satisfies the histidine requirement of these strains. Deletion 30 is the only strain examined which has imidazolyl acetol phosphate: L-glutamate aminotransferase activity, showing that the deletion in Δ30 terminates prior to the *his*C gene. Other deletions do not have any of the enzymatic activities tested. The deletions 16 and 17 and the eductant do not produce nitrogenase activity as expected, but the nitrogenase genes are intact and functional in Δ4 and Δ30. We conclude that the *nif* cluster of genes maps on the operator side of the *his* operon and before the *shi*A locus, a finding that is further substantiated by genetic experiments described below.

Genetic mapping of deletions

F′*his* episomal mapping of the deletions 4 and 30 was done to identify the precise point of termination of these classes of deletions. A set of F′*his*$^-$ episomes (F′*his*G$^-$ through F′*his*I$^-$; see Fig. 2.1) were used to infect the deletion strains in order to test for the presence of *his* gene activity. For example, the presence of a functional *his*D gene on the chromosome would be expected to complement an F′*his*D$^-$ episome, defective only in *his*D gene activity, yielding a *His*$^+$ prototrophic clone. Deletions 16 and 17 and the eductant possess no *his* gene activity as measured by this technique. However, Δ30 possesses *his*G through *his*H activity while Δ4 possesses only *his*G and *his*D activity. Other *his* genes were absent in terms of their complementing activity with episomes. The presence of a functional *his*O (operator) in Δ4 and Δ30 is implied since expression of any of the *his* genes requires a functional operator.

An F′*his shi*A episome derived from *E. coli* was found to restore the

Shu$^+$ phenotype of deletion 17. As discussed above, *E. coli* produces shikimate permease but presumably lacks other shikimate utilization enzymes. Since the F′ episome of *E. coli* contributes only the *shi*A gene, it is concluded that Δ17 lacks only the shikimate permease. The presence of a functional shikimate permease also overcomes the unknown nutritional requirement in Δ17.

These findings confirm the map order of phage resistance as (*rfb*?), *gnd*, *his*, *nif*, *shi*A (*shu*), and show that the *nif* cluster lies next to the *his*D, *his*G, *his*O (operator) end of the *his* operon. A summary of these findings is given in Fig. 2.1. The genetic distances between genes are not drawn to scale and only the various bacterial markers tested are included.

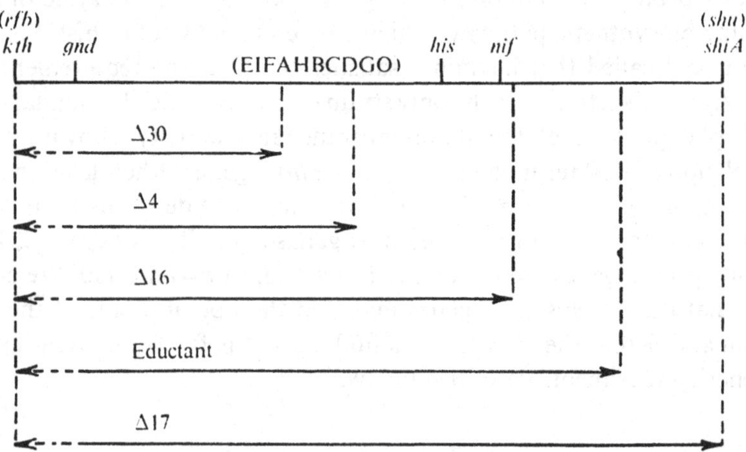

Fig. 2.1. Genetic map of the *nif* region of the *Klebsiella* chromosome; arrowed lines illustrate the extent of various deletions; dotted lines indicate that termination points of the deletions are not mapped. Symbols: *kth* (*rfb*), K3 phage resistance; *gnd*, gluconate-6-phosphate dehydrogenase; *his*, histidine operon; *nif*, nitrogen fixation; *shi A* (*shu*), shikimate permease.

Deletions will play a continuing role in our attempt to construct a fine-structure genetic map of the *nif* operon(s). The goals include the pinpointing of each structural gene for nitrogenase, and may eventually be used to produce strains forming one or more of the individual subunits of the nitrogenase complex. Such strains would represent 'pure' sources of nitrogenase subunits. Identification and genetic alteration of the *nif* operator(s) might also lead to the construction of strains which are derepressed for nitrogenase or subunit synthesis. As discussed by Brill (see Chapter 3) such strains might be of importance in agriculture. In this connection we may soon possess the necessary tools to delete old *nif* operators while replacing them with new ones designed or selected for better performance such as increased affinity for inducer. Isolation and studies of regulatory mutants of *nif* may also shed new light on the mechanism of NH_4^+ repression of nitrogenase synthesis, a general regulatory system which appears to control the synthesis of a wide array of enzymes and pathways dealing with nitrogen metabolism in the microbial and, perhaps, the plant world (Prival & Magasanik, 1971).

DNA extracted from deletion mutants of *nif* might also be useful in studies of plant cell uptake of bacterial DNA as well as in evolutionary studies of *nif* DNA homologies in different strains using techniques of DNA hybridization. Deletion mutants function as highly stable 'nonreverting' recipients in genetic crosses and may also be of importance for studies on the maintenance and origin of *nif*-bearing plasmids or episomes.

Evolutionary aspects

The presence of *nif* in coliform bacteria (*Klebsiella*) which have evolved mechanisms (sexduction and transduction) for infectious transfer of hereditary traits, such as drug resistance, raises the possibility that *nif* itself may be capable of being transferred among the coliform organisms under natural conditions. Sex factors are also known with a much broader host range (Olsen & Shipley, 1973); these sex factors promote conjugation among widely different classes of bacteria (for example, *E. coli* × root nodule bacteria), although as yet there are no reports of chromosome mobilization in such crosses (Datta *et al.*, 1971). As a working hypothesis we propose that the segment of *nif* DNA, perhaps having evolved in some other soil organism, was transferred to *Klebsiella* (via transduction or sexduction) and inserted into the chromosome at a location between the *his* operon and the *shi*A locus. Such an event is illustrated in Fig. 2.2, with *E. coli* as the hypothetical recipient of *nif* DNA. An apparent similarity of the genetic linkage map of these two organisms is obvious. A survey of the genetic literature concerned with this region of the chromosome reveals that the cotransduction frequency between a *his* locus and the *shi*A marker in *E. coli* is about 70 %, with 6–7 genes between the two markers (Pittard & Wallace, 1966). On the other hand we have observed that the cotransduction frequency between *his*D and the nearest *nif* marker of the *Klebsiella* genome is about 80 % (Streicher *et al.*, 1972*a*). The minimum length of the *nif* region itself, as calculated from P1 transductional analysis, is about 10–15 average genes. This finding, and the fact that P2 eductants, long deletions of this region of the chromosome, are *His*$^-$, *Nif*$^-$, but *Shi*A$^+$ (*Shu*$^+$) makes it unlikely that the *shi*A locus occupies a site within the *nif* cluster. It seems more probable that *nif* has been inserted into the map yielding the order *gnd, his, nif, shi*A. It will be of interest to compare this genetic order with that of the *E. coli* hybrids prepared by Dixon & Postgate (1972) as well as other naturally occurring strains of *Klebsiella*; this information may provide some clues as to the transfer of the *nif* genes among these organisms.

There is currently considerable interest in 'infecting' new species with the *nif* genes. It should be stressed that in choosing new strains of bacteria as potential recipients for the *nif* genes it is important to keep in mind the basic biochemical requirements for fixation (see Streicher & Valentine,

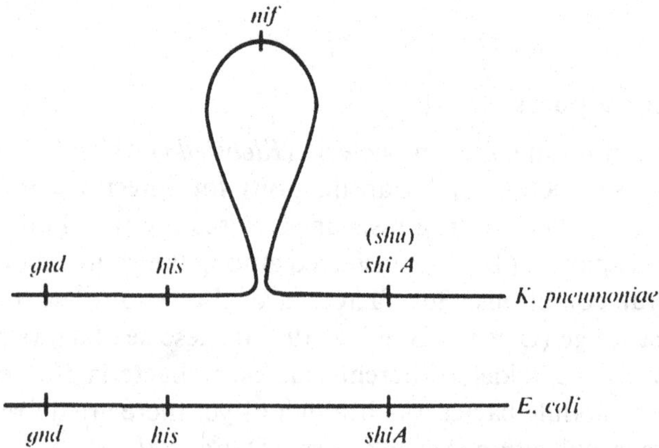

Fig. 2.2. Comparative genetic maps of the *nif* region of the *Klebsiella* genome versus *E. coli* illustrating the possible insertion of the *nif* operon(s) between *his* and *shi A*.

1973*b*, for detailed discussion). These are: (*a*) nitrogenase, (*b*) a supply of strong reductant usually as reduced ferredoxin or flavodoxin, (*c*) ATP as energy source, (*d*) a specialized NH_4^+ assimilation pathway, (*e*) NH_4^+ repression governing the genetic expression of *nif*, and (*f*) in aerobic bacteria protection of the nitrogen fixation system from oxygen denaturation and competition for reductant. There may also be a need for enzymes which catalyze the transformation of the trace element molybdenum into a form suitable for incorporation into nitrogenase and various active transport systems (permeases) for entry of substrates and exit of products of nitrogen fixation. The latter seems especially pertinent in root nodule symbiosis where nitrogenous products must be transported from the root nodules and photosynthetic products, as carbon and energy source, moved to the nodule for consumption.

In spite of the relatively stringent requirements for biological nitrogen fixation it would seem that a large number of free-living bacteria possess the basic biochemistry needed to support fixation. As already mentioned, Dixon & Postgate (1972) have constructed nitrogen-fixing hybrids of *E. coli* by sexual transfer of the *nif* set of genes from *Klebsiella* donors into this strain. A point by point check of the biochemical requirements for nitrogen fixation (a-f) mentioned above for wild-type *E. coli*, which does not naturally fix nitrogen, reveals that most, if not all, ancillary reactions are present and functional. If so it follows that such ancillary reactions may have other functions besides nitrogen fixation. For instance it is well-known that during fermentation of carbohydrates *E. coli* produces powerful reductants, as evidenced by the accumulation of hydrogen gas as fermentation product. *E. coli* is thus capable of generating a strong reductant which could be utilized for fixation. It is also known that the synthesis of at least one crucial enzyme in *E. coli*, glutamine synthetase, is repressed by NH_4^+ (Woolfolk, Shapiro & Stadtman, 1966), suggesting that a system of NH_4^+ control is operative. The new NH_4^+ assimilatory system involving glutamate synthase is also functional in wild-type *E. coli* (Miller & Stadtman, 1972). Also the presence of nitrate reductase, a molybdenum-containing enzyme (Glaser & DeMoss, 1971), indicates that *E. coli* possesses the necessary metabolic machinery for interconversion of molybdenum into a catalytically active form. Thus in retrospect *E. coli* was a good choice as recipient for the *nif* genes since it already possessed the necessary supporting enzymes. Hypothetically a wide variety of additional enteric bacteria and other free-living bacteria such as clostridia, bacilli, etc. should be suitable 'hosts' for *nif*.

The successful construction of *E. coli* strains capable of nitrogen fixation has stimulated many workers to think about similar experiments with plant and animal systems as well as a variety of free-living and symbiotic micro-organisms. Perhaps more attention has been given to plants because at first glance plants seem to possess more of the required biochemical (genetic) capacity for fixation. For example, during photosynthesis plants produce reduced ferredoxin, a powerful reductant required for fixation, However, there is the severe handicap of oxygen protection of nitrogenase to be dealt with in plants which evolve oxygen. It is also not known whether the NH_4^+ assimilatory reactions (and NH_4^+ control system if it exists) of plants could be easily integrated with nitrogen fixation. Likewise, the problem of permanent 'storage' and replication of the *nif* genes in the plant cell poses an enormous problem.

There is also considerable concern with the genetic improvement of root nodule bacteria with emphasis on broadening the host infectivity range of these symbionts for additional plants such as the cereal grains. Undoubtedly, more basic knowledge must be gathered before this will be possible. The genetic and biochemical basis of nitrogen fixation in these important organisms represents a major challenge for future workers.

We are indebted to Dr M. D. Kamen for his continued encouragement and to Arnold Loo for help with the biochemical analysis of deletion strains. Original work supported by the National Science Foundation (grant no. GB 35331). R.C.V. is the recipient of a Career Development Award from the National Institutes of Health (AI 16595).

References

Cánovas, J. L. Wheelis, H. L. & Stainer, R. Y. (1968). Regulation of the enzymes of the β-keto-adipate pathway in *Moraxella calcoacetica.* 2. The role of protocatechuate as inducer. *European Journal of Biochemistry,* **3,** 293–304.

Datta, N., Hedges, R. W., Shaw, E. R., Sykes, R. B. & Richmond, M. H. (1971). Properties of an R-factor from *Pseudomonas aeruginosa. Journal of Bacteriology,* **108,** 1244–9.

Dixon, R. A. & Postgate, J. R. (1971). Transfer of nitrogen fixation genes by conjugation in *Klebsiella pneumoniae. Nature, London,* **234,** 47–8.

Dixon, R. A. & Postgate, J. R. (1972). Genetic transfer of nitrogen fixation from *Klebsiella pneumoniae* to *Escherichia coli. Nature, London,* **237,** 102–3.

Glaser, J. H. & DeMoss, J. A. (1971). Phenotypic restoration by molybdate of nitrate reductase activity in *chl* D mutants of *Escherichia coli. Journal of Bacteriology,* **108,** 854–60.

Martin, R. G., Berberich, M. A., Ames, B. N., Davis, W. W., Goldberger, R. F. & Yourno, J. D. (1971). Enzymes and intermediates of histidine biosynthesis in *Salmonella typhimurium.* In *Methods in enzymology* (ed. H. Tabor & C. W. Tabor), vol. 17B, pp. 3–44. Academic Press, New York & London.

Matsumoto, H. & Tazaki, T. (1970). Genetic recombination in *Klebsiella pneumoniae*; an approach to genetic linkage mapping. *Japanese Journal of Microbiology,* **14,** 129–41.

Matsumoto, H. & Tazaki, T. (1971). Genetic mapping of *aro*, *pyr*, and *pur* markers in *Klebsiella pneumoniae. Japanese Journal of Microbiology,* **15,** 11–20.

Miller, R. E. & Stadtman, E. R. (1972). Glutamate synthetase from *Escherichia coli. Journal of Biological Chemistry,* **247,** 7407–19.

Nagatani, H., Shimizu, M. & Valentine, R. C. (1971). The mechanism of ammonia assimilation in nitrogen-fixing bacteria. *Archiv für Mikrobiologie,* **79,** 164–75.

Olsen, H. & Shipley, P. (1973). Host range and properties of the *Pseudomonas aeruginosa* R factor R1822. *Journal of Bacteriology,* **113,** 772–80.

Peyru, G. & Fraenkel, D. G. (1968). Genetic mapping of loci for glucose-6-phosphate dehydrogenase, gluconate-6-phosphate dehydrogenase, and gluconate-6-phosphate dehydrase in *Escherichia coli*. *Journal of Bacteriology*, **95**, 1272–8.

Pittard, J. & Wallace, B. J. (1966). Gene controlling the uptake of shikimic acid by *Escherichia coli*. *Journal of Bacteriology*, **92**, 1070–5.

Prival, M. J. & Magasanik, B. (1971). Resistance to catabolite repression of histidase and proline oxidase during nitrogen limited growth of *Klebsiella pneumoniae*. *Journal of Biological Chemistry*, **246**, 6288–96.

Spudich, J. A., Horn, V. & Yanofsky, C. (1970). On the production of deletions in the chromosome of *Escherichia coli*. *Journal of Molecular Biology*, **53**, 49–67.

Streicher, S. L., Gurney, E., Robinson, C. & Valentine, R. C. (1972*b*). Chromosomal location of nitrogenase genes of *Klebsiella pneumoniae*. *Federation Proceedings*, **31**, 910. (Abstract.)

Streicher, S. L., Gurney, E. & Valentine, R. C. (1971). Transduction of the nitrogen-fixation genes in *Klebsiella pneumoniae*. *Proceedings of the National Academy of Sciences, USA*, **68**, 1174–7.

Streicher, S. L., Gurney, E. & Valentine, R. C. (1972*a*). The nitrogen-fixation genes. *Nature, London*, **239**, 495–9.

Streicher, S. L. & Valentine, R. C. (1973*a*). The nitrogen-fixation (*nif*) operon(s) of *Klebsiella pneumoniae*. In: *Microbial iron metabolism* (ed. J. B. Neilands). Academic Press, New York & London. (In press.)

Streicher, S. L. & Valentine, R. C. (1973*b*). Comparative biochemistry of nitrogen fixation. *Annual Review of Biochemistry*, **42**, 279–303.

Sunshine, M. G. & Kelly, B. (1971). Extent of host deletions associated with bacteriophage P2-mediated eduction. *Journal of Bacteriology*, **108**, 695–704.

Woolfolk, C. A., Shapiro, B. & Stadtman, E. R. (1966). Regulation of glutamine synthetase. I. Purification and properties of glutamine synthetase from *Escherichia coli*. *Archives of Biochemistry and Biophysics*, **116**, 177–92.

3. Control of nitrogenase synthesis in *Azotobacter vinelandii*

W. J. BRILL

All free-living nitrogen-fixing bacteria so far examined only fix N_2 when unable to obtain sufficient nitrogen from another source (Brill, 1973). Readily utilizable nitrogen sources such as urea or NH_4^+ completely repress nitrogenase formation, whereas sources that are not as effective for growth only partially repress nitrogenase synthesis; these include aspartate, glutamate and asparagine (Wilson, Hull & Burris, 1943). This suggests that depletion of NH_4^+ pools (or nitrogenous compounds related to NH_4^+ metabolism) causes both components of the enzyme to be synthesized. After *Azotobacter vinelandii* has utilized available NH_4^+ from the medium, active nitrogenase appears within 15 minutes (Shah, Davis & Brill, 1972). This activity is not due to activation of an inactive enzyme because no trace of either protein component is detected by antigenic techniques prior to NH_4^+ depletion (Davis *et al.*, 1972). Addition of chloramphenicol to cells that are in the process of synthesizing nitrogenase immediately halts the increase in nitrogenase activity (Shah *et al.*, 1973*b*). Thus, if protein modification is required to activate nitrogenase, it must be very rapid.

The control of nitrogenase synthesis is very stringent because the difference of component levels between N_2- and NH_4^+-grown cells is greater than a factor of 2.6×10^4 (Gordon & Brill, 1972). Both component proteins of nitrogenase are synthesized co-ordinately in the optimum ratio for activity, reaching maximum activity levels 2 hours after NH_4^+ is exhausted from the medium (Shah *et al.*, 1972). Their co-ordinated synthesis may involve a common control gene. Nitrogen gas does not seem to be an inducer of nitrogenase because bacteria grown in an essentially N_2-free medium still produce normal amounts of nitrogenase (Strandberg & Wilson, 1968). Unfortunately, it has been impossible to eliminate all N_2 from the growth medium, but ecologically it would seem improbable for *Azotobacter* to have a control system that responds to the absence of N_2 as the gas is found in all its natural environments.

Addition of NH_4^+ to cells that have been growing on N_2 abruptly stops further synthesis of both nitrogenase components. Nitrogenase activity seems not to be inhibited because its loss initially follows component protein loss as determined antigenically (Davis *et al.*, 1972). The actual effector for repression might be NH_4^+, glutamate, aspartate, or serine because these pools increase in sulphate-limited, chemostat-grown *Azotobacter chroococcum* when NH_4^+ is the nitrogen source (Drozd, Tubb & Postgate, 1972). Nitrate also is known to repress nitrogenase synthesis (Wilson, Hull & Burris, 1943). It is the metabolism of nitrate, however, that is required for this repression because a mutant unable to reduce nitrate synthesizes identical quantities of nitrogenase in the presence or absence of nitrate (Sorger, 1969). During repression, both enzyme components seem to be diluted out by growth for the first half generation, but are then rapidly inactivated and more than 95 % of the

Table 3.1. *Specific activities* of components in wild-type and mutant strains*

Strain	Specific activity		
	Extract	*I*	*II*
UW (wild-type)	55.1	64.4	61.1
UW1	0.0	0.0	0.0
UW6	0.1	0.1	51.5
UW38	0.2	0.2	165.7
UW91	0.0	83.0	0.0

* nmoles acetylene reduced/min/mg protein. Cells were derepressed for 3 h. After Shah *et al.* (1973*b*).

activity is lost in two generations (Shah *et al.*, 1972). Component protein loss, detected antigenically, still follows dilution kinetics and the components are lost co-ordinately. An interesting area of study that has not been investigated is the mechanism of component inactivation during repression.

In order to learn more about nitrogenase synthesis in *A. vinelandii* mutant strains were isolated that are unable to fix N_2 (Fisher & Brill, 1969; Shah *et al.*, 1973*a*). Mutant strains that would revert spontaneously were chosen for further studies because they, presumably, are the result of a single mutation in genes specific for nitrogenase synthesis. All mutant strains grew as well as the wild-type in media containing excess NH_4^+. Component synthesis was derepressed in mutant strains by growing each strain on limiting NH_4^+ and then further incubating the strain for 3 hours after NH_4^+ was depleted from the medium. Crude extracts were tested for component activity by titrating the cell-free extract with pure components that had been isolated from the wild-type (Shah & Brill, 1973; Shah *et al*, 1973*a*). Mutant strains were I^-II^-, I^+II^-, and I^-II^+ for component activity (see Table 3.1).

To test whether a I^-II^+ mutant strain produced an inactive Component I or none at all, assays were required to identify the inactive components. One method utilized antisera from rabbits that had been injected with a purified component. Cross-reactions with crude extracts were noted by precipitin reactions detected on Ouchterlony plates. A negative cross-reaction, however, might be caused by a component that, because of a mutation, has a very different tertiary structure. If a mutation causes premature termination of a nitrogenase protein, the peptide formed also might not give a positive result with the Ouchterlony precipitin test. Another technique used depends on the fact that both components contain iron. Anaerobic acrylamide gel electrophoresis was

Table 3.2. *Phenotypes of mutant strains that are* I^+II^- *for activity*

Cross-reacting material	Iron stain on acrylamide gel	Typical mutant strain
I^+II^+	I^+II^+	UW91
I^+II^-	I^+II^-	UW120

performed on crude extracts of the wild-type and mutant strains and the gels treated with α,α-dipyridyl, which reacts to form a pink color with ferrous ions. Many mutant strains that lacked activity for one or both of the components still showed a positive reaction with the iron stain, while some strains that synthesized large amounts of a component as detected by the precipitin reaction did not. Perhaps such inactive components are defective in iron incorporation or are synthesized to form a tertiary structure that no longer can retain iron. Plate 3.1 shows an acrylamide gel stained with α,α-dipyridyl. Strain UW (wild-type) produces, as expected, both components when grown on N_2 but neither when grown on NH_4^+. Mutant strains UW1, UW38, UW46, and UW91 are I^-II^-, I^-II^+, I^-II^+, and I^+II^- respectively for activity.

Many mutant strains were analysed for activity, antibody cross-reaction and iron stain reaction for each of the two components. Table 3.2 classifies two types of strains that are I^+II^- for activity. Strain UW91 represents a class that presumably contains mutants with lesions in genes that specify the structure of Component II. The resulting inactive component still can be recognized by the iron stain and antigenic reaction.

The other mutant strain, UW120, may have a more drastically altered Component II or it may be the result of a mutation that controls the synthesis of that component.

Mutant strains that are I$^-$II$^+$ for activity are summarized in Table 3.3. Strain UW10 represents a mutant that presumably has a mutation in a structural gene for one of the two subunits (α and β) for Component I (Burns, Holsten & Hardy, 1970), such that the resulting structure is not drastically altered. An interesting phenotype is seen with strain UW100 that has normal amounts of Component I detected antigenically, but has no trace of it as detected with the iron stain. Perhaps this strain has a Component I altered in such a way that it no longer can hold on to the iron, or has a defect in an enzyme that specifically incorporates iron. Strain UW6 either produces no Component I or it is in a drastically altered form.

Table 3.3. *Phenotypes of mutant strains that are* I$^-$II$^+$ *for activity*

Cross-reacting material	Iron stain on acrylamide gel	Typical mutant strain
I$^+$II$^+$	I$^+$II$^+$	UW10
I$^+$II$^+$	I$^-$II$^+$	UW100
I$^-$II$^+$	I$^-$II$^+$	UW6

Of special interest are the strains that have no activity for either component. Two classes with this phenotype are described in Table 3.4. Mutant strain UW3 is typical of many mutants isolated. Component I is produced in normal amounts, but it cannot be detected with the iron stain, whereas Component II has not been detected with either technique. We have shown in preliminary experiments using electron paramagnetic resonance spectroscopy some indication of accumulation of an iron-containing substance in these strains. It is quite possible, therefore, that these mutant strains are the result of a lesion in a gene that is responsible specifically for iron uptake or incorporation into the two nitrogenase components. Component I that has no iron can still be detected with antiserum (for example, strain UW100 in Table 3.3). However, it is possible that Component II that has lost its iron cannot be detected antigenically. The only support for this idea is that no mutant that is I$^+$II$^-$ for activity and iron stain has yet been found that is I$^+$II$^+$ for cross-reacting material.

The other class of mutant strains that has no activity for either component is represented by UW2 that has no trace of either component when tested antigenically or with the iron stain. This pleiotropic-negative phenotype might be caused by a mutation in a control gene for nitrogenase synthesis, possibly analogous to the *lac* i^s mutation in the lactose-degrading system in *Escherichia coli* (Willson *et al.*, 1964). Revertants of *Lac* mutants are commonly found to be constitutive for the *lac* genes. Many revertants of strains such as UW2 also are defective in control of nitrogenase. Twenty-one spontaneous revertants of UW2 were picked and purified (Gordon & Brill, 1972); of these, fourteen did not produce nitrogenase activity in the presence of excess NH_4^+. The remaining seven were derepressed for nitrogenase synthesis. This is the first example of nitrogen fixation that is not completely repressed by excess NH_4^+. Strain UW59, a spontaneous revertant of UW2 has been studied in greater detail (Gordon & Brill, 1972). This derepressed mutant, in the presence and absence of NH_4^+ in the growth medium, produces the two components in the optimum ratio for maximum activity. Therefore, the control mutation still allows co-ordinated synthesis of the components.

Table 3.4. *Phenotypes of mutant strains that are* I^-II^- *for activity*

Cross-reacting material	Iron stain on acrylamide gel	Typical mutant strain
I^+II^-	I^-II^-	UW3
I^-II^-	I^-II^-	UW2

The specific activity of nitrogenase in strain UW59 when grown on NH_4^+ is one-quarter of the specific activity of the wild-type on N_2. When UW59 is grown on N_2 it has twice the specific activity of the same strain grown on NH_4^+. These data support the model that both nitrogenase components have at least one control gene in common, and that this control gene produces a product that is required to activate the synthesis of the components. Presumably, NH_4^+ or a related effector inactivates the activator; this type of control is known as positive control (Englesberg, Squires & Meronk, 1969). The mutation that caused the phenotype of strain UW2 could be in the control gene, the modified activator produced no longer being able to activate the various operons. Strain UW59, therefore, might be caused by a second mutation in the control gene allowing reactivation of the activator, but at the same time there is partial loss of the effector binding site(s). Such derepressed strains are being tested for agronomic applications because they will synthesize active nitrogenase in the presence of fertilizer nitrogen.

Most mutants that lack activity of only one component produce about the same amount of the active component that is normally present in the wild-type. An exception is strain UW38 (Table 3.1) which produces more than twice the amount of Component II than is made by other mutants. When this strain is derepressed for 8 hours, more than five times as much Component II is synthesized than is detected in the wild-type (Shah *et al.*, 1973*b*). The increase in activity is caused by increased synthesis rather than activation of Component II because greater amounts of its protein can be detected antigenically (Shah *et al.*, 1973*b*). It is possible that UW38 is the result of two mutations, one causing hyperproduction of Component II and a second inability to synthesize detectable Component I. To test this, four spontaneous revertants of UW38 that had regained the ability to fix N_2 were isolated. All produced lower quantities of each component than are normally found in derepressed wild-type. This

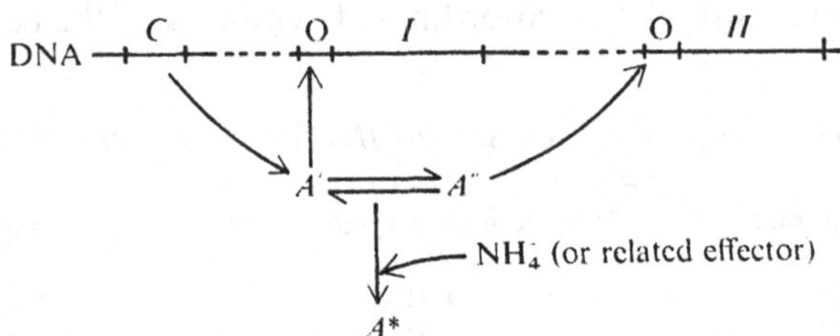

Fig. 3.1. Simplified model for positive control of nitrogenase. *C*, the control gene, *nif* C; *O*, operator; *I*, gene for component I; *II*, gene for Component II; *A′*, activator protein in a state that only activates the gene for Component I; *A″*, activator protein in a state that only activates the gene for Component II; *A**, inactive activator protein.

indicates that a single mutation causes the phenotype of strain UW38. It might be possible that the components are specified by more than one operon, having operators that respond to a single activator. This activator could be modified, by mutation, so that it no longer will bind to an operator in a gene specifying Component I, but that this same activator somehow is more efficient in activating Component II transcription.

A simplified model that can explain the data is described in Fig. 3.1. This requires that at least two operons code for Components I and II. It is possible of course, that more than one operon is necessary for active Component I synthesis, an operon being required, for example for each of its subunits. A gene, *nif*C, constitutively produces an activator protein, *A* that can exist in two structural forms, *A′* and *A″*. The conformation, *A′*, binds to the operator for Component I, but not to that for Component II, while the reverse is true for *A″*. Binding of the activator to an operator is necessary for transcription of the gene. Accumulation of the effector (perhaps NH_4^+) in the cytoplasm inactivates the activator protein by changing its conformation to A^* which can no longer interact with either operator and thus prevents synthesis of either component. This model predicts that the mutations causing the phenotypes of strains UW2 and UW38 are linked in the same gene, *nif*C. Gene mapping by transduction or transformation (Sen & Sen, 1965) should confirm or reject this model.

References

Brill, W. J. (1973). Regulation of nitrogenase. In: *Dinitrogen fixation* (ed. R. W. F. Hardy). John Wiley & Sons, New York. (In press.)

Burns, R. C., Holsten, R. D. & Hardy, R. W. F. (1970). Isolation by crystalization of the Mo–Fe protein of *Azotobacter* nitrogenase. *Biochemical and Biophysical Research Communications*, **39**, 90–4.

Davis, L. C., Shah, V. K., Brill, W. J. & Orme-Johnson, W. H. (1972). Nitrogenase. II. Changes in the EPR signal of Component I (iron–molybdenum protein) of *Azotobacter vinelandii* nitrogenase during repression and derepression. *Biochimica et Biophysica Acta*, **256**, 512–23.

Drozd, J. W., Tubb, R. S. & Postgate, J. R. (1972). A chemostat study of the effect of fixed nitrogen sources on nitrogen fixation, membranes and free amino acids in *Azotobacter chroococcum. Journal of General Microbiology*, **73**, 221–32.

Englesberg, E., Squires, C. & Meronk, F., Jr (1969). The arabinose operon in *Escherichia coli* B/r: a genetic demonstration of two functional states of the product of a regulator gene. *Proceedings of the National Academy of Sciences, USA*, **62**, 1100–7.

Fisher, R. J. & Brill, W. J. (1969). Mutants of *Azotobacter vinelandii* unable to fix nitrogen. *Biochimica et Biophysica Acta*, **184**, 99–105.

Gordon, J. K. & Brill, W. J. (1972). Mutants that produce nitrogenase in the presence of ammonia. *Proceedings of the National Academy of Sciences, USA*, **69**, 3501–3.

Sen, M. & Sen, S. P. (1965). Interspecific transformation in *Azotobacter. Journal of General Microbiology*, **41**, 1–6.

Shah, V. K. & Brill, W. J. (1973). Nitrogenase. IV. Simple method of purification to homogeneity of nitrogenase components from *Azotobacter vinelandii. Biochimica et Biophysica Acta*, **305**, 445–54.

Shah, V. K., Davis, L. C. & Brill, W. J. (1972). Nitrogenase. I. Repression and derepression of the iron–molybdenum and iron proteins of nitrogenase in *Azotobacter vinelandii. Biochimica et Biophysica Acta*, **256**, 498–511.

Shah, V. K., Davis, L. C., Gordon, J. K., Orme-Johnson, W. H. & Brill, W. J. (1973*a*). Nitrogenase. III. Nitrogenaseless mutants of *Azotobacter vinelandii*: activities, cross-reactions and EPR spectra. *Biochimica et Biophysica Acta*, **292**, 246–55.

Shah, V. K., Davis, L. C., Stieghorst, M. & Brill, W. J. (1973*b*). A mutant of *Azotobacter vinelandii* that hyper-produces nitrogenase Component II. *Journal of Bacteriology*, **117**, 917–19.
Sorger, G. J. (1969). Regulation of nitrogen fixation in *Azotobacter vinelandii* OP: the role of nitrate reductase. *Journal of Bacteriology*, **98**, 56–61.
Strandberg, G. W. & Wilson, P. W. (1968). Formation of the nitrogen-fixing enzyme system in *Azotobacter vinelandii*. *Canadian Journal of Microbiology*, **14**, 25–31.
Willson, C., Perrin, D., Cohn, M., Jacob, F. & Monod, J. (1964). Non-inducible mutants of the regulator gene in the 'lactose' system of *Escherichia coli*. *Journal of Molecular Biology*, **8**, 582–92.
Wilson, P. W., Hull, J. F. & Burris, R. H. (1943). Competition between free and combined nitrogen in nutrition of *Azotobacter*. *Proceedings of the National Academy of Sciences, USA*, **29**, 289–94.

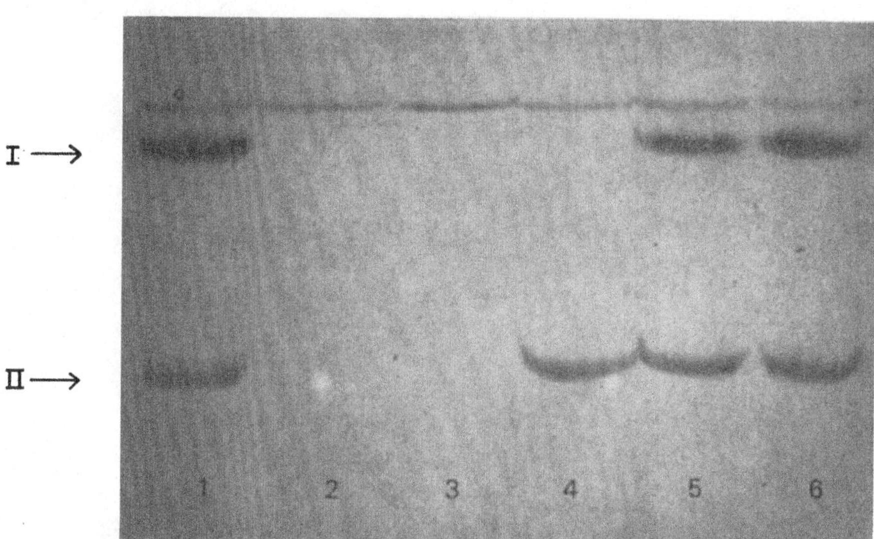

Plate 3.1. Detection of components I and II (arrowed) with the iron stain α, α-dipyridyl. Crude extracts from the wild-type and derepressed mutant strains were applied to the top of the polyacrylamide gel. After electrophoresis, the gels were stained with α, α-dipyridyl. 1, UW grown on N_2; 2, UW grown on NH_4^+; 3, UW1; 4, UW38; 5, UW46; and 6, UW91.

(*Facing p. 46*)

4. Effects of some mutations on symbiotic properties of *Rhizobium*

J. DÉNARIÉ, G. TRUCHET & B. BERGERON

In its natural environment each of the partners of the *Rhizobium*–legume symbiosis is unable by itself to fix substantial amounts of nitrogen (Bergersen, 1971). Effective nitrogen fixation occurs only when bacteria are present in their host cells: it is the phenotypic expression of two associated genomes. Cytological studies have elucidated the different steps involved in the process of effective nodule formation, while biochemical studies have dealt essentially with the working of mature nodules; however, very little is known of the biochemical nature of the bacteria – plant interactions which lead to nodule formation and nitrogen fixation. A method for genetic and biochemical analysis of the process consists of introducing into the genetic system a single point mutation and then studying the phenotypic consequences on the *Rhizobium*–legume system. A major difficulty in the analysis is that, *a priori*, the genes of *Rhizobium* involved in symbiosis cannot be selected directly *in vitro* but have to be revealed in the presence of the host plant. Our first studies, therefore, were aimed at finding *Rhizobium* mutants which showed associations between their bacterial characters *in vitro* and their symbiotic properties (*in vitro–in vivo* associations).

Schwinghamer (1975) has recently comprehensively reviewed *Rhizobium* genetics, so in this chapter we will focus on some particular aspects of the subject. We will discuss briefly the results concerning some classes of mutants (resistant to bacteriophages, antibiotics and antimetabolites) which suggest that the bacterial cell wall and/or membrane play a key role in the successful establishment of symbiosis. We will then examine work on independent auxotrophic mutants which show that leucine, purine and pyrimidine synthesis by *Rhizobium* are needed for the development of effective nodules. Finally we will discuss the isolation of mutants for late functions in symbiosis and improvement of strain efficiency.

Mutants resistant to bacteriophages, antibiotics and antimetabolites

A change of the in-vitro phenotype associated with a modification of the symbiotic properties was first described in *R. trifolii* mutants resistant to bacteriophages. Kleczkowska (1950, 1965) found that more than 50 % of phage-resistant mutants showed modified symbiotic properties. Three mechanisms of resistance to bacteriophage can be roughly differentiated; (1) phage adsorption on the bacterial surface, (2) phage nucleic acid penetration, and (3) phage nucleic acid restriction or non-replication, or defect in capsid synthesis. The first two mechanisms imply the participation of the bacterial cell wall. Recently Atkins & Hayes (1972) have studied the surface modifications of four *R. trifolii* mutants resistant to two bacteriophages. The changes concern phage adsorption, microelectrophoretic mobility, and sensitivity to lipase and lysozyme. The authors suggested that those changes were due to alterations of the composition or of the organization of cell wall polymers; the symbiotic properties of these mutants were not reported.

Schwinghamer (1964, 1967) studied the association between antibiotic resistance and modification of effectiveness in spontaneous mutants of *R. leguminosarum*, *R. meliloti* and *R. trifolii*. He distinguished three groups of antibiotics according to the modification of the symbiotic properties. In group I, resistant to chloramphenicol, spectinomycin, spiramycin and streptomycin, most of the mutants did not lose effectiveness. These antibiotics inhibit protein synthesis. For spectinomycin and streptomycin a modification of some ribosomal proteins is responsible for resistance in *E. coli* (Flacks *et al.*, 1966). These mutations can be used as genetic markers for ecological studies (Imshenetskii, Pariiskaya & Erraiz Lopez, 1970; Obaton, 1971; Schwinghamer & Dudman, 1973). In group II, resistant to novobiocin, penicillin and vancomycin, there is a partial or total loss of effectiveness in about 50 % of mutants. These antibiotics are known to inhibit cell wall or cell membrane synthesis. In group III, resistant to neomycin and viomycin, most of the mutants are completely ineffective. The mode of action of these antibiotics involves an inhibition of protein synthesis.

Jordan and co-workers have studied the biochemical basis of viomycin resistance (Vio^r) in *R. meliloti*. No antigenic difference was found between sensitive and resistant strains (Hendry & Jordan, 1969). The Vio^r mutants accumulate phospholipids in the cell wall (McKenzie & Jordan, 1970, 1972) and have fewer negatively charged sites on the envelope (Yu & Jordan, 1971). They have no N_2-fixing (C_2H_2 reducing) activity and no bacteroids could be seen in the nodules (Hendry & Jordan, 1969). Vegetative bacteria are not converted into bacteroids and degenerate (Gourret & Amarger, personal communication). There appear to be at least two classes of mutants resistant to neomycin: one is cross-resistant with kanamycin and effective, the other is sensitive to kanamycin and ineffective (Schwinghamer, 1964). Kanamycin-resistant mutants (*Kan*^{r}) are also of two kinds; effective or ineffective. Ineffective *Kan*^{r} mutants of *R. trifolii* and *R. meliloti* produce more polysaccharide than the effective parents (Damery & Alexander, 1969).

Surprisingly all the antibiotic-resistant mutants studied, except one type, are infective. The exception is among *Str*^{r} strains of *R. trifolii*, resistant to high concentrations of streptomycin (Żelazna-Kowalska, 1971). Effective strains became avirulent when transformed to *Str*^{r}. These mutations also showed pleiotrophic effects such as the modification of colonial morphology.

Schwinghamer (1968) has isolated a large number (530) of spontaneous mutants of *R. leguminosarum*, *R. meliloti*, *R. phaseoli* and *R. trifolii* resistant to metabolic inhibitors, mainly D-amino acids and amino acid analogs. Nearly half of these were defective in symbiosis, i.e. either partly effective, ineffective (*Eff*$^-$) or non-nodulating (*Inf*$^-$). There were no clearcut differences in these defects between the D-amino acid resistant and the analog resistant group. Moreover there was cross-resistance between the two groups. A number of these mutants showed altered nutritional requirements; but the ratio of *Eff*$^-$ to *Eff*$^+$ strains was similar among auxotrophic mutants and prototrophic mutants. The antimetabolites produced spheroplasts when added to wild-type strains. There was no cross-reaction between antimetabolite, viomycin and bacteriophage resistance. These results led Schwinghamer (1968, 1974) to suggest (1) that loss of cell wall or membrane integrity through mutations could, for most of the ineffective mutants, be involved in resistance to antimetabolites, group II and group III antibiotics or bacteriophages; and (2) that cell wall or membrane alterations cause a failure of rhizobia to develop into bacteroids. These are referred to as hypotheses (1) and (2) below.

Resistant mutants are easily isolated without the use of mutagenic agents. Multiple mutations are thus less likely to have occurred, however some mutations could be deletions. It is generally quite difficult to isolate sensitive revertants of resistant mutants. Genetic transfer could prove that the same gene controls both the cultural behavior and the symbiotic property, as demonstrated so far only in *R. trifolii* by means of transformation (Zelazna-Kowalska & Lorkiewicz, 1971). In the absence of such direct evidence the high frequency of a given *in vitro–in vivo* relationship supports the single mutation hypothesis. But in all the cases described the association is never absolute; mutants of different symbiotic properties are found in each gross phenotypic in-vitro class; only the proportion of *Eff*$^{+}$, *Eff*$^{-}$, *Inf*$^{-}$ varies between the different classes. This points out the need for detailed studies of individual mutants within each class.

The biochemical studies of Jordan and his co-workers on viomycin resistance support hypothesis (1), at least for the *Vio*^{r} class of mutants. Similar studies would be interesting with ineffective mutants in other classes. It is difficult to prove directly that cell wall or membrane alterations are responsible for ineffectiveness; the defects cannot be restored biochemically as with auxotrophic mutants.

Alterations of symbiotic properties have been studied in detail on viomycin-resistant mutants of *R. meliloti*; their inability to develop into bacteroids supports hypothesis (2). It would be of interest to determine at which step the process of nodule formation is blocked in other types of ineffective mutants. Conversion of vegetative cells of rhizobia into bacteroids in the host cells is associated with changes in the bacterial cell envelope structure as revealed by freeze-etching electron microscopy (McKenzie, Wail & Jordan, 1973). This envelope also undergoes some biochemical modifications as established by studies of sensitivity to lysozyme, lauryl sulfate and deoxycholate (Van Brussel, personal communication).

Auxotrophic mutants

Association between nutritional requirements and symbiotic properties

Some antimetabolite-resistant mutants isolated by Schwinghamer (1968) showed partial or multiple nutritional requirements. A D-histidine resistant mutant of *R. trifolii* was auxotrophic for riboflavin (*Rib*$^{-}$) and ineffective; the prototrophic revertants recovered effectiveness.

In the genetic analysis of symbiosis multiple mutations can lead to wrong conclusions about the association between a nutritional requirement and a defect in symbiosis. In order to avoid confusion the use of methods such as back-mutation, transformation or transduction is necessary. Accordingly Lorkiewicz & Melke (1970) studied prototrophic revertants of an ineffective histidine-requiring mutant of *R. trifolii*, and Kowalski (1971) used transduction with an ineffective lysine-requiring mutant of *R. meliloti*.

Using nitrosoguanidine (NTG) Schwinghamer (1967) obtained a *R. leguminosarum* strain with a growth requirement for adenine and thiamine which is ineffective on one pea variety and non-nodulating on another. All the prototrophic revertants were again fully effective.

Lorkiewicz and co-workers (1971) using ultraviolet irradiation of *R. trifolii* selected eighteen auxotrophic mutants; all were either ineffective or avirulent. Studies of their prototrophic revertants gave unexpected results as all were still defective, and a higher proportion were avirulent. To account for these results, Schwinghamer (1975) suggested that plasmid control of symbiotic properties was involved.

Using NTG mutagenesis and penicillin enrichment, Scherrer & Dénarié (1971) isolated *R. meliloti* auxotrophs; they found an *in vitro–in vivo* relationship. Four purine- or pyrimidine-requiring mutants (*Ade*$^-$ or *Ura*$^-$) were ineffective whereas mutants auxotrophic for the amino acids glycine, cysteine and methionine were effective (see Table 4.1).

This association between nutritional defectiveness of mutants and their symbiotic capacity has recently been confirmed (Dénarié & Bergeron, 1975); thirteen purine- or pyrimidine-requiring mutants out of fifteen were ineffective. Among thirty-four mutants auxotrophic for arginine, glycine, cysteine, methionine, phenylalanine and tryptophan, thirty-two were effective and two slightly less effective than the wild-type strain. On the contrary all the eleven leucine-requiring (*Leu*$^-$) strains were ineffective. The twenty isoleucine–valine dependent mutants (*Ilv*$^-$) gave different responses: some were effective, some ineffective and others intermediate. Statistical analysis of the acetylene-reducing activity of plants inoculated with *Ilv*$^-$ mutants showed that the variation coefficient was much higher in the group of intermediate strains than in the *Eff*$^-$ and *Eff*$^+$ group, suggesting that intermediate response may be due to the presence of revertants among some *Ilv*$^-$ strains. Two thiamine-requiring strains were fully effective.

Ten prototrophic revertants were isolated *in vitro* from each ineffective mutant (*Ade*$^-$, *Leu*$^-$, *Ura*$^-$, *Ilv*$^-$); all recovered their effectiveness. In *Leu*$^-$ strains recovery of prototrophy by transduction (Kowalski & Dénarié, 1972) restored effectiveness. One single mutation is thus responsible for both the alteration of metabolism and the loss of effectiveness.

Similar associations have been found in other systems. Cysteine- and tryptophan – requiring mutants of *R. trifolii* were effective on clover whereas mutants auxotrophic for adenine were ineffective (Barate de Bertalmio, personal communication). Three leucine-, one adenine- and one isoleucine–valine-requiring mutant of *R. leguminosarum* were ineffective on pea whereas three mutants auxotrophic for cysteine, glycine and tryptophan were effective (J. E. Beringer, personal communication). Thus in different *Rhizobium*–legume associations the host plant can supply the bacteria with some amino acids (e.g. cysteine, methionine, arginine, glycine, tryptophan, phenylalanine) during the whole symbiotic process. On the other hand the alteration of the bacterial biosynthesis of purines, pyrimidines and leucine cannot be balanced by the host during the whole process. It is worthy of note that all the auxotrophic mutants are virulent (with one exception on the variety of pea Home Freezer). Suspensions of bacteria in distilled water are used for

Table 4.1. *Association between nutritional requirements and symbiotic properties of strains of* Rhizobium

Rhizobium: species and strains	Host Plant: species and cultivars	Nutritional requirements of mutants	Number of mutants	Symbiotic properties*			
				E	PE	I	NN
R. leguminosarum	*Pisum sativum*	Adenine	1				
L4[a]	Home Freezer	+ thiamine					1
	Freezonian					1	
R. trifolii	*Trifolium pratense*	Riboflavine	1				
T1[b]	(Kenland)					1	
	Trifolium subterraneum						
	(Mt Barker)			1			
R. meliloti	*Medicago sativa*						
Sa10[c]	Du Puits	Adenine	1			1	
2011[d]		Adenine	2			2	
		Uracil	1			1	
		Glycine	3	3			
		Methionine	2	2			
		Cysteine	2	2			
L5–30[e]	Du Puits	Adenine	10	1	1	8	
		Uracil	2			2	
		Thiamine	2	2			
		Methionine	8	7	1		
		Cysteine	2	2			
		Glycine	1	1			
		Arginine	3	3			
		Ornithine	8	7	1		
		Tryptophan	1	1			
		Phenylalanine	4	4			
		Leucine	11		2(lk)	9	
		Isoleucine + valine	20	4	6	10	
		Isoleucine + valine + leucine	3			3	

* E, effective: I, ineffective; PE, partly effective; NN, non-nodulating. For *R. meliloti* E represents no significant difference with wild-type strain at $P = 0.01$, I represents no significant difference with uninoculated controls at $P = 0.01$. lk, leaky auxotrophs.

Data after [a] Schwinghamer (1967); [b] Schwinghamer (1970); [c] Dénarié (1969); [d] Scherrer & Dénarié (1971); [e] Dénarié & Bergeron (1975). Assay methods used for different strains were: visual rating of effectiveness for L4; dry weight of plants for T1, Sa10 and 2011; and dry weight of plants or acetylene-reducing assay for L5–30.

inoculation so that it can be assumed that the host can supply rhizobia with essential metabolites, including purines, pyrimidines and leucine, during the first steps of the infection processes and nodule initiation.

Restoration of effectiveness

Schwinghamer (1970) described the full restoration of effectiveness in a *Rib*$^-$ *R. trifolii* strain by adding riboflavin to the *Rhizobium*–clover system. The addition of two flavin-containing co-enzymes also restored efficiency; flavin mononucleotide was more active than flavin adenine dinucleotide. An *Ade*$^-$ mutant of *R. leguminosarum* avirulent for the pea variety Home Freezer formed nodules when adenine was added to the system, but effectiveness was not restored. Addition of L-leucine to a *Leu*$^-$ *R. meliloti*–lucerne association restored effectiveness completely (Truchet & Dénarié, 1973*b*).

Evidence of this kind demonstrates a direct biochemical relationship between the metabolic alteration of some auxotrophic mutants and the alteration of symbiotic properties. It suggests, at least for riboflavin and leucine auxotrophs, that the end-product of these pathways is involved.

Even in the cases of *in vivo–in vitro* associations, ineffective auxotrophic mutants can be expected to occur which would not be restored by an exogenous supply of the end-product. For instance those in which the mutated enzyme either (*a*) plays a role not only in the metabolic pathway concerned but in another pathway involved in symbiosis or (*b*) is situated before a branching point in a branched pathway, one branch being involved only in symbiosis or (*c*) plays a role in the transfer of fixed nitrogen compounds from bacteria to the host plant (e.g. enzymes for the synthesis of asparagine and glutamine).

The fact that among the *Ilv*$^-$ mutants some are effective and some are not, could mean that the end-products (isoleucine and/or valine) are not directly involved.

Ultrastructure of the ineffective nodules

Alteration of a *Rhizobium* gene can induce both a modification of one cultural characteristic of the bacteria and the loss of effectiveness. It is important to know at which step the nodule development is stopped; that is, which step of the process is under the control of the mutated gene.

Pankhurst, Schwinghamer & Bergersen (1972) studied the ultrastructure and acetylene-reducing activity of root nodules formed by a riboflavin-requiring mutant of *R. trifolii*. A large proportion of the bacteria in the host cells failed to develop into bacteroids; acetylene reduction was very low. Addition of riboflavin induced transformation of bacteria into bacteroids and restored nitrogen-fixing activity. The addition at different intervals after inoculation showed that the riboflavin requirement was particularly important from one to three days after nodule formation. This requirement seems greater for bacteroid formation than for vegetative growth. Two hypotheses were proposed: either riboflavin is required for increased electron transport in bacteroids or it has some specific role in bacteroid formation.

Truchet & Dénarié (1973*a*, *b*) studied ineffective nodules formed by three *Ade*$^-$, one *Ura*$^-$ and three *Leu*$^-$ mutants of *R. meliloti.* Those formed by *Ade*$^-$ mutants were very similar to those formed by *Rib*$^-$ strains, having a very low acetylene-reducing activity and containing few bacteroids (see Plate 4.1). It appears, therefore, that the demand for purines is greater for bacteroid formation and especially for nitrogen fixation. At these steps RNA and DNA content does not increase (Dilworth & Williams, 1967). It may be assumed that purine needs would be due more to synthesis of hydrogen-carrying enzymes (NAD, FMN etc.) or to energy-rich molecules (ATP etc.) than to nucleic acid synthesis. The nodules formed by the *Ura*$^-$ mutant seem to be blocked at a slightly earlier step. The multiplication of bacteria in the host cells is less, as is their conversion into bacteroids, and no ethylene is produced.

The *Leu*$^-$ mutants induce root deformations which are later converted into small spherical nodules which are completely ineffective. The bacteria are unable to leave the infection thread for the host cytoplasm (see Plate 4.2). Nuclei of cells of the medullary zone are small and the nodular meristem has a very limited activity. It is interesting to note that amino-acid-requiring mutants are frequently effective (see above). Leucine seems, therefore, to play a particular role. Some preliminary experiments have shown that the leucine content in the pool in lucerne roots is even higher than those of the sulfur and aromatic amino acids. Three working hypotheses can be considered: (i) the requirement for leucine by *Rhizobium* is much greater in the plant than for any of the other amino acids studied, (ii) membrane permeability is weaker for leucine than for other amino acids, or (iii) bacteria have to supply plants with leucine to allow nodule development. Addition of leucine completely restores effectiveness. The addition of precursors or derivatives of leucine at different steps of nodule formation could give some information on this problem.

Kinetics of nodule formation

Fig. 4.1 represents the kinetics of nodule formation for the wild-type strain, for typical strains of the ineffective *Ade*⁻ and *Leu*⁻ mutants and for a *Met*⁻ mutant representative of the effective mutants auxotrophic for various amino acids and thiamine. As far as *Ade*⁻ and *Leu*⁻ mutants are concerned, the nodules seem to appear rather late; hence the symbiotic process must be modified early. Generally the effective auxotrophs

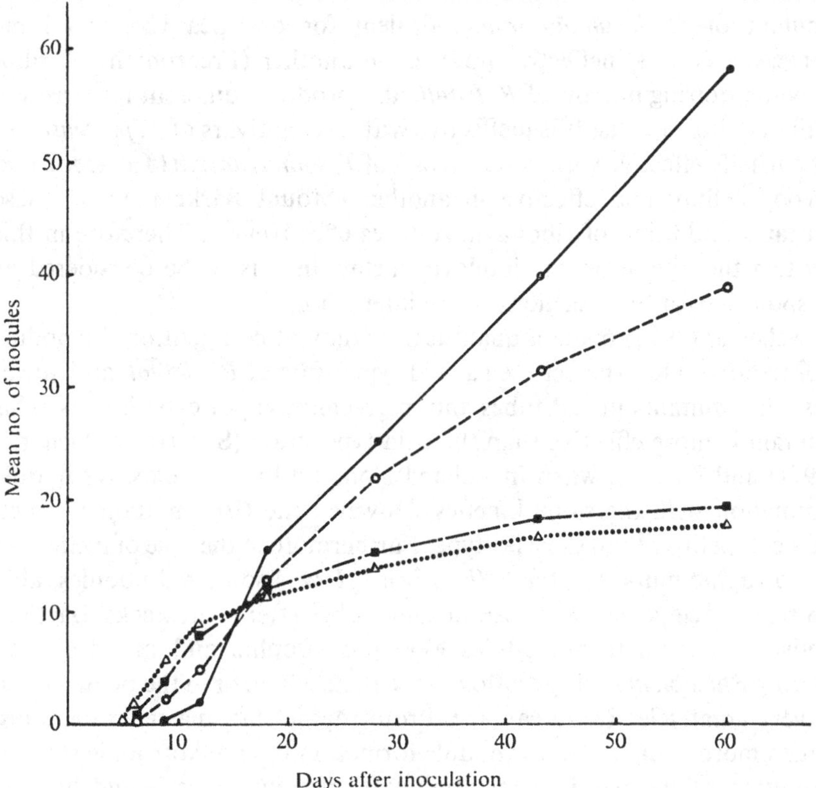

Fig. 4.1. Kinetics of nodule formation for ●——●, *Leu*⁻ mutant; ○———○, *Ade*⁻ mutant; ■—·—■, *Met*⁻ mutant; and △...△, wild-type strain.

show no significant difference as compared with the wild-type strains, but there is a greater variability between the plants for number of nodules than for dry weight. It seems, however, than the *Gly*⁻ and *Try*⁻ mutants form nodules earlier than some of the *Arg*⁻ and *Met*⁻ mutants.

It must be noted that about three weeks after inoculation the ineffective *Leu*⁻ and *Ade*⁻ strains produce many more nodules than the wild-type strain. Eight weeks after inoculation there are twice as many nodules for *Ade*$^-$ and three times as many for *Leu*$^-$ mutants. At this time the *Leu*$^-$ curve has not yet reached a plateau, as if the regulation mechanisms of nodule initiation were not working.

Specificity and competition

Schwinghamer (1967, 1970) has established a specific bacteria–host interaction at the metabolic level. An adenine and thiamine dependent mutant of *R. leguminosarum* avirulent for one pea variety (Home Freezer) induces ineffective nodules on another (Freezonian). A riboflavin-requiring mutant of *R. trifolii* also produces different responses in different host plants. It is ineffective with two cultivars of *T. pratense* but is partially effective with two cultivars of *T. subterraneum* (Tallarook and Woogenellup) and effective in another (Mount Barker). In all these instances addition of riboflavin restores effectiveness. Therefore in this system the alteration of riboflavin metabolism is to be considered as responsible for the specificity of the interaction.

Scherrer (1973) made a quantitative study of competition for nodule formation on lucerne between a wild-type strain of *R. meliloti* and one of its *Gly*$^-$ mutants in test tubes and in greenhouse pot experiments. This mutant is more effective than the wild-type strain (Scherrer & Dénarié, 1971) and induces, when inoculated alone on lucerne, effective nodule formation with the same kinetics. However the *Gly*$^-$ mutant is much less competitive than the wild-type. Furthermore in the case of ineffective auxotrophic mutants (*Ade*$^-$, *Ilv*$^-$, *Leu*$^-$, *Ura*$^-$) some red nodules, able to reduce acetylene, were seen in some tubes after eight weeks. Bacteria re-isolated from these nodules were prototrophic and probably had during *Rhizobium* multiplication in the rhizosphere or in the plants roots, a very great selective advantage. Prototrophic rhizobia seem therefore much more competitive for nodule formation than auxotrophic strains. Addition of the required metabolite at different times would help to determine the step in nodule formation at which the alteration of a metabolic pathway plays a role in the competition.

Mutants for late functions in symbiosis

Mutations for ineffectiveness described above either block nodule formation at an early stage (release from infection threads) as do the *Leu*$^-$ mutations, or at an intermediate one (multiplication of bacteria in the host cytoplasm and conversion into bacteroids) as do the *Vio*^{r}, *Ura*$^-$,

Rib⁻ and *Ade*⁻ mutations. Attempts are in progress in various laboratories to get mutations in genes controlling late functions in symbiosis, especially mutations of the nitrogenase complex itself.

Mutants of *R. meliloti* showing lack or decrease of nitrate reductase activity have been isolated (Kondorosi *et al.*, 1973). One class of these mutants is ineffective: this could be explained by the alteration of either (1) a common subunit of nitrogenase and nitrate reductase, (2) the membrane, or (3) the molybdenum mobilization system. Nitrate reductase mutants can be detected directly in Petri dishes.

Another approach consists of selecting mutants primarily on *symbiotic characters* and trying then to analyze them genetically and/or biochemically. A genetic method was used by Kowalski (1971) who found in a transducible strain of *R. meliloti* a *Lys*⁻ mutant carrying at least two mutations, one controlling lysine synthesis, the other controlling effectiveness. These two genes were not cotransducible. The mutagen NTG is thought to act preferentially at the replication point of the chromosome (Cerdá-Olmedo, Hanawalt & Guerola, 1968), and with high concentrations of NTG, it is possible to get multiple point mutations in the same chromosome region (Guerola, Ingraham & Cerdá-Olmedo, 1971). Using a high NTG concentration on *R. meliloti* we isolated a *Leu*⁻ mutant B11 whose prototrophic revertants were still ineffective. On the other hand most prototrophic transductants (more than 80 %) obtained using phages multiplied on effective prototrophic wild-type bacteria, were effective. This strain probably carries two different genes, one *leu* gene controlling leucine synthesis, one *eff-1* gene controlling effectiveness. The prototrophic ineffective revertants and transductants induce nodules which do not reduce acetylene, but nodule formation kinetics, shape and color, are the same as for the wild-type. Electron microscopy studies show that bacteria are transformed into bacteroids and that leghemoglobin is present between the membrane envelope and bacterial cell wall. Thus it is concluded that the *eff-1* gene controls a very late function in nodule formation and is situated very near a *leu* locus (less than 20 % recombination is found; see Kowalski, Chapter 5).

A biochemical approach is used by W. J. Brill and co-workers (personal communication). Mutagenized rhizobia are inoculated on homologous plants and ineffective strains detected by the acetylene reduction assay. Ineffective mutants still having nitrogenase components are then identified by immunological techniques; some nitrogenase mutants can be expected among them.

Improvement of effectiveness

Genes controlling some auxotrophic characters control expression of nitrogen fixation as well. Increase in the number of available alleles of these genes is easy to obtain by isolating prototrophic revertants of ineffective auxotrophic mutants. Schwinghamer (1970) proposed the selection of mutants from within such revertants to get strains with improved efficiency. Scherrer & Dénarié (1971) isolated three prototrophic revertants from *Ade*$^-$ and *Ura*$^-$ strains of *R. meliloti* which are significantly more effective than the wild-type strain (see Table 4.2).

Table 4.2. *Efficiency of prototrophic revertants of* R. meliloti *as measured by dry weight of plants (After Scherrer & Dénarié, 1971)*

Strains	Nutritional requirements	Phototrophic revertants	Dry weight of plants
Control			3.2
Wild-type			11.1
2011 m 32	Adenine		2.7*
		1	15.1
		2	15.7*
		3	13.8
		4	9.3
		5	10.6
2011 m 54	Adenine		2.1
		1	8.2
		2	9.0
2011 m 69	Uracil		4.9
		1	11.6
		2	16.3†
		3	13.3
		4	10.0
		5	12.5

* Significant difference from wild-type at 5 % level.
† Significant difference from wild-type at 1 % level.

A greater improvement in the level of nitrogen fixation can be expected in the future from isolation of mutants for regulation of the synthesis and activity of the enzymes playing a direct role in nitrogen fixation.

Studies of the symbiotic properties of the auxotrophs may provide information useful for inoculant improvement. Most auxotrophs appear to show defective symbiotic properties, such as a loss of effectiveness, restriction of host range, and a very strong decrease in competitiveness for nodule formation. The only auxotrophic mutants known to be more effective than the wild-type are very bad competitors. Therefore, for inoculation purposes the use of prototrophic strains ought to be more suitable. The common use of rich media for production of rhizobia may tend to select slightly auxotrophic derivatives. These observations lead us to agree with the proposal of Schwinghamer (1974), of using a synthetic medium containing as few added growth factors as possible for the industrial production of inoculant cultures.

The authors are very grateful to Professor Lorkiewicz, Dr Kowalski and P. Boistard for critically reading the manuscript and to Miss M. O. Mirae and Dr J. E. Beringer for correcting the English translation.

References

Atkins, G. J. & Hayes, A. H. (1972). Surface changes in a strain of *Rhizobium trifolii* on mutation to bacteriophage resistance. *Journal of General Microbiology*, **73,** 273–8.

Bergersen, F. J. (1971). Biochemistry of symbiotic nitrogen fixation in legumes. *Annual Review of Plant Physiology*, **22,** 121–40.

Cerdá-Olmedo, E., Hanawalt, P. C. & Guerola, N. (1968). Mutagenesis of the replication point by nitrosoguanidine: map and pattern of replication of the *Escherichia coli* chromosome. *Journal of Molecular Biology*, **33,** 705–19.

Damery, J. T. & Alexander, M. (1969). Physiological differences between effective and ineffective strains of *Rhizobium. Soil Science*, **108,** 209–16.

Dénarié, J. (1969). Une mutation provoquant l'auxotrophie pour l'adénine et la perte du pouvoir fixateur d'azote chez *Rhizobium meliloti. Comptesrendus de l'Académie des Sciences, Paris*, Ser. D, **269,** 2464–6.

Dénarié, J. & Bergeron, B. (1975). Association entre besoins nutritionnels et propriétés symbiotiques chez des mutants auxotrophes de *Rhizobium meliloti. Canadian Journal of Microbiology.* (In press.)

Dilworth, M. J. & Williams, D. C. (1967). Nucleic acid changes in bacteroids of *Rhizobium lupini* during nodule development. *Journal of General Microbiology*, **48,** 31–6.

Flacks, J. G., Leboy, P. S., Birge, E. A. & Kurland, C. G. (1966). Mutations and genetics concerned with the ribosome. *Cold Spring Harbor Symposia on Quantitative Biology*, **31,** 623–31.

Guerola, N., Ingraham, J. L. & Cerdá-Olmedo, E. (1971). Induction of closely linked multiple mutations by nitrosoguanidine. *Nature, London,* **230,** 122–5.

Hendry, G. S. & Jordan, D. C. (1969). Ineffectiveness of viomycin-resistant mutants of *Rhizobium meliloti. Canadian Journal of Microbiology*, **15,** 671–5.

Imshenetskii, A. A., Pariiskaya, A. N. & Erraiz Lopez, L. (1970). Experimental production of mutants of *Rhizobium meliloti* with increased activity. *Microbiology*, **39,** 294–7.

Jordan, D. C., Yamamura, Y. & McKague, M. E. (1969). Mode of action of viomycin on *Rhizobium meliloti*. *Canadian Journal of Microbiology*, **15**, 1005–12.

Kleczkowska, J. (1950). A study of phage-resistant mutants of *Rhizobium trifolii*. *Journal of General Microbiology*, **4**, 298–310.

Kleczkowska, J. (1965). Mutations in symbiotic effectiveness in *Rhizobium trifolii* caused by transforming DNA and other agents. *Journal of General Microbiology*, **40**, 377–83.

Kondorosi, A., Barabás, I., Sváb, Z., Orosz, L. & Sik, T. (1973). Evidence for common genetic determinants of nitrogenase and nitrate reductase in *Rhizobium meliloti*. *Nature New Biology*, **246**, 153–4.

Kowalski, M. (1971). Transduction in *Rhizobium meliloti*. *Plant and Soil*, Special Volume, 63–6.

Kowalski, M. & Dénarié, J. (1972). Transduction d'un gène contrôlant l'expression de la fixation de l'azote chez *Rhizobium meliloti*. *Comptes rendus de l'Académie des Sciences, Paris*, Ser. D, **275**, 141–4.

Lorkiewicz, Z. & Melke, A. (1970). Infectiveness of the histidine-dependent mutant of *Rhizobium trifolii*. *Acta Microbiologica Polonica*, Ser. A, **2**, 75–7.

Lorkiewicz, Z., Zurkowski, W., Kowalczuk, E. & Górska-Melke, A. (1971). Mutagenesis and conjugation in *Rhizobium trifolii*. *Acta Microbiologica Polonica*, Ser. A, **3**, 101–7.

McKenzie, C. R. & Jordan, D. C. (1970). Cell wall phospholipid and viomycin-resistance in *Rhizobium meliloti*. *Biochemical and Biophysical Research Communications*, **40**, 1008–12.

McKenzie, C. R. & Jordan, D. C. (1972). Cell wall composition and viomycin-resistance in *Rhizobium meliloti*. *Canadian Journal of Microbiology*, **18**, 1168–70.

McKenzie, C. R., Wail, W. J. & Jordan, D. C. (1973). Ultrastructure of free-living and nitrogen-fixing forms of *Rhizobium meliloti* as revealed by freeze-etching. *Journal of Bacteriology*, **113**, 387–93.

Obaton, M. (1971). Utilisation de mutants spontanés résistants aux antibiotiques pour l'étude écologique des *Rhizobium*. *Comptes-rendus de l'Académie des Sciences, Paris*, Ser. D, **272**, 2630–3.

Pankhurst, C. E., Schwinghamer, E. A. & Bergersen, F. J. (1972). The structure and acetylene-reducing activity of root nodules formed by a riboflavin-requiring mutant of *Rhizobium trifolii*. *Journal of General Microbiology*, **70**, 161–77.

Scherrer, A. (1973). Etude quantitative de la compétition pour la formation des nodules de luzerne entre une souche de *Rhizobium meliloti* et un de ses mutants auxotrophes pour la glycine. *Annales de Phytopathologie*, **5**, 105.

Scherrer, A. & Dénarié, J. (1971). Symbiotic properties of some auxotrophic mutants of *Rhizobium meliloti* and of their prototrophic revertants. *Plant and Soil*, Special Volume, 39–45.

Schwinghamer, E. A. (1964). Association between antibiotic resistance and ineffectiveness in mutant strains of *Rhizobium* spp. *Canadian Journal of Microbiology*, **10**, 221–33.

Schwinghamer, E. A. (1967). Effectiveness of *Rhizobium* as modified by mutation for resistance to antibiotics. *Antonie van Leeuwenhoek, Journal of Microbiology and Serology*, **33**, 121–36.

Schwinghamer, E. A. (1968). Loss of effectiveness and infectivity in mutants of *Rhizobium* resistant to metabolic inhibitors. *Canadian Journal of Microbiology*, **14**, 355–67.

Schwinghamer, E. A. (1969). Mutation to auxotrophy and prototrophy as related to symbiotic effectiveness in *R. leguminosarum* and *R. trifolii*. *Canadian Journal of Microbiology*, **15**, 611–22.

Schwinghamer, E. A. (1970). Requirement for riboflavin for effective symbiosis on clover by an auxotrophic mutant of *Rhizobium trifolii*. *Australian Journal of Biological Sciences*, **23**, 1187–96.

Schwinghamer, E. A. (1975). Genetic aspects of nodulation and N_2 fixation by legumes: the microsymbiont. In: *Dinitrogen fixation* (ed. R. W. F. Hardy). Wiley, New York. (In press.)

Schwinghamer, E. A. & Dudman, W. F. (1973). Evaluation of spectinomycin resistance as a marker for ecological studies with *Rhizobium* spp. *Journal of Applied Bacteriology*, **36**, 263–72.

Truchet, G. & Dénarié, J. (1973*a*). Structure et activité réductrice d'acétylène des nodules de luzerne (*Medicago sativa* L.) induits par des mutants de *Rhizobium meliloti* auxotrophes pour l'adénine et pour l'uracile. *Comptes-rendus de l'Académie des Sciences, Paris*, Ser. D, **277**, 841–4.

Truchet, G. & Dénarié, J. (1973*b*). Ultrastructure et activité réductrice d'acétylène des nodosités de luzerne (*Medicago sativa* L.) induites par des souches de *Rhizobium meliloti* auxotrophes pour la leucine. *Comptes-rendus de l'Académie des Sciences, Paris*, Ser. D, **277**, 925–8.

Yu, K. K. & Jordan, D. C. (1971). Cation content and cation-exchange capacity of intact cells and cell envelopes of viomycin-sensitive and resistant strains of *Rhizobium meliloti*. *Canadian Journal of Microbiology*, **17**, 1283–6.

Żelazna-Kowalska, I. & Lorkiewicz, Z. (1971). Transformation in *Rhizobium trifolii*. IV. Correlation between streptomycin resistance and infectiveness in *Rhizobium trifolii*. *Acta Microbiologica Polonica*, Ser. A, **3**, 11–20.

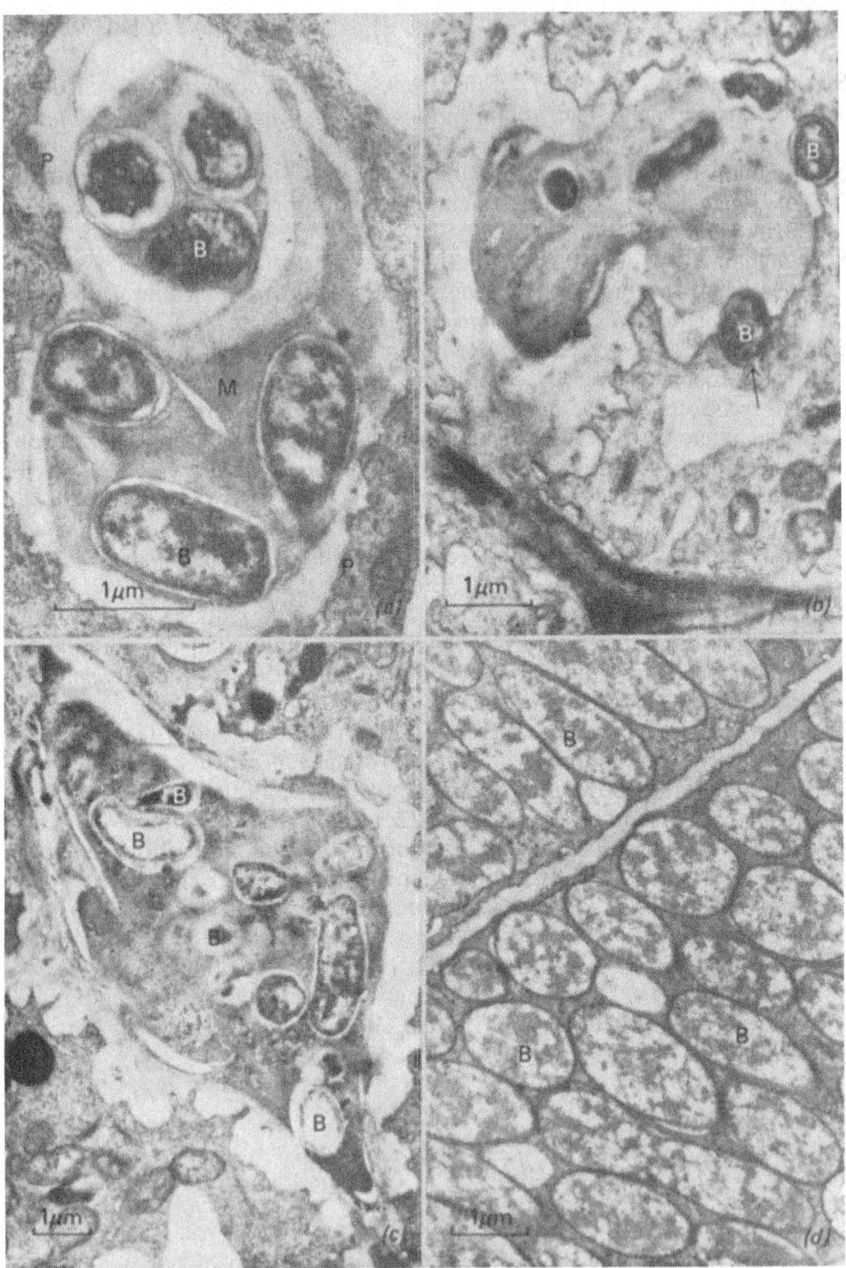

Plate 4.1. Nodule formation in auxotrophic mutants of *R. meliloti* compared with that in the prototroph. (*a*) *Ade*⁻, strain C34. Infection thread. The bacteria (B) are embedded in a matrix (M). The outer membrane of the infection thread (P) is known to be continuous with the host cell plasmalemma. (*b*) *Ade*⁻, strain C14. Release of the bacteria into the host cell cytoplasm by endocytosis. The bacteria (B) push back the plasmalemma of the host cell (arrowed). (*c*) *Leu*⁻, strain C51. Infection thread. Some bacteria (B) degenerate in the matrix of the infection thread; there is no release of bacteria into the host cytoplasm. (*d*) Normal nodule, strain L5–30. The bacteria (B) are homogenous.

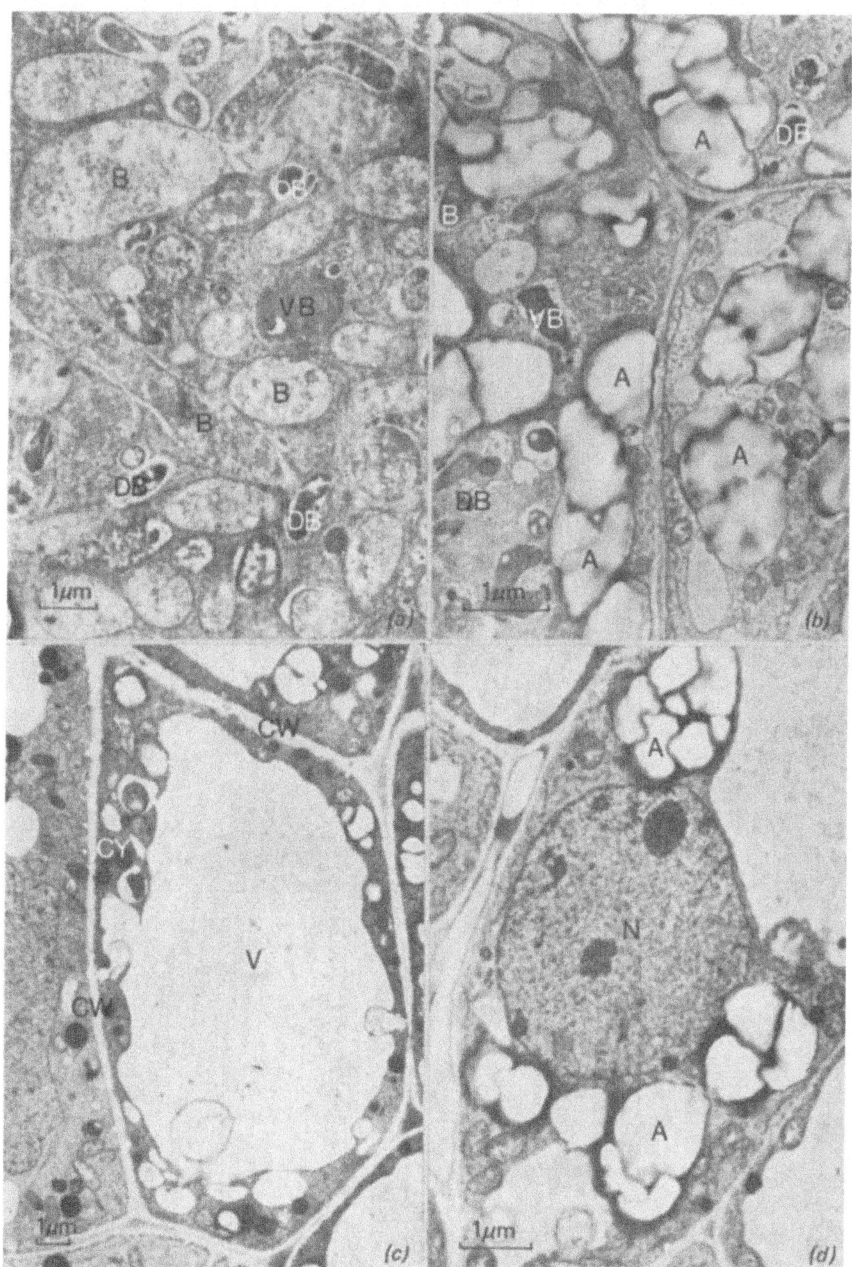

Plate 4.2. Nodule formation in auxotrophic mutants of *R. meliloti*. (*a*) *Ade*⁻, strain C62. Central zone with fully developed bacteria (B), vegetative bacteria (VB) and degenerative bacteria (DB). (*b*) *Ura*⁻, strain K307. Central zone with vegetative bacteria (VB) and degenerative bacteria (DB). Only a few bacteria are converted into bacteroids (B). The amyloplasts (A) are fully developed. (*c*) *Leu*⁻, strain K21. Distal cell of the central zone. The cell cytoplasm (CY) is reduced to a thin layer stuck to the cell wall (CW). The central vacuole (V) is developed. The cell cytoplasm is not infected by rhizobia. (*d*) *Leu*⁻, strain B13. Partial view of a proximal cell of the central zone. The nucleus (N) is small, and the amyloplasts (A) well-developed.

5. Transduction of effectiveness in *Rhizobium meliloti*

M. KOWALSKI

Studies on the genetic control of symbiotic properties of *Rhizobium* provided data on the transfer of host specificity in *R. japonicum* by transformation (Balassa, 1963) and in *R. phaseoli* by conjugation (Higashi, 1967) as well as transfer of infectivity by transformation (Marecková, 1970). Recently transformation of symbiotic effectiveness in *R. japonicum* was found (Żelazna-Kowalska, Żurkowski, Lorkiewicz & Menzies, unpublished data). In all these experiments plants were used for direct selection of host-specific, infective or effective recombinants. However, trials on the transfer of effectiveness in *R. meliloti* by transduction (Kowalski, 1970*b*) with selection of recombinants by host plants were unsuccessful. Attempts were made, therefore, to find markers easy for selection and correlated with effectiveness. Studies with auxotrophic mutants showed that leucine-dependent mutants were ineffective while prototrophic revertants and transductants became effective (Kowalski & Dénarié, 1972). This system seems to provide an opportunity for a detailed analysis of genetic control of nitrogen fixation in *Rhizobium*.

In this chapter further analysis of the relationship between effectiveness of nitrogen fixation and leucine dependence is described.

Materials and methods

Three leucine-dependent mutants, B11, B13 and B32, were described earlier (Kowalski & Dénarié, 1972). Three others, i.e. C41, C51 and E66, and one methionine-dependent, B23, were isolated by Dénarié using the same method. These mutants were isolated from the prototrophic and effective *R. meliloti* strain L5-30, a delysogenized derivative of the strain L5(L5) (Kowalski, 1967). Strain L5(L5) carrying transducing phage L5 was used in crosses as the wild-type donor. Transduction crosses were carried out with phage L5 multiplied on auxotrophic strains by the agar layer method (Adams, 1959) or UV induction of the strain L5(L5) (Kowalski, 1970*a*). Liquid cultures were grown in medium '5' (Laird, 1932) supplemented with 20 μg/ml of the required amino acid in the case of auxotrophs. For plating, the minimal medium (BS) of Bergersen modified by Sherwood (1970), without biotin and thiamine, was used and supplemented with amino acids at 20 μg/ml as required. Transducing phage lysates were prepared by the agar layer method and titrated on medium '79' (Allen, 1969). The nitrogen-free medium for inoculation tests consisted of: K_2HPO_4, 0.5 g; $MgSO_4 \cdot 7H_2O$, 0.2 g; NaCl, 0.1 g; $Ca_3(PO_4)_2$, 2.0 g; $FeSO_4$, 0.01 g; agar, 8.0 g; distilled water, 1000 ml; pH, 7.0.

For transduction, bacterial cultures (approx. 5×10^8 cells/ml) were mixed in a 1:1 ratio with transducing lysates (2×10^9 phage particles/ml), incubated for 20 min at 28 °C for phage adsorption and then plated.

In crosses where linkage of *leu*$^-$ and *met*$^-$ markers was tested, transductants were selected by plating on BS medium supplemented with the amino acid required by the donor and tested for the presence of the donor marker by subsequent plating on selective media.

Tests for effectiveness used *Medicago sativa* var. Eynsford; seeds were sterilised with ethanol and mercuric chloride according to Vincent (1970). One pre-germinated seedling per tube with nitrogen-free medium was used. After 2–4 days of growth they were inoculated with 0.2 ml of bacteria (approx. 10^7 cells/ml) suspended in saline. Results were checked after 6 weeks of plant growth in a greenhouse with a 16 hour light period. Symbiotic effectiveness was estimated on the basis of wet weight of the green part of the plants (Kowalski & Dénarié, 1972).

Results

Crosses between six *Leu*$^-$ mutants yielded recombinants with high frequency (Table 5.1), indicating that *Leu*$^-$ mutations are located in different sites. Mutations B13 and C41 seem to be closest since little or no wild-type recombinants were observed in their crosses. It is interesting to note that in both mutants 100 % cotransduction of effectiveness with *leu*$^-$ marker was observed (Table 5.2). Effectiveness of all prototrophic transductants and revertants of B13 and C41 mutants suggest that these two mutations are directly involved in the expression of effectiveness. Moreover, addition of L-leucine to plants inoculated with these two mutants caused restoration of effectiveness. This finding strongly suggests that leucine deficiency alone is a factor limiting expression of effectiveness.

In the case of the B11 mutant, loss of effectiveness seems to be caused by a mutation close to the *leu*$^-$ marker (there is less than 20 % recombination) but not identical to it, since back mutation to *Leu*$^+$ and addition of

L-leucine did not restore effectiveness. The E66 mutant has a distant mutation affecting effectiveness which is not co-transducible with the *leu*$^-$ marker.

Effectiveness of *Leu*$^+$ recombinants obtained from the crosses of *Leu*$^-$ mutants was compared with those obtained with the wild-type donor (Table 5.3). The percentage of effective clones was comparable in

Table 5.1. *Transductional crosses of* Leu$^-$ *mutants*

	Number of *Leu*$^+$ recombinants in crosses with donor*						
Recipient	B11	B13	C41	B32	C51	E66	L5(L5) wild-type
B11	0	106	44	149	280	494†	932†
B13	99	0	0	142	230	570	755
C41	310	25	0	180	445	266	656
B32	546	363	126	0	336	629	980
C51	0	18	0	0	0	27	259
E66	135†	100	53	83†	29	0	724†

* Number of recombinants from 5 plates.
† These recombinants were tested for effectiveness; results are shown in Table 5.3.

Table 5.2. *Restoration of effectiveness of leucine-dependent mutants*

Strain	Restoration of effectiveness by L-leucine*	Effectiveness of prototrophic revertants†	% frequency of *Leu*$^+$ *Eff*$^+$ cotransduction
B11	–	–	80 (59)‡
B13	+	+	100 (40)
C41	+	+	100 (28)
E66	–	–	0 (107)

* 20 μg/ml of L-leucine added 3 days after inoculation of plants with the strain tested (see Dénarié *et al.*, Chapter 4).
† Ten prototrophic revertants for each mutation were tested.
‡ The figure in parentheses gives the number of prototrophic transductants tested for effectiveness.

both crosses. These results confirm that in strains B11 and E66, leucine-dependence and effectiveness mutations are located in different sites since effective prototrophic recombinants were obtained.

Transductional crosses of the B11 mutant with auxotrophic mutants revealed that the methionine-dependent mutation (B23) was cotransduced at a frequency of 70 % with the *leu*$^-$ marker (Table 5.4).

Table 5.3. *Effectiveness of* Leu$^+$ *recombinants from crosses of B11, B32 and E66 mutants compared with those from crosses with L5(L5) wild-type donor*

Donor	Recipient	Number of *Leu*$^+$ transductants analysed	Number of transductants Effective	Number of transductants Ineffective	% effective *Leu*$^+$ recombinants
L5(L5)	B11	59	48	11	81
B32	B11	96	68	28	71
E66	B11	96	80	16	83
L5(L5)	E66	97	0	97	0
B11	E66	98	0	98	0
B32	E66	10	0	10	0

Table 5.4. *Linkage of B11 leucine-dependent mutation with B23 methionine-dependent mutation site in* Rhizobium meliloti

	Donor Recipient	Number of *Leu*$^+$ transductants analysed	Number of donor type recombinants (*Leu*$^+$ *Met*$^-$)	% cotransduction
Experiment 1	B23 × B11	104	69	70
Experiment 2	B23 × B11	393	266	68

Discussion

Leucine biosynthesis is controlled in *Salmonella typhimurium* by an operon of four structural genes plus adjacent operator region (Margolin, 1963; Sanderson, 1970). The structure of *leu*$^-$ region in *Escherichia coli* is very similar (Somers, Amzallag & Middleton, 1973). In *R. meliloti* the region controlling biosynthesis of leucine also seems to be clustered because wild-type recombinants were obtained in crosses between six mutants with lower frequency than with wild-type donor.

Results of these experiments indicate that the end-product of this operon, i.e. leucine, is necessary for symbiotic effectiveness. It was found that leucine is required in the early steps of symbiosis since cells of mutants B13 and C41 are not liberated from infection threads during infection of lucerne roots (see Dénarié *et al.*, Chapter 4).

Leucine-dependent mutants B11 and E66 showed independent mutations involved in the expression of effectiveness. Linkage of the methionine marker to the *leu*$^-$ marker of strain B11 should enable a detailed genetic analysis of this region of the chromosome and its role in determination of nitrogen fixation to be made.

Lack of cotransduction of *leu*$^-$ and effectiveness markers in the E66 mutant may be explained by the assumption that they are too distant and cannot be transferred together in the fragment of host DNA picked up by phage L5. This mutant resembles ineffective mutant L5-30 *lys*$^-$ which also did not regain effectiveness after transduction to *lys*$^+$ (Kowalski, 1970*b*).

Thanks are due to Professor Zbigniew Lorkiewicz and Dr J. Dénarié for valuable discussions. This work was supported by the Polish Academy of Sciences within the project 09.3.1.

References

Adams, M. H. (1959). *Bacteriophages.* Wiley-Interscience, New York.

Allen, O. N. (1959). *Experiments in soil bacteriology*, 3rd edition. Burges Publishing Co., Minneapolis.

Balassa, G. (1963). Genetic transformation in *Rhizobium.* A review of the work of R. Balassa. *Bacteriological Reviews*, **27**, 228–41.

Higashi, S. (1967). Transfer of clover infectivity of *Rhizobium trifolii* to *Rhizobium phaseoli* as mediated by an episomic factor. *Journal of General and Applied Microbiology*, **13**, 391–403.

Kowalski, M. (1967). Transduction in *Rhizobium meliloti. Acta Microbiologica Polonica*, **16**, 7–12.

Kowalski, M. (1970*a*). Transducing phages in *Rhizobium meliloti. Acta Microbiologica Polonica*, Ser. A, **2**, 109–14.

Kowalski, M. (1970*b*). Genetic analysis by transduction of *Rhizobium meliloti* mutants with changed symbiotic activity. *Acta Microbiologica Polonica*, Ser. A, **2**, 115–22.

Kowalski, M. & Dénarié, J. (1972). Transduction d'un géne controlant l'expression de la fixation de l'azote chez *Rhizobium meliloti. Comptes-rendus de l'Academie des Sciences, Paris*, Ser. D, **275**, 141–4.

Laird, D. C. (1932). Bacteriophage and the root nodule bacteria. *Archiv für Mikrobiologie*, **3**, 159–93.

Marecková, H. (1970). Transformation of infectivity in *Rhizobium japonicum. Zentralblatt für Bakteriologie, Parasitenkunde, Infektionskrankheiten und Hygiene*, II Abteilung, **125**, 594–6.

Margoline, P. (1963). Genetic fine structure of the leucine operon in *Salmonella. Genetics*, **48**, 441–57.

Sanderson, K. E. (1970). Current linkage map of *Salmonella typhimurium. Bacteriological Reviews*, **34**, 176–93.

Sherwood, M. T. (1970). Improved synthetic medium for the growth of *Rhizobium, Journal of Applied Bacteriology*, **33**, 708–13.

Somers, J. M., Amzallag, A. & Middleton, R. B. (1973). Genetic fine structure of the leucine operon of *Escherichia coli* K-12. *Journal of Bacteriology*, **113**, 1268–72.

Vincent, J. M. (1970). *Manual for the practical study of root nodule bacteria.* Blackwell Scientific Publications, Oxford.

6. Genetic transformation in *Rhizobium japonicum*

F. DOCTOR & V. V. MODI

Genetic transformation in *Rhizobium* was first achieved by Balassa in 1954 for antibiotic resistance (Balassa, 1957) cysteine independence (Balassa, 1960) and ability to form nodules on non-specific hosts (Balassa, 1955).

Recently in *R. japonicum* the kinetics of transformation of penicillinase and fructokinase (Gadre *et al.*, 1967) and gelatinase (Raina & Modi, 1969) markers have been studied in our laboratories. However, in this system the frequency of transformation to streptomycin resistance was found to vary between 0.002 to 0.08 % (Raina & Modi, 1972). The levels of the types of nucleases with the pH optima of 7.5 and 8.4 observed in competent and non-competent cells were the same (Peris, Modi & Raina, unpublished observations). The UV sensitive mutants isolated exhibited a lowered transformation frequency of about 10^{-5}, possibly due to the absence of enzymes concerned with excision and repair (Peris *et al.*, 1973).

Transformation frequency in *R. japonicum* is 100 to 1000 times lower than that of *B. subtilis* and *H. influenzae*. Factors responsible for this low frequency of transformation and their genetic linkage were investigated.

Materials and methods

Rhizobium japonicum Str^r strain was kindly supplied by Dr Hana Mareckova, Research Institute of Crop Production, Institute of Plant Nutrition, Prague. Different auxotrophs in this culture were obtained by mutagenesis with nitrosoguanidine (NTG). The composition of the media for the growth and development of competence has been described earlier (Raina & Modi 1971, 1972). Minimal medium with 2 % agar supplemented with required growth factors was used for selection of recombinants. Supplements were added at the following concentrations (μg/ml): L-arginine, 50; L-leucine, 50; L-isoleucine, 50; uracil, 25.

Preparation of labelled DNA

^{32}P-labelled donor DNA was prepared by growing *R. japonicum Str^r* in phosphate-free medium to which carrier-free [^{32}P] orthophosphate

(10 μci/ml) was added. The cells were collected after 18 h growth at 30 °C, washed twice with saline ethylenediamine tetra-acetic acid (EDTA) (0.15 M NaCl, 0.1 M EDTA, pH 7.0) and resuspended in 2 ml saline EDTA at pH 8.0. The suspended cells were treated with 0.1 % sodium lauryl sulphate at 55 °C for 30 min and incubated further with pronase (Calbiochem preparation, 1 mg/ml) at 37 °C for 4 h. The lysate was deproteinised with chloroform and isoamyl alcohol (24 : 1) and DNA precipitated with cold ethanol. The fibrous material was dissolved in 0.1 × SSC (0.015 M trisodium citrate, + 0.15 M sodium chloride at pH 7.0) and treated with ribonuclease (Sigma, 100 μg/ml) for 30 min at 37 °C, deproteinised after bringing the SSC strength to 1 × SSC and centrifuged. The upper layer was directly charged onto a Sephadex G-50 column (20 × 2 cm) equilibrated as well as eluted with 1 × SSC. Fractions (3 ml each) were collected at a flow rate of 0.5 ml/min. A 0.1 ml sample of each fraction was dried on filter paper (25 mm diameter) and counted in a liquid scintillation spectrometer (Packard) using a toluene-based scintillation liquid. Fractions numbered 4 to 7 contained relatively high molecular weight DNA in a concentration range of 15–20 μg of DNA per 0.1 ml, as estimated by the absorbance at 260 nm, taking 1 unit of optical density (OD) as equivalent to 50 μg of DNA.

Transformation

All transformations were performed in a final volume of 1 ml (0.4 ml of recipient cells, 0.1 ml (5 μg) of DNA and 0.5 ml of competence medium). The reaction mixture in the tubes was incubated at 30 °C on a rotary shaker (150 rev/min). After an appropriate period of incubation the reaction was terminated with deoxyribonuclease (100 μg, in 10^{-3} M $MgCl_2$), and the incubation continued for about 3 min. The reaction mixture was then chilled in an ice bath and the cells collected by centrifugation in the cold. The cell pellet was resuspended to a final volume of 1 ml in 1 × SSC. In experiments in which ^{32}P-labelled DNA was used, 0.05 ml of the suspension was dried on a paper disc and counted.

Transduction

After 16 h the culture was centrifuged and the cells suspended in a complex medium. After 2 h growth at 30 °C, the cells were mixed with phage propagated on the prototrophic wild-type strain of *R. japonicum*, incubated for 30 min at 30 °C and plated on appropriate selective media for scoring the recombinants.

Results and discussion

Irreversibly bound DNA and its inactivation

The recipient culture was exposed to donor DNA for 10 min followed by DNase treatment for 3 min and then the sample removed; the mixture was incubated further for 23 min. The 3 and 23 min samples were centrifuged, the cells washed twice with ice-cold saline and lysed in saline with 0.05 % SDS. The controls included DNA solution used in the experiments, DNA alone diluted to the same extent and assayed for transforming activity.

The amount of DNA irreversibly bound to the competent cells was determined by using [^{32}P]DNA and the percentage of bound DNA calculated from the proportion of the counts retained after treatment with DNase compared to those of the input DNA (at zero time); 14 % of the added DNA was thus shown to be irreversibly bound. Earlier Raina & Modi (1972) observed that the binding was temperature dependent and exonuclease independent.

Table 6.1. *Assay of lysates of transformed* (Str^r *transformants*) *recipient culture for donor activity in* R. japonicum

Lysates	Transformants/ml	% inactivation
Control (first transformation)	3.1×10^3	
Pre-uptake lysate	3.0×10^3	
Adjusted figure on basis of 14 % irreversibly bound DNA	4.3×10^2	
3 min	2.1×10^2	52
23 min	6.4×10	86

The transforming activity of the control donor DNA for *Str^r* marker was found to be 3.0×10^3 (Table 6.1) when exposed to recipient cells. The free DNA control also gave the same amount of activity (3.1×10^3/ml) as the pre-uptake lysate showing that the procedure of DNA addition and subsequent lysis of the recipient cells presumably does not alter the transforming activity of the added DNA. The above value of donor activity for the transformed culture is expected if the culture adsorbs irreversibly all the DNA to which it has been exposed and the donor DNA does not undergo any change in activity during transformation.

Multiplying the number of *Str*^{r} transformants obtained in the assay of the pre-uptake lysate by the percentage of irreversibly bound DNA gives 4.3×10^2 transformants/ml assay mixture. A comparison of this figure with the number of *Str*^{r} transformants/ml of the assay mixture obtained from the 3 and 23 min lysates, revealed that 48 % of the *Str*^{r} donor activity irreversibly adsorbed by the recipient culture was present after 3 min and that 14 % remained at 23 min after the termination of uptake. It may be that the progressive loss of donor activity observed after the uptake was due to the inactivation of the donor DNA. It is thus possible that one of the reasons for the low frequency of transformants observed in *R. japonicum* may be this progressive inactivation of transformable DNA.

Competence in R. japonicum

The onset of competence in *R. japonicum* occurs in the late log phase (Fig. 6.1). The sharp peak of competence observed after 8 hours at 30 °C is maintained only for a brief period of 30 minutes, indicating that competence in this organism is short-lived, and requires a specific physiological state in the recipient.

Increases in the number of transformants of approximately threefold and twofold have been observed in anaerobically (under N_2 atmosphere) and statically incubated cultures respectively (Table 6.2). A similar

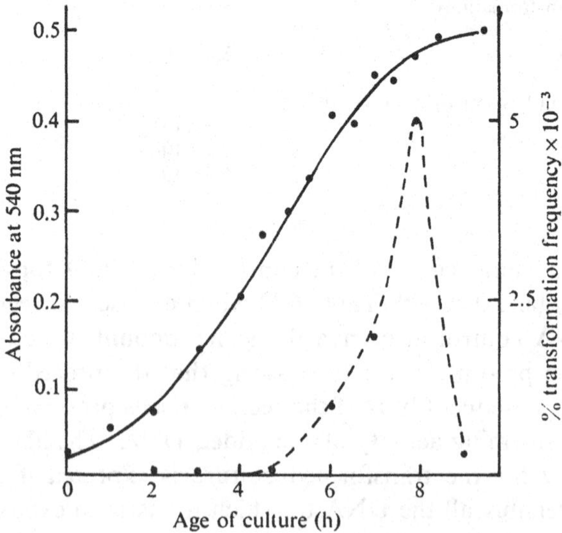

Fig. 6.1. Development of competence in *R. japonicum* in relation to age of culture. ●——●, absorbance at 540 nm; ●---●; percentage transformation frequency × 10.

observation on the stimulation of the development of competence in *H. influenzae* has been reported (Miller & Huang, 1972). The results suggest that a biochemical latency is essential for the development of competence. The threefold increase in the number of transformants may be due to arrested chromosome replication facilitating the exchange of markers.

Table 6.2. *Effect of aeration on competence in* R. japonicum

Growth condition*	Viable cell counts/ml	Transformants/ml	% frequency of transformation
Aerobic (shaken)	5.0×10^8	1.7×10^4	0.003
Aerobic (static)	2.0×10^8	1.1×10^4	0.005
Anaerobic	1.5×10^8	1.2×10^4	0.008

* After 6 h in the competence medium, with the culture shaken, it was incubated as indicated. At 8 h the culture was treated with 5 μg/ml of donor DNA. F_6 (*leu*⁻) culture was used as the recipient.

Table 6.3. *Transformation of a* ura⁻, ileu⁻ *mutant* (F_4) *with a* leu⁻ *donor* (F_6)

Selected marker	No. of transformants tested	No. carrying unselected markers	
		ura⁺	*leu*⁺
ileu	100	49	52
		ileu⁺	*leu*⁺
ura	100	56	70

Mapping of biochemical markers

Transformation In the three-point cross using F_4 (*ileu*⁻, *ura*⁻) as a recipient and F_6 (*leu*⁻) as a donor, amongst the *ileu* transformants, 49 % were *ura*⁺ and 52 % were *leu*⁺, while among the *ura* transformants selected 56 % were *ileu*⁺ and 70 % were *leu*⁺ (Table 6.3) which shows that *leu* is much closer to *ileu* than *ura*, as 48 % of the *ileu* transformants have incorporated the *leu* allele of the donor.

In another transformation cross with *R. japonicum* F_{68} (*ileu*⁻, *val*⁻) as recipient and F_6 (*leu*⁻) as donor, all the *ileu*⁺ transformants were also *val*⁺ indicating a common biosynthetic pathway. When *ileu*⁺ transformants were selected, 35 % were *ileu*⁺, *leu*⁺ which shows that 65 % of *ileu*⁺, *val*⁺ transformants have incorporated the *leu* allele along with *ileu*⁺ *val*⁺ donor allele (Table 6.4).

Table 6.4. *Transformation of a* val$^-$, ileu$^-$ *mutant* (F_{68}) *with a* leu$^-$ *donor* (F_6)

Selected markers	No. of transformants tested	No. carrying unselected markers	
		ileu$^+$	*ileu*$^+$, *leu*$^+$
val$^+$	100	100	35
		val$^+$	*val*$^+$, *leu*$^+$
ileu$^+$	100	100	34

Table 6.5. *Transformation of an* isoleu$^-$, ura$^-$, val$^-$, arg$^-$ *mutant* (F_{78}) *with wild-type* donor*

Selected marker(s)	Transformants/ml	% transformation frequency
arg$^+$	1.3×10^3	0.0021
ileu$^+$	1.5×10^3	0.0024
val$^+$	1.2×10^3	0.002
ileu$^+$, *ura*$^+$	3.4×10^2	0.00052
ura$^+$, *val*$^+$	2.0×10^2	0.00032
ileu$^+$, *val*$^+$,	1.6×10^3	0.0026
ileu$^+$, *arg*$^+$	3.2×10^2	0.00052
val$^+$, *arg*$^+$	2.1×10^2	0.00034
ura$^+$, *arg*$^+$	1.53×10^3	0.0025
ileu$^+$, *ura*$^+$, *val*$^+$, *arg*$^+$	1.8×10^2	0.00029

* *R. japonicum* 211^s: viable count, 6.12×10^7/ml, DNA concentration, 5 μg/ml.

Table 6.6. *Transduction of an* isoleu$^-$, ura$^-$ *mutant* (F_4) *with phage induced from a tryptophan revertant* (trp^r)

Selected marker(s)	% transduction frequency
ura$^+$	3.1×10^{-3}
ileu$^+$	Nil
ileu$^+$, *ura*$^+$	Nil

Viable count, 9.6×10^7; multiplicity of infection, 1.0.

Table 6.7. *Transduction of an* ileu$^-$, val$^-$, ura$^-$, arg$^-$ *mutant* (F_{78}) *with phage induced from tryptophan revertant* (trp^r)

Medium	Selected marker(s)	% transduction frequency
M* + ileu + val	*ura*$^+$, *arg*$^+$	3.1×10^{-3}
M + ileu + val + arg	*ura*$^+$	2.9×10^{-3}
M + ileu + val + ura	*arg*$^+$	Nil
M + val + ura + arg	*ileu*$^+$	Nil
M + ileu + ura + arg	*val*$^+$	Nil

Viable count, 2.6×10^8. Multiplicity of infection, 1.0. * M, minimal medium.

Mutant F_{78}, which is auxotrophic for valine, isoleucine, arginine and uracil, has a block in ornithine transcarbamylase (OTCase) and in the arginine (arg I) biosynthetic pathways. Transformation between *R. japonicum* F_{78} as recipient and *R. japonicum* wild-type 211^s as donor showed linkage between the *arg I* and *ura* loci (Table 6.5).

Transduction A temperate phage isolated from trp^r revertant of *R. japonicum* was used for transduction. Transduction crosses using *R. japonicum* F_4 (*ileu*$^-$, *ura*$^-$) as recipient showed that the *ura* marker was transduced with the frequency of 3.1×10^{-3} (Table 6.6). The absence of transduction for other markers may indicate the specialised nature of the transducing phage. The transduction of *R. japonicum* F_{78} (*ileu*$^-$, *val*$^-$, *ura*$^-$, *arg*$^-$) shows that *ura arg* markers are cotransduced to the same extent as *ura* alone (Table 6.7). However, when the same transduced recipient was plated on minimal medium containing isoleucine, valine and uracil (25 μg/ml of each), no *arg* transductants were recovered. On the assumption that uracil might be repressing the growth of *arg* transductants, the

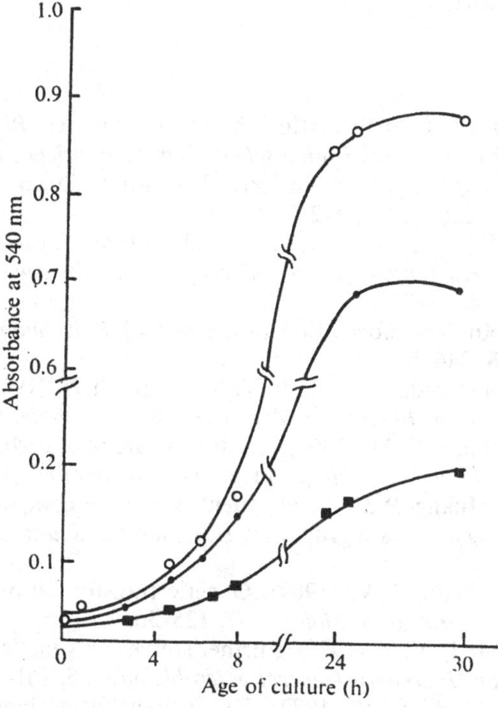

Fig. 6.2. Growth response of uracil, arginine cotransductants in the presence of ileu + val (●); ileu+val+arg (○) and ileu+val+ura (■).

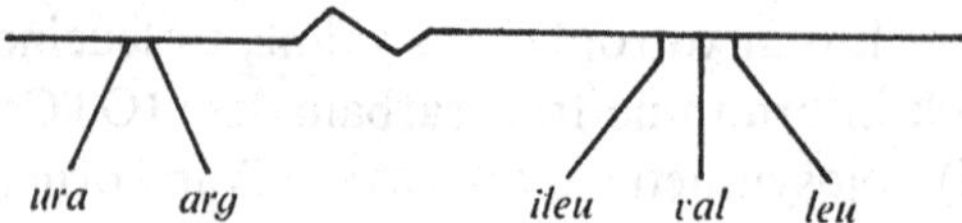

Fig. 6.3. Approximate positions of *arg*, *ura*, *val*, *ileu* and *leu* markers on the *R. japonicum* genetic map.

growth response of ura^+, arg^+ cotransductants was studied in minimal medium containing uracil and/or arginine, the addition of isoleucine and valine being common. Uracil suppresses the growth of nearly 70 % of the transductants (Fig. 6.2) while arginine stimulates it by 30 %. These results are similar to those reported in *E. coli* (Gorini & Kalman, 1963).

A genetic map constructed from the above results (Fig. 6.3) shows two clusters of genes, one being that of *ura* and *arg* and the other of *ileu*, *val* and *leu*.

We wish to thank Dr N. S. Rao and Dr N. K. Notani of Bhabba Atomic Research Centre, Bombay for providing the laboratory facilities for tracer work. This investigation was supported by grant P1-480/FG-In-394 from the Agricultural Research Service of the US Department of Agriculture.

References

Balassa, R. (1954). Transformation Mechanismen der *Rhizobium*. I–III. *Acta microbiologica Academiae scientiarum Hungaricae*, **2**, 51–78.

Balassa, R. (1955). *In vivo* induzierte Transformationen bei Rhizobien. *Naturwissenschaften*, **43**, 422–3.

Balassa, R. (1957). Durch Desoxyribonukleinsauren induzierte Veränderungen bei Rhizobien. *Acta microbiologica Academiae scientiarum Hungaricae*, **4**, 77–81.

Balassa, R. (1960). Transformation of a strain of *Rhizobium lupini*. *Nature, London*, **188**, 246–7.

Gadre, S. V., Mazumdar, L., Modi, V. V. & Parekh, V. (1967). Interspecific transformation in *Rhizobium*. *Archiv für Mikrobiologie*, **57**, 388–91.

Gorini, L. & Kalman, S. M. (1963). Control by uracil of carbamyl phosphate synthesis in *E. coli*. *Biochimica et Biophysica Acta*, **69**, 355–60.

Miller, D. H. & Huang, P. C. (1972). Identification of competence repressing factors during log-phase growth of *H. influenzae*. *Journal of Bacteriology*, **109**, 560–4.

Raina, J. L. & Modi, V. V. (1969). Genetic transformation in *Rhizobium*. *Journal of General Microbiology*, **57**, 125–30.

Raina, J. L. & Modi, V. V. (1971). Further studies on genetic transformation in *Rhizobium*. *Journal of General Microbiology*, **65**, 161–5.

Raina, J. L. & Modi, V. V. (1972). Deoxyribonucleate binding and transformation in *Rhizobium japonicum*. *Journal of Bacteriology*, **111**, 356–60.

7. Plasmid control of effectiveness in *Rhizobium*: transfer of nitrogen-fixing genes on a plasmid from *Rhizobium trifolii* to *Klebsiella aerogenes*

L. K. DUNICAN, F. O'GARA & A. B. TIERNEY

A combination of physiological and cytological interactions occurs during the symbiotic association which produces a nitrogen-fixing symbiosis in nodulated legumes. The term effectiveness is now generally applied to *Rhizobium* strains which can initiate the development of nodules on their host legume and can proceed through the several stages required to form the bacteroid-containing tissue in the nodules which fixes nitrogen. Ineffective *Rhizobium* strains can cause the production of nodules which are, however, small and since these nodules contain neither bacteroids nor haemoglobin, nitrogen is not fixed.

No method has been available to study genetically how a *Rhizobium* goes through the developmental stages required to form bacteroids. A model was proposed by Dunican & Cannon (1971) to offer a basis for the experimental examination of the genetics of symbiosis. This suggested that the genetic material in *Rhizobium* responsible for the expression of certain properties related to effectiveness was found on extrachromosomal DNA – now referred to as plasmids. Preliminary data showed that exposure of *Rhizobium trifolii* strains to plasmid-curing agents like acridine orange resulted in the loss of effectiveness in the cultures (Cannon, 1972). Interpretation of plasmid-curing data is always difficult since the only plasmid efficiently cured is F′*lac*$^+$. F-like and I-like R factors are not cured by acridines (Salisbury, Hedges & Datta, 1972).

In the last two years the plasmid model has been further examined in our laboratory using plasmid-curing experiments, physical analysis of *Rhizobium trifolii* DNA, genetic analysis and gene transfer using introduced resistance transfer factors (R factors). This chapter summarises research on the development of this model particularly with regard to the genetics of the nitrogen-fixing operon (*nif*) in *Rhizobium trifolii*.

Methods

Rhizobium trifolii strain T1 was received from Dr E. A. Schwinghamer. CSIRO, Australia. *R. trifolii* strain 712 (Coryn strain from Rothamsted collection) was obtained from C. Masterson Soils Research Centre, An Foras Taluntais, Wexford, Ireland. *Klebsiella aerogenes* strain 418 and *K. pneumoniae* strain M5a1 were obtained from Dr F. C. Cannon, ARC Unit of Nitrogen Fixation, University of Sussex. *Escherichia coli* J5-3 (RP4) and *E. coli* 22 (R1-19drd) were obtained from Dr N. Datta, Royal Postgraduate Medical School, London. *Pseudomonas aeruginosa* G23 was received from Dr J. R. W. Govan, Department of Bacteriology, University Medical School, Edinburgh. The caesium chloride – ethidium bromide and sucrose gradients used have been described previously (O'Gara & Dunican, 1973), as have the techniques used in the transformation experiments (O'Gara & Dunican, 1973; Dunican & Tierney 1973) and the methods employed in transferring the nitrogen-fixing genes from *R. trifolii* to *K. aerogenes* (Dunican & Tierney, 1974).

Results and discussion

Analysis of Rhizobium DNA by isotopic centrifugation

Plasmid DNAs can be detected in their respective host bacteria using isopycnic centrifugation involving the differential binding of ethidium bromide to supercoiled as opposed to linear or open circular DNA (Radloff, Bauer & Vinograd, 1967). Circular plasmid DNA forms have

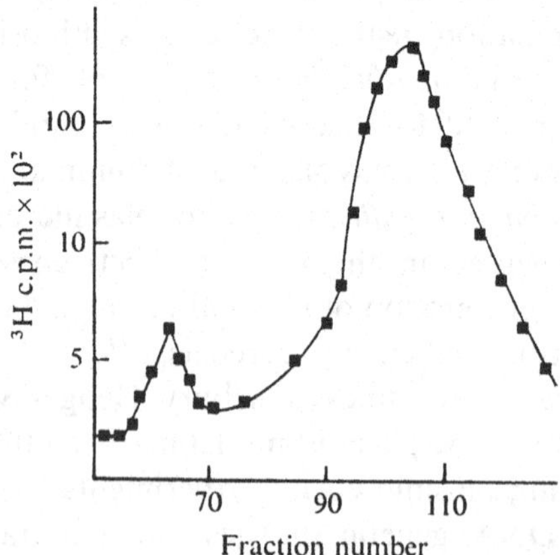

Fig. 7.1. Fractionation of an ethidium bromide–caesium chloride (3.8 g) density gradient containing ^{3}H-labelled DNA from *R. trifolii* T1. Fractions are numbered in order as collected from the bottom of the tube and two drops per fraction were collected. Note the change in scale on the y axis. (After Cannon, 1972.)

been detected for colicinogenic factors, *E. coli* sex factors and antibiotic resistance factors (Clowes, 1972). All these plasmids were detected as denser bands following centrifugation in ethidium bromide – caesium chloride density gradients. The technique, however, is not absolute since a plasmid in *Pseudomonas aeruginosa* which could be detected by genetic transfer could not be detected on such gradients (Ingram *et al.*, 1972).

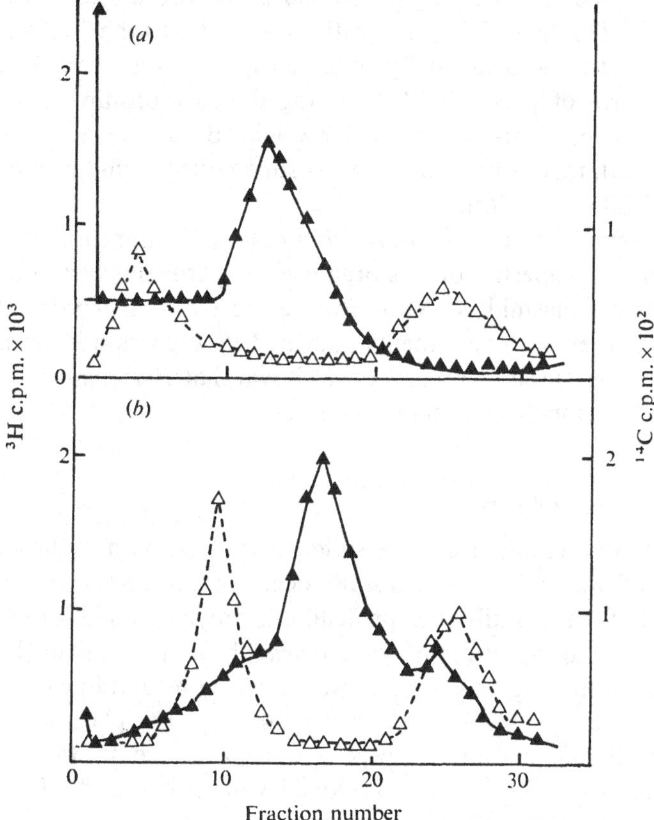

Fig. 7.2. Fractionation of 5–20 % neutral sucrose density gradients containing bulked satellite DNA fractions isolated from *R. trifolii* T1. (▲——▲). Layered on top of the gradients with this freshly prepared DNA (*a*) and DNA stored at 4 °C (*b*) was ^{14}C-labelled MS-2 phage (△——△) as a sedimentation marker. Nine drops per fraction were collected and fractions are numbered in order as collected from the bottom of the tube.

Dye-buoyant density centrifugation has been used to detect plasmid DNA in *R. trifolii* where DNA components corresponding to supercoiled DNA were detected (Cannon, Dunican & O'Gara, 1971). Fig. 7.1 shows the presence of supercoiled DNA in *R. trifolii* T1. The proportion of label in the satellite peak ranged from 0.93–1.2 % of the total (Cannon, 1972).

The collected fractions from six gradient tubes containing the denser satellite DNA component were pooled and analysed by neutral sucrose gradient centrifugation, using ^{14}C-labelled MS-2 phage as a sedimentation marker (Fig. 7.2). Sucrose gradient centrifugation indicated that freshly prepared satellite DNA from *R. trifolii* sedimented as one band with a sedimentation coefficient of 53 S (Fig. 7.2*a*). Storage of this DNA resulted in the appearance of a second peak with a sedimentation coefficient of 27 S (Fig. 7.2*b*). The ratio of the 53 S to the 27 S peak is 1.9 and this is the ratio normally observed for the supercoiled and open circular form of plasmid DNA. Using the relationship between sedimentation coefficients and molecular weight (Bazaral & Helinski, 1968) a plasmid with this value of 53 S for its supercoiled form has a molecular weight of 28×10^6 daltons.

The plasmid DNA found in *R. trifolii* cannot be immediately assigned to symbiotic properties of this organism. Experiments were undertaken to see if this plasmid could be transferred from *R. trifolii* into other genera, as previous experiments indicated that it was non-transferable. R factors were used as reports have shown that *Rhizobium* can act as a host for R factors from other organisms.

R factors in Rhizobium

The R factors are one class of plasmids which have been studied in several species of *Rhizobium*. A naturally-occurring R factor which carries resistance for the antibiotics penicillin, neomycin and chloramphenicol has been found in a strain of *R. japonicum* by Cole & Elkan (1973). The plasmid controlling these factors was removed by acridine orange treatment twice as quickly as the spontaneous rate. The plasmid containing three antibiotic resistances was also transferable to two strains of *Agrobacterium tumefaciens* at a frequency of 0.1 to 0.25 %. Transfer into *R. trifolii* was not successful.

An ineffective strain of *R. trifolii* strain 712 (Coryn) was examined and found to contain a natural plasmid controlling penicillin resistance (Dunican & O'Gara, unpublished observations). The plasmid was cured using ethidium bromide (100 μg/ml) with a frequency of approximately 50 %. An examination of the DNA of this strain (Figs. 7.3 and 7.4) before and after curing with ethidium bromide – caesium chloride density gradients shows that a covalently closed circular DNA molecule was present in the original strain 712 but was absent in the cured penicillin sensitive strain (Dunican & O'Gara, unpublished observations). The original *R. trifolii* strain 712 was ineffective and the cured strains when examined using the plant inoculation technique in Vincent (1970) on *Trifolium repens* were also found to be ineffective, showing that the presence of penicillin resistance was not inhibiting the expression of effectiveness.

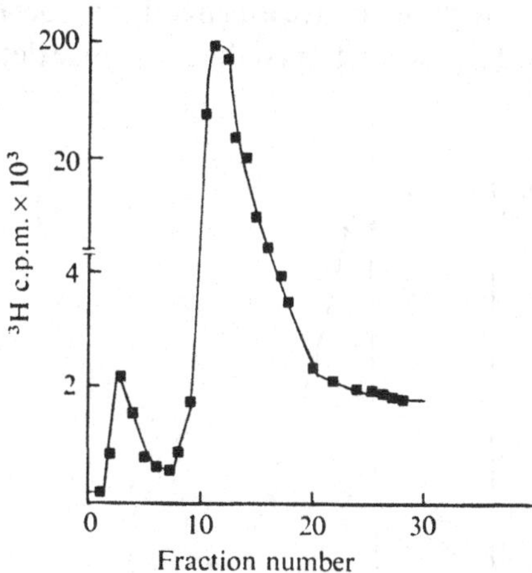

Fig. 7.3. Fractionation of an ethidum bromide–caesium chloride density gradient containing ^{3}H-labelled DNA from *R. trifolii* 712. Fractions are numbered in order as collected from the bottom of the tube and six drops per fraction were collected. Note the change in scale for the fractions which contain the chromosomal DNA, i.e. the large peak at fraction number 12.

Several reports have shown that resistance factors can be transferred into *Rhizobium* strains from other genera. Datta and co-workers (1971) showed that an R factor isolated from *Pseudomonas aeruginosa* could be transferred into several genera including *R. trifolii* and *R. lupini* at a low frequency of 10^{-7}. In a subsequent report Datta & Hedges (1972) showed that F and F-like plasmids could be transferred to *R. lupini* strain 6.2 but not into other *Rhizobium* strains or agrobacteria. Beringer (1974) has also studied R factors of the compatibility class P, introduced into *R. leguminosarum* from *E. coli* at frequencies of 10^{-4} per recipient. In an extensive study it was found that the R factor did not transfer any chromosomal genes between *R. leguminosarum* and auxotrophs of this species.

Transformation has been used to transfer R factors into *R. trifolii* in our laboratory. The transformation of the R factor RP4 from *E. coli* J5-3 into *R. trifolii* T1 has been reported (O'Gara & Dunican, 1973). Partially purified plasmid DNA of the donor strain was prepared by the

Lysozyme–EDTA–Triton–X 100 procedure and was used in transformation experiments with *Rhizobium trifolii* as recipient. Transformants of *R. trifolii* were found at a frequency of 1.3×10^{-4} per recipient and these acquired the triple resistances of the donor. The *R. trifolii* T1 (RP4) was stable when grown on MSY (mannitol salts yeast extract) agar incorporating the specific antibiotics coded by the R factor, and nodulated its host legume to give an effective association. *Rhizobium* isolates from these nodules still retained the resistance determinants of the R factor.

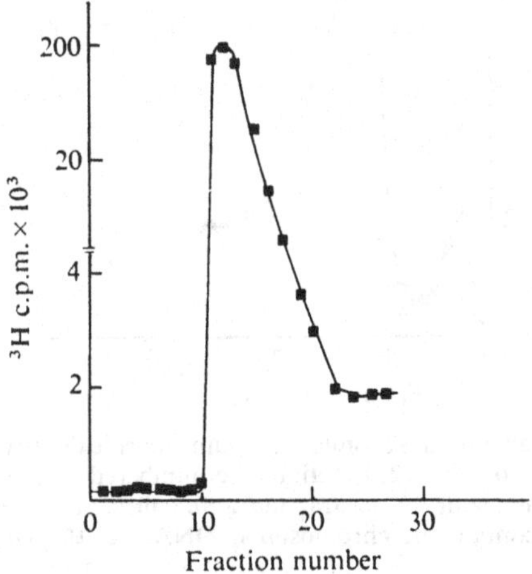

Fig. 7.4. Fractionation of an ethidium bromide–caesium chloride density gradient containing ^{3}H-labelled DNA from the 'cured' *R. trifolii* 712 strain. Fractions are numbered in order as collected from the bottom of the tube and six drops per fraction were collected. Note the change in scale for the peak which contain the chromosomal DNA.

In a parallel study in our laboratory the derepressed F-like R factor R1-19drd was transferred into *R. trifolii* (Dunican & Tierney, 1973). Since this R factor was depressed we thought it should be more efficient in mobilising *Rhizobium* genetic material to other genera. This factor was transferred from its host *E. coli* R1-19drd to *P. aeruginosa* G-23 selecting for kanamycin resistance. No supercoiled DNA could be detected in the original *Pseudomonas* strain on ethidium bromide–caesium chloride gradients, nor could acetylene reduction be detected. The DNA of the *Pseudomonas* strain containing the R factor was subsequently used to transform *R. trifolii* T1 using kanamycin resistance as a selective agent. Kanamycin resistant isolates were checked and found to have acquired chloramphenicol and amphicillin resistance – the resistances of the original R1-19drd. The transformants were found to nodulate clovers effectively, and when cultures were tested for acetylene reduction no reduction was detected, as was expected for *Rhizobium*.

Mobilisation of nitrogen-fixing genes in Rhizobium

The *R. trifolii* containing the derepressed R factor R1-19drd was used in conjugation experiments with *Klebsiella aerogenes* 418, a non-nitrogen-fixing strain. The procedures used, in outline, were as follows. Initially, conventional R factor crosses using the technique described by Datta *et al.* (1971) were employed. Two rifampicin resistant mutants of *K. aerogenes* (minimum inhibitory concentrations 50 and 500 μg/ml) were used as recipients and the selective plating procedure used rifampicin 25 μg/ml and kanamycin 25 μg/ml in Burk's nitrogen-free agar (Newton, Wilson & Burris, 1953) incubated in an N_2 atmosphere. This selective procedure permitted the detection of *K. aerogenes* hybrids which had received the R factor and the ability to fix nitrogen. Little success was achieved until it was observed that exposing the donor cells to ultraviolet light (50 % kill) prior to preparing the mating mixture was necessary. This procedure had been previously used by Anderson & Lewis (1965) to increase R factor crosses in *Salmonella typhimurium*. The rate of transfer in a typical experiment is shown in Table 7.1. Presumptive

Table 7.1. *Frequency of transfer of* nif *genes*

Donor strain	Recipient strain	Characters selected in hybrids	Frequency of transfers
R. trifolii T1K	*K. aerogenes* 418 *Rif*^{r}	*Nif*$^+$, *Kan*^{r}	1.6×10^{-6}
R. trifolii T1K	*K. aerogenes* 418 *Rif*^{r}	*Nif*$^+$	3×10^{-6}

Strain 418 *Rif*^{r} is resistant to 500 μg/ml rifamycin. Selection for *Nif, Kan*^{r} was on Burk's medium with antibiotic incorporated and incubated in an N_2 atmosphere and selection for *Nif* was on Burk's medium in N_2 atmosphere. The frequency of transfer was calculated in relation to the numbers of donors present in the mating mixture.

nitrogen-fixing colonies were detectable by their large size on plates containing the mating mixture. Control cultures of the *K. aerogenes* recipient or the UV irradiated *R. trifolii* grown separately were never found to form colonies on the selective plates in more than two dozen transfer experiments. Initial difficulties in the subsequent propagation of these colonies in Burk's nitrogen-free liquid medium were overcome by adding yeast extract (100 μg/ml) to the medium. In subsequent cultures Hino & Wilson's (1958) medium was used. Pure cultures capable of growing on Burk's nitrogen-free agar were isolated and grown on Hino & Wilson's medium and shown to be able to reduce acetylene, thus confirming that they were nitrogen-fixing hybrids (Table 7.2). Growth of the *K. aerogenes Nif* hybrids in combined nitrogen (500 μg/ml N) or under aerobic conditions repressed the production of the nitrogenase enzyme indicating that the regulatory genes had also been transferred.

Table 7.2. *Acetylene reduction by* K. aerogenes *418* Nif$^+$ *hybrids and control strains*

	Klebsiella strain	nmol C_2H_4 Formed/min/mg dry wt
K. pneumoniae	M5a1	5.9
K. aerogenes	418 *Rif^r*	0
K. aerogenes	418 hybrid B1	3.9
K. aerogenes	418 hybrid B4	10.0
K. aerogenes	418 hybrid B7	3.0

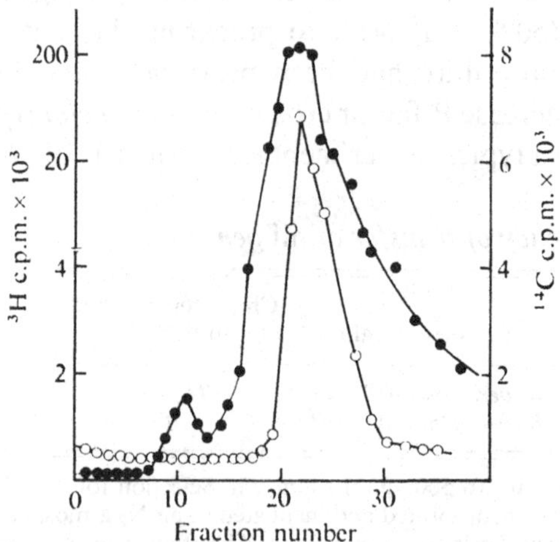

Fig. 7.5. Fractionation of an ethidium bromide–caesium chloride density gradient containing ^{3}H-labelled DNA from a *K. aerogenes Nif*$^+$ hybrid (●——●) and ^{14}C-labelled DNA from the parent *K. aerogenes* strain (○——○). Fractions are numbered in order as collected from the bottom of the tube and six drops per fraction were collected. Note the change in scale for the large peaks (fraction number 20) which contain the chromosomal DNA.

Most of the hybrids isolated were unstable since all rapidly lost the antibiotic resistances of the R factor unless maintained on the drugs. The *nif* operon was also unstable in many of the hybrids although several stable strains were isolated. All the hybrids were resistant to 500 μg/ml rifampicin, which was equal to the resistance of the original *K. aerogenes* 418. The hybrids and the parent *K. aerogenes* strain had the same base ratio (see Fig. 7.5).

Our evidence indicates that the *nif* genes in the *K. aerogenes* hybrids were derived from the *R. trifolii* strain. At least two dozen *nif* transfers were accomplished with the controls being negative in all cases. The growth peculiarities of the nitrogen-fixing *Klebsiella* hybrids and the instability of the *nif* operon in the *K. aerogenes* hybrid contrast sharply with the more rapid growth and great stability of the well-studied natural N_2-fixing *K. pneumoniae* M5a1 which was used as the basis of comparison for our hybrids (Dunican & Tierney, 1974).

Location of the nif *operon in* R. trifolii

The transferability of the *Rhizobium nif* operon to *K. aerogenes* offered the possibility of some interesting experiments which could be used to determine whether this operon was located on the chromosome or a plasmid. The efficiency of transfer of chromosomal genes by a derepressed R factor falls in the range 10^{-5}–10^{-7} per donor. Derepressed R factors transfer themselves at a frequency near unity (Cooke & Meynell, 1969). R factor mobilisation of plasmids, e.g. Col E, occurs at a frequency of 1 % of the level of R factor transfer, which is a transfer efficiency approximating that seen in Table 7.1. This would indicate that the *nif* operon was on a plasmid and that the R factor was mobilising this plasmid to the *K. aerogenes* recipient.

Some additional evidence was obtained when conjugation experiments were set up, selecting as before for kanamycin resistance and the *nif* operon with additional plates selecting for *nif* transfer only. The results in Table 7.3 show the transferability of *nif* was slightly increased when kanamycin resistance was not selected for. The results were interpreted as showing that a plasmid–plasmid association was occurring in the *R. trifolii.* Two mechanisms, termed plasmid aggregates and plasmid co-integrates, have been suggested to account for the mobilisation of one plasmid by another (Milliken & Clowes, 1973); i.e. either the plasmids undergo recombination but remain loosely attached or the two plasmids integrate to give a single large plasmid. The plasmid aggregate mechanism seems appropriate here since the evidence suggested that kanamycin resistance and the *nif* operon could be lost independently in the *K. aerogenes* hybrids, thus indicating that they were probably not a single integrated molecular structure.

The effect of ultraviolet light on the *Rhizobium* donor prior to mating was studied to elucidate this point and the effect of UV on the transfer of *nif* operon is shown in Fig. 7.6. The cotransfer of *nif* and kanamycin resistance increases to the point where 50 % of the *Rhizobium* cells are killed by the UV. Selecting for the *nif* operon only gave the same result

Table 7.3. *Transfer of* nif *genes and kanamycin resistance between* Rhizobium trifolii *and* Klebsiella aerogenes

Donor strain	Recipient strain	Characters selected in hybrids	Frequency of transfers
R. trifolii T1K	*K. aerogenes* 418 *Rif*^{r}	*Nif*$^+$, *Kan*^{r}	1.9×10^{-7}
R. trifolii T1K	*K. aerogenes* 418 *Rif*^{r}	*Kan*^{r}	2.3×10^{-5}

Strain 418 *Rif*^{r} is resistant to 500 μg/ml rifamycin. Selection for *Nif*, *Kan*^{r} was on Burk's medium incubated in an N_2 atmosphere, and selection for *Kan*^{r} was on a nutrient agar with the antibiotics added as outlined in methods. The frequency of transfer was calculated in relation to the number of donors present in the mating mixture.

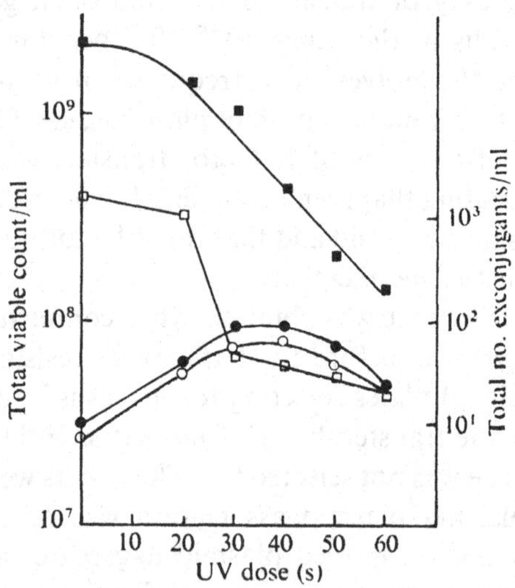

Fig. 7.6. Dose response curves of the survival of donor culture (■——■), transfer of R factor (□——□), transfer of *nif* genes (●——●) and cotransfer of R factor and *nif* genes (○——○) when *Rhizobium* was irradiated with UV light prior to mating.

with slightly higher efficiency. The transfer of kanamycin resistance alone decreased with UV dose. While these results show a similarity to data of Evenchik, Stacey & Hayes (1969) where F transfer of *pro* and *lac*$^+$ genes in *E. coli* showed essentially the same response to UV. The authors concluded that the UV was increasing gene transfer due to its stimulation of the recombination event leading to the attachment of the genes to the R factors. Although Evenchik *et al.* (1969) were studying chromosomally located genes it seems likely that recombination between two plasmids should also be stimulated by UV light by the same mechanism, but this point remains to be demonstrated.

Analysis of the nif *operon in* Klebsiella aerogenes

As shown above the *K. aerogenes* hybrids received both the *nif* operon and antibiotic resistances from *R. trifolii* T1. The two genetic elements were unstable and the antibiotic resistances and the ability to fix N_2 were lost independently of each other unless selective conditions were used to maintain them. Several *Klebsiella* hybrids which lost their antibiotic

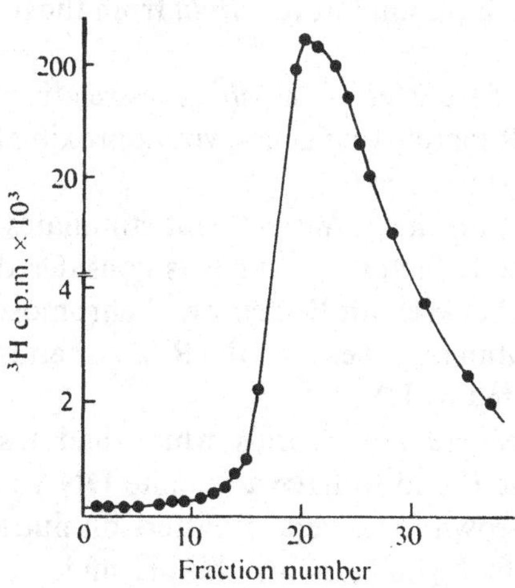

Fig. 7.7. Fractionation of an ethidium bromide–caesium chloride density gradient containing ^{3}H-labelled DNA from *K. aerogenes Nif*$^-$ segregants. Fractions are numbered in order as collected from the bottom of the tube and six drops per fraction were collected. Note the change in scale for the peak which contains the chromosomal DNA.

resistances but retained the ability to fix N_2 were examined by caesium chloride density centrifugation. The results of one such experiment are shown in Fig. 7.5 where a satellite DNA peak is seen in the *K. aerogenes Nif* hybrid but not in the wild-type strain. When the *Nif* hybrid strain shown in Fig. 7.5 was grown for several transfers on nutrient agar the ability to fix N_2 was lost and the satellite DNA peak was also lost (Fig. 7.7).

Conclusion

This chapter has examined the information available on plasmids in the genus *Rhizobium*. Naturally-occurring plasmids have been shown in two species, and these plasmids are stable showing that these strains have the ability to contain such extrachromosomal DNA. Plasmids have been transferred into *Rhizobium* from other genera and such plasmids are only maintained if the bacteria are stored on the specified antibiotics. The derepressed F-like R factor which was transferred into *R. trifolii* was the first factor which permitted experiments to study the transferability of nitrogen-fixing genes from *R. trifolii* to *K. aerogenes*. Nitrogen-fixing genes were transferable into *K. aerogenes* from *R. trifolii* T1 using this R factor, transfer being stimulated by pre-treatment of the donor with UV. Since it was shown that neither the *E. coli* nor *P. aeruginosa*, previous hosts of the R1-19drd factor, could fix nitrogen it was concluded that the nitrogen-fixing genes originated in the *R. trifolii*. We consider that the *nif* genes are located on a plasmid in *R. trifolii* from the following evidence:

(*a*) The high rate of transfer of the *nif* genes relative to the rate of co-transfer of the R factor +*nif* genes, viz. approximately 1 %, indicates that a plasmid is involved.
(*b*) The demonstration that UV treatment stimulates *nif* transfer while inactivating the R factor. UV light is considered to stimulate the recombination between an R factor and chromosomal genes. However, the recombination between the R factor and the plasmid could also be stimulated by UV.
(*c*) *Klebsiella aerogenes Nif* hybrids which had lost their antibiotic resistances were found to have a satellite DNA plasmid. When the hybrids were grown for several transfers on nutrient agar they lost their ability to fix N_2 and also lost this plasmid.

The data presented here indicate that the genetic material for nitrogen fixation is present in *Rhizobium*. This has been assumed since the symbiotic association was discovered in 1888 but no experimental procedure was available to prove it. The transfer of this genetic material to *K. aerogenes* where it resides as a plasmid offers a major opportunity for the improvement of strains of *Rhizobium* species on a truly scientific basis since it will be possible to introduce nitrogen-fixing genes into *Rhizobium* species using transformation of the *Nif* plasmid in *K. aerogenes* hybrids. Meaningful genetic experiments can also be planned to elucidate other components which are concerned in effectiveness.

We wish to thank our colleague P. R. Smith for his useful discussions on genetics, F. C. Cannon from the Nitrogen Fixation Unit, University of Sussex, for his helpful comments during the preparation of this article and Dr G. Elkan and Dr J. Beringer for making available unpublished papers. We also thank those who supplied bacterial cultures. This work was supported by the Irish Committee for IBP and by fellowships from the Department of Education (F. O'G. and A. B. T.).

References

Anderson, E. S. & Lewis, M. J. (1965). Characterization of a transfer factor associated with drug resistance in *Salmonella typhimurium*. *Nature, London*, **208**, 843–9.

Bazaral, M. & Helinski, D. (1968). Circular DNA forms of colicinogenic factors E1, E2, and E3 from *Escherichia coli*. *Journal of Molecular Biology* **36**, 185–94.

Beringer, J. E. (1974). R-factor transfer studies in *Rhizobium leguminosarum*. *Journal of General Microbiology*, **84**, 188–98.

Cannon, F. C., Dunican, L. K. & O'Gara, F. (1971). Dye-buoyant-density-gradient analysis of *Rhizobium* deoxyribonucleic acid. *Biochemical Journal*, **125**, 103.

Cannon, F. C. (1972). A study of effectiveness and viomycin resistance in relation to the genome structure of *Rhizobium*. PhD Thesis, University College, Galway.

Clowes, R. C. (1972). Molecular structure of bacterial plasmids. *Bacteriological Review*, **36**, 361–405.

Cole, M. A. & Elkan, G. H. (1973). Antibiotic resistance transfer in *Rhizobium japonicum*. *Antimicrobial Agents and Chemotherapy*, **4**, 248–53.

Cooke, M. & Meynell, E. (1969). Chromosomal transfer mediated by derepressed R factors in *Escherichia coli* K12. *Genetical Research*, **14**, 79–87.

Datta, N. & Hedges, R. W. (1972). Host range of R-factors. *Journal of General Microbiology*, **70**, 453–60.

Datta, N., Hedges, R. W., Shaw, E. J., Sykes, R. B. & Richmond, M. H. (1971). Properties of an R-factor from *Pseudomonas aeruginosa*. *Journal of Bacteriology*, **108**, 1244–9.

Dunican, L. K. & Cannon, F. C. (1971). The genetic control of symbiotic properties in *Rhizobium*: evidence for plasmid control. *Plant and Soil*, Special Volume, 73–9.

Dunican, L. K. & Tierney, A. B. (1973). Transformation of an R-factor from *Pseudomonas aeruginosa* into *Rhizobium trifolii*. *Molecular and General Genetics*, **126**, 187–90.

Dunican, L. K. & Tierney, A. B. (1974). Genetic transfer of nitrogen fixation from *Rhizobium trifolii* to *Klebsiella aerogenes*. *Biochemical and Biophysical Research Communications*, **57**, 67–72.

Evenchik, Z., Stacey, K. A. & Hayes, W. (1969). Ultraviolet induction of chromosomal transfer by autonomous sex factors in *Escherichia coli*. *Journal of General Microbiology*, **56**, 1–14.

Freidlander, D. R. & Friedlander, D. (1968). Studies on *Escherichia coli* sex factors. II. Some physical properties of F′ *lac* DNA and F DNA. *Journal of Molecular Biology*, **32,** 25–35.

Hino, S. & Wilson, P. W. (1957). Nitrogen fixation by a facultative bacillus. *Journal of Bacteriology*, **75,** 403–8.

Ingram, L., Sykes, R. B., Grinsted, J., Saunders, J. R. & Richmond, M. H. (1972). A transmissible resistance element from a strain of *Pseudomonas aeruginosa* containing no extrachromosomal DNA. *Journal of General Microbiology*, **72,** 269–79.

Milliken, C. E. & Clowes, R. C. (1973). Molecular structure of an R-factor, its component drug resistance determinants and transfer factor. *Journal of Bacteriology*, **113,** 1026–33.

Newton, J. W., Wilson, P. W. & Burris, R. H. (1953). Direct demonstration of ammonia as an intermediate in nitrogen fixation by *Azotobacter*. *Journal of Biological Chemistry*, **204,** 445–51.

O'Gara, F. & Dunican, L. K. (1973). Transformation and physical properties of an R-factor RP4 transferred from *Escherichia coli* to *Rhizobium trifolii*. *Journal of Bacteriology*, **116,** 1177–80.

Radloff, R., Bauer, W. & Vinograd, J. (1967). A dye-buoyant density method for the detection and isolation of closed circular duplex DNA: the closed circular DNA in HeLa cells. *Proceedings of the National Academy of Sciences, USA*, **57,** 1514–20.

Salisbury, V., Hedges, R. W. & Datta, N. (1972). Two modes of 'curing' transmissible bacterial plasmids. *Journal of General Microbiology*, **70,** 443–52.

Vincent, J. M. (1970). *Manual for the practical study of root nodule bacteria.* Blackwell Scientific Publications, Oxford.

8. The demonstration of conjugation in *Rhizobium leguminosarum*

J. E. BERINGER

Gene transfer in *Rhizobium* has been reported to occur by conjugation (Heumann, 1968; Lorkiewicz *et al.*, 1971; Kaushik & Venkataraman, 1972), transformation (Balassa, 1963; Zelazna-Kowalska & Lorkiewicz, 1971; Raina & Modi, 1972) and transduction (Kowalski, 1971; Sik & Orosz, 1971). Chromosomal 'mapping' has also been reported by Lester (1971) and Al-Ani (1972) using nitrosoguanidine mutagenesis of synchronously growing cultures (Cerdá-Olmedo, Hanawalt & Guerola, 1968). Genetic maps have been published for *R. lupini* (Heumann, Pühler & Wagner, 1973) and *R. trifolii* (Al-Ani, 1972).

These methods of genetic analysis have, as yet, contributed little to our knowledge of the plant–bacterium symbiosis for two main reasons. Transformation and transduction have restricted potential as only short lengths of the genome are usually transferred, making it difficult to demonstrate linkage and to perform long-range genetic mapping. Also, though conjugation is suitable for preliminary studies, the only well-documented conjugation system – that of Heumann (Heumann, 1968; Heumann, Pühler & Wagner, 1973) – has been developed using strains of *R. lupini* whose identity and symbiotic properties have not been confirmed.

Genetic analysis in prokaryotes is most easily performed by conjugation, though this is difficult to demonstrate in previously untested species, because in most conjugation systems donor ability is a property of strains carrying sex factors; for example F in *Escherichia coli* (Lederberg, Cavalli & Lederberg, 1952), FP2 in *Pseudomonas aeruginosa* (Holloway, 1969) and SCP1 in *Streptomyces coelicolor* (Vivian, 1971), though in *S. coelicolor* fertility due to SCP1 is superimposed on an already existing level of fertility not yet associated with a sex factor (Hopwood *et al.*, 1973). Therefore, in the absence of known donor strains, chromosomal transfer can be detected only if at least one parent in the cross happens to carry a sex factor, and if mating conditions are suitable for its transfer.

This chapter describes a method for demonstrating conjugation in

R. leguminosarum using self-transmissible R factors derived from *Pseudomonas aeruginosa*. Having shown that plasmid transfer occurs, mating conditions can be optimised and the screening of large numbers of *Rhizobium* strains for their ability to form recombinants in mixed culture becomes more realistic.

Intergeneric transfer of plasmids

Transfer of plasmids between members of the Enterobacteriaceae and genera of other families has only recently been recorded (Sykes & Richmond, 1970; Fullbrook, Elson & Slocombe, 1970; Datta *et al.*, 1971; Olsen & Shipley, 1973) and has to date only involved R factors of the compatibility class P. These R factors were originally observed in strains of *Pseudomonas aeruginosa* carrying high-level resistance to carbenicillin and have since been shown to be transferable to *Escherichia coli* (Sykes & Richmond, 1970) and from *E. coli* to *Rhizobium trifolii* and *R. meliloti* (Datta *et al.*, 1971). Transfer between *Rhizobium* strains has been reported by Pühler, Burkardt & Heumann (1972) and Beringer (1973*a*). Stanisich & Holloway (1971) have shown that P group R factors mediate chromosomal transfer in *P. aeruginosa*, but no reports of chromosomal transfer by these plasmids in other genera have yet appeared.

Transfer of P group R factors to *Rhizobium leguminosarum*

We have tested a large number of chromosomally marked *R. leguminosarum* strains for their ability to form recombinants in mixed culture. Because gene transfer was not demonstrated (Beringer, 1973*b*), plasmid transfer between *R. leguminosarum* strains was examined as an indicator of conjugation. It was hoped that if transfer occurred it could be used to determine optimum mating conditions and that chromosomal marker transfer might be mediated by the R factors.

Datta *et al.* (1971) transferred the P group R factor RP4 from *Escherichia coli* to *Rhizobium meliloti* and *R. trifolii* by mating in nutrient broth for 2–3 hours at 27 °C. Transfer occurred at a frequency of about 10^{-6} per donor. *E. coli* was, therefore, chosen as a donor of P group R factors in crosses with *R. leguminosarum* in an enriched liquid minimal medium – Vincent's (1970) yeast sucrose (phage) broth – at 28 °C for 20 hours. Three different P group R factors (RP4, RK2 and R6886), all carrying kanamycin/neomycin, tetracycline and carbenicillin resistance, have been transferred to *R. leguminosarum* from *E. coli* (Table 8.1). Transfer was observed in all donor–recipient combinations tested. R^+ rhizobia were selected on suitably supplemented minimal media, using kanamycin for selection of R factor transfer; all these R^+ transconjugants simultaneously became tetracycline and carbenicillin resistant.

The level of antibiotic resistance mediated by these R factors in *R. leguminosarum* was similar to that in *E. coli*, except that carbenicillin resistance was poorly expressed in *R. leguminosarum*. All three R factors were transferred back to R^- *E. coli* strains and full expression was observed, indicating that the low-level of carbenicillin resistance in *Rhizobium* was due to impaired phenotypic expression rather than

Table 8.1. *R factor crosses*

Cross No.	Donor	Recipient	Frequency of transfer (per recipient)	% NB* in mating medium
1	J5-3*nal*^{r} (RP4)	100	4×10^{-5}	5
2	100 (RP4)	J5-3	8×10^{-5}	1
3	J5-3*nal*^{r} (RP4)	41	6×10^{-5}	40
4	J5-3*nal*^{r} (R6886)	41	1×10^{-7}	40
5	J5-3*nal*^{r} (RK2)	41	6×10^{-5}	40
6	145 (RP4)	41	3×10^{-4}	40
7	145 (R6886)	41	2×10^{-4}	40
8	145 (RK2)	41	6×10^{-4}	40
9	41 (RP4)	145	6×10^{-6}	0
10	41 (RP4)	145	8×10^{-5}	40
11	41 (RP4)	145	2×10^{-5}	60

Crosses were performed by adding equal (1 ml) volumes of late log phase cultures of donors and recipients to Vincent's (1970) yeast sucrose (phage) broth and shaking the culture gently for 20 h at 28 °C. Nutrient broth was added in the proportions shown.

J5-3 and J5-3*nal*^{r} are *pro met* and *pro met nal*^{r} derivatives of *Escherichia coli* K12, kindly provided by Dr E. Meynell: 41, 100 and 145 are respectively *ade ura str*^{r}, *gly str*^{r} and *leu* derivatives of *Rhizobium leguminosarum*.

* NB, nutrient broth (Oxoid).

genetic alteration of the R factors. Further experiments (Beringer, 1973*b*) showed that R factor transfer between *R. leguminosarum* strains was insensitive to DNase and that a wide range of independent 'wild-type' isolates can act as recipients of R factors in crosses (47 out of 52 tested). These features of R factor transfer, together with the ability of *R. leguminosarum* strains to transfer the R factors back to *E. coli*, make conjugation, rather than spontaneous transformation or transduction, the most likely mode of transfer.

Transfer of R factors between *Rhizobium leguminosarum* strains

Crosses between *R. leguminosarum* strains were initially performed in Vincent's (1970) yeast sucrose (phage) broth (VS), but the frequency of R factor transfer observed was low (see cross 9 in Table 8.1). This frequency was about 100 times less than that observed by other workers for transfer of these R factors between strains of *Escherichia coli* (Datta *et al.*, 1971) or *Pseudomonas aeruginosa* (Stanisich & Holloway, 1971) and could have been due to an inherent genetic property of the *R. leguminosarum* strains or to differences in the mating technique. In an attempt to improve the frequency of R factor transfer, increasing amounts of nutrient broth (NB) were added to the liquid medium. It was found that added nutrient broth increased the frequency of transfer, but above 40 % NB in the medium the frequency declined and significant morphological changes and inhibition of growth occurred. As the addition of NB increased transfer approximately tenfold (see crosses 9–11 in Table 8.1) subsequent crosses were performed in 60 % VS × 40 % NB (see crosses 6–8 in Table 8.1). Sherwood (1972) observed that inhibition of *Rhizobium* by yeast extracts was almost entirely due to the glycine present. When crosses were performed in the presence of increasing amounts of glycine in place of NB a similar effect on the frequency of transfer was observed, suggesting that the increased frequency of transfer may be due to the formation of partial spheroplasts. Glycine is known to induce morphological alterations and spheroplast formation in Gram-negative bacteria (McQuillen, 1960; Strominger & Birge, 1965).

Controls

A serious criticism of much *Rhizobium* genetic research has been the lack of adequate genetic markers for characterising strains and the absence of data for the symbiotic properties of the bacteria used. Therefore, to ensure that R factor transfer was indeed occurring in *R. leguminosarum* the following controls were included: (*a*) genetically marked strains were used in all experiments; (*b*) R factor transfer was always selected by one marker carried on the plasmid and transconjugants were tested for the presence of the other two non-selected markers; (*c*) auxotrophic mutants were tested for their symbiotic properties on the pea (*Pisum*) using a cloned line of the variety Wisconsin Perfection; and (*d*) bacteria were isolated from nodules formed by genetically marked strains and characterised for their phenotypic properties. A limited number of R^+ derivatives of effective auxotrophic mutants were tested and found to have unaltered symbiotic properties.

Conclusion

A genetic understanding of the *Rhizobium*–legume symbiosis depends upon the genetic analysis of both symbionts. For this reason, the *Pisum–R. leguminosarum* symbiosis is very suitable for further study as the pea is genetically the best-known legume and is amenable to experimentation.

R factor transfer provides a simple method for demonstrating conjugation in *Rhizobium* and can be used to monitor the ability of rhizobia to transfer DNA when screening large numbers of strains for sex factor activity. The demonstration that *R. leguminosarum* can transfer R factors by conjugation, together with the observation of Cole & Elkan (1972) that some rhizobia carry self-transmissible R factors, suggests that rhizobia, like many other Gram-negative bacteria, have evolved the ability to transfer genetic information by this method. It remains for this to be exploited to develop a genetic analysis of *R. leguminosarum*.

I would like to thank Professor D. A. Hopwood and A. W. B. Johnston for their helpful comments during the preparation of this article.

References

Al-Ani, F. Y. (1972). Genetic investigations on the root nodule bacteria *Rhizobium*. PhD Thesis, University of Liverpool.

Balassa, G. (1963). Genetic transformation of *Rhizobium*: a review of the work of R. Balassa. *Bacteriological Reviews*, **27**, 228–41.

Beringer, J. E. (1973*a*). R-factor transfer studies in *Rhizobium leguminosarum*. *Proceedings of the Society for General Microbiology*, **1**, 10–11.

Beringer, J. E. (1973*b*). Genetic studies with *Rhizobium leguminosarum*. PhD Thesis, University of East Anglia.

Cerdá-Olmedo, E., Hanawalt, P. C. & Guerola, N. (1968). Mutagenesis of the replication point by nitrosoguanidine: map and pattern of replication of the *Escherichia coli* chromosome. *Journal of Molecular Biology*, **33**, 705–19.

Cole, M. A. & Elkan, G. H. (1972). Non-chromosomal determinants for penicillin and chloramphenicol resistance in *Rhizobium japonicum*. Paper presented to 72nd Annual Meeting of the American Society for Microbiology.

Datta, N., Hedges, R. W., Shaw, E. J., Sykes, R. B. & Richmond, M. H. (1971). Properties of an R-factor from *Pseudomonas aeruginosa*. *Journal of Bacteriology*, **108**, 1244–9.

Fullbrook, P. D., Elson, S. W. & Slocombe, B. (1970). R-factor mediated beta-lactamase in *Pseudomonas aeruginosa*. *Nature, London*, **226**, 1054–6.

Heumann, W. (1968). Conjugation in star-forming *Rhizobium lupini*. *Molecular and General Genetics*, **102**, 132–44.

Heumann, W., Pühler, A. & Wagner, E. (1973). The two transfer regions of the *Rhizobium lupini* conjugation. II. Genetic characterization of the transferred chromosomal segments. *Molecular and General Genetics*, **126,** 267–74.

Holloway, B. W. (1969). Genetics of *Pseudomonas. Bacteriological Reviews*, **33,** 419–43.

Hopwood, D. A., Chater, K. F., Dowding, J. E. & Vivian, A. (1973). Advances in *Streptomyces coelicolor* genetics. *Bacteriological Reviews*, **37,** 371–405.

Kaushik, B. D. & Venkataraman, G. S. (1972). Induced variations in microorganisms. III. Genetic transfer between intact cells of the mutants of *Rhizobium trifolii. Folia Microbiologica*, **17,** 393–5.

Kowalski, M. (1971). Transduction in *Rhizobium meliloti. Plant and Soil*, Special Volume, 63–6.

Lederberg, J., Cavalli, L. L. & Lederberg, E. M. (1952). Sex compatibility in *E. coli. Genetics*, **37,** 720–30.

Lester, L. P. (1971). Development of a method for genetic mapping in *Rhizobium trifolii.* PhD Thesis, Purdue University.

Lorkiewicz, Z., Zurkowski, W., Kowalczuk, E. & Górska-Melke, A. (1971). Mutagenesis and conjugation in *Rhizobium trifolii. Acta Microbiologica Polonica*, Ser. A, **3,** 101–7.

McQuillen, K. (1960). Bacterial protoplasts. In: *The Bacteria* (ed. I. C. Gunsalus & R. Y. Stanier), vol. 1, pp. 249–359. Academic Press, New York & London.

Olsen, R. H. & Shipley, P. (1973). Host range and properties of the *Pseudomonas aeruginosa* R-factor R1822. *Journal of Bacteriology*, **113,** 772–80.

Pühler, A., Burkardt, H. J. & Heumann, W. (1972). Genetic experiments with the *Pseudomonas aeruginosa* R-factor RP4 in *Rhizobium lupini. Journal of General Microbiology*, **73,** xxvi.

Raina, J. L. & Modi, V. V. (1972). Deoxyribonucleate binding and transformation in *Rhizobium japonicum. Journal of Bacteriology*, **111,** 356–60.

Sherwood, M. T. (1972). Inhibition of *Rhizobium trifolii* by yeast extracts or glycine is prevented by calcium. *Journal of General Microbiology*, **71,** 351–8.

Sik, T. & Orosz, L. (1971). Chemistry and genetics of *Rhizobium meliloti* phage 16–3. *Plant and Soil.* Special Volume, 57–62.

Stanisich, V. A. & Holloway, B. W. (1971). Chromosome transfer in *Pseudomonas aeruginosa* mediated by R-factors. *Genetical Research*, **17,** 169–72.

Strominger, J. L. & Birge, C. H. (1965). Nucleotide accumulation induced in *Staphylococcus aureus* by glycine. *Journal of Bacteriology*, **89,** 1124–7.

Sykes, R. B. & Richmond, M. H. (1970). Intergeneric transfer of a beta-lactamase gene between *Ps. aeruginosa* and *E. coli. Nature, London*, **226,** 952–4.

Vincent, J. M. (1970). *Manual for the practical study of root nodule bacteria.* Blackwell Scientific Publications, Oxford.

Vivian, A. (1971). Genetical control of fertility in *Streptomyces coelicolor* A3(2): plasmid involvement in the interconversion of UF and IF strains. *Journal of General Microbiology*, **69**, 353–64.

Zelazna-Kowalska, I. & Lorkiewicz, Z. (1971). Conditions for genetical transformation in *Rhizobium meliloti*. *Acta Microbiologica Polonica*, Ser. A, **3**, 21–8.

9. Identification and classification of root nodule bacteria

P. H. GRAHAM

The rhizobial taxonomist serves two masters. Onc is interested primarily in the ability of rhizobia to form root nodules, especially with agronomically important legumes; the other wishes to examine a range of characteristics and to identify strains according to this wider spectrum of properties. Commonly he is interested in comparing groups of *Rhizobium* with apparently related organisms such as *Agrobacterium* spp.

As the taxonomic groupings resulting from DNA studies (Wagenbreth, 1961; de Ley, 1968; Gibbins & Gregory, 1972), serology (Kleczkowski & Thornton, 1944; Graham, 1963*a*; Vincent & Humphrey, 1970; Vincent, Humphrey & Skrdleta, 1973) and Adansonian analysis (Graham, 1964; Moffett & Colwell, 1968) often have conflicted with those obtained from host plant studies, *Rhizobium* taxonomy has become polarised to the point where some scientists no longer use specific names for *Rhizobium* strains. This review will consider the advantages and disadvantages of the major approaches to *Rhizobium* classification and identification and will suggest areas of work necessary to improve the existing situation.

The infective classification of *Rhizobium*

The present classification of *Rhizobium* stems largely from the work of Baldwin & Fred (1927, 1929) Eckhardt, Baldwin & Fred (1931) and Fred, Baldwin & McCoy (1932). As given by Breed, Murray & Smith (1957) the genus *Rhizobium*, together with *Agrobacterium* and *Chromobacterium* comprise the family Rhizobiaceae. Only six species of *Rhizobium* are recognised, and are distinguished by differences in infectiveness coupled with some biochemical tests (see Table 9.1).

This classification assumes that each species of *Rhizobium* will nodulate only those plants which fall within a particular 'infective' or 'cross-inoculation' group (for definition see Nutman, 1956) and that within such groups rhizobia from one plant will nodulate all other plants, and vice versa (Fred *et al.*, 1932). The classification lost credibility with the repeated demonstration of symbiotic promiscuity and non-reciprocal nodulation (Leonard, 1932; Walker & Browne, 1935; Allen

& Allen, 1939; Wilson, 1939, 1944; Lange, 1961, 1966). Its practical value declined with increased awareness of effectiveness subgroups, such as occur in clovers (Nutman, 1965), soybean (Hamatova, 1965) *Centrosema* (Bowen & Kennedy, 1961) and *Stylosanthes* (Souto, Coser & Dobereiner, 1970) and it is now widely criticised (Wilson, 1944; Lange, 1961, 1966; Norris, 1956, 1965; Graham, 1964*a*; de Ley, 1968).

Table 9.1. *Species of* Rhizobium *and their characteristics.* (After Breed *et al.*, 1957)

Species	Serum zone in litmus milk	Acid reaction in litmus milk	Growth rate	Host nodulated	Common name of group
R. trifolii	+	–	Fast	*Trifolium*	Clover
R. leguminosarum	+	–	Fast	*Pisum, Lens, Lathyrus, Vicia*	Pea
R. phaseoli	+	–	Fast	*Phaseolus vulgaris*	French bean
R. meliloti	+	+	Fast	*Melilotus, Medicago, Trigonella*	
R. japonicum	–	–	Slow	*Glycine max*	Soybean
R. lupini	–	–	Slow	*Lupinus, Ornithopus*	Lupin
Rhizobium spp.	Variable	Variable	Variable	*Vigna, Desmodium, Arachis, Centrosema, Stylosanthes* etc.	Cowpea miscellany

The major limitations of the infective classification are as follows:

1. Symbiotic promiscuity

Symbiotic promiscuity is the nodulation of plants from one infective group by rhizobia belonging to another. Today only one species, *R. meliloti*, still satisfies the original nodulation criteria, most strains from this species nodulating only with species of *Medicago*, *Melilotus*, and *Trigonella*. Even here promiscuity has been observed (Wilson, 1944) especially with rhizobia from *Leucaena* (Trinick, 1965, 1968) though not to an extent which would invalidate the species.

Cross-infection is most common in the cowpea, soybean and lupin cross-inoculation groups (for review see Lange, 1966). Lange (1961), for example, examined eighty-five isolates from legumes indigenous to the south of Western Australia. Forty-five of these strains nodulate a lupin, cowpea, soybean and French bean (*Phaseolus vulgaris* L.) and could therefore have been assigned to each of four different rhizobial species.

Only three of the strains were specific for hosts of the cowpea group. Though the majority of strains nodulated *Lupinus digitatus*, few nodulated *L. albus* or *L. angustifolius*. The report by Trinick (1973) of cowpea rhizobia able to nodulate a non-legume, only extends this problem.

Cross-infection also occurs between the pea and clover inoculation groups (Kleczkowska, Nutman & Bond, 1944; Norris, 1956, 1959), with some strains isolated from African clover species showing greater affinity for *Pisum* and *Vicia* than for clovers of European origin.

2. *Insufficient nodulation data*

The order Leguminosae contains between 431 genera with 8887 species (Norris, 1956) and 700 genera with 14 000 species (Tutin, 1958). Only 8–12 % of these have been examined for nodulation (Allen & Allen, 1961) and a much smaller percentage included in infectiveness trials. Moreover, nodulation studies generally have been strongly biased toward agriculturally important legumes. This is evident in Table 9.1, where the six *Rhizobium* specific names are derived from only eleven genera of plants.

The tendency is to regard all rhizobia associated with previously untested legumes as similar to the slow-growing cowpea organisms (Norris, 1956). While fast-growing organisms isolated from such legumes must always be suspected of contamination with *Agrobacterium* (Lange, 1960) and be tested accordingly, there are numerous instances where rhizobia similar to *R. meliloti* or *R. trifolii* have been isolated from root nodules of plants included in the cowpea cross-inoculation group. In this category are isolates from *Leucaena* (Trinick, 1965), *Sesbania* (Johnson & Allen, 1952) *Samanea, Andura, Albizzia* and *Cytisus* (Allen & Allen, 1936) and *Psoralea* (Norris, 1965). Norris (1965) regarded these faster growing organisms as forms intermediate between *R. trifolii* and the cowpea rhizobia. Further testing might instead permit them to be included with existing or proposed species. What subgroupings might be obtained from such legume–*Rhizobium* associations where the microsymbiont is slow-growing or in the *Rhizobium*–non-legume system is impossible to predict.

3. *Lack of supporting biochemical data*

Biochemical and serological differences among rhizobia are not used to advantage in the current classification. Only three tests are significant.

(*a*) Strains of *R. trifolii*, *R. leguminosarum*, *R. phaseoli* and *R. meliloti* are faster growing than cowpea, soybean and lupin rhizobia (Cameron & Sherman, 1935; Graham & Parker, 1964). Whereas the first group produces visible colonies on agar in 3–5 days, the second group takes 7–15 days to achieve comparable development.

(*b*) While the rhizobia are all small Gram-negative bacilli which do not form endospores, they differ in flagellation. Thus the fast-growing rhizobia normally have peritrichous flagellation, while cowpea, lupin or soybean rhizobia have a subpolar monotrichous flagellation uncommon in other organisms (Lohnis & Hansen, 1921; Bushnell & Sarles, 1937; Leifson & Erdman, 1958).

(*c*) Norris (1965) tested 717 strains of rhizobia for growth on a mannitol yeast-extract medium containing bromthymol blue. Strains of *R. trifolii*, *R. meliloti*, *R. leguminosarum* and *R. phaseoli* produced acid in this media while most other strains gave an alkaline reaction. Norris used this data to propose differences in *Rhizobium* evolution (Norris, 1965, 1970, 1972) drawing conclusions which have been strongly criticised (Parker, 1968). The test does, however, underline differences in carbohydrate utilisation by rhizobia (see also Graham, 1964*b*) and provide a practical means for showing culture purity.

The litmus milk test has been used in *Rhizobium* characterisation (Stevens, 1925; Breed *et al.*, 1957) but the long incubation times necessary, coupled with the variable results obtained make it of doubtful taxonomic value for species separation (Hofer, 1941; Jensen, 1942; Graham & Parker, 1964).

Because these tests are of such limited value the only technique by which presumptive rhizobia can be screened is through their nodulation ability. This is time-consuming, and can be difficult when seed of the appropriate host is not readily available. Non-infective mutants cannot be identified by biochemical tests.

Alternative proposals for *Rhizobium* classification

Rhizobium taxonomy is following the pattern set by the plant-pathogenic Psuedomonadaceae (de Ley, 1968). As with these organisms, an infective classification has proved unsatisfactory, and alternative methods of obtaining a meaningful classification are being sought. These have included Adansonian analysis, DNA base ratio studies and hybridisation. With *Rhizobium*, despite differences in sample size and testing procedures, the results obtained from such studies have shown a surprising degree of agreement.

In an early investigation, Graham (1964*a*) studied 100 features of 121 strains of *Rhizobium*, *Agrobacterium*, *Beijerinckia*, *Bacillus* and *Chromobacterium* and concluded:

(i) that the present species *R. meliloti* continue without change.

(ii) that the species *R. leguminosarum*, *R. trifolii* and *R. phaseoli* be pooled to form a single species, *R. leguminosarum*.

(iii) that the species *Agrobacterium radiobacter* and *A. tumefaciens* be united and included as *R. radiobacter* in the genus *Rhizobium*.

(iv) that the species *R. japonicum* and *R. lupini* be pooled with organisms of the cowpea miscellany to form a single species. Since these organisms appeared distinct from other rhizobia, further studies were suggested to determine the merits of placing them in a separate genus; *A. rhizogenes* and *A. rubi* were not included in this study.

't Mannetje (1967) challenged some of those conclusions. Using different sorting techniques to re-analyse Graham's data, he suggested (*a*) that *A. radiobacter* and *A. tumefaciens* be combined, but not included in the genus *Rhizobium* until further studies had been made, and (*b*) that slow-growing root nodule bacteria be maintained as a single species in the genus *Rhizobium*. In this recommendation he followed closely the proposals of Norris (1956, 1965) based on host plant studies and carbohydrate utilisation tests.

de Ley (1968) reviewed the information obtained from infective groupings, computer taxonomy (Graham, 1964*a*; 't Mannetje, 1967; Moffet & Colwell, 1968), DNA base ratios (Wagenbreth, 1961; de Ley & Rassel, 1965; de Ley *et al.*, 1966) and hybridisation data (Herberlein, de Ley & Tijtgat, 1967). He pointed out that three of the *Agrobacterium* spp. (*A. pseudotsugae*, *A. gypsophilae* and *A. stellulatum*) were atypical in G+C percentages or in numerical analysis (Moffett & Colwell, 1968) and suggested that they, together with the genus *Chromobacterium*, be removed completely from the Rhizobiaceae. For those organisms which remained, he proposed the formation of a single genus *Rhizobium* which would contain both *Rhizobium* and *Agrobacterium* species. Species designations intermediate between those of Graham (1964*a*) and 't Mannetje (1967) were chosen, and characterised as in Table 9.2.

While the system proposed by de Ley (1968) provides a more logical classification of the known rhizobia, and thus an interim solution, it does not solve two major problems:

(i) As with the infective classification, it does not consider the rhizobia

Table 9.2. *Species of* Rhizobium *and their characteristics.* (After de Ley, 1968)

Species	Relation to species of Breed *et al.* (1957)	Flagellation	Per cent G+C	Serum zone litmus milk	Acid reaction litmus milk	Growth rate	Nodule forming characteristics, special features
R. leguminosarum	*R. phaseoli* + *R. trifolii* + *R. leguminosarum*	Peritrichous	59.0–63.5	+	–	Fast	Forms nodules on one or more of *Trifolium*, *P. vulgaris*, *Vicia*, *Pisum*, *Lathyrus*, *Lens*
R. meliloti	Unchanged	Peritrichous	62.0–63.5	+	+	Fast	Forms nodules with *Melilotus*, *Medicago*, *Trigonella*
R. rhizogenes	*A. rhizogenes*	Peritrichous	61.0–63.0	+	–	Fast	Causes hairy root disease of apples and other plants
R. radiobacter	*A. tumefaciens* + *A. radiobacter* + *A. rubi*	Peritrichous	59.5–63.0	+	–	Fast	Frequently produce galls on angiosperms Produce 3-ketoglycosides
R. japonicum	*R. japonicum* + *R. lupini* + Cowpea miscellany	Subpolar	59.5–65.5	–	–	Slow	Nodulates many different legumes including one or more of *Vigna*, *Glycine*, *Lupinus*, *Ornithopus*, *Centrosema*, etc.

associated with the many as yet untested legume species. Only an extensive investigation will show if these organisms conform to the proposed species and the extent of differences which exist.

(ii) No new information is advanced by which strains could be identified in the absence of host plant data. The DNA base ratio data suggested by de Ley are useful, but the technique is one not readily applicable in many soil microbiology laboratories. This need for better identification methods is discussed below.

Identification and characterisation of rhizobia

In this section the species names used are those proposed by de Ley (1968). Irrespective of the method of classification adopted, rhizobia and agrobacteria will continue to be identified by host plant reactions and by morphological features. There is need, therefore, to find host varieties that have a wide spectrum of susceptibilities, and to standardise the system of testing. Methods which could be used are well-documented by Vincent (1970) and Kleczkowska *et al.* (1968). Some morphological features of the rhizobia are described earlier in this chapter. Ideally, however, additional characteristics should be available to permit strain identification in the absence of infectiveness data, e.g. in the case of non-infective mutants or *A. radiobacter*. Pointers to such characteristics are evident in the papers of Hofer (1941), Graham & Parker (1964), Graham (1964*b*), Kleczkowska *et al.* (1968) and in the species descriptions of Breed *et al.* (1957). These include:

1. Biochemical reactions

Production of 3-ketoglycosides. Strains of *R. radiobacter* (= *A. radiobacter* and *A. tumefaciens*) metabolise lactose and sucrose with the production of 3-ketoglycosid es,substances which are readily detected using Benedict's reagent (Bernaerts & de Ley, 1963). The reaction is not known to occur in other organisms, and provides a simple means for distinguishing between rhizobia and *R. radiobacter* occurring as contaminants in nodules.

Utilisation of carbohydrates. Though carbohydrate utilisation by rhizobia has been extensively studied (Baldwin & Fred, 1927; Georgi & Ettinger, 1941), strain differences have often been obscured by the inclusion of yeast extract in the test media. As pointed out by Graham

(1964*b*) and Parker (1971), rhizobia can use amino acids as energy sources, the media often becoming alkaline as a result. Methods must be used which overcome this difficulty. Mini-tests as used for the identification of Enterobacteriaceae could be valuable.

Table 9.3 shows utilisation differences obtained by Graham (1964*b*). Most strains of *Rhizobium* utilised glucose, xylose and arabinose, but

Table 9.3. *Utilisation of selected carbohydrates by* Rhizobium *spp.* (After Graham, 1964*b*)

	No. strains tested			
Sugar	*R. leguminosarum*	*R. meliloti*	*R. radiobacter*	*R. japonicum*
	35	5	13	55
Glucose	35	5	13	37
Xylose	35	5	13	51
Arabinose	34	4	13	54
Dextrin	4	—	1	—
Sucrose	35	5	13	—
Lactose	33	3	6	3
Raffinose	32	3	8	3
Rhamnose	33	5	13	1
Maltose	35	4	13	6

lactose and sucrose were utilised by few slow-growing rhizobia. Additional studies are needed to extend the range of carbohydrates which can be used, and to apply enzymatic studies such as those of Martínez- de Drets & Arias (1972) and Elkan (1971) to a greater range of *Rhizobium* strains.

Vitamin requirements. Biotin, thiamine and pantothenate are required by most strains of *R. leguminosarum* (=*R. trifolii*, *R. phaseoli* and *R. leguminosarum*) (West & Wilson, 1939; Ljunggren, 1961; Graham, 1963*b*). Biotin is also required by some strains of *R. meliloti*, slow-growing rhizobia and agrobacteria, but only *R. leguminosarum* strains require the other growth factors mentioned (McBurney, Bollen & Williams, 1935; Graham, 1963*b*).

Growth on calcium glycerophosphate medium. Hofer (1941) and Graham & Parker (1964) compared the growth of *Rhizobium* spp. in calcium glycerophosphate agar. Only strains of *R. meliloti* and *R. radiobacter* produced a precipitate in this medium, though *R. leguminosarum* strains also grew well. Slow-growing rhizobia grew poorly or not at all.

Graham & Parker (1964) used the 'G' term proposed by Sneath (1962) to determine features of taxonomic value in rhizobia. In addition to those tests outlined above, a number of other features showed promise. These included production of penicillinase, growth at pH 9.5 and in 2 % sodium chloride, and production of hydrogen sulphide in certain media. Differences in antibiotic sensitivity also warrant consideration (Elkan, 1971). Better definition is needed for these tests, but they could also be of value in future identification procedures.

2. Serology

Though early serological studies using agglutination reactions showed cross-reactions in line with the classification proposed by de Ley (1968) – that is of three broad serological groups in the rhizobia with some cross-reaction between agrobacteria and fast-growing rhizobia (Bushnell & Sarles, 1939; Kleczkowski & Thornton, 1944; Graham, 1963*a*, 1971) – serological methods have been little used in the classification of *Rhizobium* (see review by Graham, 1969). A major reason was the variety of somatic antigens found in these organisms, Purchase, Vincent & Ward (1951) postulating sixteen different O antigens in only fifteen strains of *R. meliloti*.

More recent evidence using agar diffusion techniques and disrupted bacterial cells suggests internal group-specific antigens which could be used in strain classification. Thus Vincent & Humphrey (1970) found two internal antigens common to all eighteen *R. leguminosarum* strains studied, but could not detect the antigens in nine slow-growing cultures. Some, but not all, *R. meliloti* and *R. radiobacter* also possessed one or both of the antigens. Similarly, Vincent *et al.* (1973) found at least one and generally two common antigens in twenty-three slow-growing rhizobia tested against antisera from three soybean strains, but fifty-one fast-growing rhizobia failed to give any reaction. Common internal antigens have also been found in *R. radiobacter* (Graham, 1971).

Serological methods including crushed nodule preparations (Means, Johnson & Date, 1964; Parker & Grove, 1970) and fluorescent antibody techniques (Trinick, 1969) are already widely used in strain identification of rhizobia from nodules formed as a result of inoculation (Dudman & Brockwell, 1968).

3. Susceptibility to bacteriophage

As suggested by Kleczkowska *et al.* (1968), knowledge of *Rhizobium* susceptibility to bacteriophage has not yet reached the point where such techniques could be used in *Rhizobium* classification.

Conclusions

Kleczkowska *et al.* (1968) end their review of *Rhizobium* classification with 'More needs to be known about the nodule bacteria of legumes generally before classification of *Rhizobium* can be rationalised and completed.' 't Mannetje (1967) is similarly pessimistic, calling for detailed studies of additional rhizobia and agrobacteria. The present author is concerned by the declining number of *Rhizobium* taxonomic papers published since 1968, and feels that a stimulus to continued research in this field is necessary. Therefore, while acknowledging that any classification erected now will have deficiencies, he proposes adoption of the classification suggested by de Ley (1968) and shown in Table 9.2.

A number of future goals should be set, the most important of which would be a detailed examination of the organisms at present grouped in the cowpea miscellany. Taxonomic subgroupings (if any) need to be established. A second need is for improved and standardised biochemical tests which could be used to classify and identify *Rhizobium* strains. Some tests of potential value are suggested; others need further investigation. Finally, studies on group-specific antigens in *Rhizobium* are necessary to determine whether serological methods can be applied routinely to genus or species rather than to strain identification.

References

Allen, E. K. & Allen, O. N. (1961). The scope of nodulation in the Leguminosae. In: *Recent advances in botany*, pp. 585–8. University of Toronto Press, Toronto & New York.

Allen, O. N. & Allen, E. K. (1936). Root nodule bacteria of some tropical leguminous plants. I. Cross inoculation studies with *Vigna sinensis* L. *Soil Science*, **42,** 61–77.

Allen, O. N. & Allen, E. K. (1939). Root nodule bacteria of some tropical leguminous plants. II. Cross inoculation tests within the cowpea group. *Soil Science*, **47,** 63–76.

Baldwin, I. L. & Fred, E. B. (1927). The fermentation characters of the root nodule bacteria of the Leguminosae. *Soil Science*, **24,** 217–30.

Baldwin, I. L. & Fred, E. B. (1929). Nomenclature of the root nodule bacteria of the Leguminosae. *Journal of Bacteriology*, **17,** 141–50.

Bernaerts, M. J. & de Ley, J. (1963). A biochemical test for crown gall bacteria. *Nature, London*, **197,** 406–7.

Bowen, G. D. & Kennedy, M. M. (1961). Heritable variation in nodulation of *Centrosema pubescens. Queensland Journal of Agricultural Science*, **18**, 161–70.

Breed, R. S., Murray, E. G. D. & Smith, N. R. (1957). *Bergey's manual of determinative bacteriology*, 7th edition. Williams & Wilkins, Baltimore.

Bushnell, O. A. & Sarles, W. B. (1937). Studies on the root nodule bacteria of wild leguminous plants in Wisconsin. *Soil Science*, **44**, 409–23.

Bushnell, O. A. & Sarles, W. B. (1939). Investigations upon the antigen relationships among the root-nodule bacteria of the soybean, cowpea and lupin cross inoculation groups. *Journal of Bacteriology*, **38**, 401–10.

Cameron, G. N. & Sherman, J. M. (1935). The rate of growth of rhizobia. *Journal of Bacteriology*, **30**, 647–50.

de Ley, J. (1968). DNA base composition and hybridisation in the taxonomy of phytopathogenic bacteria. *Annual Review of Phytopathology*, **6**, 63–90.

de Ley, J., Bernaerts, M., Rassel, A. & Guilmot, J. (1966). Approach to an improved taxonomy of the genus *Agrobacterium. Journal of General Microbiology*. **43**, 7–17.

de Ley, J. & Rassel, A. (1965). DNA base composition, flagellation and taxonomy of the genus *Rhizobium. Journal of General Microbiology*, **41**, 85–91.

Dudman, W. F. & Brockwell, J. (1968). Ecological studies of root-nodule bacteria introduced into field environments. I. A survey of field performance of clover inoculants by gel immune diffusion serology. *Australian Journal of Agricultural Research*, **19**, 739–47.

Eckhardt, M. M., Baldwin, I. L. & Fred, E. B. (1931). Studies on the root nodule organism of *Lupinus. Journal of Bacteriology*, **21**, 273–85.

Elkan, G. H. (1971). Biochemical and genetical aspects of the taxonomy of *Rhizobium japonicum. Plant and Soil*, Special Volume, 85–104.

Fred, E. B., Baldwin, I. L. & McCoy, E. (1932). *Root nodule bacteria and leguminous plants*. University of Wisconsin Press, Madison.

Georgi, C. E. & Ettinger, J. M. (1941). Utilisation of carbohydrates and sugar acids by the rhizobia. *Journal of Bacteriology*, **41**, 323–40.

Gibbins, A. M. & Gregory, K. F. (1972). Relatedness among *Rhizobium* and *Agrobacterium* species determined by three methods of nucleic acid hybridisation. *Journal of Bacteriology*, **111**, 129–41.

Graham, P. H. (1963*a*). Vitamin requirements of root nodule bacteria. *Journal of General Microbiology*, **30**, 245–8.

Graham, P. H. (1963*b*). Antigenic affinities of the root nodule bacteria of legumes. *Antonie van Leeuwenhoek, Journal of Microbiology and Serology*, **29**, 281–91.

Graham, P. H. (1964*a*). The application of computer techniques to the taxonomy of the root nodule bacteria of legumes. *Journal of General Microbiology*, **35**, 511–7.

Graham, P. H. (1964*b*). Studies on the utilisation of carbohydrates and Kreb's cycle intermediates by rhizobia, using an agar plate method. *Antonie van Leeuwenhoek, Journal of Microbiology and Serology*, **30**, 68–72.

Graham, P. H. (1969). Serology of the Rhizobiaceae. In: *Analytical serology of Microorganisms* (ed. J. B. Kwapinski), vol. 2, pp. 353–78. John Wiley & Sons, New York.

Graham, P. H. (1971). Serological studies with *Agrobacterium radiobacter*, *A. tumefaciens* and *Rhizobium* strains. *Archiv für Mikrobiologie*, **78,** 70–5.

Graham, P. H. & Parker, C. A. (1964). Diagnostic features in the characterisation of the root nodule bacteria of legumes. *Plant and Soil*, **20,** 383–96.

Hamatova, E. (1965). The selecting of strains of *Rhizobium japonicum* capable of reliable nodulation in Czechoslovak soya varieties. *Rostlinne Vyroby*, **9,** 193–209.

Herberlein, G. T., de Ley, J. & Tijtgat, R. (1967). Deoxyribonucleic acid homology and taxonomy of *Agrobacterium*, *Rhizobium* and *Chromobacterium*. *Journal of Bacteriology*, **94,** 116–24.

Hofer, A. W. (1941). A characterisation of *Bacterium radiobacter* (Beijerinck and van Delden) Löhnis. *Journal of Bacteriology*, **41,** 193–224.

Jensen, H. L. (1942). Nitrogen fixation in leguminous plants. I. General characters of root nodule bacteria isolated from species of *Medicago* and *Trifolium*. *Proceedings of the Linnean Society of New South Wales*, **67,** 98–108.

Johnson, M. D. & Allen, O. N. (1952). Nodulation studies with special reference to strains isolated from *Sesbania* species. *Antonie van Leeuwenhoek, Journal of Microbiology and Serology*, **18,** 12–22.

Kleczkowska, J., Nutman, P. S. & Bond, G. (1944). Note on the ability of certain strains of rhizobia from peas and clover to infect each others host plants. *Journal of Bacteriology*, **48,** 673–5.

Kleczkowska, J., Nutman, P. S., Skinner, F. A. & Vincent, J. M. (1968). The identification and classification of *Rhizobium*. In: *Identification methods for microbiologists* (ed. B. M. Gibbs & D. A. Shapton), pp. 51–65. Academic Press, New York & London.

Kleczkowski, A. & Thornton, H. G. (1944). A serological study of the root nodule bacteria from the pea and clover inoculation groups. *Journal of Bacteriology*, **48,** 661–73.

Lange, R. T. (1960). *Rhizobium* of South Western Australia. PhD Thesis, University of Western Australia.

Lange, R. T. (1961). Nodule bacteria associated with the indigenous Leguminosae of South Western Australia. *Journal of General Microbiology*, **26,** 351–9.

Lange, R. T. (1966). Bacterial symbiosis with plants. In: *Symbiosis* (ed. S. M. Henry), vol. 1, pp. 99–170. Academic Press, New York & London.

Leifson, E. & Erdman, L. W. (1958). Flagellar characteristics of *Rhizobium* species. *Antonie van Leeuwenhoek, Journal of Microbiology and Serology*, **24,** 97–110.

Leonard, L. T. (1923). Nodule production kinship between the soybean and the cowpea. *Soil Science*, **15,** 277–83.

Ljunggren, H. (1961). Transfer of virulence in *Rhizobium trifolii*. *Nature, London*, **191,** 623.

Lohnis, F. & Hansen, R. (1921). Nodule bacteria of leguminous plants. *Journal of Agricultural Research*, **20,** 543–56.

McBurney, C. H., Bollen, W. B. & Williams, R. J. (1935). Pantothenic acid and the nodule bacteria – legume symbiosis. *Proceedings of the National Academy of Sciences, USA*, **21,** 301–4.

Martínez de Drets, G. M. & Arias, A. (1972). Enzymatic basis for differentiation of *Rhizobium* into fast and slow growing groups. *Journal of Bacteriology*, **109,** 467–70.

Means, U. M., Johnson, H. W. & Date, R. A. (1964). Quick serological method of classifying strains of *Rhizobium japonicum* in nodules. *Journal of Bacteriology*, **87,** 547–53.

Moffett, M. L. & Colwell, R. R. (1968). Adansonian analysis of the Rhizobeaceae. *Journal of General Microbiology*, **51,** 245–66.

Norris, D. O. (1956). Legumes and the *Rhizobium* symbiosis. *Empire Journal of Experimental Agriculture*, **24,** 247–70.

Norris, D. O. (1959). *Rhizobium* affinities of African species of *Trifolium*. *Empire Journal of Experimental Agriculture*, **27,** 87–97.

Norris, D. O. (1965). Acid production by *Rhizobium*. A unifying concept. *Plant and Soil*, **22,** 143–66.

Norris, D. O. (1970). Nodulation of pasture legumes. In: *Australian grasslands* (ed. R. M. Moore), pp. 339–48. Australian National University Press, Canberra.

Norris, D. O. (1972). Leguminous plants in tropical pastures. *Tropical Grasslands*, **6,** 159–70.

Nutman, P. S. (1956). The influence of the legume in the root nodule symbiosis. *Biological Reviews*, **31,** 109–51.

Nutman, P. S. (1965). Symbiotic nitrogen fixation. In: *Soil nitrogen* (ed. W. V. Bartholomew & F. E. Clark), pp. 363–79. American Society of Agronomy, Madison.

Parker, C. A. (1968). On the evolution of symbiosis in legumes. In: *Festkrift til Hans Laurits Jensen*. Gadgaard Nielsens Bogtrykkeri, Lemvig.

Parker, C. A. (1971). The significance of acid and alkali production by rhizobia on laboratory media. Fourth Australian *Rhizobium* Conference, Canberra.

Parker, C. A. & Grove, P. L. (1970). Rapid serological identification of rhizobia in small nodules. *Journal of Applied Bacteriology*, **33,** 248–52.

Purchase, H. F., Vincent, J. M. & Ward, L. M. (1951). Serological studies of the root nodule bacteria. IV. Further studies of isolates from *Trifolium* and *Medicago*. *Proceedings of the Linnean Society of New South Wales*, **76,** 1–6.

Sneath, P. H. A. (1962). The construction of taxonomic groups. *Symposia of the Society of General Microbiology*, **12,** 289–332.

Souto, S. M., Coser, A. C. & Dobereiner, J. (1970). Especifidade de una variedade native de 'Alfalfa de Nordeste' (*Stylosanthes gracilis* N.B.K.) na symbiosis con *Rhizobium* sp. In: Proc. V. Reunion Latinoamericana de *Rhizobium*, pp. 78–91. Departamento Nacional de Pesquisa Agropecuária, Rio de Janeiro, Brazil.

Stevens, J. W. (1925). A study of various strains of *Bacillus radicicola* from nodules of alfalfa and sweet clover. *Soil Science*, **20,** 45–66.

't Mannetje, L. (1967). A re-examination of the taxonomy of the genus *Rhizobium* and related genera using numerical analysis. *Antonie van Leeuwenhoek, Journal of Bacteriology and Serology*, **33,** 477–91.

Trinick, M. J. (1965). *Medicago sativa* nodulation with *Leucaena leucocephala* root nodule bacteria. *Australian Journal of Science*, **27,** 263–4.

Trinick, M. J. (1968). Nodulation of tropical legumes. I. Specificity in the *Rhizobium* symbiosis of *Leucaena leucocephala. Experimental Agriculture*, **4,** 243–53.

Trinick, M. J. (1969). Identification of legume nodule bacteria by the fluorescent antibody technique. *Journal of Applied Bacteriology*, **32,** 181–6.

Trinick, M. J. (1973). Symbiosis between *Rhizobium* and the non-legume, *Trema aspera. Nature, London*, **244,** 459–60.

Tutin, T. G. (1958). Classification of the legumes. In: *Nutrition of legumes* (ed. E. G. Hallsworth), pp. 3–14. Butterworth, London.

Vincent, J. M. (1970). *Manual for the practical study of root nodule bacteria.* IBP Handbook No. 15. Blackwell Scientific Publications, Oxford.

Vincent, J. M. & Humphrey, B. A. (1970). Taxonomically significant group antigens in *Rhizobium. Journal of General Microbiology*, **63,** 379–82.

Vincent, J. M., Humphrey, B. A. & Skrdleta, V. (1973). Group antigens in slow-growing rhizobia. *Archiv für Mikrobiologie*, **89,** 79–82.

Wagenbreth, D. (1961). Ein Beitrag zur systematischen Einordnung der Knollchenbaktieren durch Bestimmung des relativen Basengehaltes ihrer Desoxyribonukleosauren. *Flora Jena*, **151,** 219–30.

Walker, R. H. & Brown, P. E. (1935). The nomenclature of the cowpea group of root nodule bacteria. *Soil Science*, **39,** 221–5.

West, P. M. & Wilson, P. W. (1939). Growth factor requirements of the root nodule bacteria. *Journal of Bacteriology*, **37,** 161–85.

Wilson, J. K. (1939). Symbiotic promiscuity in the Leguminosae. In: *IIIrd International Soil Science Society Congress*, **A,** pp. 49–63.

Wilson, J. K. (1944). Over five hundred reasons for abandoning the cross inoculation groups of legumes. *Soil Science*, **58,** 61–9.

10. The nodulation profile of the genus *Cassia*

ETHEL K. ALLEN & O. N. ALLEN

Cassia is the fourth largest genus of the family Leguminosae, the largest genus in the subfamily Caesalpinioideae, and amongst the twenty-five largest genera of dicotyledonous plants. The 600-odd species of *Cassia* abound in warm regions throughout the world; few are found in temperate areas. Species range in habit from prostrate, annual herbs bearing flat, linear pods 2 cm in length to tall forest trees with cylindrical, woody, club-like pods 100 cm or more long. *Cassia* 'shower trees' are tropical ornamentals unexcelled in floral brilliance. Tree bark is a source of tannin. Dried leaflets of members of subgenus *Senna* yield the cathartic senna. *C. acutifolia* Del. and *C. angustifolia* Vahl are commercial sources of Alexandria and Tinnevelly senna, respectively. Herbaceous cassias are green manuring and cover crops in tea, coffee and cacao plantations.

This genus was defined originally by Linnaeus in the first edition of his *Species Plantarum*. Parts of this large genus were later given generic consideration, but most botanists have been content to retain the many species in a single genus. Bentham (1871) referred to *Cassia* as 'an excellent instance of a large, widely distributed, much varied, but well-defined group'. His system recognizing the division of *Cassia* into subgenera *Fistula*, *Senna* and *Lasiorhegma* is still accepted to a large extent.

This chapter presents the results of a global survey of nodule occurrence on *Cassia* species superimposed upon Bentham's infrageneric organization of the genus.

Results and discussion

Nodule formation is more commonly present than absent in the subfamilies Mimosoideae and Papilionoideae; the reverse is true in the subfamily Caesalpinioideae. A current estimate (Allen & Allen, unpublished data) that about 70 % of caesalpinioid species lack nodules is in general agreement with an earlier report (Allen & Allen, 1961). Nodulated and non-nodulated genera often occur within the same tribe. Nodulated species are present in about 20 % of the genera; 10 % of the genera have both nodulated and non-nodulated species.

Discontinuity of nodulation within Caesalpinioideae occurs commonly at the generic level; within *Cassia* it is abruptly evident at the subgeneric level. Lack of nodules within subgenera *Fistula* and *Senna* (Table 10.1*a*) is in marked contrast to nodule prevalence in subgenus *Lasiorhegma*, section *Chamaecrista* (Table 10.1*e*). Reports of nodulation on five species in *Fistula* and *Senna* are unchallenged (Table 10.1*b*), however the reports for four other species are conflicting (Table 10.1*c*). Since the negative evidence for the latter outweighs the positive, these species probably lack nodulating ability. Conversely, it is conjectured that *C. wrightii* (Table 10.1*f*) may nodulate under growth conditions

Table 10.1. *Nodulated and non-nodulated* Cassia *species*

SUBGENERA *Fistula* and *Senna*

(a) Non-nodulated species (50)

C. abbreviata Oliv. subsp. *abbreviata*, subsp. *beareana* (Holmes) Brenan, Rhodesia (Corby, 1974); *aculeata* Pohl., Venezuela (Barrios & Gonzalez, 1971); *acutifolia* Del., France (Lechtova-Trnka, 1931); *alata* L., Hawaii (Allen & Allen, 1936*b*), Philippines (Bañados & Fernadez, 1954), Trinidad (DeSouza, 1966); *antillana* (Britt. & Rose) Alain, Hawaii (personal observation, 1962); *artemisoides* Gaud., USDA (Leonard, 1925), Australia (Beadle, 1964); *bacillaris* L.f., Philippines (Bañados & Fernandez, 1954); *bauhinioides* Gray, Arizona (Martin, 1948); *bicapsularis* L., Rhodesia (Corby, 1974), USDA (Leonard, 1925), Venezuela (Barrios & Gonzalez, 1971); *biflora* L., Trinidad (DeSouza, 1966); *circinnata* Benth., Australia (N. C. W. Beadle (personal communication, 1963); *coluteoides* Collad., Rhodesia (Corby, 1974); *corymbosa* Lam., S. Africa (Grobbelaar, van Beijma & Saubert, 1964), USDA (Leonard, 1925); *covesii* Gray, Arizona (Martin, 1948); *desolata* F. Muell., Australia (Beadle, 1964); *emarginata* L., USDA (Leonard, 1925); *eremophila* A. Cunn., Australia (Beadle, 1964); *fistula* L., Hawaii (Allen & Allen, 1936*b*), Philippines (Bañados & Fernandez, 1954), Brazil (Campêlo & Döbereiner, 1969); *floribunda* Cav., Rhodesia (Corby, 1974); *fruticosa* Mill., Trinidad (DeSouza, 1966); *glauca* L.f., Hawaii (Allen & Allen, 1936*b*); *italica* (Mill.) Lam., Rhodesia (Corby, 1974), subsp. *arachoides* (Burch.) Brenan, S. Africa (Grobbelaar *et al.*, 1967); *javanica* L., Java (Muller & Frémont, 1935), Hawaii (personal observation, 1962), S. Africa (Grobbelaar *et al.*, 1964), Brazil (Campêlo & Döbereiner, 1969); (*laevigata* Willd.) = *floribunda* Cav., Java (Steinmann, 1930; Muller & Frémont, 1935), Hawaii (Allen & Allen, 1936*b*); *latifolia* G. F. W. Mey., British Guiana (Norris, 1969); *leptocarpa* Benth., Arizona (Martin, 1948); *marilandica* L., USDA (Leonard, 1925); *marksiana* Domin, Hawaii (personal observation, 1962); *medsgeri* Shafer, Illinois, USA (Burrill & Hansen, 1917), USDA (Leonard, 1925); *moschata* H.B. & K., Hawaii (Allen & Allen, 1936*b*), Philippines (Bañados & Fernandez, 1954), Venezuela (Barrios & Gonzalez, 1971); *multijuga* Rich., Java (Steinmann, 1930; Muller & Frémont, 1935), Trinidad (DeSouza, 1966); *nodosa* Buch.-Ham., Hawaii (Allen & Allen, 1936*b*); *obovata* Collad., S. Africa (Mostert, 1955); *obtusifolia* L., widespread (many reports); *oracle*, New York (Wilson, 1939); *petersiana* Bolle, Rhodesia (Corby, 1974); *phyllodinea* R. Br., *pleurocarpa* F. Muell., Australia (Beadle, 1964); *quadrifoliolata* Pitt., *saeri* Britt. & Rose, *sericea* Sw., Venezuela (Barrios & Gonzalez, 1971); *siamea* Lam., Java (Muller & Frémont, 1935), Hawaii (Allen & Allen, 1936*b*), Philippines (Bañados & Fernandez, 1954), S. Africa (Grobbelaar *et al.*, 1964); *singueana* Del., Rhodesia (Corby, 1974); *sophera* L. var. *schinifolia* Benth., S. Africa (Grobbelaar, personal communication, 1971); *sturtii* R. Br., Australia (Beadle, 1964); *timorensis* DC., Java (Steinmann, 1930); *tomentosa* L.f., USDA (Leonard, 1925), Java (Steinmann, 1930); *tora* L., widespread (many reports); *wislizeni* A. Gray, Arizona (Martin, 1948).

(b) Nodulated species (5)

C. leiandra Benth. var. *guianensis* Sandw., British Guiana (Norris, 1969); *pilifera* Vog.,

Hawaii (Allen & Allen, 1936*a*); *pteridophylla* Sandw., British Guiana (Norris, 1969); *sophera* L., S. Africa (Grobbelaar *et al.*, 1967); *speciosa* Schrad., Brazil (E. S. Lopes, personal communication, 1966).

(c) Conflicting reports (4)

C. didymobotrya Fres. (+), Ceylon (W. R. C. Paul, personal communication, 1951); (−), Java (Steinmann, 1930; Muller & Frémont, 1935), Rhodesia (Corby, 1974); *grandis* L.f. (+), Ceylon (Wright, 1903); (−), Hawaii (Allen & Allen, 1936*b*); *hirsuta* L. (+), Ceylon (Wright, 1903), Java (Keuchenius, 1924); (−), Ceylon (Bamber & Holmes, 1911; Holland, 1924), Java (Steinmann, 1930; Muller & Frémont, 1925), Philippines (Bañados & Fernandez, 1954), Trinidad (DeSouza, 1966); *occidentalis* L. (+), Java (Keuchenius, 1924), Philippines (Bañados & Fernandez, 1954); (−), Java (Steinmann, 1930; Muller & Frémont, 1935), Russia (Lopatina, 1931), USDA (Leonard, 1925), Japan (Itano & Matsuura, 1936), Argentina (Rothschild, 1970), Hawaii (Allen & Allen, 1936*b*), Venezuela (Barrios & Gonzalez, 1971), Rhodesia (Corby, 1974).

SUBGENUS *Lasiorhegma*

SECTION *Absus*

(d) Nodulated species (2)

C. absus L., widespread (many reports); *hispidula* Vahl, Trinidad (DeSouza, 1966).

SECTION *Chamaecrista*

(e) Nodulated species (37)

C. bauhiniaefolia Kunth, Venezuela (Barrios & Gonzalez, 1971); *biensis* (Stey.) Mend. & Torre, S. Africa (N. Grobbelaar, personal communication, 1972), Rhodesia (Corby, 1974); *calycioides* DC., Venezuela (Barrios & Gonzalez, 1971); *capensis* Thunb. var *keiensis* Stey., S. Africa (Grobbelaar & Clarke, 1972); *chamaecrista* L., USA (Burrill & Hansen, 1917; Leonard, 1925; Wilson, 1939); *chamaecristoides* Collad., Mexico (personal observation, 1970); *comosa* Vog., S. Africa (Grobbelaar *et al.*, 1967); *cultrifolia* H.B. & K., Venezuela (Barrios & Gonzalez, 1971); *diphylla* L., Puerto Rico (Kinman, 1916); *falcinella* Oliv. var. *parviflora* Stey., Rhodesia (Corby, 1974); *fasciculata* Michx., widespread (many reports); *fenarolii* Mend. & Torre, Rhodesia (Corby, 1974); *flexuosa* L., Venezuela (Barrios & Gonzalez, 1971); *glandulosa* L., Mexico (personal observation, 1970); *gracilior* (Ghesq.) Stey., Rhodesia (Corby, 1974); *gracilis* Kunth., Venezuela (Barrios & Gonzalez, 1971); *hochstetteri* Ghesq., Rhodesia (Corby, 1974); *katangensis* (Ghesq.) Stey., Belgian Congo (Steyaert, 1952); *kirkii* Oliv., S. Africa (N. Grobbelaar, personal communication, 1970); var. *kirkii*; *leptadenia* Greenm., Rhodesia (Corby, 1974); *leschenaultiana* DC., Java (Keuchenius, 1924; Steinmann, 1930; Muller & Frémont, 1935), Ceylon (Paul, 1949), Taiwan (personal observation, 1962); *lucesiae* Pitt., Venezuela (Barrios & Gonzalez, 1971); *mimosoides* L., widespread (many reports); *nictitans* L., USA (Burrill & Hansen, 1917; Shunk, 1921; Leonard, 1923; Conklin, 1936); *parva* Stey., Rhodesia (Corby, 1974); *patellaria* DC., Java (Keuchenius, 1924; Steinmann, 1930; Muller & Frémont, 1935), Trinidad (DeSouza, 1966), Argentina (Rothschild, 1970); *polytricha* Brenan var. *polytricha*, Rhodesia (Corby, 1974); *pumila* Lam., Java (Helten, 1915; Muller & Frémont, 1935), W. India (Y. S. Kulkarni, personal communication, 1972); *quarrei* (Ghesq.) Stey., Rhodesia (Corby, 1974); *rotundifolia* Pers., Australia (Norris, 1959), S. Africa (University of Pretoria, 1955–6; N. Grobbelaar, personal communication, 1963), Rhodesia (Corby, 1974); *serpens* L., Venezuela (Barrios & Gonzalez, 1971); *simpsoni* Pollard, USDA (Leonard, 1925); *swartzii* Wickstr., Puerto Rico (Dubey, Woodbury & Rodríguez, 1972); *tagera* L., Venezuela (Barrios & Gonzalez, 1971); *wittei* Ghesq., Rhodesia (Corby, 1974); *zambesica* Oliv., Rhodesia (Corby, 1974).

(f) Non-nodulated species (1)

C. wrightii A. Gray, Arizona (Martin, 1948).

more favorable than in arid Arizona (Martin, 1948). This assumption stems from the recent observation in Rhodesia (Corby, 1974) of nodules on *C. leptadenia*, the only other member of the *Chamaecrista* complex reported not to nodulate (Martin, 1948).

Information on nodule morphology and rhizobia characterization is lacking for the five nodulated species in *Fistula* and *Senna* (Table 10.1*b*). Nodules on species in section *Chamaecrista* have apical meristems that become dichotomously branched with age (personal observation). Rhizobia from these nodules are slow-growing strains of the soybean–lupin–cowpea type. *C. chamaecrista* L. (Burrill & Hansen, 1917; Wilson, 1939), *C. fasciculata* Michx. (Bushnell & Sarles, 1937), *C. mimosoides* L. (Allen & Allen, 1936*a*), and *C. nictitans* L. (Conklin, 1936) in the *Chamaecrista* complex, and *C. pilifera* Vog. (Allen & Allen, 1936*a*) in subgenus *Senna* are aligned symbiotically within the cowpea miscellany.

Why certain *Cassia* species are nodulated and others are not is an enigma. Morphological factors in members of *Fistula* and *Senna* militate against nodule formation in that root hairs are uncommon and, when present, are sparse and thick-walled. Rhizobia from a range of host species have the ability to enter root hairs of *C. tora* and *C. grandis*, but they are immobilized at the base of the hairs (personal observation). Cortical root tissue is much reduced. Roots are ordinarily dark brown, or black, and wiry; rootlet production is meager. Plant parts yield antibacterial compounds (Fowler & Srinivasian, 1921: Itano & Matsuura, 1936; Wasicky, 1942; George, Venkataraman & Pandalei, 1947; Robbins, Kavanagh & Thayer, 1947; Anchel, 1949; Hauptmann & Nazario, 1950). Anthraquinones and their derivatives are rare in the family Leguminosae, but occur widely in *Cassia* species (Harborne, 1971). Phenolic compounds, tannins, quinones and derivatives are demonstrable microchemically in overlapping cortical root cells of *C. fistula*, *C. grandis* and *C. tora* (Plate 10.1). These cell layers presumably form a physico-chemical barrier to infection by rhizobia; they are known to thwart nematode gall formation (personal observation), a unique function that accounts for the successful use of *C. tora* as a nematode-trap crop.

Favorable returns from the use of *C. tora* and *C. occidentalis* (Leonard & Reed, 1930) as green manure gave rebirth to the idea that symbiotic nitrogen-fixing bacteria may occur and function within root tissues without forming nodules (Fehér & Bokor, 1926; Friesner, 1926). The absence of micro-organisms in all plant parts of *C. tora* (Allen & Allen,

1933) does not support this supposition. Species in *Fistula* and *Senna*, moreover, do not enrich the soil; this undoubtedly accounted for the cautionary statement from the International Institute of Agriculture (1936) that: 'As a rule, all leguminous plants of *Cassia* genus, without nodules, are harmful to the main crops with which they are associated. Their use either as a cover crop or as a shade plant is not recommended.'

Taxonomic aspects

Bentham (1871) deduced that *Fistula*, *Senna* and *Lasiorhegma* as subgenera 'were differentiated . . . before the areas they occupied had become broken up by the intervention of apparently insurmountable obstacles . . .' and that some of the subordinate groups 'are exclusively limited to one hemisphere, showing the possibility of their formation since the interposition of impassable barriers'. This comment is especially applicable to members of section *Chamaecrista* which are almost exclusively indigenous to the western hemisphere, and thus of more recent lineage.

Table 10.2. *Nodulation profile of Cassia*

Subgenera	Nodulated	Non-nodulated	+/− reports	Total
Fistula and *Senna*	5	50	4	59
Lasiorhegma				
Section *Apoucouita*		None examined		
Section *Absus*	2	—	—	2
Section *Chamaecrista*	37	1	—	38
Total	44	51	4	99

Symbiotaxonomy (Norris, 1965), a term analogous to chemotaxonomy, denotes physiological specialization of leguminous species relative to rhizobial affinities. Inasmuch as the plant is the dominant agent in the symbiosis (Allen & Allen, 1958), nodule formation *per se* merits recognition. Demarcation between nodule prevalence within the *Chamaecrista* complex and nodule paucity in *Fistula* and *Senna* is well-defined (Table 10.2), and suggests that nodulation in *Cassia* is a relatively recent physiological adaptation. Whether the requisite genetic ingredients for symbiosis within the primitive components of *Fistula* and *Senna* ever existed is questionable.

Some botanists have considered *Cassia* heterogeneous. Section *Chamaecrista* has had a varied taxonomic history relative to its proper status either as a section within *Cassia* or as a separate genus (Senn, 1938*a*, *b*; Irwin & Turner, 1960; Irwin, 1964). A nodulation survey of about twenty-six native and introduced *Cassia* species in Rhodesia prompted Corby (1974) to view *Chamaecrista* as a separate genus. Nodulation data presented here corroborate this conclusion and support nodulating ability as a useful criterion for defining phylogenetic relationships.

Summary

This nodulation survey of very many specimens of ninety-nine randomly sampled, geographically dispersed *Cassia* species has been presented against Bentham's infrageneric organization of the genus. Paucity of nodulation in subgenera *Fistula* and *Senna* contrasts with prevalence of nodulation in section *Chamaecrista*, subgenus *Lasiorhegma*. Nodulation data support *Chamaecrista* as a subgenus and perhaps even as a generic segregate.

References

Allen, Ethel K. & Allen, O. N. (1933). Attempts to demonstrate symbiotic nitrogen-fixing bacteria within the tissues of *Cassia tora*. *American Journal of Botany*, **20,** 79–84.

Allen, Ethel K. & Allen, O. N. (1958). Biological aspects of nitrogen fixation. In: *Handbuch der Pflanzenphysiologie* (ed. W. Ruhland), vol. 8, pp. 48–118. Springer-Verlag, Berlin.

Allen, Ethel K. & Allen, O. N. (1961). The scope of nodulation in the Leguminosae. *Recent Advances in Botany*, **1,** 585–8.

Allen, O. N. & Allen, Ethel K. (1936*a*). Root nodule bacteria of some tropical leguminous plants. I. Cross-inoculation studies with *Vigna sinensis* L. *Soil Science*, **42,** 61–77.

Allen, O. N. & Allen, Ethel K. (1936*b*). Plants in the sub-family Caesalpinioideae observed to be lacking nodules. *Soil Science*, **42,** 87–91.

Anchel, Marjorie (1949). Identification of the antibiotic substance from *Cassia reticulata* as 4, 5-dihydroxyanthraquinone-2-carboxylic acid. *Journal of Biological Chemistry*, **177,** 169–77.

Bamber, M. K. & Holmes, J. A. (1911). Green manuring. *Circulars and Agricultural Journal of the Royal Botanic Gardens*, Ceylon, **5,** 217–30.

Bañados, L. L. & Fernandez, W. L. (1954). Nodulation among the Leguminosae. *The Philippine Agriculturist*, **37,** 529–33.

Barrios, S. & Gonzalez, V. (1971). Rhizobial symbiosis on Venezuelan savannas. *Plant and Soil*, **34,** 707–19.

Beadle, N. C. W. (1964). Nitrogen economy in arid and semi-arid plant communities. III. The symbiotic nitrogen-fixing organisms. *Proceedings of the Linnean Society of New South Wales*, **89,** 273–86.

Bentham, G. (1871). Revision of the genus *Cassia. Transactions of the Linnean Society of London, Botany*, **27**, 503–91.

Burrill, T. J. & Hansen, R. (1917). Is symbiosis possible between legume bacteria and non-legume plants? *Illinois Agricultural Experimental Station Bulletin*, **202**, 115–81.

Bushnell, O. A. & Sarles, W. B. (1937). Studies on the root nodule bacteria of wild leguminous plants in Wisconsin. *Soil Science*, **44**, 409–23.

Campêlo, A. B. & Döbereiner, Johanna (1969). Estudo sôbre inoculacão cruzada de algumas leguminosas florestais. *Pesquisa agropecúaria brasileira*, **4**, 67–72.

Conklin, Marie E. (1936). Studies of the root nodule organisms of certain wild legumes. *Soil Science*, **41**, 167–85.

Corby, H. D. L. (1974). Systematic implications of nodulation among Rhodesian legumes. *Kirkia*, **9**, 301–29.

DeSouza, D. I. A. (1966). Nodulation of indigenous Trinidad legumes. *Tropical Agriculture*, Trinidad, **43**, 265–7.

Dubey, H. D., Woodbury, R. & Rodríguez, Rita L. (1972). New records of tropical legume nodulation. *Botanical Gazette*, **133**, 35–8.

Fehér, D. & Bokor, R. (1926). Untersuchungen über die bakterielle Wurzelsymbiose einiger Leguminosenhölzer. *Planta: Archiv für wissenschaftliche Botanik* (*Abteil E der Zeitschrift für wissenschaftliche Biologie*), **2**, 406–13.

Fowler, G. J. & Srinivasian, M. (1921). The biochemistry of the indigenous indigo dye vat. *Journal of the Indian Institute of Science*, **4**, 205–21.

Friesner, G. M. (1926). Bacteria in the roots of *Gleditsia triacanthos* L. *Proceedings of the Indiana Academy of Science*, **34**, 215–24.

George, M., Venkataraman, P. R. & Pandalei, K. M. (1947). Investigations on plant antibiotics. II. A search for antibiotic substances in some Indian medicinal plants. *Journal of Scientific and Industrial Research*, **6B**, 42–6.

Grobbelaar, N., Beijma, M. C. van & Saubert, S. (1964). Additions to the list of nodule-bearing legume species. *South African Journal of Agricultural Science*, **7**, 265–70.

Grobbelaar, N., Beijma, M. C. van & Todd, C. M. (1967). A qualitative study of the nodulating ability of legume species: List 1. Publications of the University of Pretoria, New Series, No. 38.

Grobbelaar, N. & Clarke, Brenda (1972). A qualitative study of the nodulating ability of legume species: List 2. *Journal of South African Botany*, **38**, 241–7.

Harborne, J. B. (1971). Terpenoid and other low molecular weight substances of systematic interest in the Leguminosae. In: *Chemotaxonomy of the Leguminosae* (ed. J. B. Harborne, D. Boulter & B. L. Turner), pp. 257–83. Academic Press, London & New York.

Hauptmann, H. & Nazario, L. L. (1950). Some constituents of the leaves of *Cassia alata* L. *Journal of the American Chemical Society*, **72**, 1492–5.

Helten, W. M. van (1915). Resultaten verkregen in den cultuurtuin met eenige nieuwe groenbemesters. *Department van Landbouw, Nijverheid en Handel. Mededeelingen Cultuurtuin*, **2**, 28–32.

Holland, T. H. (1924). Some green manures and cover plants. Department of Agriculture, Ceylon, Leaflet No. 30.

International Institute of Agriculture (1936). Use of leguminous plants in tropical countries as green manure, as cover and as shade. International Institute of Agriculture, Rome.

Irwin, H. S. (1964). Monographic studies in *Cassia* (Leguminosae–Caesalpinioideae). I. Section Xerocalyx. *Memoirs of the New York Botanical Garden*, **12,** 1–114.

Irwin, H. S. & Turner, B. L. (1960). Chromosomal relationships and taxonomic considerations in the genus *Cassia. American Journal of Botany*, **47,** 309–18.

Itano, A. & Matsuura, A. (1936). Studies on nodule bacteria. V. Influence of plant extract as accessory substance on the growth of nodule bacteria. *Berichte des Ohara Instituts für landwirtschaftliche Forschungen*, **7,** 185–214.

Keuchenius, A. A. M. N. (1924). Botanische kenmerken en cultuurwaarde als groenbemester van een 60-tal nieuwe soorten van Leguminosen. Department van Landbouw, Nijverheid en Handel. Mededeelingen van het Proefstation voor Thee, No. 90.

Kinman, C. F. (1916). Cover crops for Porto Rico. Porto Rico Agricultural Experiment Station Bulletin No. 19.

Lechtova-Trnka, Mara (1931). Etude sur les bactéries des légumineuses, et observations sur quelques champignons parasites des nodosités. *Le Botanise*, Série **23,** 301–530.

Leonard, L. T. (1923). Nodule-production kinship between the soybean and the cowpea. *Soil Science*, **15,** 277–83.

Leonard, L. T. (1925). Lack of nodule formation in a subfamily of the Leguminosae. *Soil Science*, **20,** 165–7.

Leonard, L. T. & Reed, H. R. (1930). A comparison of some nodule forming and non-nodule forming legumes for green manuring. *Soil Science*, **30,** 231–6.

Lopatina, G. V. (1931). Investigations of the nodule-forming bacteria. II. Observations on nodule formation on leguminous plants. *Bulletin of the State Institute of Agricultural Microbiology, USSR*, **4,** 105–10. (In Russian with English summary.)

Martin, W. P. (1948). Observations on the nodulation of leguminous plants of the Southwest. *USDA Regional Bulletin*, **107,** *Plant Study Series* 4. Mimeographed, 10 pp. Albuquerque, New Mexico.

Mostert, J. W. C. (1955). Observations on the nodulation of some leguminous species. *Farming in South Africa*, **30,** 338–40.

Muller, H. R. A. & Frémont, T. (1935). Observations sur l'infection mycorhizienne dans le genre *Cassia* (Caesalpiniaceae). *Annales Agronomiques, Paris*, **5,** 678–90.

Norris, D. O. (1959). The role of calcium and magnesium in the nutrition of *Rhizobium*. *Australian Journal of Agricultural Research*, **10,** 651–98.

Norris, D. O. (1965). Rhizobium relationships in legumes. In: *Proceedings of the 9th International Grassland Congress, São Paulo, Brazil*, vol. 2, pp. 1087–92. Edições Limitada, São Paulo, Brazil.

Norris, D. O. (1969). Observations on the nodulation status of rainforest leguminous species in Amazonia and Guyana. *Tropical Agriculture, Trinidad*, **46**, 145–51.

Paul, W. R. C. (1949). On the value of some legumes. *Tropical Agriculturist*, **105**, 109–16.

Robbins, W. J., Kavanagh, F. & Thayer, J. D. (1947). Antibiotic activity of *Cassia reticulata* Willd. *Bulletin of the Torrey Botanical Club*, **74**, 287–92.

Rothschild, Delia I. de (1970) Nodulacion en leguminosas subtropicales de la flora Argentina *Revista del Museo Argentino de Ciencias Naturales 'Bernardino Rivadavia' e Instituto Nacional de Investigacion de las Ciencias Naturales, Botánica*, **3**, 267–86.

Senn, H. A. (1938*a*). Cytological evidence on the status of the genus *Chamaecrista* Moench. *Journal of the Arnold Arboretum*, **19**, 153–7.

Senn, H. A. (1938*b*). Chromosome number relationships in the Leguminosae. *Bibliographia Genetica*, **12**, 175–336.

Shunk, I. V. (1921). Notes on the flagellation of the nodule bacteria of Leguminosae. *Journal of Bacteriology*, **6**, 239–48.

Steinmann. (1930). De Bergcultures. *Orgaan van het algemeen Landbouw Syndicaat*, **4**, 1099–1103.

Steyaert, R. (1952). Cassieae. In: *Flore du Congo belge et du Ruanda-Urundi*, vol. 3, pp. 496–545. Institute National pour l'Etude Agronomique de Congo Belge (INEAC), Brussels.

University of Pretoria (1955–6). Annual report on the research activities of the Plant Physiological Research Institute. 45 pp.

Wasicky, R. (1942). Pharmacological investigations of leaves of *Cassia fistula* L. and some Brazilian species of *Cassia*. *Boletim da Academia Nacional de Farmacia, Rio de Janeiro*, **4**, 133–5.

Wilson, J. K. (1939). Leguminous plants and their associated organisms. Cornell University Agricultural Experiment Station Memoir No. 221.

Wright, H. (1903). Nitrogenous plants. *Circulars and Agricultural Journal of the Royal Botanic Gardens, Ceylon*, **2**, 77–81.

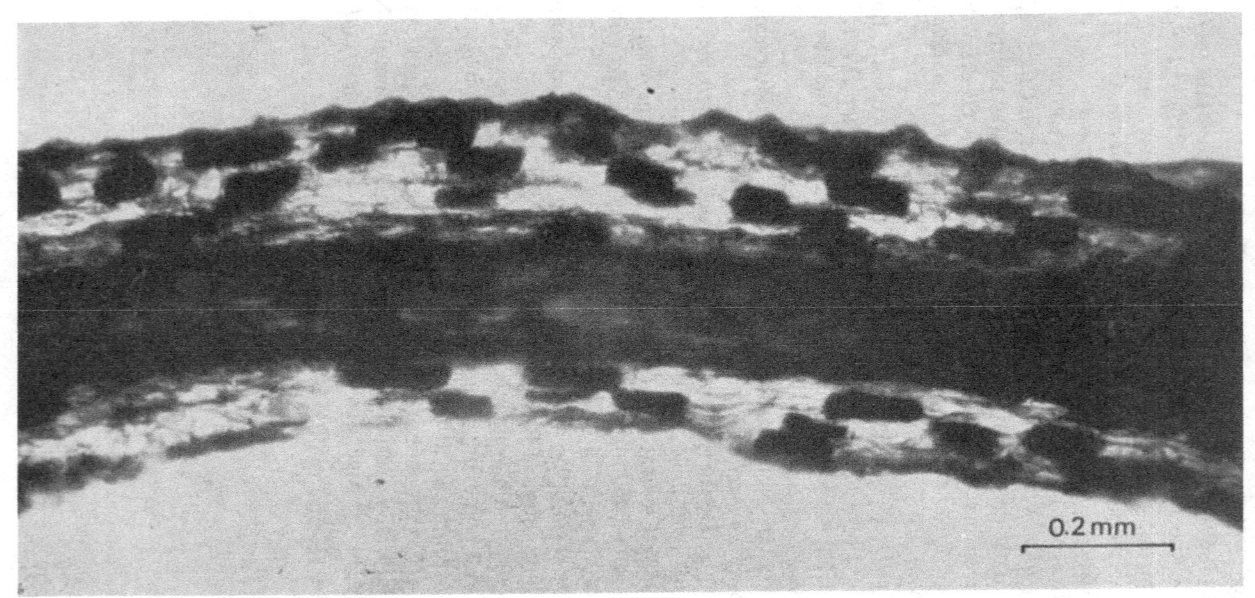

Plate 10.1. Longitudinal section of *C. tora* rootlet showing cortical cell inclusions inhibitory to nodulation.

PART II

Quality of legume inoculants

11. The production of high quality inoculants and their contribution to legume yield

R. J. ROUGHLEY

Nitrogen fixed by legumes is an essential input for high protein foods produced in both crop and pasture systems. Establishment of effectively nodulated legumes cannot in many situations be left to chance but requires the introduction of effective strains of rhizobia into the soil at sowing.

This was once done by transfer of soil from an established legume stand to a new site. At the end of the nineteenth century pure cultures of rhizobia in broth were used to inoculate seed. Later the broth was used to impregnate powdered peat which formed a convenient carrier and it is this type of inoculant that is almost universally used today. Some inoculants are still prepared on agar slopes or as freeze-dried cultures, but in neither of these forms do the rhizobia survive well on the seed coat.

Many of the failures to establish rhizobia in soil successfully may be attributed either to insufficient or unsuitable rhizobia in the inoculant.

Quality of inoculants

The quality of legume inoculants depends on both the number of rhizobia they contain and their effectiveness in fixing nitrogen with the intended host. Standards by which inoculants are judged are ultimately determined by field performance in different situations and because these differ widely, it would be unrealistic to set a rigid standard to apply for a wide range of environments. Where legumes are to be established in *Rhizobium*-free soil with good conditions, 100 rhizobia per seed provides a satisfactory inoculum level, but where large numbers of ineffective rhizobia occur and/or conditions are adverse for *Rhizobium* survival, numbers in excess of 10^6 per seed may be required. Ireland & Vincent (1968) showed that lime-pelleting seed improved the relative performance of inoculants for a given number of rhizobia per seed. This made it possible to establish rhizobia in soil containing large numbers of rhizobia which otherwise would have required extremely heavy inoculation. Within the bounds of inoculum technology, therefore, it is the agronomic demands made on cultures which determine the standards by which they are judged.

The ability of a strain to form nodules and fix nitrogen in particular localised conditions is an important requirement of inoculum quality e.g. D. L. Chatel & R. M. Greenwood (personal communication) selected *R. trifolii* strain WU290 on its ability to colonise the rhizosphere and multiply, forming large populations in sandy soils of Western Australia; this allowed it so survive high temperatures during summer in the absence of the host. It is important to stress the dangers inherent in the export of inoculants prepared in one country or continent for use in another, particularly when such inoculants are for a legume programme in a developmental stage. It should be a matter of top priority to select strains of rhizobia suited to local conditions and if inoculants cannot be made locally these strains should be sent to an inoculant manufacturer for inclusion in the cultures he supplies.

To reduce the number of different inoculants made, manufacturers often compromise by grouping legumes for inoculation, and using strains that have a wide spectrum of effectiveness but which may not be the best available for all of the legumes it is intended to inoculate. Such compromises and principles of strain selection are discussed later (see Date, Chapter 12) but I would stress the disadvantages in this practice especially in areas where there are no naturalised rhizobia. In such circumstances it is desirable to use the best strain available for a particular legume in order to avoid developing problems of competition from less effective strains.

Stages in the production of inoculants and their relation to quality

Methods used in making legume inoculants are described in detail by Burton (1967) and Roughley (1970). Further details on the effect of preparation and sterilising the carrier material are discussed by Strijdom & Deschodt (see Chapter 13). I will only briefly outline here the sequence of steps in manufacture and their significance with regard to quality.

The broth inoculum

Following the selection of suitable strains these should be maintained as freeze-dried cultures or on agar in screw-capped tubes to reduce strain variation. Mother cultures are prepared from an ampoule or tube from the stock culture collection. Purity and trueness to type of the rhizobia used to inoculate the broth, later to be used to inoculate the carrier, is important for both pure culture inoculants and those that do not claim to exclude all other micro-organisms. In the latter case the number of viable rhizobia per millilitre is particularly important but when the carrier has been sterilised is less so (Roughley, 1968).

The carrier materials

Peats and soils high in organic matter are most commonly used as carriers. The ability of such materials to prevent deterioration and even to promote growth of rhizobia, can vary widely within the one deposit and may also be affected by their treatment prior to inoculation (Roughley & Vincent, 1967). The carrier material is usually dried and milled and may then be sterilised either by steam under pressure or gamma irradiation.

Different carrier materials vary widely in their ability to absorb moisture. This determines how much broth should be added at inoculation. Expressing moisture content on a percentage basis is misleading and has caused confusion. It is preferable to express it on the basis of water potential (pF); this may conveniently be calculated according to Fawcett & Collis-George (1967). The optimum pF for most strains of rhizobia lies between 3.9 and 4.0 (Steinborn & Roughley, 1974). The peat is adjusted to this level by injecting broth culture into sterilised peat in 0.038 mm polyethylene bags or glass containers or by mixing with unsterilised peat before packaging.

Packaging material

Suitable packaging must allow some exchange of oxygen but restrict the passage of water. The literature on this aspect is confusing probably because rhizobia have a small but definite demand for some gas exchange through the packaging material (Roughley, 1968). Incubation for two weeks at 28 °C is desirable before use. Storage after incubation should, where possible, be at 4 °C. High temperatures have been shown by many workers to affect adversely the survival of rhizobia in inoculants (Vincent, 1958). This poses a real hazard during transport unless inocula are kept under refrigeration.

Problems in inoculant production in Australia

During the past seventeen years since inoculants made in Australia have been under official control, four unforeseen and critical difficulties have occurred, namely (i) toxin production in peat caused by overheating during drying (Roughley & Vincent, 1967); (ii) variability in the suitability of peat from within the one deposit (Roughley & Vincent, 1967); (iii) salt accumulation in the peat deposit (Steinborn & Roughley, 1974); and (iv) strain variability.

Salt accumulation in the peat deposit

A rapid decline in numbers of rhizobia, particularly *R. trifolii*, occurred in 1970. The peat deposit at Mount Gambier, South Australia,

Table 11.1. *Effect of salt concentration on survival of* R. trifolii *WU290 (iii)*. (After Steinborn & Roughley, 1974)

	Log mean viable cells after (weeks)			
% chloride	0	1	4	12
0.14	8.26	9.15	8.46	7.63
0.26	8.29	8.86	8.15	7.45
0.59	8.21	6.90	7.83	–
0.88	8.20	7.53	7.40	–
1.20	8.13	6.00	–	–
0.19*	8.32	8.72	7.99	7.60

* Original salt content 1.20, but leached with water.

Table 11.2. *Concentration of salt in uninoculated peat from Mt Gambier, S. Australia.* (After Steinborn & Roughley, 1974)

Year first harvested	Mean % chloride when harvested	Mean % chloride in 1972 at depth	
		10 cm	46 cm
1969	0.29	0.11	0.29
1970	0.58	0.15	0.33
1971	0.46	0.13	0.31

is situated one mile from the coast and is subject to prevailing salt-bearing winds. Deposited salt is normally leached by heavy winter rain but a series of drier years caused it to accumulate in the top 10 cm.

Table 11.1 shows the effect of salt concentration expressed as per cent chloride on growth of *R. trifolii* WU290 (iii). Its effect was evident after one week and a progressive decline followed. Peat with a high concentration of salt was reclaimed by leaching with water. Table 11.2 shows the mean percentage of chloride in peat harvested in 1969, 1970 and

Table 11.3. *Effect of carrier material on numbers of rhizobia*

	Mean log number/g at (weeks)										
	R. trifolii TA1				*R. meliloti* SU47			Cowpea strain CB1024			
	0	1	4	12	0	1	4	0	1	4	12
Mt Gambier peat	8.391	8.262	9.081	7.933	8.236	8.962	9.664	7.556	9.405	9.568	9.270
Pulverised briquette brown coal	7.997	6.287	5.421	0.000	8.510	8.276	6.792	8.029	5.544	0.000	0.000
Yallourn 3 brown coal	8.078	6.292	5.468	0.575	8.433	7.742	6.736	8.152	6.238	4.314	0.000
Yallourn standard brown coal	8.215	8.068	8.131	6.877	8.461	9.034	8.721	8.155	8.234	8.596	8.249
Morwell brown coal	8.232	8.022	7.520	6.833	8.464	10.130	8.690	7.971	8.372	8.340	8.394

1971 compared with peat harvested from the same sites in 1972. The amount of salt declined in 1972 in the surface 10 cm, but there was a corresponding accumulation deeper down the profile.

These results show the need to test constantly the carrier material particularly if it is exposed to changes in weather patterns.

We are investigating large and apparently uniform deposits of brown coal, the next stage in the metamorphosis of peat, as a carrier for rhizobia. Preliminary results (see Table 11.3) show that two out of four brown coals tested have some promise as carriers of three groups of rhizobia, but require nutrient amendment to improve rhizobial growth.

Variation in strain effectiveness

It is commonly accepted that strains of *R. trifolii* are prone to produce ineffective mutants while those of *R. meliloti* are more stable; few comparable studies of other species of *Rhizobium* have been done. Both frequent subculturing and high levels of some amino acids in the medium increase the chances of strain variation. The effect of the

Table 11.4. *Variation in strains used commercially**

Strain	Species	Variation found
WU290	*R. trifolii*	Loss of effectiveness and failure to form nodules
CC2480a	*R. trifolii*	Reduction in effectiveness
CB782	*R. trifolii*	Variation in effectiveness
CB756	Cowpea	Variation in effectiveness and colony type
WU425	*R. lupini*	Variation in colony type and time to form nodules
U45	*R. meliloti*	Variation in colony type
SU47	*R. meliloti*	Variation in colony type

* Issued by Australian Inoculants Research and Control Service.

medium has been described in papers by Strijdom & Allen (1967, 1969) and Staphorst & Strijdom (1971). *R. meliloti* serially cultivated on media supplemented with D-alanine became less effective; some strains lost their ability to form nodules. The effectiveness of *R. trifolii* strains decreased when grown on media supplemented with alanine, histidine, phenylalanine and glycine. Neither *R. meliloti* or *R. trifolii* regained their effectiveness when cultivated on normal media. Cultures in our laboratory have been maintained in freeze-dried ampoules and cultivated on a basal yeast-water mannitol medium. Despite this, we have been troubled with variation.

The response of *Macroptilium atropurpureum* cv Siratro to seven different sources of CB756 was tested (R. J. Roughley & C. F. Herridge, unpublished data). Table 11.6 shows the increase in time required by the less effective strains to form nodules and the dry weight of the plants at forty days. There was a fourfold range in plant dry weight between the least and the most effective strain. Seventeen single colony isolates from one source were similarly tested (Table 11.7). There was an eightfold variation in plant dry weight and a range of <1 mg to 15 mg in mean nodule weight. Nodule efficiency ranged from <1 mg to 12.3 mg of plant material/mg of nodule, with the largest nodules tending to be most efficient.

These results strikingly demonstrate both the need for regular tests of strain effectiveness and the possible danger of using single colony isolates as mother cultures for inoculant production. They are relevant also to the maintenance of any culture collection and to the proposal of a world *Rhizobium* bank.

Table 11.7. *Growth of Siratro inoculated with seventeen cultures of CB756 derived from the one source*

Isolate of CB756	Mean nodule wt* (mg)	Mean plant wt* (mg)	Increase/mg nodule (mg)
1	15	209	12.1
2	11	156	11.6
3	11	138	10.0
4	9	135	11.9
5	10	126	9.8
6	8	119	11.4
7	8	119	11.4
8	8	106	9.8
9	9	102	8.2
10	7	93	9.3
11	6	72	7.3
12	3	65	12.3
13	6	65	6.2
14	6	57	4.8
15	6	58	5.0
16	2	37	4.5
17	–	26	—
Uninoculated	0	28	—

* Ten replicates.

Predicting the need to inoculate

The need to inoculate seed must be judged on the basis of the size of the natural or previously introduced population of rhizobia occurring in the soil and its general level of effectiveness for the legumes in question. Whereas it was once considered that a problem soil was one in which rhizobia were absent, the greatest challenge is presented by sites which are environmentally suited to colonisation by rhizobia but contain many ineffective strains. Extensive surveys to assess the effectiveness of native rhizobia are few, but one in southeastern Australia indicated a general level of effectiveness of *R. trifolii* of approximately 50 % of the inoculum strain TA1 (Bergersen, 1970). Failure to establish a more effective strain in such a situation will not lead to crop failure, but does not achieve maximum production.

Two attempts have been made in New South Wales, Australia, to predict the need for inoculation of pasture legumes. Brockwell & Robinson (1970) attempted to relate the occurrence of rhizobia in soil to climate, vegetation, topography and soil type. None of these characteristics except vegetation had any consistent relationship except in so far as they affected the occurrence of leguminous plants. Consistent relationships did occur between presence of legume host and the number and type of rhizobia in soil.

Roughley & Walker (1973) attempted to relate soil type, colour, pH, previous history of the site including cropping data and fertiliser applications, and vegetation before sowing, with the need to inoculate subterranean clover at thirty-six sites in New South Wales. Only the previous occurrence of the host provided a partial guide to the presence of sufficient, effective native rhizobia, but even so it was impossible to predict consistently the need to inoculate.

Benefits resulting from inoculation

The amount of nitrogen fixed by legumes nodulated by effective strains of rhizobia is strongly affected by the nitrogen status of the soil. Different legumes behave differently in the same soil when inoculated with the same strain of rhizobia. For example, results from our laboratory show that when nitrogen is mineralised on the north coast of New South Wales following cultivation, nodulation of *Lablab purpureum* is delayed and at times prevented while *Macroptilium atropurpureum* forms effective nodules. Estimates of the large amount of nitrogen that can be fixed by legumes growing in nitrogen-free media has fostered the concept that legumes always fix sufficient or at least the major part of the nitrogen for their own requirements. This can be far from the truth when soil nitrogen levels are high or when nitrogen is added during the growing season. Variations in nitrogen status of the medium may account for some of the large discrepancies in the literature on the amount of nitrogen fixed by different *Rhizobium*/host combinations. Such data is summarised by Thomas (1973) for results on the African continent and by Subba Rao (1972) and Nutman (1974) for a range of legumes in different localities.

Zackharchenko & Pirozhenko (1970) grew lupins, vetches and peas in pots with added nitrogen. At harvest, the percentage of the total plant nitrogen derived from fixation was 82 % for lupin, 74 % for vetch and 57 % for peas. Half of the mineral nitrogen added to the pots was taken up by these legumes. Gukova, Lavrova & Ageeva (1971) showed that when nitrogen was applied to lupins grown in pots at 150 or 900 mg/pot the percentage of fixed nitrogen in the plant material fell from 80 to 20 %.

When nitrogen was applied at 16.8 or 84 mg/kg soil, the proportion of the total nitrogen derived from fixation by lucerne plants was 95 and 67 % respectively (Dorosinskii, Lazareva & Afanaseva, 1969). In field conditions when soil nitrogen is very low plants may suffer a period of nitrogen starvation before nodules begin to function. In these circumstances there may be a synergistic effect of adding small amounts of inorganic nitrogen.

In conclusion, it is safe to predict that increased demands will be made by the agriculturists for larger numbers of rhizobia to be applied per seed in inocula as harsher environments are sown with legumes or established areas providing competitive situations are re-sown. Also the proportion of the seed which will be pre-inoculated, often well before sale to the farmer, will increase. Both these situations must be met either by increasing the number of rhizobia in peat or improving their survival on seed. It is general experience that numbers of rhizobia in peat are regularly in the range $1\text{–}5 \times 10^9$/g; a tantalising few reach 10^{10}/g. Nevertheless prospects for increasing inoculant quality significantly are not good, but the number of rhizobia which survive on seed can be much improved by seed pelleting and use of adhesives. Further work on these aspects seems to offer the greatest opportunity to meet the challenges posed above.

References

Bergersen, F. J. (1970). Some Australian studies relating to the long-term effects of the inoculation of legume seeds. *Plant and Soil*, **32**, 727–36.

Brockwell, J. & Robinson, A. C. (1970). Observations on the natural distribution of *Rhizobium* spp relative to physical features of the landscape. In: *XIth International Grasslands Congress* (ed. M. I. T. Norman), pp. 438–41. University of Queensland Press.

Burton, J. C. (1967). *Rhizobium* culture and use. In: *Microbial technology* (ed. H. J. Peppler), pp. 1–33. Van Nostrand Rheinhold, New York.

Dorosinksii, L. M., Lazareva, N. M. & Afanaseva, L. M. (1969). The amount of biological fixation of nitrogen by lucerne. *Agrokhimiya*, No. 8, 59–63; *Herbage Abstracts*, **40,** 199 (1970).

Fawcett, R. G. & Collis-George, N. (1967). A filter-paper method for determining the moisture characteristics of soil. *Australian Journal of Experimental Agriculture and Animal Husbandry*, **7,** 162–7.

Gukova, M. M., Lavrova, E. K. & Ageeva, V. S. (1971). Utilization of N from atmosphere and fertilizers by lupin. *Doklady Moskovskoi Sel' skokhozyaistvennoi Akademii im K.A. Timiryazeva*, No. **162,** 189–93; *Herbage Abstracts*, **42,** 80.

Ireland, J. A. & Vincent, J. M. (1968) A quantitative study of competition for nodule formation. *IXth International Congress of Soil Science Transactions*, **2,** 85–93.

Nutman, P. S. (1974). The potential of legumes for protein production. *Proceedings of the Indian Academy of Science*, in press.

Roughley, R. J. (1968). Some factors influencing the growth and survival of root nodule bacteria in peat culture. *Journal of Applied Bacteriology*, **31,** 259–65.

Roughley, R. J. (1970). The preparation and use of legume seed inoculants. *Plant and Soil*, **32,** 675–701.

Roughley, R. J. & Vincent, J. M. (1967). Growth and survival of *Rhizobium* in peat culture. *Journal of Applied Bacteriology*, **30,** 362–76.

Roughley, R. J. & Walker, M. (1973). A study of inoculation and sowing methods for *Trifolium subterraneum* in New South Wales *Australian Journal of Experimental Agriculture and Animal Husbandry*, **13,** 284–91.

Staphorst, J. L. & Strijdom, B. W. (1971). Infectivity and effectiveness of colony isolates of ineffective glycine-resistant *Rhizobium meliloti* strains. *Phytophlactica*, **3,** 131–6.

Steinborn, Julia & Roughley, R. J. (1974). Sodium chloride as a cause of low numbers of *Rhizobium* in legume inoculants. *Journal of Applied Bacteriology*, in press.

Strijdom, B. W. & Allen, O. N. (1967). Studies with strains of *Rhizobium meliloti* cultivated on media supplemented with amino acids. *South African Journal of Agricultural Science*, **10,** 623–30.

Strijdom, B. W. & Allen, O. N. (1969). Properties of strains of *Rhizobium trifolii* after cultivation on media supplemented with amino acids. *Phytophylactica*, **1,** 147–52.

Subba Rao, N. S. (1972). A case for production of bacterial fertilizers in India. *Fertiliser News*, **17,** 37–43.

Thomas, D. (1973). Nitrogen from tropical pasture legumes on the African continent. *Herbage Abstracts*, **43**, 33–9.

Vincent, J. M. (1958). Survival of the root nodule bacteria. In: *Nutrition of the legumes* (ed. E. G. Hallsworth), pp 108–23. Butterworth & Co., London.

Zakharchenko, I. G. & Pirozhenko, G. S. (1970). On the nitrogen fixation by legumes. *Agrokhimiya* No. 5, 28–33; *Herbage Abstracts*, **41**, 84.

12. Principles of *Rhizobium* strain selection

R. A. DATE

Assessment of the need for inoculation

Most discussions of selection of strains of *Rhizobium* concentrate on one or more of the criteria used for their evaluation. Obviously these criteria are important, but there are other aspects which must also be considered in production of legume inoculants. The preparation of high-quality inoculants often assumes that inoculation of legume seed is essential or desirable for satisfactory growth; the need for inoculation should first be established. Awareness of the benefits of inoculation has increased in recent years, but in my experience, in many instances, considerable effort has been channelled into selection of 'efficient' strains of rhizobia without knowing whether inoculation is necessary, or whether the strain used makes any contribution to yield or even forms nodules. This was well-illustrated by Ham, Cardwell & Johnson (1971) where inoculation of soybeans with a strain known to be effective failed to increase grain yield over that of uninoculated plots or to account for many of the nodules in the inoculated treatments. Where no suitable naturally occurring strains are present in the soil, inoculation with an effective strain is essential, e.g. in the case of *Lotononis bainesii* (Norris, 1958) or various tropical pasture legumes (Bryan & Andrew, 1955). Response to inoculation with effective strains of *Rhizobium* varies between these two extremes.

Surveys of the success of inoculation from routine sowings, conducted by means of questionnaires completed by farmers or researchers whose direct interest is not legume bacteriology, are of little value since they tend to be subjective and inaccurate. Such surveys may be more useful when done by a single researcher or group of workers employing well-standardised criteria of evaluation. Better information can come from trials similar to, but simpler than that described by Nutman & Vincent (1968) and by Nutman in this volume (Chapter 19). Three treatments are required: (*a*) uninoculated control plots to check for the presence of naturally occurring rhizobia, (*b*) plots inoculated with a potentially useful strain known to be effective in controlled-environment or glasshouse tests, and (*c*) plots inoculated with the same strain and supplied with combined nitrogen. From such plots the following results and explanations are possible:

(i) Uninoculated control treatment plants are without nodules, small and nitrogen deficient. Those of the inoculated plot may also be without nodules, indicating that the test strain is non-infective, or contain nodulated plants which show no improvement in yield over the uninoculated controls, indicating an ineffective test strain. Alternatively if inoculated plants are nodulated and show a substantial increase in yield the test strain is effective. Such strains may vary widely in their degree of effectiveness, those causing the largest dry weight (or nitrogen) yield being selected as potential inoculum strains. In this situation where uninoculated controls are free of nodules, it can be assumed that the nodules formed were due to the test strain. This situation is common in Australia for *Lotononis bainesii*, some African *Trifolium* species and in certain situations also *Leucaena leucocephala*, *Lablab purpureus*, *Desmodium intortum* and *D. uncinatum*.

(ii) Uninoculated control treatment plants have nodules but are judged as ineffective because growth is poorer than in plants that were inoculated and given nitrogen treatment. A low-yielding inoculated treatment similar to the uninoculated control suggests either that the inoculant strain was ineffective in nitrogen fixation or was unable to compete with the naturally occurring strains for nodule sites. Serological analysis, or other strain-identifying techniques would indicate which was true. Alternatively, the inoculated treatment may be high-yielding, significantly better than the uninoculated control and similar to the inoculated plus nitrogen treatment. Such plants from inoculated treatments may bear both effective and ineffective nodules, suggesting that the inoculum strain was able to compete with the naturally occurring strains for nodules sites.

(iii) Uninoculated control plants have nitrogen-fixing nodules with plant and nitrogen yields similar to the inoculated treatments. Here it is essential to determine how successful the applied strain has been in forming nodules and to make some estimate of their contribution to the total nitrogen fixed.

These, then, are the various possibilities that may be encountered. The inoculated plus combined nitrogen treatment is to avoid either classing a strain as ineffective in nitrogen fixation when some other factor such as phosphorus deficiency is a major factor for limiting plant growth, or discarding a legume in species evaluation or plant introduction programmes, which does not nodulate with either the test or naturally occurring strains of *Rhizobium*, but which otherwise yields satisfactorily.

Criteria used in the selection of *Rhizobium* strains

A strain must be able to form effective nitrogen-fixing nodules on the hosts for which it is recommended under a wide range of field conditions (Vincent, 1956). This has long been used as a selection criterion and is perhaps still of primary importance because only effective, nitrogen-fixing strains should be used when assessing other characters. Chief among these are competitiveness in nodule formation and survival and multiplication in the soil, particularly in the absence of the host. Brockwell *et al.* (1968) proposed additional criteria including prompt effective nodulation over a range of root temperatures, ability to grow well in culture, to grow and survive in peat and to survive on the seed. Others include pH tolerance (Norris, 1973), pesticide tolerance and nodulation in the presence of combined nitrogen.

Specificity and effectiveness in nitrogen fixation

Specificities in nodule formation and nitrogen fixation involve genotype interactions between the host legume and the invading *Rhizobium* in which the plant plays a dominant role. There are, therefore, two sources of variation (i.e. that between plants and that between strains of *Rhizobium*) which may be exploited in selection programmes. The microbiologist is usually confronted with an already selected plant line for which he must find a suitable *Rhizobium* strain, and has only the variation between strains of rhizobia in which to search for an effective combination. The amount of work and number of strains screened could be reduced substantially if plant breeders paid more attention to effective nodulation in exploitation of plant variability in selection and breeding programmes. The influence of host genotype on the effectiveness of the symbiotic association has been described for *Trifolium* species (Nutman, 1969), *Glycine max* (Weber *et al.*, 1971) and *Medicago sativa* (Aughtry, 1948). Similar, but not genetically defined, specificities have been described for African clovers (Norris & 't Mannetje, 1964) *Medicago* and *Melilotus* species (Brockwell & Hely, 1966), *Centrosema pubescens* (Bowen & Kennedy, 1961) and *Lotononsis bainesii* (Norris, 1958). Specific interactions with the environment also play a role. For example soil (root) temperatures for soybean (Gibson, 1971) and clover (Roughley, 1970*a*; Pankhurst & Gibson, 1973).

Generally specificity is used to refer to effectiveness in nitrogen fixation rather than ability to form nodules. Confusion in definition of this point has led to the breakdown and misunderstanding of the concept of cross-inoculation groups. There is no good scientific reason for maintaining these groups and their abandonment has been strongly recommended (Wilson, 1944; Norris, 1956). However, host groups based on effectiveness of association of a given strain of *Rhizobium* with a number of host species, which may be from different genera and even different tribes, is convenient and of practical value.

To accommodate these specificities three alternatives are possible in selecting strains for inoculant production: (*a*) to provide numerous inoculants each with a highly effective strain for an individual species, (*b*) to use 'wide-spectrum' strains that vary from good to excellent in nitrogen fixation with a range of legumes, or (*c*) to make multiple-strain inoculants containing the best strain for each of the host species recommended. These alternatives present a dilemma for the microbiologist 'who must decide between the practical aspects of inoculant production and marketing and scientific excellence in providing the best available material for inoculant manufacture' (Date & Roughley, 1975). In Australia 'wide-spectrum' strains alone are used when available. However, there has been a recent trend toward the specialised culture with the specific strain for each host. The use of multi-strain inoculants should be avoided because of possible antagonistic and competitive effects in the culture (Marshall, 1956) and the likelihood of competition in nodule formation from the less effective strains in the inoculant for a particular host (Caldwell, 1969).

Competition for nodule formation

The ability of a strain of *Rhizobium* to compete successfully with other rhizobia for nodule (infection) sites is an important characteristic, and marked qualitative and quantitative differences have been demonstrated. For example, Brockwell & Dudman (1968) have shown large differences in competitive ability between known effective strains on *Trifolium subterraneum* when applied to the seed in approximately equal numbers and in competition with the native rhizobial population. In such situations the more competitive inoculum strains may account for 80 to 90 % of nodules formed. In these experiments, numbers of rhizobia applied per seed were lower than the number of native rhizobia per gram of soil but presumably the applied rhizobia had an advantage in being closer to the seed and seedling root than the more widely distributed native rhizobia. Competition may also be influenced by the number of rhizobia added in the inoculum and the number present in the soil. Such qualitative relations are affected by the influence of the host in the formation of nodules and whether the competing strains are effective or ineffective in nitrogen fixation. Ireland & Vincent's (1968) results suggest that in the presence of a large soil population of ineffective strains the number of inoculant rhizobia per seed should exceed the number of ineffective rhizobia per gram of soil for effective nodulation of 90 to 100 % of plants. Increasing the number of effective rhizobia on the seed has about twice the influence of a proportional increase in the number of ineffective strains in the soil.

Although the host plant has a dominant role in determining which strain or strains forms nodules (Vincent & Waters, 1953; Means, Johnson & Erdman, 1961) this is determined by genotypic interactions between the competing strains and the host. For example Means *et al.* (1961) found as little as 1 % of strain 76 for soybeans in a mixture with strain 38 accounted for 85 % of the nodules, whereas strain 94 in mixture with strain 38 accounted for only 19 % of nodules. These two strains varied from 0 to 100 % successful when in competition individually with five other effective strains. Success also depended on the proportion of each in the mixture. Similar qualitative and quantitative differences in success of soybean inoculation in competitive situations are reported by Johnson & Means (1964), Johnson, Means & Weber (1965), Caldwell (1969) and Ham, Cardwell & Johnson (1971).

Survival and multiplication in the soil

Rhizobia introduced into the soil on seed must multiply rapidly in the soil solution and rhizosphere to allow prompt nodulation of the seedling plant. They must persist also in the soil in the absence of the host and compete for nodule sites on the roots of regenerating annuals and new roots on perennials. The ability to colonise roots and soil is described by Chatel, Greenwood & Parker (1968) as saprophytic competence. Population densities measured at four soil depths at approximately monthly intervals over the establishment and growth periods of the plants revealed marked differences between strains of *Rhizobium* for a single host species (*Trifolium subterraneum*) and even greater differences between strains from *T. subterraneum* and *Ornithopus sativus* (Chatel & Parker, 1973*a*, *b*; Chatel & Greenwood, 1973). Corresponding differences in proportion of nodules due to the inoculum strains have also been recorded (Chatel, *et al.*, 1968). In plots sampled one or two years after sowing, one strain failed to form nodules, whilst the other was still able to account for most (97–100 %) of the nodules formed and had even spread from one plot to the next (30–68 % of nodules). Some of the unidentified nodule isolates were tested for colonising ability and in many instances were superior to the original inoculum strain.

Other criteria

Several strain characteristics may be useful in specific situations. For example, Norris (1973) recommended an alkali-producing strain for *Leucaena leucocephala* in place of an acid-producing strain for sowings in acid (pH 5.0) soils. A particular strain or strains of a single serotype can account for a majority of nodules according to soil pH (Damargi, Frederick & Anderson, 1967).

Selection for nitrogen fixation at low (or high) temperatures may also be desirable; e.g. Brockwell, Bryant & Gault (1973), demonstrated better nitrogen fixation by white clover with a particular strain in alpine regions. Similarly strains can be selected to nodulate in the presence of combined nitrogen (Gibson *et al.*, 1971). In my own experience, host species tend to be influenced more than strains by combined nitrogen; for example, the nodulation of red and white clovers in newly cleared land was not affected by 19 kg N/ha, whereas nodulation of subterranean clover by the same *Rhizobium* was completely inhibited until the soil nitrogen was depleted, as indicated by associated grass turning yellow. Similarly, nodulation of a species such as *Lablab purpureus* is inhibited by combined nitrogen but not that of *Macroptilium atropurpureum*.

Other characteristics such as tolerance to pesticides (mostly fungicides), ease of culture, growth and survival in peat and survival on seed can also be used as criteria for selection of strains.

Collection, isolation and authentication of strains of *Rhizobium*

Pure cultures of authenticated strains of *Rhizobium* are necessary for assay purposes and for preparation of legume inoculants. They may originate from apparently effective nodules on a healthy plant, peat cultures, or pure cultures of a recommended strain in a culture collection.

If a strain is isolated from nodules or peat (for methods see Vincent, 1970; or Roughley, 1970*b*) it is essential that it originate from a single well-isolated colony to avoid problems of mixed cultures. Cultures received from other collections should also be checked for purity and authenticated as *Rhizobium*. A collection of such cultures can then be tested for nitrogen-fixing ability on suitable host species.

There are two approaches to collection of suitable strains for new legumes or the establishment of a new collection of strains:

(*a*) If the legume species is a new introduction, plant collectors should be requested to collect nodules at the same time so that the microbiologist can isolate the *Rhizobium* and subsequently provide it as the inoculum. This approach has been used extensively in this laboratory and in most instances has been successful; e.g. with the species *Lotononis bainesii*. However, in a minority of cases, e.g. some new species of *Stylosanthes*, the nodules yielded strains highly effective in nitrogen fixation in controlled-environment tests but unsuccessful as field inoculants as they failed to form nodules.

(*b*) Uninoculated seed of the new species can be sown at a number of sites in the hope, which is not always realised, that native strains will form effective nodules from which suitable *Rhizobium* strains can be isolated. This system also has the disadvantage of reducing the genetic base within which selection can be made.

Evaluation of strains

Rhizobium for legume inoculants should be evaluated in two or three phases: first a screening for effectiveness in nitrogen fixation, second an assay of selected effective strains for nitrogen-fixing ability under field conditions including competition for nodule sites, persistence and colonising abilities, and third, when necessary, coverage of other aspects such as pH, pesticide tolerance, survival in culture and on seed etc. These are arbitrary but convenient divisions that streamline actual testing procedures by reducing the number of strains at each phase. Controlled-environment facilities provide reliable comparative information of the ability of a large number of strains to fix nitrogen, but are not suitable for the second and third phases. Field trials, on the other hand, can evaluate only a limited number of strains because of demands on time, labour and facilities. In special circumstances glasshouse trials in pure culture (Johnson & Means, 1964) can be used to forecast field performance (Caldwell, 1969).

Methods

For controlled environment and glasshouse conditions, aseptic culture of plants in tubes or sand-jars as described by Norris (1964), Vincent (1970) or Roughley (1970*b*) provides a satisfactory means of ranking strains for effectiveness. A simple method, suited to small-seeded legumes and used in Australia for some time, utilises an undisturbed core of soil, collected by driving an empty can into the ground and then removing it to the glasshouse where the can serves as a pot with soil. This method is described by Roughley (1970*b*) and Vincent (1970) and has the advantage that host–strain combinations can be tested with a microflora environment similar to that in the field whilst eliminating the hazards of field trials. Additionally, several 'sites' can be compared simultaneously.

Field trials provide the ultimate test of a strain's ability, since the proportion of nodules formed and amount of nitrogen fixed are the end result of many interacting factors. For example, differences in nitrogen-fixing ability between strains in nitrogen-free tube or sand-jar culture may be increased or decreased because of interaction with fluctuating soil temperatures, presence of combined nitrogen, deficiencies of mineral nutrients, pH or differential competition with other rhizobia and micro-organisms in the rhizosphere.

Aspects of design and conduct of field trials are important. The simplest and most convenient design is a series of single rows of uniformly spaced plants, each row representing a host–strain combination. The advantages of spacing plants are that it permits maximum expression of yield potential and facilitates harvest of nodules when competitive and persistence characters are being assayed. The main disadvantage is that legumes are rarely found as such, but rather as swards, either alone or in association with grasses.

Experience suggests that spaced plants on a 30–50 cm grid or 30–50 cm spacing in rows 50–100 cm apart in a completely randomised design is satisfactory for most temperate species and the erect non-runner-type tropical legumes. For tropical legumes with a strong runner habit a grid of 3–5 m is required. If grown as pure swards, plots 2×3.5 m are sufficient for most clover and medic species, but should be at least 5×5 m for most crop and tropical legumes. Field layout should permit construction of suitable drainage channels between plots or rows to prevent cross-contamination. The drainage system should have a capacity, in tropical areas, to cope with frequent and highly intensive rainfall. Yield and nodule assessment can be made at 10–14 weeks, or sooner in some species. If spaced plants or rows are used, the same sample should be used for dry weight and nitrogen determinations and for assessment of nodulation. In swards a quadrat of one m^2 is suggested for cutting.

Effectiveness is measured either directly by determining the amount of nitrogen fixed or indirectly by measuring plant dry weight; these two parameters are highly correlated (Erdman & Means, 1952). However, some differentiation between strains may be lost due to differences in nitrogen concentration at very high or low levels of effectiveness (Haydock & Norris, 1967). Thompson, Roughley & Herridge (1974) suggest that visual ratings of yield (in plots or rows) provides an accurate (relative) assessment of dry matter yield. Effectiveness measured in terms of grain yield for crop legumes (pulses) is more difficult to assess since nitrogen fixation can be severely reduced at the flowering and pod-filling stages. Factors such as frost, moisture and temperature affect grain yield more than a limited supply of nitrogen at those stages. The acetylene reduction technique has limited use in *Rhizobium* strain comparisons since total rather than rate of nitrogen fixation is the criterion of interest. Competitiveness, persistence and saprophytic competence are usually evaluated by determining the proportion of nodules formed by the inoculum using serological, antibiotic resistance or other strain identifying techniques. There is a strong argument which claims that assays of this kind may not have practical significance. If, for example, a particular strain accounts for only 60 % of nodules compared with 100 % for another, there may still be no difference in the amount of nitrogen contributed by it, since fixation is related to volume of active fixing tissue and its rate of fixation rather than to number of nodules.

Attempts to correlate effectiveness in nitrogen fixation with physiological (Fred, Baldwin & McCoy, 1932; Gupta & Sen, 1965; Sarma, Lakshmi-Kumari, Apte & Subba Rao, 1973), serological (Koontz & Faber, 1961), cultural (Fred *et al.*, 1932) or other characteristics have been unsuccessful, although recent studies suggest some correlation with enzyme activity (e.g. Romeiko, Dubovenko & Malinskaya, 1972).

Recommendation of strains for use in inoculant production

Responsibility for recommending strains varies from country to country, depending on the general level of experience and the degree of control exercised by official bodies. Four measures are strongly recommended:

(*a*) that the final responsibility for strain recommendation lies with a central laboratory, but allowing others involved in testing strains the opportunity to contribute.

(*b*) that recommendations be uniform unless local specific conditions demand variation.

(*c*) that changes to recommendations be made only when absolutely necessary and that a replacement strain has adequate support data of at least two years' field information.

(*d*) that the responsible body be physically, and preferably institutionally, separated from the inoculant manufacturing group(s).

In Australia each of the various research groups, e.g. CSIRO, Departments of Agriculture and Universities, conduct independently their own strain selection programmes. The results are generally considered at a workshop-type meeting and final recommendations agreed upon for the next season. The central laboratory (in this case AIRCS, see following section) then assumes responsibility for recommendation and issue of strains to inoculant manufacturers and other interested groups. It also regularly checks recommended strains for effectiveness and other pertinent characters in field trials at several sites.

Maintenance and supply of recommended strains

Freeze-drying in small ampoules is a satisfactory and convenient way of storing cultures to minimise variation and contamination over a period of many years. Ideally this should be combined with a series of checks on specificity, effectiveness and purity each time an ampoule is removed from storage. Other forms of storage, such as porcelain beads (Norris, 1963) or screw-cap tube or bottle cultures are satisfactory for shorter periods, but subculturing of parent stocks must be reduced to a minimum to reduce chances of variation and contamination. A combination of freeze-dried and screw-cap tube cultures is used by the Australian Inoculant Research and Control Service (AIRCS) in a system which meets the above conditions and provides cultures for distribution to manufacturers of inoculants, with only two transfers necessary between an ampoule and any two cultures received by a manufacturer (Date, 1969). Using this system each manufacturing unit is provided with a fresh mother culture from the parent stock at regular intervals. The AIRCS also tests the quality of inoculants at two stages during manufacture and once during distribution.

The responsibility of the central control laboratory should include a programme of testing designed to check recommended strains in comparative trials with reserve strains and potential new strains, i.e. those submitted by the various research groups. In this way the central control laboratory can maintain a bank of tested strains available for immediate replacement of a recommended strain in the event of strain failure. In the past we have experienced loss of invasiveness, effectiveness and failure to survive in peat culture for various reasons, and have been able to replace such errant strains from the bank of tested strains.

Summary

Before selecting *Rhizobium* strains for inoculant production, the need for inoculation should be determined and then the criteria of selection based on the requirements of the locality. After collection and checks for purity and authenticity, *Rhizobium* strains are compared in controlled environment and field studies using appropriate criteria. Suitable strains are then recommended, a supply being maintained and regularly checked.

Financial assistance from the Commonwealth Foundation is gratefully acknowledged.

References

Aughtry, J. D. (1948). Effect of genetic factors in *Medicago* on symbiosis with *Rhizobium.* Cornell University Agricultural Experiment Station *Memoir* No. 280. 18 pp.

Bowen, G. D. & Kennedy, M. M. (1961). Heritable variation in nodulation of *Centrosema pubescens. Queensland Journal of Agricultural Science,* **18,** 161–70.

Brockwell, J., Bryant, W. G. & Gault, R. R. (1972). Ecological studies of root-nodule bacteria introduced into field environments. 3. Persistence of *Rhizobium trifolii* in association with white clover at high elevations. *Australian Journal of Experimental Agriculture and Animal Husbandry,* **12,** 407–13.

Brockwell, J. & Dudman, W. F. (1968). Ecological studies of root-nodule bacteria introduced into field environments. II. Initial competition between seed inocula in the nodulation of *Trifolium subterraneum* seedlings. *Australian Journal of Agricultural Research*, **19,** 749–57.

Brockwell, J., Dudman, W. F., Gibson, A. H., Hely, F. W. & Robinson, A. C. (1968). An integrated program for the improvement of legume inoculant strains. In: *Transactions of the Ninth International Congress Soil Science* (ed. J. W. Holmes), vol. 2, pp. 103–14. ISSS/Angus & Robertson, Sydney.

Brockwell, J. & Hely, F. W. (1966). Symbiotic characteristics of the systematic treatment of nodulation and nitrogen fixation interactions between hosts and rhizobia of diverse origins *Australian Journal of Agricultural Research*, **17,** 885–99.

Bryan, W. W. & Andrew, C. S. (1955). Pasture studies on the coastal lowlands of subtropical Queensland. II. The interrelation of legumes, *Rhizobium* and calcium. *Australian Journal of Agricultural Research*, **6,** 291–8.

Caldwell, B. E. (1969). Initial competition of root-nodule bacteria on soybeans in a field environment. *Agronomy Journal*, **61,** 813–15.

Chatel, D. L. & Greenwood, R. M. (1973). The colonization of host root and soil by rhizobia. II. Strain differences in the species *Rhizobium trifolii*. *Soil Biology and Biochemistry*, **5,** 433–40.

Chatel, D. L., Greenwood, R. M. & Parker, C. A. (1968). Saprophytic-competence as an important character in the selection of *Rhizobium* for inoculation. In: *Transactions of the Ninth International Congress of Soil Science* (ed. J. W. Holmes), vol. 2, pp. 65–73. ISSS/Angus & Robertson, Sydney.

Chatel, D. L. & Parker, C. A. (1973*a*). Survival of field grown rhizobia over the dry summer period in Western Australia. *Soil Biology and Biochemistry*, **5,** 415–23.

Chatel, D. L. & Parker, C. A. (1973*b*). The colonization of host-root and soil by rhizobia. I. Species and strain differences in the field. *Soil Biology and Biochemistry*, **5,** 425–32.

Damargi, S. M., Frederick, L. R. & Anderson, I. C. (1967). Serogroups of *Rhizobium japonicum* in soybean nodules as affected by soil types. *Agronomy Journal*, **59,** 10–12.

Date, R. A. (1969). A decade of legume inoculant quality control in Australia *Journal of the Australian Institute of Agricultural Science*, **35,** 27–37.

Date, R. A. & Roughley, R. J. (1975). Legume inoculant production. In: *Dinitrogen fixation* (ed. R. W. F. Hardy), vol. 2. John Wiley & Sons, New York. (In press.)

Erdman, L. W. & Means, U. M. (1952). Use of total yield for predicting N content of inoculated legumes grown in sand culture. *Soil Science*, **23,** 231–5.

Fred, E. E., Baldwin, I. L. & McCoy, E. (1932). *Root nodule bacteria and leguminous plants*. University of Wisconsin Press, Madison, Wisconsin.

Gibson, A. H. (1971). Factors in the physical and biological environment affecting nodulation and nitrogen fixation by legumes. *Plant and Soil*, Special Volume, 139–52.

Gibson, A. H., Dudman, W. F., Weaver, R. W., Horton, I. C. & Anderson, I. C. (1971). Variation within serogroups 123 of *Rhizobium japonicum. Plant and Soil,* Special Volume, 33–7.

Gupta, B. M. & Sen, A. (1965). The relationship between glucose consumption by *Rhizobium* species from some common cultivated legumes and their efficiencies. *Plant and Soil,* **22,** 229–38.

Ham, G. E., Cardwell, V. B. & Johnson, H. W. (1971). Evaluation of *Rhizobium japonicum* inoculants in soils containing naturalized populations of rhizobia. *Agronomy Journal,* **63,** 301–3.

Haydock, K. P. & Norris, D. O. (1967). Opposed curves for nitrogen percent on dry weight given by *Rhizobium* dependent and nitrate dependent legumes. *Australian Journal of Science,* **29,** 426–7.

Ireland, J. A. & Vincent, J. M. (1968). A quantitative study of competition for nodule formation. In *Transactions of the Ninth International Congress Soil Science* (ed. J. W. Holmes), vol. 2, pp. 85–93. ISSS/Angus & Robertson, Sydney.

Johnson, H. W. & Means, U. M. (1964). Selection of competitive strains of soybean nodulating bacteria. *Agronomy Journal,* **56,** 60–2.

Johnson, H. W., Means, U. M. & Weber, C. R. (1965). Competition for nodule sites between strains of *Rhizobium japonicum* applied as inoculum and strains in the soil. *Agronomy Journal,* **57,** 179–85.

Koontz, F. P. & Faber, J. E. (1961). Somatic antigens of *Rhizobium japonicum. Soil Science,* **91,** 228–32.

Marshall, K. C. (1956). Competition between strains of *Rhizobium trifolii* in peat and broth cultures. *Journal of the Australian Institute of Agricultural Science,* **22,** 137–40.

Means, U. M., Johnson, H. W. & Erdman, L. W. (1961). Competition between bacterial strains effecting nodulation in soybeans. *Proceedings of the Soil Science Society of America,* **25,** 105–8.

Norris, D. O. (1956). Legumes and the *Rhizobium* symbiosis. *Empire Journal of Experimental Agriculture,* **240,** 247–70.

Norris, D. O. (1958). A red strain of *Rhizobium* for *Lotononis bainesii. Australian Journal of Agricultural Research,* **9,** 629–32.

Norris, D. O. (1963). A porcelain bead method for storing *Rhizobium. Empire Journal of Experimental Agriculture,* **31,** 255–8.

Norris, D. O. (1964). Techniques used in work with *Rhizobium.* In: *Some concepts and methods in subtropical pasture research,* Commonwealth Agricultural Bureau Bulletin No. 47 (ed. CSIRO Cunningham Laboratory), pp. 186–98. Commonwealth Agricultural Bureau, Harpenden, England.

Norris, D. O. (1973). Seed pelleting to improve nodulation of tropical and subtropical legumes. 5. The contrasting response to lime pelleting of two *Rhizobium* strains on *Leucaena leucocephala. Australian Journal of Experimental Agriculture and Animal Husbandry,* **13,** 98–101.

Norris, D. O. & 't Mannetje, L. (1964). The symbiotic specialization of African *Trifolium* spp. in relation to their taxonomy and their agronomic use. *East African Agriculture and Forestry Journal,* **29,** 214–35.

Nutman, P. S. (1969). Symbiotic effectiveness in nodulated red clover. *Heredity*, **23,** 537–51.

Nutman, P. S. & Vincent, J. M. (1968). Experiments on nitrogen fixation by nodulated legumes, International Research Theme No. 3. Royal Society Special Document IBP/69(66) ammended design 1968.

Pankhurst, C. & Gibson, A. H. (1973). *Rhizobium* strain influence on disruption of clover nodule development at high root temperature. *Journal of General Microbiology*, **74,** 219–31.

Romeiko, I. N., Dubovenko, E. K. & Malinskaya, S. M. (1972). Dehydrogenase and nitrogen-fixing activity of pea nodule bacteria. *Microbiology, USSR*, **61,** 247–50.

Roughley, R. J. (1970*a*). The influence of root temperature, *Rhizobium* strain and host selection on the structure and nitrogen fixing efficiency of the root nodules of *Trifolium subterraneum. Annals of Botany*, **34,** 631–46.

Roughley, R. J. (1970*b*). The preparation and use of legume seed inoculants. *Plant and Soil*, **32,** 675–701.

Sarma, K. S. B., Lakshmi-Kumari, M., Apte, R. & Subba Rao, N. S. (1973). Some physiological characteristics of *Rhizobium meliloti* and *R. trifolii* in relation to efficiency of symbiosis with lucerne and Egyptian clover. *Plant and Soil*, **38,** 299–305.

Thompson, J. A., Roughley, R. J. & Herridge, D. F. (1974). Criteria and methods for comparing the effectiveness of *Rhizobium* strains for pasture legumes under field conditions. *Plant and Soil*, **40,** 511–24.

Vincent, J. M. (1956). Strains of rhizobia in relation to clover establishment. In: *Seventh International Grassland Congress, Palmerston North, New Zealand*, pp. 179–89.

Vincent, J. M. (1970). *A Manual for the practical study of root-nodule bacteria.* IBP Handbook No. 15. Blackwell Scientific Publications, Oxford.

Vincent, J. M. & Waters, L. M. (1953). The influence of host on competition amongst root-nodule bacteria. *Journal of General Microbiology*, **9,** 357–70.

Weber, D. F., Caldwell, B. E., Sloger, C. & Vest, H. H. (1971). Some USDA studies on the soybean – *Rhizobium* symbiosis. *Plant and Soil*, Special Volume, 293–304.

Wilson, J. K. (1944). Over five hundred reasons for abandoning the cross-inoculation groups of legumes. *Soil Science*, **58,** 61–9.

Table 11.4 summarises our experience with a number of strains during the past three years. Three *R. trifolii* strains have varied in effectiveness or invasiveness but not in colony type. The cowpea strain CB756 varied in both effectiveness and colony type, the lupin strain WU425 varied in colony type and time required to form nodules, and two medic strains U45 and SU47 varied in colony type only.

Table 11.5. *Growth of* Trifolium semipilosum *inoculated with CB782 obtained from different sources*

Isolate of CB782	Plant dry wt (mg)*
1	40
2	58
3	106
4	258
5	360
Uninoculated	61

* Mean of ten replicates

C. Labanderas (personal communication) selected a single colony of CC2480a which gave an unusual type of nodulation on *Trifolium subterraneum*. Early nodules were ineffective but among the later nodules a few were effective. Similarly with WU290 (iii) one subculture produced early ineffective nodules followed by two to three effective nodules. Another colony isolate of this strain failed to nodulate plants in our field trials. The growth of *T. semipilosum* inoculated with CB782 obtained from five different sources is shown in Table 11.5. The response varied from a 30 % depression in growth to a sixfold increase above the uninoculated control.

Table 11.6. *Growth and time for nodulation of Siratro inoculated with seven isolates of CB756 obtained from different sources*

Isolates of CB756	Time to nodulate (days)	Shoot wt (mg)
1	14	158
2	14	145
3	16	68
4	11	74
5	19	37
6	>19	32
7	>19	37
Uninoculated	–	6

13. Carriers of rhizobia and the effects of prior treatment on the survival of rhizobia

B. W. STRIJDOM & C. C. DESCHODT

Carriers of rhizobia

The first artificial inoculants were liquid bacterial cultures added to seed or directly to soil (Fred, Baldwin & McCoy, 1932). Although the results were frequently unsatisfactory, broth cultures or *Rhizobium* suspensions washed from solid media continued to receive attention over the years and good results have been reported by some workers (Van Schreven *et al.*, 1953; Van Schreven, 1958, 1963; Schiffmann & Alper, 1968). A progressive increase in the use of liquid inoculants for soybeans occurred in the USA because of ease of application by machinery (Burton, 1967). However, critical comparisons of liquid cultures with peat inoculants, consistently confirmed the superiority of the latter (Fellers, 1918; Vincent, Thompson & Donovan, 1962; Brockwell & Phillips, 1965; Burton, 1965; Burton & Curley, 1965; Vincent, 1968). Liquid cultures seem to lack the protective effect afforded by peat to the rhizobia on seed following inoculation.

Despite the advantages that a suitable synthetic carrier of rhizobia could have over natural materials, no synthetic carrier is in general use. Fraser (1966) devised an inoculant consisting of calcium sulphate granules impregnated with rhizobia but this is not widely used. Lyophilised cultures had been used in Australia for several years (Vincent, 1958, 1965, 1968; McLeod & Roughley, 1961), and commercial mixtures of talc and lyophilised cultures were also tested in the USA (Burton, 1967). A major disadvantage which lyophilised inoculants have in common with liquid cultures is their inability to prolong post-inoculation survival of rhizobia on seed (Date, 1968).

The indisputable capability of soil to support survival of rhizobia (Jensen, 1961) directed most efforts to obtain carriers superior to peat, around a neutralised soil-peat base enriched with nutrients. Among these were peat containing carbon black (Hedlin & Newton, 1948), the so-called humus inoculants of Newbould (1951) and Gunning & Jordan (1954), carriers consisting of peat or soil, supplemented with materials such as lucerne meal, ground straw, yeast and sugar (Temple, 1916;

Van Schreven, 1958, 1963, 1970; Van Schreven, Otzen & Lindenberg, 1954; Wrobel & Ziemiecka, 1960), Nile silt supplemented with nutrients (Afify *et al.*, 1968), soil plus coir dust or soybean meal (John, 1966; Iswaran, 1972), soil plus wood charcoal (Newbould, 1951; Gunning & Jordan, 1954; Iswaran, 1972), and peat amended with nutrients and an adhesive to improve its ability to adhere to the seed (Hastings, Greenwood & Proctor, 1966).

Various materials other than soil have also received attention as possible carriers of rhizobia (Fred *et al.*, 1932; Vincent, 1965). These include vermiculite, decomposed sawdust, perlite and rice husk compost (Bonnier, 1960), ground rock phosphate (R. A. Date, personal com-

Table 13.1. *Survival of* R. meliloti *strain U45 and* Rhizobium *strain CB756 of the cowpea group in various inoculants*

Inoculant formulation	Log number rhizobia/g after (days)			
	15	30	120	180
	Strain U45			
Coal*	9.88	9.43	9.10	7.70
Coal+lucerne*	10.59	10.59	8.44	8.51
Coal+lucerne+bentonite*	10.75	10.07	9.64	9.68
Peat	10.38	10.48	9.23	9.21
	Strain CB756			
Coal	7.37	7.31	7.45	6.60
Coal+lucerne	7.20	7.68	7.84	6.57
Coal+lucerne+bentonite	8.48	9.58	8.27	6.94
Peat	8.59	9.37	8.80	7.61
LSD at				
$P = 0.01$	0.21	0.29	1.16	0.97
$P = 0.05$	0.17	0.23	0.92	0.77

* With 1 % sucrose. Approximately 4 parts coal to 1 part lucerne meal or 2 parts coal, 1 part lucerne meal and 2 parts bentonite.

munication) and coffee husk compost. According to B. K. Pugashetti, R. B. Patil, D. J. Bagyaraj & H. P. Srinivas (N. S. Subba Rao, personal communication) the number of rhizobia in a coffee husk compost inoculant remained constant when stored at 10 °C for 30 days.

A so-called cob-earth carrier consisting of milled, decomposed, ground maize cobs supplemented with nutrients is currently used for inoculants in Rhodesia (see Corby, Chapter 14). The material is sterilised before being inoculated aseptically. Survival of some *Rhizobium* strains in the cob-earth carrier compared favourably to that in peat used in South African inoculants (see Table 13.2).

Leiderman (1971) found finely ground bagasse suitable as a carrier of soybean rhizobia for periods up to 100 days after preparation. According to P. H. Graham (personal communication) bagasse is a good carrier at 28 °C especially when soaked before use to eliminate excess sugar; at 37 °C numbers of *Rhizobium* rapidly decline, falling

Table 13.2. *Survival of* R. meliloti (*U45*), R. japonicum (*WB61*) *and a cowpea strain* (*CB756*) *in each of three carriers*

Rhizobium strain	Log number rhizobia/g after (days)		
	40	60	140
		Peat*	
U45	9.89	9.71	9.83
CB756	9.56	9.79	9.02
WB61	9.61	9.89	9.11
Mean	9.67	9.80	9.32
		Coal–bentonite†	
U45	10.15	10.56	10.00
CB756	9.30	10.53	9.80
WB61	10.01	10.13	8.73
Mean	9.82	10.41	9.51
		Cob-earth‡	
U45	10.15	10.02	11.10
CB756	9.49	10.36	7.84
WB61	9.06	9.87	6.90
Mean	9.57	10.08	8.60
LSD (means)			
$P = 0.01$	0.47	0.71	0.96
$P = 0.05$	0.34	0.56	0.69

* Putfontein peat as used in inoculants in South Africa.
† Coal–bentonite–lucerne meal (2:2:1).
‡ Decomposed maize cobs supplemented with nutrients.

below 10^6/g in two weeks. Fungus contamination seems to be an important problem with this carrier but is reduced when the bagasse is washed.

The suitability of a carrier of rhizobia based on coal, was investigated in South Africa (Deschodt & Strijdom, unpublished results). Three formulations were studied; i.e. coal alone, coal with lucerne meal and coal with bentonite and lucerne meal. Lucerne meal was included because of the good results obtained by Van Schreven *et al.* (1954) with inoculants containing lucerne. The protective effect which clay minerals have on rhizobia (Marshall, 1964) prompted the inclusion of bentonite. The ingredients were ground to pass a 170 mesh (British Standard) sieve, mixed and supplemented with 1.0 % sucrose. The pH was adjusted to 7.0 with calcium carbonate and the carrier sterilised for 2 hours at 121 °C. Survival of *Rhizobium meliloti* strain U45 and of the cowpea strain CB756 in the coal–bentonite–lucerne carrier was comparable to that in peat (Table 13.1).

Subsequent experiments showed that best results with each of three *Rhizobium* strains were obtained in a carrier (designated CBL carrier) consisting of coal, bentonite and lucerne meal in a ratio of 2:2:1 (mass basis) and a moisture content of 50 %. Survival of each strain in this carrier over a 140-day period was similar to that in sterilised Putfontein peat used in inoculants in South Africa (Table 13.2). Survival of *R. meliloti* was better in the cob-earth carrier used in Rhodesia than in either the peat or CBL carrier, but *R. japonicum* (WB61) and the cowpea strain CB756 survived relatively poorly in the cob-earth carrier.

Results indicated that the type and source of the coal used in the carrier had no marked effect on growth or survival of the rhizobia. In

Table 13.3. *Survival of* R. meliloti *strain U45 and* Rhizobium *strain CB756 in carriers containing coal from different sources*

	Log number rhizobia/g after (days)				
Type of coal*	10	15	30	120	180
			Strain U45		
Anthracite A†	9.59	10.07	10.57	8.34	7.C0
Anthracite B†	9.62	10.39	9.34	8.71	7.47
Large nuts‡	9.30	9.70	10.18	8.33	7.47
			Strain CB756		
Anthracite A	7.94	7.97	7.65	6.89	7.29
Anthracite B	7.97	7.71	7.80	6.77	6.81
Large nuts	7 74	7.29	7.44	6.45	6.70
LSD at					
$P = 0.01$	0.27	0.39	0.33	0.70	1.21
$P = 0.05$	0.22	0.31	0.26	0.56	0.96

* Coal carriers consisted of 80 % coal + 19 % lucerne meal + 1 % sucrose.
† High grade coals from different sources.
‡ Low grade coal.

carriers consisting of 80 % coal, 20 % lucerne meal and 1 % sucrose (Table 13.3), no significant differences in survival of two *Rhizobium* strains tested were obtained, irrespective of the grade or source of the coal used. This is undoubtedly a promising feature of this carrier.

The ability of inoculated and uninoculated CBL carriers to adhere to glass beads was compared with that of peat (Table 13.4). Two per cent of an adhesive was added to some inoculants during preparation to determine if it would improve the eventual ability of the carrier to stick to beads moistened with water. Although the adhesive had no effect on either inoculated or uninoculated carriers it was found that CBL carrier

Table 13.4. *Amounts (g) of peat or a coal carrier adhering to 100 g quantities of glass beads*

Treatment of carrier	Age of carrier when applied (days)			Dry mass (mean)
	2	10	30	
Coal+pino.*	1.45	1.46	1.42	1.44
Peat+pino.	0.52	0.83	0.92	0.76
Coal+U45†	1.35	1.41	1.44	1.40
Peat+U45	1.05	1.14	0.93	1.04
Coal+pino.+U45	1.36	1.40	1.42	1.39
Peat+pino.+U45	0.96	0.99	0.87	0.94
Coal+CB756‡	1.32	1.42	1.43	1.39
Peat+CB756	1.22	1.22	1.10	1.18
Coal+pino.+CB756	1.27	1.38	1.43	1.36
Peat+pino.+CB756	1.09	1.21	0.92	1.07

LSD of means: $P = 0.01$, 0.49; $P = 0.05$, 0.43.
* Non-inoculated carrier+2 % pinolene (poly-L-*p*-menthene-8, 9-diyl).
† Carrier inoculated with *R. meliloti* strain U45.
‡ Carrier inoculated with *Rhizobium* strain CB756 of the cowpea group.

tended to adhere to beads in larger quantities than did peat. In practice this would mean that more rhizobia would be obtained on seed treated with CBL inoculant than on seed treated with a peat inoculant containing the same number of rhizobia per gram.

Conclusion

An assessment of the literature on carriers of rhizobia leaves the firm impression that (*a*) peat is still unchallenged as a carrier and (*b*) that it is relatively easy to devise a substrate from a variety of materials that would support satisfactory growth and survival of rhizobia. Most of these potential carriers have not been studied in detail, but they do not seem to possess superior properties that would justify their replacement of peat. Some of the materials also suffer from the disadvantage of being unavailable in many countries. However, it is clear that the search for new carriers has often been prompted by the lack of suitable local peat, and that the materials investigated were cheap and readily available in those areas. Each of these carriers may have considerable merit for local use and should not be judged strictly as a general replacement for peat.

Marked *Rhizobium*–carrier interactions are prevalent. This was emphasized by very large numbers of a *R. meliloti* strain in the cob-earth inoculant after 140 days, and the relatively poor survival of a cowpea and *R. japonicum* strain in the same carrier (Table 13.2). Survival studies with a range of strains are essential before a potential carrier can be considered for inoculant production.

The CBL carrier warrants further attention. It is easy to produce from materials that are relatively cheap and obtainable in many countries. In South Africa 250 g of this carrier would cost about 3 cents (US). The source and type of coal do not seem to be as important as with peat; survival is as good as or better than in peat and it apparently adheres to seed in larger quantities when used as inoculant. The CBL carrier is currently being investigated in more detail to determine if it would protect rhizobia on seed as well as peat does.

Prior treatment of carriers

Because few materials have been subjected to critical studies as possible carriers of rhizobia, attention will be restricted to the prior treatment of peat.

Source and character of peat

The most important factors to consider before selecting a particular peat as carrier are the source and character of the material. Whereas the choice may be aided by a chemical analysis, actual multiplication and survival studies of *Rhizobium* strains in the peat are essential (Burton, 1965; Roughley & Vincent, 1967; Roughley, 1970). This is emphasised on the one hand by marked differences in chemical composition and physical properties of peats used with success in inoculants

Table 13.5. *Composition of peat from two sources used as carriers of rhizobia in South Africa*

Peat source	Nitrogen content			Moisture content (%)	Organic matter (%)	Ash (%)	P (%)	Fe (%)	Al (%)	Cu (ppm)	Exchangeable (mM/100 g)				pH (water)	Microbial population		
	NH_4 (%)	NO_3 (%)	Total (%)								Ca	Mg	K	Na		Bacteria $\times 10^4$	Actino-mycetes $\times 10^4$	Fungi $\times 10^4$
Putfontein																		
10–40 cm	0.2	0	1.9	8.4	58.3	33.1	0.1	1.52	2.10	17.8	19.8	9.7	0.2	0.4	5.4	2.1	0.6	3.3
40–50 cm	0.2	0	1.9	9.3	69.2	21.5	0.07	1.03	1.51	15.0	21.4	13.3	0.2	0.5	5.1			
Barrydale																		
10–40 cm	0.2	0	0.9	6.8	45.7	47.5	0.2	0.42	1.00	435.0	2.6	6.5	1.1	4.0	3.1	32.0	6.7	3.1

in South Africa (Table 13.5) and elsewhere (Burton, 1965; Roughley & Vincent, 1967; Roughley, 1970) and, on the other, by the differences in the abilities of peats of similar composition to support growth and survival of rhizobia. Even peat samples from a single peat deposit may vary greatly in ability to promote survival of *Rhizobium* strains (Roughley & Vincent, 1967), especially when taken at different depths (Van Schreven, 1970).

The peat carriers used in most inoculants in the USA and Australia are characterised by a relatively high organic matter content (Burton, 1967; Roughley, 1970), but this is not an essential feature of peat carriers in good inoculants. In Colombia six peats, with organic matter content varying between 35 and 78 %, were studied by P. H. Graham (personal communication). The most suitable had an organic matter content of only 40 %. A peat also used with reasonable success in South Africa has 45 % organic matter (Table 13.5).

Drying of peat and particle size

Drying of the harvested peat is an important step in its preparation. It has been shown with Australian peat that the final drying temperature should not exceed 100 °C in order to avoid the formation of toxic degradation products and excessive rises in temperature when the peat is subsequently wetted by the addition of broth culture (Roughley & Vincent, 1967; Roughley, 1970). However, Burton (1967) described the production of high-quality inoculants by flash drying a sedge peat carrier with hot air at 650 °C to a moisture content of approximately 7 % (wet basis). Van Schreven (1958, 1963, 1970) produced soil-peat inoculants without completely drying the peat before milling. The type of peat and the eventual particle size desired will determine to what extent drying is required. There appears to be no practical advantage in drying peat below 12 % moisture content (Roughley & Vincent, 1967). Even at a moisture content of 15–19 % (dry weight basis) the Putfontein peat used in South Africa is easily ground to allow 50–60 % of the particles to pass a 200 mesh (British Standard) sieve (C. H. Jacobs, personal communication).

Dried peat is usually ground in a hammer mill. In Australia best results are obtained with peat carriers that pass through a 200 mesh sieve (Roughley & Vincent, 1967; Roughley, 1970). Fifty to sixty per cent of the peat particles in inoculants in South Africa will pass through a 200 mesh (British Standard) sieve. Whereas a small particle size is considered important by many inoculant producers, good results have been reported by Van Schreven (1963, 1970) with peat sieved through a 2 mm screen. Peat and soil carriers studied by Indian workers, were ground to pass a 60 mesh sieve (Iswaran *et al.*, 1969; Iswaran, 1972). The seed inoculation procedure will determine to some extent the requirements with respect to particle size of the carrier.

pH and moisture content

Most peat deposits, with the exception of the Mt Gambier peats used in Australia (Stephens, 1943; Roughley & Vincent, 1967) have a low pH and require the addition of a neutralising agent before being suitable as carriers. According to P. H. Graham (personal communication) strongly acid peats in Colombia with the pH adjusted to 6.5 will support good survival of rhizobia at temperatures of 28 °C or below. At 37 °C the rhizobial population drops below 10^6/g in 2 weeks and only a Medellin peat with an initial pH of 5.1–5.3 before addition of calcium carbonate, supported adequate rhizobial survival for up to 7 weeks at 37 °C. Considering the satisfactory results obtained with neutralised acid peats in many countries, further studies are warranted to determine if the superior quality of the Medellin peat could be attributed to its relatively high initial pH.

Although *Rhizobium* strains may differ somewhat with respect to optimum pH conditions (Jensen, 1942), most strains grow well at 6.0–7.0 and peat carriers are usually adjusted to pH 6.5–7.0 (Burton, 1965; Iswaran *et al.*, 1969; Roughley, 1970; P. H. Graham, personal communication; Deschodt & Strijdom, unpublished results). Calcium carbonate (lime) is the neutralising agent most frequently used, although magnesium carbonate may also be used with certain peats; sodium and potassium carbonates are unfavourable for rhizobial survival (Fulmer, 1918; Roughley & Vincent, 1967). Van Schreven (1970) confirmed the superiority of calcium carbonate over calcium carbonate with di-potassium hydrogen phosphate; ammonium hydroxide was found to be extremely harmful for rhizobia in peat.

It is essential that the moisture content of a peat carrier be reduced sufficiently in order that the eventual addition of inoculum brings the moisture content of the inoculant to the desired level (Hedlin & Newton, 1948; Van Schreven *et al.*, 1954; Roughley & Vincent, 1967). A final moisture content of c. 40–55 % appears to be favourable for most peats used as carriers (Gunning & Jordan, 1954; Van Schreven, 1958, 1963,

1970), although Burton (1967) reported good survival in peat inoculants in the USA with a moisture content of 35–40 %. Obviously, a factor such as organic matter content could have a marked effect on the optimum moisture content of a particular peat carrier. Expression of optimum moisture status in terms of pF (water potential) values would therefore be more acceptable for comparative purposes. Steinborn & Roughley (1974) found an optimum pF of 3.9–4.0 for all strains tested.

Rhizobia in non-sterile peat may be more susceptible to harmful effects of high moisture levels than rhizobia in sterile peat (Roughley & Vincent, 1967; Vincent, 1968; Roughley, 1968, 1970; Date, 1968). Forty to fifty per cent moisture was found suitable in unsterilised Australian peat whereas moisture below 30 % or above 60 % was unfavourable for survival. Conversely, a moisture content of 60 % was recommended by Roughley (1968, 1970) for sterilised peat.

Containers

The containers in which carriers are packed are important since their gas exchange and moisture retention properties may have a marked effect on the quality of the inoculant. Whereas Newbould (1951), Spencer & Newton (1953) and Gunning & Jordan (1954) claimed satisfactory survival of rhizobia in tightly sealed cans or jars, free access of air to carriers is recommended by Hedlin & Newton (1948), Van Schreven *et al.* (1954) and Roughley (1968), and to this end bottles plugged with cotton wool and covered with cellophane are used in Holland (Van Schreven, 1958, 1963) and bags of low density polythene (0.038–0.051 mm gauge) in Australia (Roughley, 1970). Polythene is also employed with success in the USA (Burton, 1965). Iswaran (1971) claimed better survival of a *R. japonicum* strain in sealed polythene bags than in open aerated ones.

Although a critical examination of the results reported leaves little doubt that some degree of aeration is necessary for satisfactory survival of rhizobia in peat carriers, the amount required is small. This is stressed by survival studies with four strains representing *R. meliloti*, *R. trifolii*, *R. japonicum* and the cowpea group, in steam-sterilised Putfontein peat in sealed high density polypropylene bags of 0.31–0.32 mm gauge (Strijdom, unpublished results). After 150 days storage at 27 °C the numbers of each strain still ranged between 3.5×10^8/g and 62×10^8/g. Good survival was also obtained in New Zealand, in experiments using heat-sterilised peat packed in high-density polythene bags (A. Hastings, personal communication).

Sterilisation of peat

Most inoculants are probably still being produced in unsterilised or partially sterilised peat. Many of these are of high quality (Burton, 1967; Vincent, 1968) and there appears to be no advantage for the inoculant producers concerned to switch to the difficult and more expensive process of aseptic inoculation of sterilised peat in packets or other containers. Unfortunately this situation does not hold true for all peats used in inoculants. A rapid loss of viability of rhizobia in non-sterile peat stored at 25 °C was reported by Vincent (1958), Roughley & Vincent (1967) and Roughley (1968), especially if the moisture content was below 30 % or above 60 %. Survival in sterile peat was not affected by temperatures up to 26 °C and high moisture level was not detrimental to survival.

In South Africa commercial inoculants are frequently withdrawn from the market because of a rapid decline in numbers in the unsterilised peat carrier (Strijdom, unpublished results); the same strains survived in steam-sterilised peat in numbers of 10^8–10^9/g six months after production. Many experiments showed that sterilisation of the peat was beneficial for the initial multiplication of the rhizobia and for survival during storage. There is also evidence that certain slow-growing *Rhizobium* strains, for example, those from cowpea, soybean and *Lotononis*, may survive poorly in a non-sterile peat that satisfactorily supports survival of fast-growing strains (Roughley & Vincent, 1967; Vincent, 1968). Whereas sterilisation may, therefore, not be essential for all peats, it is a prerequisite for others in order to ensure success with the various *Rhizobium* strains used. There is little doubt that, in general, a carrier free of contaminants is superior to a non-sterilised one (Van Schreven, 1958; Hamatova, 1962; Roughley & Vincent, 1967).

Steam is most commonly used to sterilise carriers. Sterilisation times and temperatures range from 1 to 4 hours at around 120 °C to 3 hours at 125 °C (Newbould, 1951; Van Schreven, 1958, 1963, 1970; Roughley & Vincent, 1967; Iswaran, 1972). In the large-scale production of inoculants, the nature of the peat, the amount in each packet and the number and arrangement of packets in the autoclave may all affect the minimum time required to obtain complete sterility. Dry heat was used by Gunning & Jordan (1954) who treated peat for 1 hour at 170 °C.

Australian workers found excessive heat treatment of peat detrimental to subsequent growth and survival of rhizobia (Roughley & Vincent, 1967). The deleterious effect on rhizobia of overnight exposure of two South African peats to a temperature of 160 °C prior to inoculation emphasised this fact. Conversely, improved survival of rhizobia was consistently obtained in South Africa when peat was sterilised in an autoclave, although Roughley & Vincent (1967) showed that steam-sterilisation for 4 hours at 121 °C may also render peat less favourable for rhizobia than, for example, gamma irradiation at a dose of 5.0×10^6 rad.

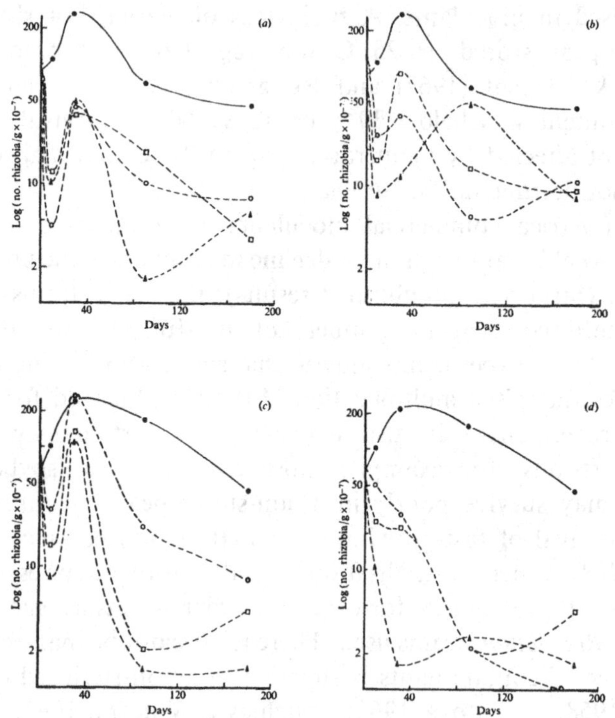

Fig. 13.1. Survival of *Rhizobium meliloti* strain U45 in peat treated with ethylene oxide. Barrydale peat treated with etox for 8 hours (*a*) and 16 hours (*b*). Putfontein peat treated with etox for 8 hours (*c*) and 16 hours (*d*). ○---○, Etox at 500 mg/l; □---□, etox at 750 mg/l; △---△, etox at 1000 mg/l; ●——●, steam (2 hours at 121 °C).

Because of the undesirable effect of excessive heat on certain peats, attention may either be directed towards methods of sterilisation other than by heat treatment, or to the use of carriers that have been partially sterilised by the restricted application of heat. Should heat be excluded, the means of sterilisation are limited to either irradiation or gaseous treatments, neither of which seems as effective as heat for providing complete sterilisation of soil or peat in routine practice. Roughley &

Vincent (1967) reported excellent survival of rhizobia in gamma-irradiated peat (5.0×10^6 rad) and this method is currently used for all inoculants produced in Australia (R. J. Roughley, personal communication). However, gamma irradiation does not completely sterilise the peat in routine practice (Vincent, 1968) and certain micro-organisms seem remarkably resistant to doses up to 2.4–2.5×10^6 rad (Frances Parker, personal communication; Anellis, Berkowitz & Kemper, 1973).

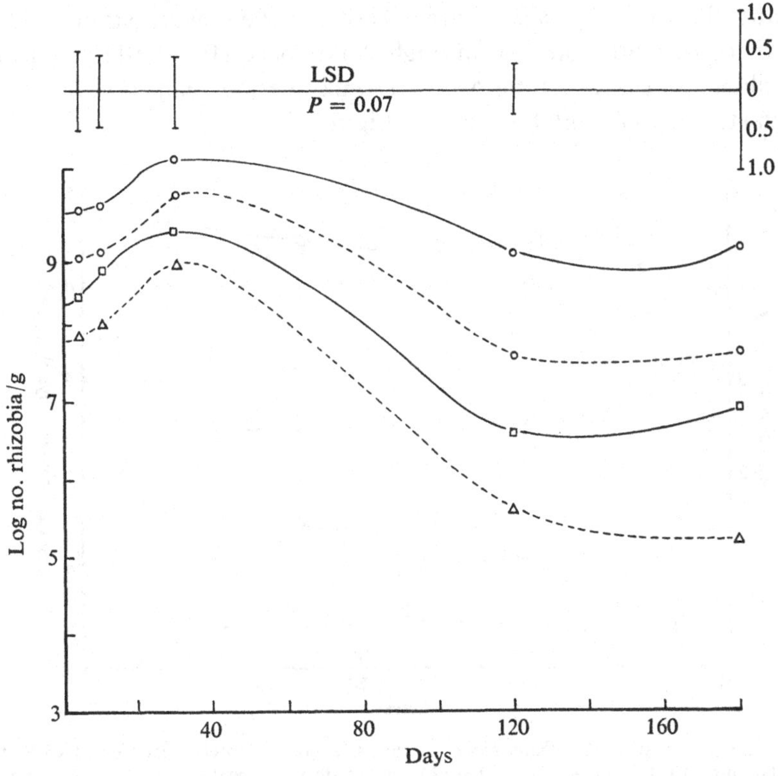

Fig. 13.2. Survival of *Rhizobium meliloti* strain U45 and the cowpea strain CB756 in steam-sterilised peat and in peat treated with etox at 1000 mg/l for 8 hours. ○——○, U45 in steam-sterilised peat; □——□ U45 in etox-treated peat; ○---○, CB756 in steam-sterilised peat; △---△, CB756 in etox-treated peat.

The effect of gaseous treatment of peat on the subsequent survival of rhizobia in inoculants was investigated with the two peats used in inoculants in South Africa. Survival of *R. meliloti* strain U45 was compared in peat which was sterilised for 2 hours at 121 °C with peat treated with etox (90 % ethylene oxide + 10 % CO_2) at concentrations of 500, 750 and 1000 mg/l respectively, for periods of 8 and 16 hours (Fig. 13.1). A common characteristic of all the etox-treated peats was the marked decrease in the numbers of rhizobia shortly after inoculation, followed by a subsequent increase in most treatments after 30 days. Although contaminants were not demonstrated in the peat at the time of inoculation, various contaminants were present on plates when counts were made after 30 days. The numbers of these increased progressively as the rhizobia decreased and approximately equalled the rhizobial numbers after 180 days. In subsequent experiments it was shown that both *R. meliloti* strain U45 and the cowpea strain CB756, were consistently present in higher numbers (P = 0.01) in steam-sterilised peat at any time over a 180-day period than in peat treated with etox at 1000 mg/l for 8 hours (Fig. 13.2).

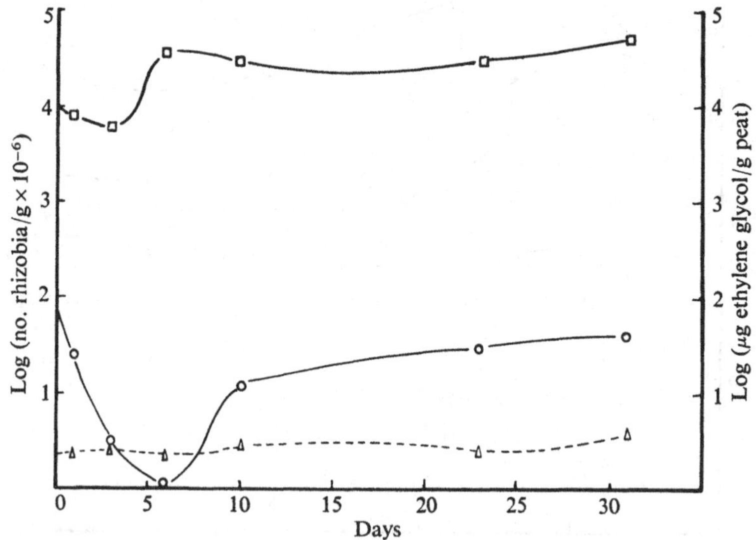

Fig. 13.3. Survival of *Rhizobium meliloti* strain U45 in Putfontein peat treated with etox at 1000 mg/l for 8 hours (○——○), and in steam-sterilised peat (□——□). The amount of ethylene glycol in the etox-treated peat is shown (△--- △).

Chromatographic analysis of gas samples in ethylene oxide treated peat showed that the ethylene concentration did not at any stage exceed that in steam-sterilised peat over the first 26 days. The marked decrease in rhizobial numbers over the first 10 days in survival experiments with etox-treated peats, could thus not be attributed to free ethylene oxide in the peat. Although the peat contained approximately 12 μg/g ethylene glycol, the concentration of this compound did not change significantly over the first 30 days and could not account for the 'dip' in the survival curve immediately after inoculation (Fig. 13.3). Propylene glycol was not formed in significant quantities in the etox-treated peat.

Conclusion

It is clear from the literature cited and the problems encountered with South African peats, that sterilisation is one of the most rewarding of the various carrier treatments considered beneficial for the survival of rhizobia in inoculants. Satisfactory results may be obtained with some non-sterile peats but a contaminant-free carrier is superior. It is also apparent that a peat which had supported satisfactory survival of rhizobia in unsterilised inoculants for decades might eventually become unsuitable for some *Rhizobium* strains unless sterilised. This is perhaps best illustrated by the increased death rate of clover strains in non-sterile Mt Gambier peat recently experienced in Australia (R. J. Roughley, Chapter 11).

Whereas excessive heat applications render peat less favourable for rhizobia, steam-sterilisation is acceptable and practiced with good results by many inoculant producers. Gamma irradiation is considered superior to steam-sterilisation, but irradiation facilities are not always available, especially in those countries which have the greatest need for cheap, biologically-fixed nitrogen.

The finding that gamma irradiation does not sterilise the peat completely in routine practice, in conjunction with the fact that peat, flash-dried at more than 600 °C (Burton, 1967) supports satisfactory survival of rhizobia, indicates that complete sterility is not essential for success. One nevertheless hesitates to generalise on this aspect. Logically, the most important step in the preparation of inoculants from partially sterilised peat, is the addition of a large inoculum before surviving contaminants in the peat have had the opportunity to increase in number.

At present our experience with gaseous sterilisation of peat is limited. Results with ethylene oxide showed that this treatment requires special equipment, that sterilisation is incomplete and that it renders the peat less favourable for rhizobia than steam-sterilisation. Although survival of rhizobia in South African peats treated with ethylene oxide was better than in unsterilised peat, it does not seem to warrant consideration as an alternative to steam-sterilisation for the peats studied.

References

Afify, M. N., Moharram, A. A., El-Nady, M. A. L., Hamdi, Y. A., El-Sherbini, M. F. & Lofti, M. (1968). Studies on the carriers of root nodule bacteria. I. Effect of organic materials on the longevity of rhizobia in the carriers. *Agricultural Research Review*, **46,** 2–5.

Anellis, A., Berkowitz, D. & Kemper, D. (1973). Comparative resistance of non-sporogenic bacteria to low-temperature gamma irradiation. *Applied Microbiology*, **25,** 517–23.

Bonnier, C. (1960). Symbiose *Rhizobium*-légumineuses: aspects particuliers aux régions tropicales. *Annales De L'Institut Pasteur*, **98,** 537–56.

Brockwell, J. & Phillips, L. J. (1965). Survival at high temperatures of *Rhizobium meliloti* in peat inoculant on lucerne seed. *Australian Journal of Science*, **27,** 332–3.

Burton, J. C. (1965). The *Rhizobium*-legume association. In: *Microbiology and soil fertility* (ed. G. M. Gilmour & O. N. Allen), pp. 107–34. Oregon State University Press, Corvallis.

Burton, J. C. (1967). *Rhizobium* culture and use. In: *Microbial technology* (ed. H. J. Peppler), pp. 1–33. Van Nostrand-Reinhold, New York.

Burton, J. C. & Curley, R. L. (1965). Comparative efficiency of liquid and peat-base inoculants on field-grown soybeans (*Glycine max*). *Agronomy Journal*, **57,** 379–81.

Date, R. A. (1968). Rhizobial survival on the inoculated legume seed. In: Transactions of the *Ninth International Congress of Soil Science Transactions* (ed. J. W. Holmes), vol. 2, pp. 75–83. ISSS/Angus & Robertson, Sydney.

Fellers, C. R. (1918). Report on the examination of commercial cultures of legume-infecting bacteria. *Soil Science*, **6,** 53–67.

Fraser, Margaret E. (1966). Pre-inoculation of lucerne seed. *Journal of Applied Bacteriology*, **29,** 587–95.

Fred, E. B., Baldwin, I. R. & McCoy, Elizabeth (1932). *Root nodule bacteria and leguminous plants.* Studies in Science No. 5. University of Wisconsin Press, Madison, Wisconsin.

Fulmer, H. L. (1918). Influence of carbonates of magnesium and calcium on bacteria of certain Wisconsin soils. *Journal of Agricultural Research*, **12,** 463–504.

Gunning, C. & Jordon, D. C. (1954). Studies on humus type legume inoculants. II. Preparational effectivity. *Canadian Journal of Agricultural Science*, **34,** 225–33.

Hamatova, E. (1962). Storage of nitrazon. *Rostlinná výroba*, **35,** 825–38.

Hastings, A., Greenwood, R. M. & Proctor, M. H. (1966). Legume inoculation in New Zealand. New Zealand Department of Scientific and Industrial Research Information Series, No. 58.

Hedlin, R. A. & Newton, J. D. (1948). Some factors influencing the growth and survival of rhizobia in humus and soil culture. *Canadian Journal of Research*, **26,** 174–87.

Iswaran, V. (1971). Effect of aeration on survival of *Rhizobium japonicum* in peat contained in polythene bags. *Mysore Journal of Agricultural Science*, **5**, 230–2.

Iswaran, V. (1972). Growth and survival of *Rhizobium trifolii* in coir dust and soybean meal compost. *Madras Agricultural Journal*, **59**, 52–3.

Iswaran, V., Sundara Rao, W. V. B., Magu, S. P. & Jauhri, K. S. (1969). Indian peat as a carrier of *Rhizobium. Current Science*, **38**, 468–9.

Jensen, H. L. (1942). Nitrogen fixation in leguminous plants. I. General characters of root-nodule bacteria isolated from species of *Medicago* and *Trifolium* in Australia. *Proceedings of the Linnean Society of New South Wales*, **67**, 98–108.

Jensen, H. L. (1961). Survival of *Rhizobium meliloti* in soil culture. *Nature, London*, **192**, 682–3.

John, K. P. (1966). A coir-dust soil compost for *Rhizobium. Journal of the Rubber Research Institute of Malaya*, **19**, 173–5.

Leiderman, J. (1971). El bagazo como exipiente de inoculante para leguminosas. *Revista Industrial y Agricola de Tucumán*, **48**, 51–8.

McLeod, R. W. & Roughley, R. J. (1961). Freeze-dried cultures as commercial legume inoculants. *Australian Journal of Experimental Agriculture and Animal Husbandry*, **1**, 29–33.

Marshall, K. C. (1964). Survival of root-nodule bacteria in dry soils exposed to high temperatures. *Australian Journal of Agricultural Research*, **15**, 273–81.

Newbould, F. H. S. (1951). Studies on humus type legume inoculants. I. Growth and survival in storage. *Scientific Agriculture*, **31**, 463–9.

Roughley, R. J. (1968). Some factors influencing the growth and survival of root nodule bacteria in peat culture. *Journal of Applied Bacteriology*, **31**, 259–65.

Roughley, R. J. (1970). The preparation and use of legume seed inoculants. *Plant and Soil*, **32**, 675–701.

Roughley, R. J. & Vincent, J. M. (1967). Growth and survival of *Rhizobium* spp. in peat culture. *Journal of Applied Bacteriology*, **30**, 362–76.

Schiffmann, J. & Alper, Y. (1968). Inoculation of peanuts by application of *Rhizobium* suspension into the planting furrows. *Experimental Agriculture*, **4**, 219–26.

Spencer, J. F. T. & Newton, J. D. (1953). Factors influencing the growth and survival of rhizobia in humus and soil cultures. II. *Canadian Journal of Botany*, **31**, 253–64.

Steinborn, Julia & Roughley, R. J. (1974). Sodium chloride as a cause of low numbers of *Rhizobium* in legume inoculants. *Journal of Applied Bacteriology*, in press.

Stephens, C. G. (1943). The pedology of a South Australian fen. *Transactions of the Royal Society of South Australia*, **67**, 191–9.

Temple, J. C. (1916). Studies of *Bacillus radicicola*. 1. Testing commercial cultures. II. Soil as a medium for growing *B. radicicola. Georgia Agricultural Experimental Station Bulletin*, **120**, 65–80.

Van Schreven, D. A. (1958). Methods used in the Netherlands for the production of legume inoculants. In: *Nutrition of the legumes* (ed. E. G. Hallsworth), pp. 328–33. Butterworth & Co., London.

Van Schreven, D. A. (1963). Ontwikkeling van de methodiek van entstoffen voor vlinderbloemige gewassen en de controle van rhizobium-stammen in het microbiologisch laboratorium. *Van Zee tot Land, Zwolle*, **36**, 88–108.

Van Schreven, D. A. (1970). Some factors affecting growth and survival of *Rhizobium* spp. in soil-peat cultures. *Plant and Soil*, **32**, 113–30.

Van Schreven, D. A., Harmsen, G. W., Lindenbergh, D. J. & Otzen, D. (1953). Experiments on the cultivation of *Rhizobium* in liquid media for use on the Zuiderzee polders. *Antonie van Leeuwenhoek, Journal of Microbiology and Serology*, **19**, 300–8.

Van Schreven, D. A., Otzen, D. & Lindenbergh, D. J. (1954). On the production of legume inoculants in a mixture of peat and soil. *Antonie van Leeuwenhoek, Journal of Microbiology and Serology*, **20**, 33–57.

Vincent, J. M. (1958). Survival of the root nodule bacteria. In: *Nutrition of the legumes* (ed. E. G. Hallsworth), pp. 108–23. Butterworth & Co., London.

Vincent, J. M. (1965). Environmental factors in the fixation of nitrogen by the legume. In: *Soil nitrogen* (ed. W. V. Bartholomew & Frances E. Clark), pp. 384–435. American Society of Agronomy Inc., Madison, Wisconsin.

Vincent, J. M. (1968). Basic considerations affecting the practice of legume seed inoculation. In: *Festskrift til Hans Laurits Jensen*, pp. 145–58. Gadgaard Nielsens Bogtrykkeri, Lemvig, Denmark.

Vincent, J. M., Thompson, J. A. & Donovan, K. A. (1962). Death of root-nodule bacteria on drying. *Australian Journal of Agricultural Research*, **13**, 258–70.

Wrobel, T. & Ziemiecka, M. J. (1960). Results of field experiments with the inoculation of leguminous plants. *Roczniki Nauk Rolniczych*, **82**-A-1, 201–9.

14. A method of making a pure-culture, peat-type, legume inoculant, using a substitute for peat

H. D. L. CORBY

Current commercial methods of making peat-based legume inoculants, as described by Date (1965, 1969), Roughley (1970), and Vincent (1970), are inherently septic. Various methods of producing pure-culture inoculants have been developed (Vincent, 1970), all of which depend on sterilising the carrier in its container and then adding *Rhizobium* aseptically.

With septic methods the prime means of restricting the growth of contaminants is to swamp them by adding very large numbers of *Rhizobium*. This practice has been continued, seemingly unnecessarily, with the various aseptic methods to reduce the multiplication of chance contaminants.

Peat, the common base of commercial legume inoculants for the past half-century (Fred, Baldwin & McCoy, 1932), does not occur everywhere, and where it does occur is of unpredictable value as a base for inoculants (see Roughley, Chapter 11). Accordingly, attempts have been made, for example by Strijdom & Deschodt (see Chapter 13), to find local substitutes having the protective power ascribed to peat.

Commercial production of legume inoculants tends to be seasonal to match the seasonal sowing of the leguminous seed to be inoculated. This leads to inefficient use of labour and equipment and probably to hurried work – all to be avoided if possible.

This Chapter describes a method of making a legume inoculant that is (*a*) a pure culture of *Rhizobium*, (*b*) made with a local substitute for peat, (*c*) made with small rather than large additions of *Rhizobium*, and (*d*) more or less free of seasonal crises of production.

The method

The bacterial carrier

The earth resulting from rotting down maize cobs has been chosen, somewhat arbitrarily, as a local substitute for peat. Cobs are mixed with fertilisers as follows (Wood, 1937): maize cobs (dry matter), 1 tonne; ground limestone, 27 kg; single superphosphate, 9 kg; and ammonium nitrate, 12 kg. The mixture is heaped on a concrete floor, moistened well, covered with plastic sheeting, turned and re-moistened occasionally, and left to rot for about six months. The resulting earth is air-dried, hammer-milled and sifted.

A test-batch allowed to rot for 30 weeks yielded earth having 19 % of the dry matter weight of the original cobs, an organic content of 76 %, a pH of 7.2 and an absorptive capacity for water of about three times its own dry weight. One tonne of air-dried cobs yields enough earth for about 4000 hectare-units of inoculant.

Water and added nutrients

The aim is to produce an inoculant which, on completion of manufacture, contains about 40 % of dry matter and 60 % of water by weight, with the water containing the following *final* concentration (g/l) of added nutrients: sucrose, 7.0; concentrated yeast extract, 3.5; K_2HPO_4, 0.35; NaCl, 0.15; $MgSO_4 \cdot 7H_2O$, 0.075; and $FeCl_3 \cdot 6H_2O$, 0.0075.

Bagging and sterilisation

The inoculant is made, and sold, in a bag of high-density polythene film, this being the cheapest locally made film that will withstand steam-sterilisation without losing its permeability to respiratory gases or its impermeability to water and micro-organisms. Film varying in thickness from 30 to 130 μm has been used successfully, the thicker grades being preferred for their greater strength and stiffness. The polythene bags, each containing 100 g of moist cob-earth, are heat-sealed after tucking in a slender plug of non-absorbent cotton wool to allow for gas expansion during sterilisation (Fig. 14.1, Steps 1 and 2).

The sealed bag is steam-sterilised at 1.4×10^5 N m^{-2} pressure for 1 hour. The vent is sealed off as the bag is withdrawn from the autoclave (Fig. 14.1, Step 3). A loss of about 5 % of the water present in the bag occurs during sterilisation, but once the bag has been fully sealed further loss is negligible.

Inoculation, storage and incubation

A 5 ml culture of *Rhizobium* in yeast–mannitol broth is diluted with yeast–sugar broth (as already described) to produce sufficient inoculum to inoculate, at 5 ml per bag, the number of bags to be inoculated in one day. The inoculum is injected via one corner of the bag and the corner immediately sealed off (Fig. 14.1, Steps 4 and 5). A diluted inoculum containing as few as 10^3 live rhizobial cells per ml has been used successfully, but in practice the amount required for one day's work never calls for such a degree of dilution.

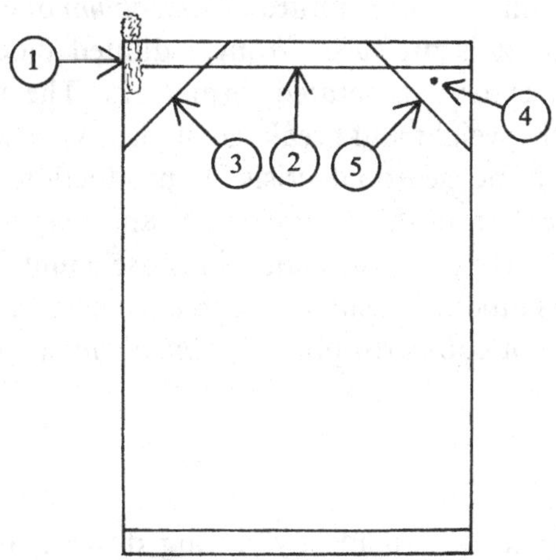

Fig. 14.1. Steps in sealing and inoculating the plastic bag. Step no. 1, tuck wisp of cotton wool in open corner to provide vent; 2, seal open end of bag; 3, seal off vent after sterilisation; 4, inject inoculum near corner of bag; 5, seal off injection-puncture after inoculation.

Inoculated bags are stored at 4 °C until required. They are then incubated at 28 °C, for one week in the case of the fast-growing strains of *Rhizobium* and for two weeks with slow-growing strains. Bags are then returned to cold storage until used.

Potency and longevity of the inoculant

Inoculated bags may be held in cold storage for at least a month before being incubated. During incubation, numbers of viable rhizobial cells normally rise to 10^9 or more per gram of inoculant dry matter, populations of this order having been achieved with strains of *Rhizobium* for *Arachis hypogaea*, *Glycine max*, *Lotus corniculatus*, *Medicago sativa*, *Phaseolus vulgaris*, and *Pisum sativum*. The method, however, seems to be better suited to the fast-growing strains of *Rhizobium* than to the slow-growing ones. Incubated bags may be kept in cold storage for at least two months without the rhizobial population falling below 10^9

viable cells per gram of inoculant dry matter, and Deschodt & Strijdom (Chapter 13) in testing the method, found that they may be held at 28 °C with similar result.

Discussion

The method has still to be proven commercially but has a number of merits. The inoculant is a pure culture of *Rhizobium* of consistently high potency, which, as it is produced from a diluted inoculum does not require large and expensive aerated fermenters. The use of a plastic container avoids the weight and breakage of glassware, and by deferring incubation there is no seasonal crisis of production. The labour of producing an inoculant in this way is great, and may well be too great for some countries. On the other hand, in those countries most in need of good inoculants labour is usually cheap and plentiful. We do not yet know whether or not cob-earth protects *Rhizobium* as peat is known to do.

Summary

A neutral organic earth is made by rotting down maize cobs. This is moistened with a nutrient solution, sealed in high-density polythene bags, and steam-sterilised. A diluted broth culture of *Rhizobium* is injected into the bags, which are held at 4 °C until required, and then incubated at 28 °C. This produces an inoculant containing at least 10^9 viable rhizobial cells per gram of inoculant dry matter. The method seems better suited to the fast-growing strains of *Rhizobium* than to slow-growing strains.

Thanks are gratefully due to Mr P. A. Donovan of the Rhodesian Department of Research and Specialist Services for commissioning this project, Dr R. A. Date of the Australian Commonwealth Scientific and Industrial Research Organisation for extensive advice, Mrs W. Petrie, Mrs J. B. Walker and Mrs A. A. Bullock for all the laboratory work, Mr T. C. D. Kennan and his colleagues of the Marandellas Research Station for experimental production of cob-earth and Mr J. F. Douse of Surrey Downs Farm for providing cob-earth.

References

Date, R. A. (1965). Legume inoculation and legume inoculant production. Report (No. 2012) to the Government of Uraguay, Appendix II. Food and Agriculture Organisation, Rome.

Date, R. A. (1969). A decade of legume inoculant quality control in Australia. *Journal of the Australian Institute of Agricultural Science*, **35,** 27–37.

Fred, E. B., Baldwin, I. L. & McCoy, E. (1932). *Root nodule bacteria and leguminous plants*. Studies in Science No. 5. University of Wisconsin Press, Madison, Wisconsin.

Roughley, R. J. (1970). The preparation and use of legume seed inoculants. *Plant and Soil*, **32**, 675–701.

Vincent, J. M. (1970). *Manual for the practical study of root-nodule bacteria*. IBP Handbook No. 15. Blackwell Scientific Publications, Oxford.

Wood, R. C. (1937). *Notebook of tropical agriculture*, 2nd edition. Imperial College of Tropical Agriculture, Trinidad.

15. Methods of inoculating seeds and their effect on survival of rhizobia

J. C. BURTON

The potential benefit to agriculture of inoculation of leguminous seed with rhizobia was realized early, but the practice developed slowly because of variable and disappointing results. Various forms of inocula were applied in many ways with sugars, glues, milk, lime, phosphate, manure and other substances, but only occasional success was realized. The merit of certain treatments was probably masked by improper matching of nodule bacteria and leguminous host or too few rhizobia for good nodulation.

Leguminous seeds differ widely in size, shape, nature of seed coat and other characteristics. The range in size of the seeds of legumes used for feed or forage production is illustrated by large hop clover (*Trifolium procumbens* L.) with more than 4000 seeds per gram compared with broadbean (*Vicia faba* L.) with 1 seed per gram. If one postulates that all leguminous seeds need a uniform number of rhizobia for good nodulation, a kilogram of hop clover would need 4000 times the number of rhizobia required for 1 kilogram of broadbean. However, this theorem has not been proven.

Small seeds must be planted on or close to the soil surface in order to obtain good stands, and here the rhizobia may be exposed to drought, high temperature, irradiation, high salt concentrations or other hazardous conditions. Considering this, small seeds should have more rhizobia than large seeds, but there are other considerations. Large seeds are usually planted deeper but nodulation may be delayed until the nitrogen in the seed has been used. Rhizobia on the seed must survive and multiply in a highly competitive environment for several weeks until their host becomes susceptible to nodulation. Large inocula give the best assurance of success under these conditions also.

Seed coats of leguminous seeds vary greatly in structure and composition and may contain substances toxic to rhizobia (Thompson, 1960; Bowen, 1961). Species and strains of rhizobia may also differ in their ability to survive on seed and in the soil (Brockwell & Phillips, 1965; Norris, 1971). Inoculation methods effective on one kind of seed are not necessarily effective on another.

This chapter discusses inoculation methods and their effect on extending the longevity of rhizobia on seed.

Experimental methods

Rhizobial survival may be determined by plate counting or by the 'most probable number' (MPN) method (Date & Vincent, 1962; Weaver & Frederick, 1972).

Inoculated seeds may be planted immediately after inoculation or following storage under prescribed conditions. A particular treatment may then be evaluated by counting, measuring, or weighing nodules, by rating plant growth visually or by determining yield. While these criteria are objective, the results are often confounded by variable, indefinable soil and climatic conditions.

The data reported here were obtained largely by the plating method. Seeds were initially screened to exclude those with micro-organisms which might interfere with plate counts. The medium used contained yeast water, mannitol and mineral salts; it was reinforced before pouring with 30 ppm of rose bengal and 30 ppm of actidione or oligomycin to suppress growth of molds and actinomycetes. Ten-gram samples of the inoculated seed were agitated briskly in 100 ml saline solution for 2 minutes in a blender, and 0.1 ml samples of appropriate dilutions spread uniformly over agar plates which were then incubated at 30 °C for 5 days for 'fast-growers' and 7 days for 'slow-growers'.

Some samples were also tested in field trials at the appropriate season. Seeds from the various inoculation treatments were planted in a rhizobia-free soil where measurement of nodulation, growth, and yield could be made. The most effective treatments in the laboratory usually proved best in the field.

Results and discussion

Survival of different forms of inocula

Broth cultures or suspensions of rhizobia washed from agar surfaces are often used to inoculate leguminous seeds despite the fact that this form of inoculum has frequently given poor results. Survival curves for *Rhizobium japonicum* applied to Chippewa soybeans in peat and broth culture are shown in Fig. 15.1.

The broth inoculum supplied 10^6 rhizobia per seed compared to 6×10^5 for the peat-base inoculum, but after storage for 1 week the seed treated with peat-base inoculum had twice as many viable rhizobia as those inoculated with the broth. After 4 weeks, the seed with the peat-base inoculum had 4×10^4 viable rhizobia per seed or about 80 times as many as those inoculated with broth culture.

Gum arabic, an exudate from *Acacia* species, has proven very beneficial in increasing longevity of rhizobia in combination with a peat-base inoculum (Brockwell, 1962*a*), but had no beneficial effect on *R. japonicum* applied to soybean seed in broth (Fig. 15.1). Mesquite gum, an

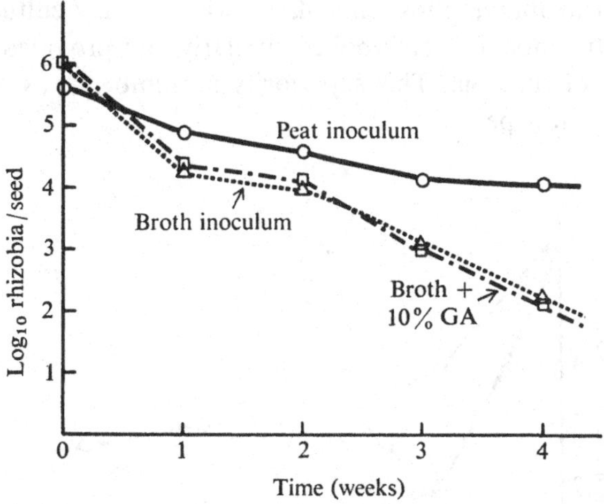

Fig. 15.1. Survival of *R. japonicum* on Chippewa soybean seed as influenced by form of inoculum and gum arabic (GA).

exudate from *Prosopis* sp., a leguminous plant native to North and South America, is equal to gum arabic in increasing longevity of rhizobia in combination with peat inoculum, but unfortunately it is not available commercially (Nitragin Co., unpublished results).

Poor results with broth inocula have been reported by many workers. Nodulation of subterranean clover was improved only when a peat inoculum was used (Radcliffe, McGuire & Dawson, 1967). Only 2 % of rhizobia in a broth inoculum survived on soybean seeds for 2 days but 30–70 % of the rhizobia survived 7 days when applied in peat (Dadarwal & Sen, 1971).

Shipton & Parker (1967) studied the effect of lime pelleting on survival of rhizobia applied to yellow lupin (*Lupinus luteus* L.) and serradella (*Ornithopus compressus* L.) as a water suspension washed from agar and as a peat inoculum. Lime coating greatly reduced nodulation where the rhizobia were applied as a suspension but had no adverse effect with the peat inoculum.

Lyophilized or freeze-dried cultures of rhizobia survive well when kept under vacuum in glass containers but die more rapidly on seeds than do rhizobia in broth culture (Vincent, 1965).

Superiority of peat-base over liquid and lyophilized inocula has been established but little attention has been given to age of the peat inoculum. Survival of *R. meliloti* on alfalfa seed is shown in Fig. 15.2. After 1 week, seed inoculated with a 28-day culture had 18 times as many viable rhizobia as those inoculated with a 7-day culture, despite the fact that the inocula were applied similarly and provided about the same number of rhizobia. This superiority in numbers was maintained throughout the test period.

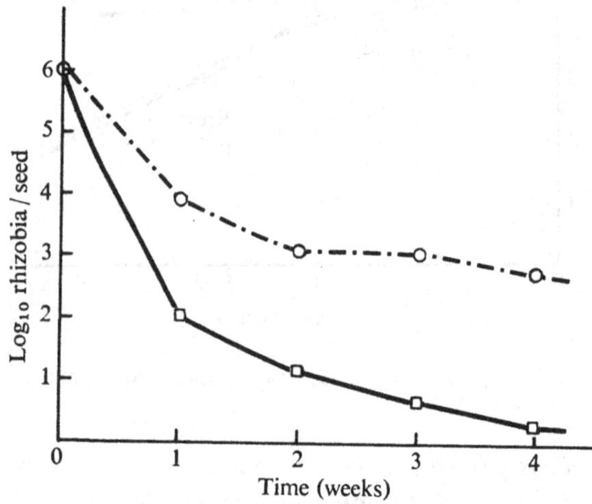

Fig. 15.2. Survival of *R. meliloti* on Vernal alfalfa seed as influenced by 'maturity' of the peat inoculum. □——□, 7-day-old inoculum; ○-.-○, 28-day-old inoculum.

The reasons for this are not understood but presumably rhizobia grown with the minimal supply of water provided by the peat become acclimatized and better conditioned to withstand drying on the seed. Peat-base inoculant kept at 20–7 °C reach optimum maturity in 4 to 5 weeks.

Influence of sugar on rhizobia longevity on seed

Early methods of inoculating sometimes entailed use of molasses or sugar for sticking the inoculant to the seed (Dobson & Lovvorn, 1949;

Smith, Blaser & Thornton, 1945). Further study has shown that they served other important functions (Vincent, 1958; Burton, 1964). Rhizobia die quickly on freshly inoculated seed, particularly with broth inocula, but sugars dissolved in the aqueous inoculum before application to the seed substantially reduce this loss. Disaccharides are more effective than monosaccharides. Sucrose and maltose are superior to mannitol and sorbitol, alcohol sugars used often in growing rhizobia (Nitragin Co., unpublished results). The effect of sucrose concentration on longevity of rhizobia on alfalfa seed is shown in Fig. 15.3. The amount of peat inoculum applied was uniform in all treatments.

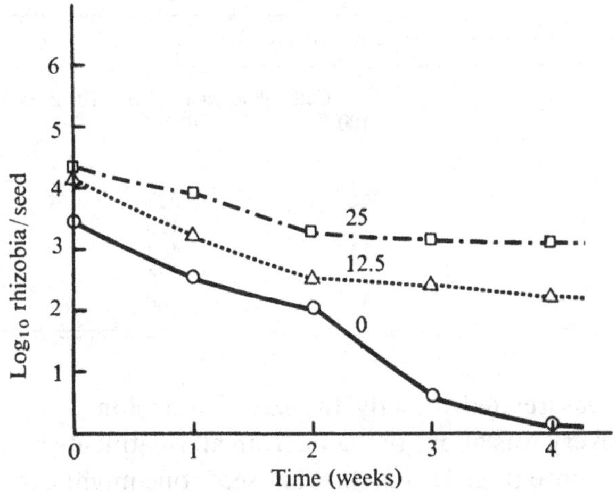

Fig. 15.3. Survival of *R. meliloti* on California common alfalfa seed as influenced by percentage sucrose in the peat inoculum slurry.

The protective action of sucrose improved with sugar concentration up to 25 % and differences between treatments increased with age.

Size of inoculum

It is generally conceded that large numbers of rhizobia on seed favor survival before planting and that large numbers of rhizobia on seed at planting favor rhizobia multiplication in the rhizosphere and early nodulation. On an acid silicious sand in southern Australia, White (1967) obtained larger proportions of nodulated plants, increased numbers of nodules and greater yields when rhizobia on the seed were increased from 4×10^3 to 4×10^6/seed. These increases in nodulation and yield were almost logarithmic on a field that had received 2.28 tonnes lime per hectare.

Quality of legume inoculants

In studies with *R. meliloti*, Burton (1964) reported that recovery of rhizobia from alfalfa seed was related directly to the amount of inoculum supplied. Differences between treatments increased with time. The fate of *R. japonicum* on Chippewa soybeans inoculated with three levels of rhizobia/seed as a peat slurry containing 10 % sucrose is shown in Table 15.1.

Table 15.1. *Survival of* R. japonicum *on Chippewa soybeans as influenced by inoculum level*

	Level of peat-base inoculum (g/kg seed)		
Time after inoculating	2.6	5.2	7.8
	Cells of *R. japonicum* × 10^3/g seed		
Applied	5100	10 100	15 300
Recovered after			
1 hour	366	527	989
1 week	98	126	164
2 weeks	73	103	123
3 weeks	37	46	47
4 weeks	33	46	39

Survival was related directly to size of inoculum for the first two weeks. Recovery was about the same from all treatments in the final two weeks. With more than 3×10^4 rhizobia/seed, one might expect excellent nodulation. However, Burton & Curley (1965) showed that a minimum of 2×10^5 rhizobia/seed was needed for good nodulation of soybeans planted under optimum moisture conditions in central Wisconsin. Alfalfa or clover seed planted under similar conditions is nodulated effectively by 5×10^2 rhizobia/seed. The number of rhizobia needed for effective nodulation varies with the species of legume as well as conditions. Working with tropical soils, Cloonan (1966) found that *Dolichos lablab* required higher levels of inoculum than did *Vigna sinensis*; approximately 1×10^6 rhizobia/seed were required to give 50 % or better crown nodulation.

Species of rhizobia

It has been suggested that soybean rhizobia (*R. japonicum*) die more rapidly on seeds than do other species of nodule bacteria (Sears & Hershberger, 1931; Iswaran, 1971). In studying this question, inocula of

R. meliloti and *R. japonicum* were applied to soybean and alfalfa seed. The peat-base inocula (2.2 g/kg seed) were applied as a water slurry. The *R. meliloti* culture provided 6×10^6 rhizobia and the *R. japonicum* culture 5×10^6 rhizobia/g seed. Survival was measured 1 hour, 7 days, and 14 days after inoculation (Table 15.2).

Table 15.2. *Survival of two* Rhizobium *species applied as a peat base inoculum slurry to alfalfa and soybean seeds*

Time after inoculating	Vernal alfalfa	Chippewa soybean
	Cells of *R. meliloti* $\times 10^3$/g seed	
Applied	6660	6660
Recovered after		
1 hour	103	1380
7 days	66	1323
14 days	11	580
	Cells of *R. japonicum* $\times 10^3$/g seed	
Applied	4814	4814
Recovered after		
1 hour	507	1243
7 days	36	118
14 days	10	115

The alfalfa and soybean rhizobia survived in greater numbers on soybean seed than on alfalfa seed. This is surprising since the alfalfa seed has a larger hilum scar and groove where rhizobia might be protected. The alfalfa rhizobia survived in greater numbers than soybean rhizobia on soybean seed.

There are indications that clover rhizobia die more rapidly on seeds than do alfalfa rhizobia and that greater numbers of clover rhizobia are needed to bring about effective nodulation. Brockwell & Phillips (1970) sowed seed of four leguminous species into hot, dry soil where they lay dormant for 7 to 9 weeks before germinating. Rhizobial survival was measured by examining young seedlings for nodulation. Ninety per cent of the alfalfa seedlings were nodulated compared with only 50 % of the red clover plants from the same inoculation treatment. The relatively poor showing of pre-inoculated clover seed as compared to alfalfa in the USA gives further support to the postulate that clover rhizobia are more delicate than alfalfa rhizobia (Carter, 1963).

Table 15.3 gives the results of an experiment in which *R. meliloti* and *R. trifolii* were applied separately to alfalfa, and subterranean and arrowleaf clover seeds, and survival determined over a 6-week period.

Table 15.3. *Survival of two species of rhizobia applied as a peat-base inoculum slurry to alfalfa* (Medicago sativa) *and to arrowleaf* (Trifolium vesiculosum) *and subterranean* (T. subterranean) *clovers*

Time after inoculating	Vernal alfalfa	Yuchi arrowleaf clover	Woogenellup subterranean clover
		Cells of *R. meliloti* × 10³/g seed	
Applied	23 000	23 000	23 000
Recovered after			
1 hour	21 800	14 200	11 800
7 days	1 360	1 940	1 350
14 days	480	750	550
21 days	310	600	400
28 days	230	320	290
42 days	66	89	90
		Cells of *R. trifolii* × 10³/g seed	
Applied	9 500	9 500	9 500
Recovered after			
1 hour	1 440	1 530	870
7 days	390	670	550
14 days	150	208	163
21 days	123	85	60
28 days	17	23	19
42 days	6	10	15

In all treatments, the peat-base inoculum (10 g/kg seed) was applied as a slurry with 10 % sucrose. The alfalfa inoculum supplied 23×10^6 and the clover inoculum 10×10^6 rhizobia per gram of seed. Survival of the *R. meliloti* was excellent on all samples. One hour after inoculating, 95 % of the applied rhizobia were recovered from alfalfa, 61 % from arrowleaf clover, and 51 % from the subterranean clover seed. The recoveries after one week were 9 % for alfalfa, 8 % for arrowleaf clover, and 6 % for subterranean clover. Numbers of rhizobia varied from 66×10^3 to 90×10^3 per gram 6 weeks after inoculation.

Death of clover rhizobia on the seeds was faster. Recovery of *R. trifolii* after 1 hour was 16 % for alfalfa, 9.6 % for subterranean clover, and 17 % for arrowleaf clover. After one week, these recoveries had decreased to 4 %, 6 % and 7 % respectively. After 6 weeks numbers of rhizobia varied from 6×10^3 for alfalfa to 15×10^3 for subterranean clover.

Effect of temperature

A high temperature is detrimental to rhizobia in the inoculum package, on the seed and in the soil, but the effects are most acute with inoculated seed. The intensity of the effect is reduced when peat inoculum is applied in a slurry with sucrose or gum arabic, but it is not eliminated (Burton, 1964; Vincent, 1965; Date, 1970). Soybeans inoculated with 10^6 rhizobia per seed as peat slurry containing 10 % sucrose had only 1.5×10^3 rhizobia per seed (1.5 % of those applied) after storage for 30 days at 32 °C. A peat inoculum stored under similar conditions maintained 50 % viability. Inoculated soybeans kept at 22 °C lost all but 5 % viable rhizobia in 30 days. A peat inoculum kept under identical conditions maintained 87 % viability (Nitragin Co., unpublished results).

The influence of high temperatures on packaged inoculants varies with species of rhizobia and with age. Peat inoculants containing *R. meliloti*, *R. leguminosarum*, and *R. japonicum* were matured for 2 weeks, packaged in polyethylene and stored at 27 °C for 4 weeks. Viable rhizobia were determined by plating at weekly intervals (Fig. 15.4).

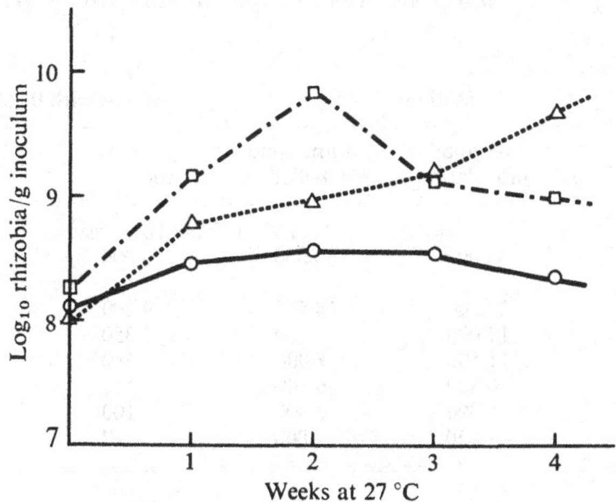

Fig. 15.4. Growth and survival of three species of rhizobia at 27 °C in a peat medium in moisture-proof polythene packages as determined by plating. □-.-□, *R. meliloti*; △ . . . △, *R. japonicum*; ○——○, *R. leguminosarum*.

The alfalfa rhizobia increased during the first 2 weeks and then decreased, the soybean rhizobia increased at each sampling period and the pea rhizobia remained fairly constant.

Pellet inoculation of leguminous seeds

The practice of simultaneously inoculating and coating leguminous seeds with pulverized limestone originated in Australia (Loneragan *et al.*, 1955; Cass-Smith & Goss, 1958). Clover seeds wetted with a broth inoculum were rolled in pulverized limestone until they were coated. When sown in acid soils or mixed with acid superphosphate, these seeds produced a greater percentage of effectively nodulated plants than did inoculated uncoated seed.

When pellet inoculation spread to areas without acid soils, various other coating materials were substituted for the limestone. Bentonite clay mixed with various organic supplements proved very successful on subterranean clover (Bergersen, Brockwell & Thompson, 1958). Thompson (1961) claims that the main objective in pelleting legume seed is to physically separate the rhizobia from the seed coat and the toxic substances it contains rather than protect against acidity. Brockwell (1962*b*) tested a variety of organic and inorganic materials on subterranean clover. A high percentage of effectively nodulated plants was

Table 15.4. *Effect of limestone pelleting on the survival of* R. trifolii *on seed*

Time after mixing	Without fertilizer		Mixed with 0–20–0 pH 3	
	Regular inoculation	Limestone pellet	Regular inoculation	Limestone pellet
	Cells of *R. trifolii* × 10^3/g seed			
Applied	50 000	50 000	50 000	50 000
Recovered after				
1 hour	22 500	18 500	9 000	13 000
2 hours	12 000	7 700	3 350	1 250
3 hours	11 500	7 000	680	1 180
4 hours	10 300	6 900	450	1 120
24 hours	4 000	2 500	100	980
6 days	1 650	800	2	450

obtained with all of the pelleting materials and also with the seed receiving only the inoculum in the gum arabic solution. Seed inoculated with peat suspended in water without any adhesive or coating material produced only about 40 % effectively nodulated plants.

Norris (1967, 1971) cautioned that limestone coatings on seeds of tropical legumes may be harmful and recommended powdered rock phosphate.

Information in the literature on seed pelleting is fragmentary; important factors such as inoculum size, physical form of coating materials and soil conditions are often not specified. The effect of coating materials on rhizobia will vary with their physical form, impurities and soil conditions. Fine limestone is beneficial when seed are to be planted in acid soils or mixed with acid fertilizers, but it has little, if any, effect on the life of rhizobia on the seed. On the other hand, precipitated calcium carbonate as a coating on inoculated seed can be detrimental. Calcium phosphate and calcium sulfate may have a similar effect (Nitragin Co., unpublished results).

The survival of *R. trifolii* on subterranean clover seed inoculated by a water slurry and by pelleting with gum arabic and limestone is shown in Table 15.4.

In pelleting, 1 kg of seed was inoculated by wetting with 100 ml of a 48 % gum arabic solution containing 16 g of peat inoculum and then coating with finely pulverized limestone. The control sample inoculated with the water slurry of peat and 10 % sucrose received 4 g peat/kg seed. The inoculated seed samples were then split; one half was not treated further and the other mixed with superphosphate. At 1-hour intervals, seed and fertilizer were separated by screening and viable rhizobia determined by plating. Survival of rhizobia on the uncoated clover seed was directly proportional to the inoculum added. After 4 hours, there were 9×10^4 rhizobia per pelleted seed as compared with 3×10^4 for the uncoated inoculated seed. When mixed with superphosphate for 4 hours, the uncoated inoculated seed had no viable rhizobia whereas the lime pelleted seed had 5×10^4 rhizobia/seed.

One advantage of pellet inoculation is that it permits application of a large inoculum without fear of overwetting the seed because the coating powder has a drying effect. Brockwell (1962*b*) and others (Date, 1970; Norris, 1971) give the advantages and disadvantages of limestone, rock phosphate and other pelleting powders. In the USA alfalfa and clovers are often planted in neutral or slightly acid soils where the pH is not critical for rhizobial survival. Larger inocula could be more beneficial to legume establishment than lime or phosphate coating particularly during the hot, dry Fall planting season.

Alfalfa and arrowleaf clover were pellet-inoculated using three coating powders: limestone, bonemeal and a special inoculum powder prepared with 76 μm peat enriched with 10 % calcium carbonate (0.2 μm particle size). In the first two treatments, 1 kg of seed was inoculated with a slurry of 20 g peat inoculum suspended in 100 ml of 48 % gum arabic solution. The seeds were then rolled in limestone (600 g) or bonemeal (290 g) until they were thoroughly coated and free-flowing. In the inoculum powder treatment, 1 kg of seed was inoculated with a slurry of 15 g of peat inoculum in 45 ml of 48 % gum arabic solution and the wet seeds were then dried by mixing with 30 g of moist inoculum powder. Pelleted seeds were kept in moisture-proof polythene bags at 24 °C. Plate counts were made weekly (Table 15.5).

Table 15.5. *Survival of rhizobia on pellet-inoculated seed as influenced by coating material*

Time after inoculating	Coating material		
	Limestone	Bonemeal	Powder inoculant
	Cells of *R. meliloti* × 10^3/g alfalfa seed		
Applied	16 000	16 000	46 000
Recovered after			
1 hour	2 240	3 180	14 700
7 days	1 150	1 960	8 500
14 days	536	1 660	5 200
21 days	420	1 120	4 000
28 days	310	770	1 600
42 days	177	580	1 000
	Cells of *R. trifolii* × 10^3/g arrowleaf clover seed		
Applied	12 600	12 600	35 900
Recovered after			
1 hour	7 700	10 300	33 000
7 days	1 510	5 200	22 000
14 days	740	2 600	12 200
21 days	570	1 700	8 400
28 days	500	1 000	7 400
42 days	250	560	3 800

Recovery of rhizobia from the alfalfa seed 1 hour after inoculating was 14 % for the limestone, 20 % for the bonemeal and 32 % for the inoculant powder. At 7 days recovery decreased to 7 %, 12 % and 18 % respectively. Only 1 to 3 % of the rhizobia remained viable for as long as 6 weeks, but seed coated with the inoculum powder had 10^6 viable rhizobia/gram (about 2000/seed) because of the very large inocula applied. In contrast, seed with the limestone coating retained only 350 rhizobia/seed.

Survival of *R. trifolii* on arrowleaf clover seed was superior to that obtained with alfalfa. Recovery 1 week after inoculating was 12 % for limestone, 41 % for bonemeal and 61 % with the powder inoculant treatment. Viable rhizobia decreased with time in all treatments but the death rate was much higher in the limestone and bonemeal treatments. When the experiment ended at 42 days, the inoculant powder treatment had 38×10^5 rhizobia/g; this was about seven times as many viable rhizobia as the bonemeal treatment and fifteen times as many as the limestone treatment. Other advantages of the inoculant powder are its light weight, adaptability for transport and lack of abrasiveness on drill equipment. Also, sufficient calcium carbonate can be incorporated in the powder inoculant to neutralize the micro-environment of seed planted in acid soils.

Summary

The evidence presented leaves some questions on legume inoculation unanswered, but certain guidelines are clear. Properly prepared peat-base inocula are superior to other forms. Sucrose or maltose in the inoculant slurry decreases the death rate of rhizobia on seeds. Natural and synthetic gums such as gum arabic and mesquite in the inoculant slurry bind the inoculum to the seed and extend longevity. Large inocula are not always essential but there is safety in numbers regardless of whether inoculated seeds are stored or planted immediately. An abundance of rhizobia is particularly important when seeds are planted under adverse conditions or in soils which harbor a large population of ineffective rhizobia.

Pelleting of leguminous seeds with peat-base inocula and pulverized limestone introduced by Australian workers as a safe method to inoculate seeds for mixing with superphosphate or planting in highly acid soils was a major contribution, but caution in use of limestone is indicated as certain limestones are toxic to rhizobia. Also, nodule bacteria from some of the tropical legumes produce alkaline by-products. Powdered rock phosphate or other coating powders are more advantageous. The results from pelleting seeds with inoculant powders are promising. They enable application of larger inocula and ensure greater survival of rhizobia on the seeds. They are also easily transported add little weight or abrasiveness to seeds and are thus easily drilled.

References

Bergersen, F. J., Brockwell, J. & Thompson, J. A. (1958). Clover seed pelleted with bentonite and organic material as an aid to inoculation with nodule bacteria. *Journal of the Australian Institute of Agricultural Science*, **24,** 158–60.

Bowen, G. D. (1961). The toxicity of legume seed diffusates toward rhizobia and other bacteria. *Plant and Soil*, **15,** 155–65.

Brockwell, J. (1962*a*). Incorporation of peat inoculant in seed pellets for inoculation of *Medicago tribuloides* Desr. sown in dry soil. *Australian Journal of Science*, **24,** 458.

Brockwell, J. (1962*b*). Studies on seed pelleting as an aid to legume seed inoculation. 1. Coating materials, adhesives, and methods of inoculation. *Australian Journal of Agricultural Research*, **13,** 638–49.

Brockwell, J. & Phillips, L. J. (1965). Survival at high temperatures of *Rhizobium meliloti* in peat inoculant on lucerne seed. *Australian Journal of Science*, **27**, 332–3.

Brockwell, J. & Phillips, L. J. (1970). Studies on seed pelleting as an aid to legume seed inoculation. 3. Survival of *Rhizobium* applied to seed sown into hot, dry soil. *Australian Journal of Experimental Agriculture and Animal Husbandry*, **10,** 739–44.

Burton, J. C. (1964). The *Rhizobium*–legume association. In: *Microbiology and soil fertility* (ed. C. M. Gilmour & O. N. Allen), pp. 107–34. Oregon State University Press, Corvallis, Oregon.

Burton, J. C. & Curley, R. L. (1965). Comparative efficiency of liquid and peat-base inoculants on field-grown soybeans (*Glycine max*). *Agronomy Journal*, **57**, 379–81.

Carter, A. S. (1963). Preinoculating legume seed. *Soybean News*, **14,** No. 2, 4 pp.

Cass-Smith, W. P. & Goss, Olga, M. (1958). A method of inoculating and lime-pelleting leguminous seeds. *Journal of Agriculture of Western Australia*, 3rd Series, **7,** 119–21.

Cloonan, M. J. (1966). The root-nodule bacteria as factors in the establishment of tropical legumes. *Journal of the Australian Institute of Agricultural Science*, **32,** 284.

Dadarwal, K. R. & Sen, A. N. (1971). Survival of *Rhizobium japonicum* on soybean seeds. *Indian Journal of Agricultural Science*, **41,** 564–8.

Date, R. A. (1970). Microbiological problems in the inoculation and nodulation of legumes. *Plant and Soil*, **32,** 703–25.

Date, R. A. & Vincent, J. M. (1962). Determination of the number of root-nodule bacteria in the presence of other organisms. *Australian Journal of Experimental Agriculture and Animal Husbandry*, **2,** 5–7.

Dobson, S. H. & Lovvorn, R. L. (1949). Inoculation of legumes. *North Carolina State University Extension Circular*, No. 309.

Iswaran, F. (1971). Survival of *Rhizobium japonicum* on soybean seeds. *Indian Journal of Agricultural Research*, **21,** 79–80.

Loneragan, J. F., Meyer, D., Faucett, R. G. & Anderson, A. J. (1955). Lime-pelleted clover seeds for nodulation in acid soils. *Journal of the Australian Institute of Agricultural Research*, **21,** 264–5.

Norris, D. O. (1967). The intelligent use of inoculants and lime pelleting of tropical legumes. *Tropical Grassland*, **1,** 107–21.

Norris, D. O. (1971). Seed pelleting to improve nodulation of tropical and subtropical legumes. 2. The variable response to lime and rock phosphate pelleting of eight legumes in the field. *Australian Journal of Experimental Agriculture and Animal Husbandry*, **11.** 282–9.

Radcliffe, J. C., McGuire, W. S. & Dawson, M. D. (1967). Survival of rhizobia on pelleted seeds of *Trifolium subterraneum* L. *Agronomy Journal*, **59,** 56–8.

Sears, O. H. & Hershberger, M. F. (1931). Nodule bacteria die fast after applied to soybeans. Annual Report of the Illinois Agricultural Experiment Station, pp. 45–6.

Shipton, W. A. & Parker, C. A. (1967). Nodulation of lime-pelleted lupines and serradella when inoculated with peat and agar cultures. *Australian Journal of Experimental Agriculture and Animal Husbandry*, **26,** 259–62.

Smith, F. B., Blaser, R. E. & Thornton, G. D. (1945). Legume inoculation. *University of Florida Bulletin*, No. 417.

Thompson, J. A. (1960). Inhibition of nodule bacteria by an antibiotic from legume seed coats. *Nature London*, **187,** 619–20.

Thompson, J. A. (1961). Studies on nodulation responses to pelleting of subterranean clover seed. *Australian Journal of Agricultural Research*, **12,** 578–92.

Vincent, J. M. (1958). Survival of the root-nodule bacteria. In: *Nutrition of the legumes* (ed. E. G. Hallsworth), pp. 108–23. Academic Press, New York & London.

Vincent, J. M. (1965). Environmental factors in the fixation of nitrogen by the legume. In *Soil nitrogen* (ed. W. F. Bartholomew & F. E. Clark), pp. 384–435. American Society of Agronomy Monograph No. 10. ASA, Madison, Wisconsin.

Weaver, R. W. & Frederick, L. R. (1972). A new technique for most-probable-number counts of rhizobia. *Plant and Soil*, **36,** 219–22.

White, J. G. H. (1967). Establishment of lucerne on acid soils. In: *The lucerne crop* (ed. R. H. M. Langer), pp. 105–14. A. H. & A. W. Reed, Wellington (Auckland) & Sydney (Australia)

16. Some studies on the necessity of legume inoculation in Serbia (Yugoslavia)

Z. D. VOJINOVIĆ

Occurrence and activity of naturalised strains of *Rhizobium* spp.

Investigations were carried out on the economically more important legumes, belonging to the genera *Medicago*, *Trifolium*, *Phaseolus*, *Vicia*, *Pisum* and *Lathyrus*, on the principal soil types of Serbia – chernozems (phaeozems*), smonitza soils (vertisols), brown forest soils (cambisols), alluvial soils (fluvisols) and pseudogley soils (planosols).

Distribution of root nodule bacteria and activity of symbiotic nitrogen fixation were estimated by observing nodulation of legumes in the field. Nodulation was graded on a scale from 0–4 (0, without nodules; 4, very effective nodules).

Table 16.1. *Spontaneous nodulation of some legumes on the main types of soil in Serbia**

	Legumes			
Soil type	*Medicago sativa*	*Trifolium pratense*	*Phaseolus vulgaris*	*Vicia* and *Pisum* spp.
Chernozems	1–4	3	1–2	3–4
Smonitza	2–4	1–3	1–3	3
Brown forest	1–4	1–4	0–3	3–4
Alluvial	2–4	2–4	0–3	3
Pseudogley	0–1	2–3	0–2	–

* Minima and maxima ratings, based on a scale from 0 to 4 (0, without nodules; 4, very effective nodules).

The results, summarised in Tables 16.1 and 16.2, are from some 300 records. No nodules were observed on lucerne grown on pseudogley, or on beans grown on brown forest soil, alluvial sandy soils or pseudogley. Very poor nodulation (rating 1) occurred on lucerne (*Medicago sativa*), red clover (*Trifolium pratense*) and beans (*Phaseolus vulgaris*)

* Names of soils given in parentheses are according to FAO (1970).

on several different soils. Plants from the group *Vicia–Pisum* were generally well-nodulated.

Table 16.2 presents the frequency of nodulation rated for each legume, regardless of soil type. Using type of nodulation as a criterion for assessing nitrogen-fixing activity, nitrogen fixation was limiting (ratings 0, 1 and 2) in about 80, 46, 43 and 4 % of samples for beans, lucerne, red clover and the *Vicia–Pisum* group respectively.

Table 16.2. *Frequency of nodulation ratings (as % total no. observations)*

	Nodulation ratings				
Legumes	4	3	2	1	0
M. sativa	11.1	42.9	22.2	20.6	3.2
T. pratense	8.1	48.6	29.8	8.0	5.5
P. vulgaris	0.0	19.7	32.0	38.7	10.6
Vicia spp. + *Pisum* spp.	37.5	58.3	4.2	0.0	0.0

Numbers of naturally occurring rhizobia in soils

The number of naturally occurring rhizobia in soils from Serbia was counted by the 'plant infection' method, with plants of small-seeded legume hosts grown on agar in tubes and of large-seeded ones in sand culture in pots.

Table 16.3 (summarised data from 138 determinations) shows that *R. meliloti* was not detected in pseudogleys and in some brown forest soils, nor was *R. trifolii* in some chernozems and brown forest soils, or *R. phaseoli* and *R. leguminosarum* in pseudogleys. Therefore, the greatest effect of inoculation of corresponding legumes should be expected in these soils. Numbers of *R. phaseoli* and *R. leguminosarum* were underestimated by the method used (Vincent, 1970). The most important finding is the lack of *R. meliloti* in pseudogleys. These soils are acid, poor in humus, available phosphorus and potassium, and are alternately water-logged and dry. Artificial inoculation of lucerne and addition of lime and fertilisers may enable lucerne to be grown on these soils; field tests support this conclusion (see below).

Low numbers of *R. meliloti* in brown forest soils are a consequence of the long absence of the host. Numbers of *R. meliloti* decreased (in the absence of the plant) more quickly in leached smonitza and brown forest soils, than in chernozems and alluvial soils. Particularly characteristic were the leached and calcareous chernozems in Stig (East Serbia) in which large numbers of effective *R. meliloti* were found. Lucerne, however, nodulated poorly under field conditions. Field tests demonstrated a need for molybdenum and lime on the leached chernozem, and for phosphorus, potassium and stable manure on calcareous chernozem. Poor nodulation of lucerne in this area may also be influenced by the arid climate.

Table 16.3. *Population density of* Rhizobium *spp. in the main soil types of Serbia (expressed as log. no. bacteria/g dry soil)*

		Rhizobium spp.			
Soil type		*meliloti*	*trifolii*	*phaseoli*	*legum-inosarum*
Chernozems	Mean	5.2	2.6	2.5	2.8
	Range	4.7–6.1	<1.0–5.7	2.3–3.4	1.8–3.4
Smonitza	Mean	4.4	4.2	3.4	4.2
	Range	2.1–6.7	2.1–5.7	2.8–4.1	3.8–4.8
Brown forest	Mean	3.5	3.4	3.5	3.9
	Range	<1.1–5.1	<1.0–5.1	1.8–4.8	3.4–4.4
Alluvial	Mean	4.8	3.7	3.3	3.1
	Range	4.1–5.7	2.1–4.7	2.8–3.8	2.8–3.4
Pseudogley	Mean	1.5	4.1	1.4	2.0
	Range	<1.0–3.1	2.7–5.7	<1.0–1.8	<1.0–3.4

R. trifolii has been found in small numbers only in some samples of chernozem and brown forest soil, probably caused by prolonged absence of the host on the arable fields from which the samples were taken. This is particularly so with chernozems in Vojvodina, where red clover is only rarely sown. Wild species of the genus *Trifolium* on waste lands nodulate well. A field test with red clover in the region of Srem (Zemun Polje) did not show any positive effect of inoculation. Uninoculated plants were also well-nodulated, illustrating the ability of naturalised *R. trifolii* to multiply quickly in the presence of the host.

Beans (*Phaseolus* spp.) are frequently poorly nodulated. *R. phaseoli* was not found in some pseudogley samples; causes of poor nodulation in other soils has not been determined. In tests carried out so far inoculation does not appear to be important, although in some cases improvement in nodulation and plant development was achieved.

R. leguminosarum is widespread in the soils investigated, but in some pseudogleys it may occur in small numbers. Nitrogen fixation by plants belonging to this inoculation group is normally effective.

Quality of legume inoculants

The effectiveness of strains of rhizobia was estimated, at first, from the appearance of nodules on the roots of legumes in the field. A better measure was obtained from the growth of plants used in determining the population density by the dilution method. Finally, the activity of pure cultures of isolated strains was measured by inoculating the test plants under aseptic conditions.

The strains of *R. meliloti* present in chernozem and alluvial soils were effective; in brown forest soils and smonitza there were also some effective strains, but their activity decreased with prolonged absence of lucerne. In pseudogleys, strains of *R. meliloti* were less active. The most active strains of *R. trifolii* were found in pseudogleys, smonitza, and alluvial soils, isolates from chernozem and brown forest soils being less effective. Strains of *R. phaseoli* isolated from alluvial soils and smonitza were more effective than those from chernozem and pseudogley; most of the other isolates were of low effectiveness. The strains of *R. leguminosarum* have not been examined individually yet, but the appearance of nodules on host plants suggests that active strains are widespread in the soils of Serbia.

Field response of legumes to inoculation and fertilisation

Effect of inoculation

From the above results, pseudogleys are distinguished by low numbers of root nodule bacteria, particularly *R. meliloti*. Because there is considerable interest in growing lucerne in pseudogleys, the effect of

Table 16.4. *Effect of inoculation on lucerne*

Soil and locality	Year	No. of cuts made	Increase of hay yield (%)	Surplus N in hay (kg/ha)
Pseudogley, Varna	1st	1	120	127.2 (1st–3rd years)
	2nd	1	40	
	3rd	2	15	
Pseudogley, Metriš	1st	1	379	13.8
	2nd	2	136	148.4
Brown forest soil, Mladenovac	1st	1	13	14.4
Chernozem, Trnjane	1st	1	7	
Chernozem, Bradarac	1st	1	22	
Chernozem, V. Crniće	1st	1	9	
Chernozem, V. Crniće	2nd	2	3	

inoculation under these conditions was investigated. The field tests were sown at Varna (near Sabac, West Serbia) and at Metriš (near Negotin, East Serbia). In addition to inoculation, the tests included fertilisation with phosphorus, potassium, stable manure and lime (CaO). The effect of inoculation of lucerne was investigated also on brown forest soil (at Mladenovac) and chernozem (in the region of Stig). Some inoculation tests were also made using beans and red clover.

Table 16.4 shows that lucerne benefits most from inoculation when grown on pseudogleys. In pseudogley from Metriš (pH in KCl, 4.5), *R. meliloti* was not detected by the 'plant dilution' method; that from Varna (pH 4.7) contained about 25 bacteria/g soil under lucerne in its second year. In other soils naturally occurring *R. meliloti* were considerably more numerous; here the effect of inoculation on the yield was small, but an increase in nitrogen content of the hay from inoculated lucerne was still observed. Consequently, on these soils too, inoculation may be justified.

The effect of inoculation of red clover and beans in our tests was less spectacular. On pseudogley at Varna, inoculation increased the yield of red clover by about 25 % in two years. In another test using the same soil, the increase in yield was only 7 %. In tests with beans, inoculation improved nodulation to a certain extent, which, in some cases, increased yield, but the effect was not great.

Effect of fertilisers

All fertilisers, applied either individually or in combinations on lucerne at Varna increased yield. The effect of inoculation was positive in all fertiliser combinations, but the greatest increase in yield from inoculation was on unfertilised plots. Fertilising stimulated nitrogen fixation by indigenous strains. Maximum hay yield was achieved with the combination phosphorus, potassium and calcium+stable manure+inoculation (243, 135 and 43 % increase of hay yield in the first, second and third years, respectively).

On pseudogley at Metriš the effect on inoculation of different quantities of mineral nitrogen (0, 40, 80 and 160 kg N/ha, together with 120 kg P/ha and 150 kg K/ha) was investigated. In this test, which lasted one year, hay yield increased with increasing quantities of nitrogen fertilisers, and inoculation showed a positive effect at all nitrogen levels. Uninoculated lucerne did not nodulate. The maximum increase of nitrogen in hay by inoculation occurred with the fertiliser combination containing 40 kg N/ha. The effect of inoculation on the yield of hay was estimated as equivalent to the effect of fertilising with a quantity of 50–60 kg N/ha.

A positive effect of molybdenum (3 kg sodium molybdate/ha before sowing), lime (1000 kg calcium oxide/ha) and a small quantity of nitrogen (30 kg/ha) occurred on leached chernozem in Stig (Veliko Crniće). The hay yield for three years amounted to (in kg/ha): 15640, 17710, 21790 and 19360 with P+K, N+P+K, N+P+K+Mo and P+K+CaO, respectively. The change in nodulation and the increase in the nitrogen content of hay with molybdenum indicated that it stimulated nitrogen fixation in this soil.

In the tests on calcareous chernozem in Stig (Bubušinac) there was a positive effect of P+K (increase in yield of about 36 % in the course of 2 years) and of stable manure (55 % in the first and 41 % in the second year). Nitrogen had less effect than on leached chernozem, and the mixture of micro-elements which contained molybdenum also, had little effect.

Summary

The occurrence, population density and activity of *Rhizobium meliloti*, *R. trifolii*, *R. phaseoli* and *R. leguminosarum* have been investigated in the main soil types of Serbia – chernozems, smonitza, brown forest, alluvial and pseudogley soils. Field tests were sown to determine the benefits of inoculating lucerne, red clover and beans.

Nodulation of legumes was unsatisfactory in many instances. In some cases it was due to the lack of specific rhizobia (e.g. for lucerne on pseudogleys), and in others to unfavourable environmental conditions (for lucerne on chernozem in Stig, to arid climate and lack of phosphorus and molybdenum in the soil). Usually, pseudogley soils have the lowest numbers of root nodule bacteria, except *R. trifolii.* In other soil types the low number of nodule bacteria was most frequently due to prolonged absence of the host plant.

Artificial inoculation had the greatest effect on lucerne yields on pseudogley soils; it had little effect on lucerne yields on chernozem and brown forest soil, but there was a definite increase in nitrogen content of the hay. There was also a positive but smaller effect of inoculation on red clover grown on pseudogley. Inoculation of beans, in some cases, improved nodulation and development of plants, but the effect was comparatively small.

The original papers on which this report is based are listed below.

References

FAO (1970). *Key to soil units for the soil map of the world.* Food and Agricultural Organisation, Rome.

Gutschy, Lj. (1931). Results of field tests with inoculated and uninoculated soybeans. *Journal of the Ministry of Agriculture, Beograd,* **35.**

Mickovski, M. & Micev, N. (1962). Distribution of nodule bacteria of soybeans, lucerne and clover in Macedonia. *Annual Review of Research Works, Faculty of Agronomy and Forestry, Skopje,* **15,** 231–40.

Mickovski, M. & Mickovska, V. (1966). Distribution of nodule bacteria of bean (*Phaseolus vulgaris*) in some areas of Macedonia. *Annual Review of Research Works, Faculty of Agronomy and Forestry, Skopje,* **17.**

Mihalić, V. & Modrić, A. (1954). Investigation on inoculation of soybeans. *Soil and Plant, Beograd,* **3,** 43–57.

Modrić, A. (1970). Production of legume inoculants in Croatia. *Agrochemie, Beograd,* **11/12,** 501–5.

Petrović, V. (1967). Effect of inoculation, P, K, and lime on lucerne (*Medicago sativa*) grown on parapodzol. *Agrochemie, Beograd,* **1/2,** 23–32.

Petrović, V. & Vojinović, Ž. (1963). Effect of inoculation of lucerne and red clover in field tests. *Soil and Plant, Beograd,* **12,** 295–300.

Petrović, V. & Vojinović, Ž. (1967*a*). Symbiotic N-fixation of bean (*Phaseolus vulgaris*). I. Nodulation by naturalized rhizobia and effect of inoculation. *Microbiology, Beograd,* **4,** 37–44.

Petrović, V. & Vojinović, Ž. (1967*b*). Symbiotic N-fixation of bean (*Phaseolus vulgaris*). II. Nodulation ability and response to inoculation of different bean varieties. *Microbiology, Beograd,* **4,** 229–34.

Petrović, V. & Vojinović, Ž. (1968). Contribution to the study of nodule bacteria distribution in different soil types of Serbia. *Review of Research Works of Institute of Soil Science, Beograd,* **1,** 127–45.

Radulović, V. (1960). Contribution to the study of nodule bacteria (*Rhizobium* spp.) on legumes in Bosnia and Hercegovina. *Review of Research Works, Faculty of Agronomy, Sarajevo,* **10/11,** 269–88.

Radulović, V. (1966). Investigation of some legume nodule bacteria. *Review of Research Works, Faculty of Agronomy, Sarajevo,* **16,** 105–52.

Radulović, V. (1969). Investigations of the nitrogen-fixing power of symbiotic root-nodule bacteria in high mountain pastures III. *Congress of Yugoslav Biologists,* Ljubjlana.

Sarić, Z. (1953*a*). Effect of inoculation on the yield of soybeans. *Soil and Plant, Beograd,* **2,** 157–68.

Sarić, Z. (1953*b*). Effectiveness of inoculation and some factors which influence it. *Journal for Scientific Agricultural Research, Beograd,* **6,** 107–19.

Vincent, J. M. (1970). *A Manual for the practical study of root-nodule bacteria.* IBP Handbook No. 15. Blackwell Scientific Publications, Oxford.

Vojinović, Ž. & Petrović, V. (1957). Virulence and effectiveness of indigenous strains of *R. meliloti* and *R. trifolii. Journal for Scientific Agricultural Research, Beograd,* **10,** 96–107.

Vojinović, Ž. & Petrović, V. (1961). Distribution of some important root-nodule bacteria in soils of Serbia. *Journal for Scientific Agricultural Research, Beograd,* **14,** 34–50.
Vojinović, Ž., Šević, N. & Sarić, Z. (1956). Effect of inoculation of soybeans grown on different soils. *Soil and Plant, Beograd,* **5,** 89–108.
Vojinović, Ž. & Popović, Ž. (1967). Some factors influencing growth of lucerne in region of Stig. *Journal for Scientific Agricultural Research, Beograd,* **20,** 76–85.

17. Agar and peat inoculation efficiency in Bulgaria

L. RAICHEVA

Since 1971 agar inoculant has been produced for soybeans in Bulgaria and used in agricultural practice. Its introduction was preceded by pot and field experiments done to establish its efficiency. Because peat inoculants are produced in a number of countries, and are considered to be cheaper and easier to apply than agar inoculants, we examined the efficiency of peat inocula under our conditions (Roughley & Vincent, 1967; Date, 1969, 1970; Manninger, Bakondi & Soos, 1969; Roughley, 1970; Vincent, 1970; Sundara Rao, 1971). To this end ten field trials were set up in 1971 and 1972 at the following five experimental stations: A, Trastenik, calcareous chernozem soil; B, Alvanovo, calcareous chernozem soil; C, Gramada, leached chernozem soil; D, Rakovski, podzolized chernozem soil; and E, Belogradchik, grey forest soil. The field trials were laid out in a Latin rectangle, the size of the plot being 30 m^2. We compared the efficiency of the following inoculants of *R. japonicum*: (1) an agar inoculant prepared from strain No. 646, a standard strain of Russian origin; (2) an agar inoculant prepared from strain D-2 isolated from a Bulgarian leached chernozem soil; (3) a peat inoculant prepared from strain No. 646; and (4) a peat inoculant containing the trace elements boron, molybdenum, manganese and cobalt (Manninger *et al.*, 1969). The peat inoculants were prepared from a peat near the village of Baykal and neutralised with calcium carbonate.

A mineral nitrogen treatment was also included in the study, and all treatments were given a basal dressing of phosphorus and potassium fertilisers; superphosphate at the rate of 80 kg/ha P_2O_5, and potassium sulphate, at the rate of 80 kg/ha of K_2O. For the mineral nitrogen treatments ammonium nitrate was applied at the rate of 80 kg/ha of nitrogen. The protein content of the grain was determined according to the method of Kjeldahl, and nodulation was recorded.

Table 17.1 shows the initial numbers of *Rhizobium japonicum* in the different inocula. The trace elements boron, molybdenum, manganese and cobalt caused about a tenfold reduction in the count.

Figs. 17.1 and 17.2 show the soybean yields for 1971 and 1972 respectively. The lowest yields were obtained without inoculation. All the inoculants increased yields, the increases differing with soil type and location. In 1971 No. 2 agar inoculant prepared from the Bulgarian strain D-2 was highly effective as was the No. 3 peat inoculant. The No. 1 agar inoculant was intermediate. The lowest efficiency was obtained with the No. 4 peat inoculant, which could be explained by its lower content of *R. japonicum*.

Table 17.1. *Initial numbers of* Rhizobium japonicum *(in 1 ml or 1 g) in the inoculants tested*

Year of application	Type of inoculant*			
	1	2	3	4
1971	235×10^6	378×10^6	590×10^6	70×10^6
1972	303×10^6	442×10^6	1187×10^6	97×10^6

* See text for details.

In 1972 all the inoculants tested produced large increases in yield, especially the No. 3 peat inoculant. A good efficiency was established for the No. 2 agar inoculant, No. 1 being slightly less effective. The No. 4 peat inoculant was also effective, particularly on the calcareous chernozem (at Trastenik) and grey forest soil (at Belogradchik). These soils evidently responded to the trace elements included in the inoculant.

Table 17.2. *Number of nodules per ten plants in 1971*

Treatment	Site				
	A	B	C	D	E
Without inoculant	–	142	5	–	–
Agar inoculant No. 1	212	1126	229	1751	151
Agar inoculant No. 2	135	1586	173	947	187
Peat inoculant No. 3	174	1116	308	909	202
Peat inoculant No. 4	9	1046	60	520	38
Fertiliser nitrogen	–	63	2	–	–

The very low yields in Belogradchik were due to damage caused by hailstorms. It is clear from the two years' data that no considerable differences were seen between the different inoculants tested. The efficiency of mineral nitrogen treatments for both years was less effective than inoculation.

Inoculation also improved the quality of the crop in terms of protein content. Tables 17.2 and 17.3 show that few or no nodules were found on the control plants and some of the plants given mineral nitrogen. Many nodules were formed on inoculated plants, with a tendency towards a maximum number in treatments with highest yields.

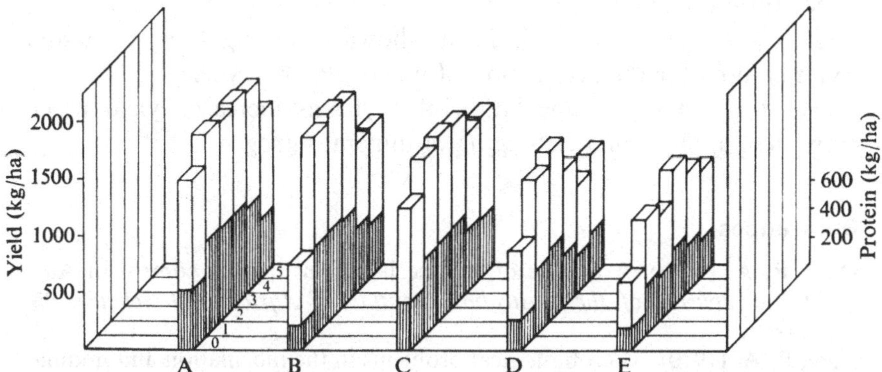

Fig. 17.1. Soybean yield at sites A–E (see text for details) using the following treatments: 0, no inoculant; 1–4, inoculant Nos. 1–4 (see text for details); 5, mineral nitrogen.

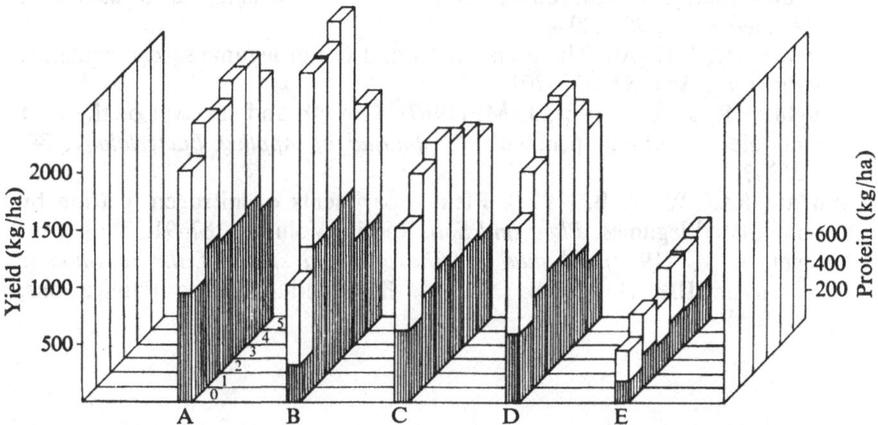

Fig. 17.2. Soybean yield in 1972. Experimental conditions as described for Fig. 17.1.

Table 17.3. *Number of nodules per ten plants in 1972*

	Site				
Treatment	A	B	C	D	E
Without inoculant	94	2	12	187	–
Agar inoculant No. 1	113	234	165	165	203
Agar inoculant No. 2	64	305	177	355	150
Peat inoculant No. 3	62	293	197	690	105
Peat inoculant No. 4	10	51	23	126	55
Fertiliser nitrogen	35	4	56	4	–

The following conclusions were drawn from the results:

(1) Peat and agar inoculants were highly efficient under the conditions of test and should, therefore, be produced and introduced in agricultural practice.

(2) The Bulgarian strain D-2 was shown to be highly effective and could be used for the production of inoculant for soybeans.

(3) At all sites the inoculation of soybeans increased yield much more than application of 80 kg/ha of mineral nitrogen.

References

Date, R. A. (1969). A decade of legume inoculant quality control in Australia. *Journal of the Australian Institute of Agricultural Science*, **35,** 27–37.

Date, R. A. (1970). Microbiological problems in the inoculation and nodulation of legumes. *Plant and Soil*, **32,** 703–25.

Manninger, E., Bakondi, E. & Soos, T. (1969). A Hazai *Rhizobium* Kutatások néhány gyakorkali eredménye-és problémája. *Agrartudomanyi Közlemenyek*, **28,** 229–45.

Roughley, R. J. (1970). The preparation and use of legume seed inoculants. *Plant and Soil*, **32,** 675–701.

Roughley, R. J. & Vincent, J. M. (1967). Growth and survival of the root nodule bacteria in peat culture. *Journal of Applied Bacteriology*, **30,** 362–76.

Sundara Rao, W. V. B. (1971). Field experiments on nitrogen fixation by nodulated legumes. *Plant and Soil*, Special Volume, 287–91.

Vincent, J. M. (1970). *Manual for the practical study of the root-nodule bacteria.* IBP Handbook No. 15. Blackwell Scientific Publications, Oxford.

18. Yield responses of soybean, chickpea, pea and lentil to inoculation with legume inoculants

J. N. DUBE

Croplands in the Jabalpur region may be classified on the basis of usage of legume inoculants into two ecosystems: (1) virgin land where rhizobia for introduced host(s) are either absent, or if present, are ineffective or poor in nitrogen-fixing capacity; and (2) cultivated land where there is an established population of rhizobia with varying levels of efficiency on different hosts.

Nodulation in virgin land is often clearly correlated with grain yield, which is a fairly good indicator of the effectiveness of a particular legume inoculant. Nodulation is a poor measure of the effectiveness of the inoculants in cultivated land, because endemic rhizobia are generally abundant and heterogeneous with respect to nitrogen-fixing capacity over a range of environmental conditions and competitive ability.

Trials in virgin land

The introduction of soybeans in India has formed an integral part of USAID programme since 1968. In collaboration with the Indian Council of Agricultural Research, the University of Illinois, Jawaharlal Nehru Agricultural University (JNAU) and Govind Ballabh Pant University of Agriculture and Technology (GBPUAT) began a series of experiments to test responses of exotic soybean varieties such as 'Bragg' to inoculation. The results convinced crop scientists and agricultural advisors that inoculation was essential for soybeans.

Because Indian soils have few *R. japonicum*, effective strains of rhizobia were first isolated from nodules on soybeans previously inoculated with the Nitragin Co. inoculant. Such acclimatised strains derived from the Nitragin inoculant are now used to produce soybean inoculants by manufacturers and national institutions. These inoculants are usually multistrain cultures, but the strains included differ between individual institutions. The national institutions, including agricultural universities and institutes, are primarily concerned with research and development of inoculants, and only produce sufficient to meet the demands of the Departments of Agriculture of the States.

During 1969–73 the American and Indian legume inoculants listed in Table 18.1 were tested. The number of viable rhizobia and contaminants

Table 18.1. *Legume inoculants tested*

Trade name	Carrier	Address of manufacturer
American		
Nitragin	Peat	The Nitragin Co. Inc., Milwaukee, Wisconsin 53209
Urbana	Humus	Urbana Laboratories, Urbana, Illinois 61801
Legume Aid	Peat	Agricultural Laboratories Inc., 1145 Chesapeake Avenue, Columbus, Ohio 43212
RP	Peat	The Rudy-Patric Co., Inoculant Laboratories, Princeton, Illinois 61356
Brazil	Peat	Geverno Do Estado. Co Rio Do sul Secretaria da Agricultural Departmento da Producao Vegetal
Noctin	-	
Dormal	-	-
Indian		
Jeevankhad	Lignite	The Biofertilizers Co., Amar Hill, Saki Vihar Rd., Bombay 400074
Bactogin	Lignite	The Bactogin Labs., near Gour River, Mandla Rd., Jabalpur 482002
IARI (Rhizogin)	Peat	Indian Agricultural Research Institute, New Delhi 110012
Kanpur	Peat	UP Institute of Agricultural Sciences, Kanpur–2
Calcutta	Lignite	Bose Institute, Calcutta–9
Hissar	Soil–compost	Haryana Agricultural University, Hissar–1
TDC	Lignite	Govind Ballabh Pant University of Agriculture & Technology, Pantnagar, Nainital
Krishinagar (Jawahar)	Lignite	Jawaharial Nehru Agricultural University, Krishinagar, Jabalpur–482004

were counted in all inoculants by plate counts during the shelf-life of up to 6–9 months. The field test was sown in a nine-row plot 6 m long with 0.45 m between rows. The outer two rows on either side were not inoculated. The central five rows were either not inoculated to act as controls or inoculated; three of these five rows were reserved for dry weight yield, and the other two to assess nodulation. The plots were surrounded by channels to drain rain water. However, heavy sporadic rains during some years caused contamination among the plots.

All inoculants had many viable rhizobia per gram of carrier. The American inoculants were free of fungal contamination but contained bacterial colonies atypical of rhizobia. The inoculant of one of the Indian manufacturers had recurrent fungal contamination during the first three years, but was free of fungal contamination subsequently. Rhizobia survived well in storage in all brands, but when subjected to field tests after a nine-month storage at room temperature the differences between brands were greater than at the beginning of the storage period. Noctin, Dormal and Urbana, and most of the Indian inoculants were poor; Nitragin, RP and Legume Aid were excellent.

The yield responses of soybeans inoculated with high-quality inoculants increased yield by up to 100 % when planted at the normal time; if planting was advanced by a few weeks yield was increased to about 125 %. Experiments under IBP Theme No. 2 using isolines of Lee soybeans proved that the best inoculants increased yield by more than the equivalent of 210 kg/ha fertiliser nitrogen. Tables 18.2 and 18.3 assess inoculant performance in 1972 and 1973.

Table 18.2. *Effect of commercial inoculants on yield of soybean (var. 'Bragg') in 1972*

Commercial inoculants	% increase over control*
Hissar, Kanpur	27
Jeevankhad, Calcutta, IARI, Noctin Bactogin, JNKVV (agar base)	53
Nitragin 70, Nitragin 72, RP, Nitragin 71, Dormal	78
Urbana, Krishnagar (Jawahar)	84

* Yield of control, 1651 kg/ha.

Table 18.3. *Effect of commercial inoculants on yield of soybean (var. 'Bragg') in 1973*

Commercial inoculants	% increase over control*
IARI 1973, Nitragin 72, RP	41
JNKVV, TDC, Bactogin	66
Brazil, Jeevankhad	78
Urbana, Nitragin 73	92

* Yield of control, 1439 kg/ha. Legume Aid, Hissar, Kanpur and IARI 1972 and Calcutta gave similar yields to the control.

Quality of legume inoculants

With a few exceptions, the effectiveness of Indian inoculants varied from year to year but was generally lower than those from the USA. Plants inoculated with low-quality inoculants also nodulated late. Strains of rhizobia which were good in strain trials of the All India Coordinated Research Project on Soybeans (AICRPS) were only poor or mediocre, when used in commercial inoculants.

Trials in cultivated land

In this ecosystem chickpea, pea and lentil are grown principally under dry farming conditions after the monsoon rains. Typically, nodulation, even by native rhizobia, was not observed until four to eight weeks after planting, although in greenhouse tests the selected strains nodulated after two weeks. Single strain inoculants were used and varied in effectiveness from 39 % for chickpea, 46 % for pea to 0–67 % for lentil. Effectiveness tended to be lower when seed was sown early, and characteristically in chickpea the variety × strain interaction was significant. The effectiveness of commercial inoculants was 14 % less than when the same strain was used in inoculants prepared in the laboratory. The potential of inoculants to increase yield of legumes for this ecosystem was demonstrated, but the selection of rhizobial strains for high level of effectiveness and prompt nodulation under wide range of environmental conditions warrants special attention. The consistent quality of some brands of American inoculants compared with the variability in the quality of Indian inoculants suggests a deficiency in large-scale production in India and there is a need to initiate a systematic programme of quality control. Satisfactory numbers of rhizobia in the inoculant do not guarantee optimum field performance; effectiveness tests remain the ultimate criterion of quality.

Competitive ability of *Rhizobium* strain

Multistrain inocula applied to soybeans growing in soils with few *R. japonicum* strains show differential success of the applied strains. Thirty nodule isolates obtained from a single plant inoculated with a high-quality inoculant were classified into serogroups for four consecutive years. Two serogroups nodulated only the primary roots, five only the secondary roots, and another five both primary and secondary roots. Only a few serogroups of *R. japonicum* out of twelve persisted after a lapse of three years. Repeated inoculation every season led to the predominance of four serogroups. Strains from acclimatised groups having the ability to persist are best for inclusion in commercial inoculants.

I thank Dr C. N. Hittle, Department of Agronomy, University of Illinois, Urbana, Illinois and Dr D. P. Motiramani, Director of Research Services, Jawaharlal Nehru Agricultural University, Jabalpur for technical and financial assistance, and collegues for help in the field trials.

PART III

Field experiments on nitrogen fixation by nodulated legumes

19. IBP field experiments on nitrogen fixation by nodulated legumes

P. S. NUTMAN

The factors determining the growth of a field-grown legume dependent wholly or in part on nitrogen fixation are too complex to be fully analysed in a single field experiment. The symbiosis is both highly integrated and to a considerable degree self-regulating, but yet sensitive to external influences. Nevertheless, something can be attempted towards a partial analysis because the more important factors fall into two categories: those intrinsic to host and bacteria and those of nutrition and of the environment, including climate. Climate can only be monitored, but the effects of weeds, disease etc. can be much reduced or eliminated and nutrition adjusted to the needs of the plant. The intrinsic factors can also be controlled to some extent by using known plant cultivars and *Rhizobium* strains, employing the latter at very heavy inoculation rates so as to favour strongly the introduced strains.

In the IBP field experiments proposed by Professor J. M. Vincent and myself (IBP 69 (66), amended 57/68), and approved for feasibility studies, we examined the host's nutrition by looking separately at requirements for nitrogen and for other nutrients. This was done by using nitrogen fertilisers and a general 'soil-amelioration' treatment to improve the chemical status and available nutrients. This varied according to local conditions, and often included liming and fertilisers containing phosphorus, potassium, sulphur and minor elements. Because we wished to estimate the amount of soil nitrogen available to the plant we used two further sets of treatments: (i) plots sown with lucerne heavily inoculated with an ineffective (non-fixing) strain of bacteria, and (ii) other plots sown with a non-legume (usually a grass or cereal). Ineffective strains were first used for this purpose by Virtanen & Holmberg (1958) in pot experiments with peas in which they showed that effectively nodulated plants obtained from 69 to 90 % of their nitrogen content from the atmosphere, depending on the amount of combined nitrogen available in the soil. In our experiments the amount of nitrogen fixed by the legume and retained in its herbage was estimated against the ineffectively inoculated and non-legume treatments in the light of their responses to nitrogenous fertiliser. The four nitrogenous fertilisers, two 'soil amelioration' and four inoculation treatments (including the non-legume as a second control) were combined factorially to give thirty-two treatments (Fig. 19.1).

To reduce cross-contamination, the plant type and inoculation treatments were set out in strips, randomised in four blocks, and the other treatments distributed at random within blocks, but using only two replicates of each treatment; certain of the less important interactions were confounded with block effects.

In addition to the field trials, pot experiments were done at some centres, using undisturbed cores of soil removed from the sites with precautions to avoid contamination, and given identical fertiliser or inoculation treatments.

	N_0	N_1	N_2	N_3	
O					U
					M
I					U
					M
E					U
					M
G					U
					M

Fig. 19.1. IBP factorial field experiment, summary of thirty-two treatments. O, uninoculated legume; I, legume inoculated with ineffective (non-nitrogen-fixing) strain; E, legume inoculated with effective strain; G, non-legume (grass or cereal); N_0, no nitrogenous fertiliser; N_1–N_3, nitrogenous fertiliser; U, no other fertiliser; M, Phosphorus, potassium, lime etc. as required.

Our own experiments for the period 1967–70, and the details of the methods used by Bell & Nutman (1971) were fully reported and it is not proposed to deal with these and collaborators' work in such detail. Instead the effects of the main factors and major interactions in all experiments will be summarised, without at first looking at their actual magnitudes. Estimates will then be made of the amounts of nitrogen fixed in these experiments and in ones of simpler design. Collaborators names are given in footnotes.

Results

Experiments with lucerne and grass or cereal

Table 19.1 summarises the first year's results in terms of the nitrogen contents of the cut herbage, of all the field experiments with lucerne and ryegrass in which the full range of thirty-two treatments was used. The + and – signs indicate larger or smaller yields than the control treatments with respect to the named factors; o indicates no effect. The following standards for significance are used: a single + (or –) where differences exceed the standard error (S.E.) and double signs where differences exceed twice the S.E. In Tables 19.1–19.9 the influence of the level of nitrogen fertiliser will not be considered; for example a double + indicates a significant response at any one of the three levels used.

Table 19.1. *Lucerne and ryegrass: main effects of treatments on nitrogen content of first year's herbage*

Place		Roth. A†	Roth. B	Wob. C	Wob. D	Stops.	Abery.
Soil type		Clay	Clay	Sand	Sand	Chalk	Loam
pH		6.0	6.0	5.5	6.0	7.5	5.0
Soil nitrogen (%)		0.18	0.16	0.07	0.06	0.64	—
R. meliloti/g		<1	<1	10^3–10^4	<1	10–10^2	<1
Inoculation	*Fertiliser*						
O	N	+	++	o	++	+	o
	M	++	+	++	o	++	++
I	None	– –	–	– –	o	o	–
	N	++	++	+	++	+	++
	M	+	+	+	++	+	++
E	None	++	++	o	++	+	++
	N	o	o	+	+	+	–
	M	++	++	+	+	+	++
G	N	++	++	++	++	+	++
	M	o	o	o	o	++	o

† For details of sites etc. see Tables 19.2, 19.3 and 19.4.
N, nitrogenous fertiliser; M, other fertilisers (lime, phosphorus, potassium etc.). For explanation of symbols see text.

Table 19.1 shows that the responses have an overall pattern. Thus the first two columns are clearly similar and differ strikingly from the next pair which differ somewhat from the next and markedly from the last. These differences corresponded to soil type and location – clay loam at Rothamsted, loamy sand at Woburn, a rendzina derived from Cretaceous limestone (chalk) at Stopsley, and silty clay loam at Aberystwyth.

Table 19.2 shows in more detail the first-year results of the Rothamsted experiments on slightly acid soils of rather low nitrogen content, and containing less than one cell of *Rhizobium meliloti* per gram of soil. The responses are qualitatively almost identical though done in different years. Inoculation with an effective strain markedly increased the nitrogen content, whereas inoculation with the ineffective strain strongly reduced it. The uninoculated and ineffectively inoculated lucerne and the grass responded strongly to nitrogen fertiliser but the effectively inoculated lucerne did not do so to any extent. Lime, phosphorus and potassium increased uptake by the lucerne, but only slightly the nitrogen uptake of ineffectively inoculated plants and still less that of the grass. Counts of *Rhizobium* done during the experiments (see Bell & Nutman, 1971) showed that soil amelioration promoted multiplication of the few lucerne bacteria present naturally, or introduced, and that this effect masked the nitrogen responses, indicated by the negative interaction. This interaction, however, was ill-defined in the ineffectively

Table 19.2. *Lucerne and ryegrass grown at Rothamsted: main effects and interactions on nitrogen content of first year's herbage*

Site	Rothamsted A	Rothamsted B
Year	1967	1968
Climate*	154 mm, 15.4 °C	219 mm, 14.9 °C
Previous cropping	Arable	Arable/pasture
Soil type, pH	Clay loam, 6.0	Clay loam, 6.0
Soil nitrogen (%)	0.18	0.16
Fertilisers (kg N/ha)	66–198 (+Ca, P, K)	99–297 (+Ca, P, K)
R. meliloti/g	<1	<1

Inoculation	*Fertiliser*	*Effects*	*Nitrogen yield* (kg/ha)	*Effects*	*Nitrogen yield* (kg/ha)
O	None	.	106	.	73
	N	+		++	
	M	++	127	+	126
	N×M	−		−	
I	None	−−	27	−	66
	N	++		++	
	M	+		+	
	N×M	O		O	
E	None	++	169	++	165
	N	O		O	
	M	++	260	++	237
	N×M	−−		−−	
G	None	.	37	.	73
	N	++		++	
	M	O		O	
	N×M	O		O	

* Rainfall and mean air temperature for June, July and August.

inoculated plots, due no doubt to the swamping effect of the non-fixing strain; it was absent for nitrogen uptake by grass.

Also shown are the nitrogen uptakes in kg/ha in some of the treatments not given nitrogen fertiliser, from which we can estimate the amount of nitrogen fixed and accumulated in the cut herbage.

At site A the ineffectively inoculated lucerne and the grass had uptakes that did not differ significantly from each other, so that either can be used to estimate the available soil nitrogen, whereas the uninoculated lucerne had clearly fixed some nitrogen, viz. 60–70 kg N/ha, and would be unsuitable for this purpose. Fixation of about 75 kg N/ha in the uninoculated plots was increased to about 100 kg N/ha by liming and addition of phosphorus and potassium and to about 230 kg N/ha by inoculating with an effective strain and adding lime, phosphorus and potassium.

Table 19.3. *Lucerne and ryegrass grown at Woburn: main effects and interactions on nitrogen content of first year's herbage*

Site, year		Woburn C, 1967		Woburn D, 1968	
Climate*		107.4 mm, 15.4 °C		261.4 mm, 14.9 °C	
Previous cropping		Arable		9-year fallow	
Soil type, pH		Sandy loam, 5.5		Sandy loam, 6.0	
Soil nitrogen (%)		0.07		0.06	
Fertilisers (kg N/ha)		66–198 (+Ca, P, K)		99–297 (+Ca, P, K)	
R. meliloti/g		10^3–10^4		<1	
Inoculation	*Fertiliser*	*Effects*	*Nitrogen yield* (kg/ha)	*Effects*	*Nitrogen yield* (kg/ha)
O	None	.	104	.	58
	N	O		++	
	M	++	133	O	60
	N×M	−−		O	
I	None	−−	76	O	56
	N	+		++	
	M	+		++	
	N×M	O		O	
E	None	O	103	++	112
	N	+		+	
	M	+	187	+	145
	N×M	O		O	
G	None	.	11	.	26
	N	++		++	
	M	O		O	
	N×M	O		O	

* Rainfall and mean air temperature for June, July and August.

At site B the grass and ineffectively inoculated lucerne again took up about the same amount of nitrogen, but more than at site A, and here the influence of the indigenous rhizobia was not evident so that either of these three values can serve as a datum for estimating the amount of nitrogen fixed. Soil amelioration without inoculation promoted nodulation and allowed about 50 kg N/ha to be fixed; with effective strain inoculation this was increased to about 160 kg N/ha.

Table 19.3 shows the results of experiments at two sites at Woburn on sandy soil but with different cropping histories, one being an arable rotation that included *Medicago lupulina* undersown in barley in 1964, and the other a clean fallow since 1959; these differences are reflected in the *Rhizobium* populations. The soil nitrogen was low at each site.

Treatment responses are clearly related to these differences. The uninoculated plots only responded to nitrogen fertiliser at site D where rhizobia were absent, and responded to lime, phosphorus+potassium most strongly where rhizobia were present. The introduced ineffective bacteria suppressed the naturally occurring effective strains at site C, but at D, they had no influence, because here there were no naturally occurring rhizobia.

The nitrogen uptakes of the grass and ineffectively inoculated lucerne were very different in this soil, especially at site C in 1967, which was a very dry year when the grass suffered much from drought. Using the ineffectively inoculated lucerne to assess available soil nitrogen, about 100 kg N/ha was fixed at both sites in the first year by the effectively nodulated lucerne given lime, phosphorus+potassium.

The remaining two lucerne–ryegrass experiments summarised on Table 19.4 are especially interesting because they represent almost the extreme conditions for lucerne cultivation in the UK – dry chalky downland in southeast England and a wet somewhat acid area near Aberystwyth in West Wales. The latter region is pre-eminent for grass–clover production, where grass responds well to heavy dressings of nitrogen fertiliser; this was the reason for the very large amounts used. The Aberystwyth site was also known to be very potassium deficient so that a basal dressing of 100 kg K/ha was applied in addition to the potassium given in the soil amelioration treatment.

At the chalkland site the lucerne and the grass responded to phosphorus and potassium fertiliser; further analysis showed this to be due to correction of potassium deficiency which was not suspected at this site. There were also small responses to nitrogen fertiliser and to effective strain inoculation, but not to ineffective strain inoculation. These effects are evidently related to the high nitrogen content of the soil and the presence of a small indigenous population of bacteria, and were the probable reason why only 40–50 kg N/ha was fixed by the lucerne given phosphorus and potassium, with or without inoculation by the effective strain. The grass was more affected than the lucerne by the dry conditions and had there been no ineffectively inoculated treatment for comparison, the nitrogen available to the lucerne from the soil would have been underestimated by nearly 40 kg.

Table 19.4. *Lucerne and ryegrass grown at Stopsley and Aberystwyth: main effects and interactions on nitrogen content of first year's herbage*

Place, year		Stopsley, 1969	Aberystwyth, 1969†
Climate*		203 mm, 13.7 °C	409 mm, 13.8 °C
Previous cropping		Pasture/arable	(+100 kg K/ha)
Soil type, pH		Chalky loam, 7.5	Silt clay loam, c. 5.0
Soil nitrogen (%)		0.64	—
Fertilisers (kg N/ha)		99–297 (+P, K)	168–673 (+Ca, P, K)
R. meliloti/g soil		10–10^2	<1

Inoculation	*Fertiliser*	*Effects*	*Nitrogen yield* (kg/ha)	*Effects*	*Nitrogen yield* (kg/ha)
O	None	.	125	.	105
	N	+		O	
	M	++	164	++	229
	N×M	−		−−	
I	None	O	124	−	70
	N	+		++	
	M	O		++	
	N×M	O		−	
E	None	+	144	++	248
	N	+		−	
	M	++	179	++	316
	N×M	O		−−	
G	None	.	85	.	113
	N	+		++	
	M	++		O	
	N×M	O		O	

* Rainfall and mean air temperature for May–August.
† Experiment conducted by W. Ellis Davies, Welsh Plant Breeding Station, Plas Gogerddan, Aberystwyth, Wales.

In the Aberystwyth experiment a large response to effective inoculation was obtained and a smaller one to ineffective inoculation. Grass responded more to nitrogen than lucerne. Lime and supplementary phosphorus+potassium were important for lucerne but not for the grass; the interactions between nitrogen and lime, phosphorus+potassium were negative for lucerne. Estimates of available soil nitrogen by uninoculated or ineffectively inoculated lucerne and by the grass did not much differ, indicating that about 200 kg/ha was fixed by inoculated lucerne given phosphorus+potassium; this is a large amount for first year production in an area which is thought generally not to be suitable for this legume.

Reviewing these six experiments it is clear that the grass and ineffectively inoculated lucerne give similar estimates of the quantity of soil nitrogen available for plant growth in those experiments where grass growth was not seriously limited by soil or climatic factors,

Table 19.5. (*a*) *Estimates of nitrogen fixed by lucerne in the second year* (*kg N/ha*)

Inoculation	*Fertiliser*	Roth. A	Wob. C (2 cuts)	Stops.	Abery.
O	U	220	190	90	249
	M	306	213	90	249
E	U	235	157	97	234
	M	342	239	116	246

(*b*) *Treatments showing maximum dry matter* (*t/ha*) *and maximum nitrogen yields* (*kg N/ha*) *irrespective of fertiliser applied*

	Treatment			
	EMN_2	EMN_0	EMN_3	UMN_3
t/ha	11.3	7.4	10.6	16.3
kg N/ha	434	258	332	430

especially water shortage. In the other experiments on chalky or sandy soil the grass uptakes were appreciably lower and in one case very much so. The ineffectively inoculated lucerne in these cases provided the best datum from which to estimate fixation in other treatments. This conclusion is supported, as indicated below, by the nitrogen response curves.

Towards the end of the first year and during the second year at each site, the ineffective strains were shown by differential counts to be overgrown or supplanted by effective strains, either those occurring naturally or migrating from the effectively inoculated plots. The ineffectively inoculated plots were thus no longer of any use and only ryegrass was used to estimate soil nitrogen. However, as an established cover it was less susceptible to unfavourable conditions than in the seedling year and therefore more satisfactory.

Using grass uptake, Table 19.5 gives estimates of nitrogen fixed at four of these sites in the *second year*. The established lucerne achieved much higher rates of fixation in the second year compared with the first, except at Stopsley where the soil nitrogen content was large.

Table 19.6. *Lucerne and ryegrass* (*at Putnok*) *or oats* (*at Keszthely and Bicserd*)*: fresh weight of cut herbage* (*t/ha*)*

		Putnok				Keszthely				Bicserd			
		1969		1970		1969		1970		1969		1970	
Inoculation	*Fertiliser*	A†	B‡	A	B	A	B	A	B	A	B	A	B
O	None	.	34	.	35	.	19	.	38	.	40	.	67
	N	++		O		++		O		+		+	
	M	++		+		++		+		++		O	
	N×M	+		O		O		O		O		O	
I	None	O		O		−−	15	O		O		+	
	N	+		O		++		O		O		O	
	M	++		O		++		+		O		O	
	N×M	+		O		O		O		O		O	
E	None	O		O		O		O		++	48	O	
	N	O		+		+		+		O		+	
	M	++	40	O	36	++	24	+	51	O		O	75
	N×M	O		O		O		O		O		O	
G	None	.	36	.	20	.	20	.		.	40	.	
	N	+		O		++		.		O		.	
	M	++		O		+		.		O		.	
	N×M	O		O		O		.		O		.	

* Experiments done by Eva Bakondi-Zámory, National Institute for Agricultural Quality Testing, Keletí Károly utca 24, Budapest II, Hungary.
†A, effects. ‡ B, yield.

Also shown are the maximum nitrogen yields irrespective of treatment; the nitrogen content of lucerne can sometimes be increased by nitrogenous fertiliser, though the efficiency of its use is very poor.

Table 19.6 summarises experiments at three sites in Hungary in terms of herbage fresh weight. Results at Putnok and Keszthely were similar in showing large first-year responses, in legume and grass, to nitrogenous and other fertilisers and small responses to effective strain inoculation, but only in ameliorated soil. At Keszthely yields were

Table 19.7. *Lucerne and grass or cereal: main effects and interactions on nitrogen content of herbage*

Treatments		IARI[a] New Delhi (Saline soil pH 8.9)	Ksar Gheris[b] Tunisia (Serozem)	Tamworth[c] NSW Site 1 1967	Site 2 1967	Site 2 1968	Sekirovo[d] Bulgaria (Podzol cinnamon soil)	Trustenik[d] Bulgaria (Calcareous chernozem)	Bozhourishte[d] Bulgaria 1967	1968	1969	G. Dubnik[d] Bulgaria (Leached chernozem)	Dijon[e] France (Sandy soil pH 4.7, %N 0.89) 1968	1969	1970	Shardlow[f] UK
O	N	○	○	○	○	+	+	+	○	○	○	○	+	.	.	+
	M	+	+	+	+	+	+	+	○	○	+	○	.	.	.	+
	N×M	○	○	○	○	○	○	○	○	○	○	○	.	.	.	○
E	None	○	○	○	○	○	+	○	○	○	○	○	++	.	.	○
	N	○	○	○	○	+	+	○	○	○	○	○	○	++	+	○
	M	+	+	+	+	○	−	○	−	+	+	○	.	.	.	○
	N×M	○	○	○	○	○	○	○	○	○	○	○	.	.	.	○
G	N	+	+	○	○	+	+	++	++	+	+	++	++	++	++	++
	M	○	+	+	+	○	+	○	○	○	+	○	.	.	.	.
	N×M	+	○	○	○	○	○	○	○	○	○	○	.	.	.	.

Data of: [a] W. V. B. Sundara-Rao, IARI, New Delhi 12, India; [b] E. Bouaziz, R. Combremont, J. Dejardin, M. Obaton & Y. Dommergues, CRUESI, Tunis, ORSTOM, Bondy, CNRS, Vandoeuvre-les-Nancy & INRA, Dijon, France; [c] J. A. Thompson, New South Wales Department of Agriculture, Tamworth, Australia; [d] L. Raicheva, N. Poushkarov Institute of Soil Science, Sofia, Bulgaria; [e] M. Obaton, G. Sommer and [f] J. Harrison, Shardlow Hall, Derby, England.

depressed by ineffective strain inoculation and on such plots response to nitrogen fertiliser was large.

Only at Bicserd were the yields increased in unamended soil by effective strain inoculation and the uninoculated lucerne responded to lime, phosphorus, potassium and molybdenum. Second-year yields were everywhere larger than first-year yields but showed fewer and smaller treatment differences. No nitrogen analyses were done from which estimates of fixation could have been made.

Table 19.8. *Estimates of nitrogen fixed by lucerne*

Place	Non-legume (G)	Nitrogen fertiliser (kg/ha)	M treatment	Year	Nitrogen fixed (kg N/ha) O, U*	E, M†
Ksar Gheris, Tunisia	Maize	0–99	P, K etc.	1st	14	21
New Delhi, India	Wheat	0–30	P, S	1st	37	25
Sekirovo, Bulgaria	Maize	0–160	Ca, P, K, S	1st	0	10
Trustenik, Bulgaria	Wheat	0–160	P, K, S	1st	106	110
Bozhourishte, Bulgaria	Wheat	0–102	P, S	1st	73	98
Bozhourishte, Bulgaria	Maize	0–102	P, S	1st	84	202
Shardlow, UK	Ryegrass	0–102	P, K	1st	198	225
Dijon, France	Cocksfoot	0–210	Ca, P, K etc.	1st	no U	65
Dijon, France	Cocksfoot	0–300	Ca, P. K etc.	2nd	no U	463
Dijon, France	Cocksfoot	0–300	Ca, P, K etc.	3rd	no U	388

* Control without inoculation or fertiliser.
† Effective inoculation and with soil amelioration treatment.

Fifteen further abridged IBP experiments, summarised on Table 19.7, on fixation by lucerne were reported by six other collaborators. These did not include plots inoculated with an ineffective strain and usually had fewer nitrogen level treatments; some also omitted the control plots without lime, phosphorus and potassium. Appreciable responses to effective strain inoculation occurred at five sites, on light soils and podzols. Lucerne benefited from nitrogen fertilisers and soil amelioration (M) at most sites but at none was there a significant N×M interaction. In nearly all experiments the non-legume (ryegrass, cocksfoot, wheat or maize) responded strongly to nitrogen, but only at four sites to other fertilisers. At one site only (a saline soil) was there a significant N×M interaction with the non-legume. The unfertilised grass or cereal gave yields equal to or smaller than the uninoculated and unfertilised lucerne except at two sites – a sandy soil at Dijon,

France, and a podzol at Sekirovo, Bulgaria; an effect which was not repeated at the Dijon site in the following two years, which were drier. Nitrogen analyses were done for ten of these experiments from which fixation was estimated; these are shown in Table 19.8.

In the first year the amounts of nitrogen fixed and retained in the herbage ranged up to 225 kg N/ha and except at New Delhi were increased by soil amelioration. Lowest rates were for sites in hot climates, or at Sekirovo where the non-legume yielded about twice as much as at other Bulgarian sites and may therefore have led to an underestimate of fixation. It would have been interesting to have made comparisons with ineffectively inoculated plots at this site.

At Dijon low first-year estimates were also associated with very high grass yields favoured by the abundant rainfall; the following years were drier. Here again a response of ineffectively inoculated lucerne to fertiliser nitrogen would have been of interest. Because control plots were omitted in this trial it is also not known how far lucerne growth may have been limited by soil or nutritional factors; the area was, however, considered to be quite unsuitable for lucerne.

Still simpler lucerne experiments were reported from France and Poland. Field trials at Lusignan from 1962–7 and at La Miniere from 1956–68* without any nitrogen fertiliser being applied, gave average nitrogen yields of 412 and 425 kg N/ha per annum respectively, which were about twice that of cocksfoot grown at the same sites and given nitrogen fertiliser at 211 kg N/ha. The Polish pot experiments with soils from forty-five different sites used three treatments only: control, effective strain inoculation and inoculation plus nitrogenous fertiliser.† Rhizobia were counted in the soils before inoculation and nodulation was later scored. Only one soil had populations of *R. meliloti* in the low range 200–20 000/g; this soil responded most to inoculation. Other soils had more than 20 000 *R. meliloti*/g and showed only small responses to inoculation and nitrogenous fertiliser. The patterns of nodulation were similar in the control and inoculated pots and were usually much reduced by nitrogenous fertiliser. Without knowing more about the actual distribution of the rhizobia in undisturbed soil (which is altered by collecting and placing in pots) or about nutrient deficiencies or other soil factors, it is not possible to draw wider conclusions from this work.

* M. Obaton and G. Sommer, Laboratoire de Microbiologie des Sols, 7 rue Sully, 21 Dijon, France.

† J. Golębiowska, WSR Wolyńska 35, Poznan, Poland.

Influence of amount of nitrogenous fertiliser

So far the effect of nitrogenous fertiliser has been considered without regard to the quantities used. Table 19.9 and 19.10 show first-year nitrogen response curves obtained in all experiments done with lucerne and ryegrass, but for two sets of contrasted treatments only.

Table 19.9. *Response to nitrogenous fertiliser of ryegrass and of lucerne inoculated with an ineffective strain of* Rhizobium

	Site	Rates of application of fertiliser (kg N/ha)			Uptake by crop (kg N/ha)			
		N_1	N_2	N_3	N_0	N_1	N_2	N_3
Ryegrass (G, U)	Rothamstead A	66	132	198	37	74	137	161
	Rothamsted B	66	132	198	73	169	227	343
	Woburn C	66	132	198	11	71	109	151
	Woburn D	99	198	297	26	110	172	286
	Stopsley	99	198	297	85	124	182	154
	Aberystwyth	168	336	673	113	133	140	263
Lucerne (I, U)	Rothamsted A	66	132	198	27	80	116	128
	Rothamsted B	66	132	198	66	115	198	251
	Woburn C	66	132	198	76	80	116	140
	Woburn D	99	198	297	56	87	139	147
	Stopsley	99	198	297	124	160	175	173
	Aberystwyth	168	336	673	69	81	120	226

Table 19.9 compares, over the range of 0–200 kg N/ha of fertiliser, the nitrogen contents of grass and of ineffectively inoculated lucerne, both grown in soil without lime, phosphorus and potassium. The large variation mostly depends on site and season. Within a trial, the responses of the grass and ineffectively inoculated lucerne to nitrogenous fertiliser run parallel and, except as already noted, the values at zero nitrogen lie close together. Where they do not, other circumstances, such as rainfall on the more responsive grass, may assist the experimenter in choosing the best treatment for estimating uptake of nitrogen from the soil. Most responses are linear, with, however, generally less than half of the applied nitrogen being utilised in top growth. Whether one uses

the grass or lucerne to measure soil nitrogen makes an appreciable difference in some experiments to the estimate of amount of nitrogen fixed, although the average difference of 16.5 kg N, was small in relation to the large nitrogen content of most effectively nodulated crops.

Table 19.10 shows the first year's response of lucerne to fertiliser either when uninoculated and grown without lime, phosphorus and potassium or when provided with an effective inoculum and given lime, phosphorus, potassium etc.

Table 19.10. *First year response of lucerne to nitrogenous fertiliser, either when uninoculated and without lime, P, K etc. or when inoculated with effective strain of* Rhizobium *and given lime, P, K etc.*

Treatment	Site	Rates of application of fertiliser (kg N/ha)			Uptake in lucerne (kg N/ha)			
		N_1	N_2	N_3	N_0	N_1	N_2	N_3
Uninoculated and without lime, P, K etc.								
(O, U)	Aberystwyth	168	336	673	105	115	119	191
	Dijon	.	.	.	.	.	.	.
	Shardlow	33	66	99	207	251	241	236
	Rothamsted B	66	132	198	73	134	203	245
	Rothamsted A	66	132	198	106	81	152	156
	Stopsley	99	198	297	125	140	177	196
	Woburn D	99	198	297	58	119	111	172
	Woburn C	66	132	198	104	124	89	119
	Trustenik	36	72	144	117	106	134	170
	Sekirovo	36	72	144	50	54	108	128
	Bozhourishte	33	66	99	95	100	111	101
	New Delhi	10	20	30	44	40	49	48
	Ksar Gheris	99	.	.	9	10	.	.
Effective inoculation and with lime, P, K etc.								
(E, M)	Aberystwyth	168	336	673	316	344	351	309
	Dijon	.	.	.	227	265	320	284
	Shardlow	33	66	99	228	251	284	268
	Rothamsted B	66	132	198	237	251	274	266
	Rothamsted A	66	132	198	260	225	252	231
	Stopsley	99	198	297	178	193	192	228
	Woburn D	99	198	297	145	207	177	191
	Woburn C	66	132	198	187	205	176	187
	Trustenik	36	72	144	119	107	108	132
	Sekirovo	36	72	144	51	51	114	140
	Bozhourishte	33	66	99	96	96	88	82
	New Delhi	10	20	30	58	63	65	69
	Ksar Gheris	99	.	.	10	10	.	.

It illustrates the well-known effect of inoculation in much reducing or eliminating the response to nitrogen, and the slight tendency for higher rates of application to depress yield of nodulated plants and for low rates sometimes to stimulate yield. It also records the judgement of different experimenters on what each thought a suitable maximum rate of nitrogen fertiliser to use; this varied from 30 to 673 kg/ha. The figure also shows the enormous range in nitrogen yield between experiments, from a few tens of kilograms at New Delhi and Ksar Gheris to more than 300 at Aberystwyth. It is of special interest that these last two relationships are connected; thus where low yields were obtained they were evidently anticipated and small amounts of nitrogen fertiliser were given and, vice versa, experimenters expecting large yields applied very large quantities of nitrogen fertiliser.

Lucerne, which originated in Persia, is considered to be a crop of the temperate and warm temperate regions, extending under irrigation into the subtropics, but these results indicate that it is poorly adapted to become established and yield well in the first year in warm and hot climates and performs best under cool-temperate conditions, with yields at other sites falling between these extremes. The sequence of yields from the lowest is as follows: Tunisia, India, S.E. Europe, W. Europe and Wales. This is broadly a climatic sequence from hot and dry to cool and moist conditions, suggesting that high soil temperature may be a crucial limiting factor. This important aspect of the environment will be considered in detail in Part IV of this volume. Established lucerne is, of course, known to yield well in warmer climates.

The variety, De Puits, used in these experiments, though widely grown, was bred in France and the results may partly reflect this origin; it would be of interest to investigate cultivar differences in this connection.

Soil-core experiments

Only one collaborator undertook parallel soil-core experiments. This used a fairly fertile UK soil.* No ryegrass treatments were included but effective and ineffective strain inoculations were used. Extrapolating to field scale, a basal fixation by uninoculated lucerne of only 34 kg N/ha was increased by effective strain inoculation to 58 kg N/ha and by inoculation and soil amelioration to 85 kg N/ha.

The results of the Rothamsted and Woburn field and soil-core

* J. W. Egdell, Westbury-on-Trym, Bristol.

experiments with lucerne and ryegrass were highly correlated (Bell & Nutman, 1971). Most of the experimental variation was accounted for by this correlation, with only two or three of the thirty-two pairs of treatment means in each experiment diverging significantly from a linear relationship.

Soil-core experiments can provide, therefore, in relative terms as good information as the field experiments.

Experiments with beans and wheat

Six IBP-type experiments, four without ineffective inoculation, were reported using field beans (*Vicia faba*) and wheat; these are summarised in Table 19.11. The Rothamsted and Woburn experiments were assessed on nitrogen contents of samples cut green in early July and early August. Beans and Spring wheat responded more to nitrogenous fertiliser at Rothamsted than at Woburn, especially at the second cut.

Table 19.11. *Field beans* (Vicia faba) *and wheat: nitrogen content and yield*

		Nitrogen content					Grain yield			
		Rothamsted	Stackyard	Woburn	Horsepool	Year	Lukavec[a] 1967	Lukavec[a] 1968	Barovo[a]	Cockle Park[b]
						Soil	Podzol	Podzol	Brown forest	
Treatments		1st cut	2nd cut	1st cut	2nd cut	pH	4.4	4.4	5.6	
O	N	+	+	O	+		+	+	+	O
	M	O	O	O	O		O	O	O	O
	N×M	.	.	.	.		O	O	O	O
E	None	O	O	O	O		++	+	O	O
	N	O	+	+	+		+	+	+	O
	M	O	+	O	O		O	O	O	O
	N×M	.	.	.	.		O	O	O	O
I	None	–	–	O	O		.	.	.	.
	N	+	+	+	O		.	.	.	.
	M	O	+	O	O		.	.	.	.
	N×M	.	.	.	.		.	.	.	.
G	N	++	++	++	++		+	.	O	O
	M	–	O	O	O		++	.	+	O
	N×M	.	.	.	.		O	.	++	O

Experiments of [a] J. Vondrys, Ceské Budejovice, Czechoslovakia, [b] A. W. Cooper, MAAF, Newcastle-upon-Tyne, England.

At both sites the cereals' responses were much larger than the legumes' responses. Wheat nitrogen yields were increased by nitrogen fertiliser by up to 200 % and 121 % at the two cuts in the Rothamsted experiment and the bean nitrogen yields by only 19 % and 16 %. At Woburn the corresponding increases were 120 % and 186 % for wheat and 14 % and 12 % for beans.

More than 10^5 *Rhizobium meliloti* were counted at each site and there was no significant response to inoculation with an effective strain, but at Rothamsted the heavy inoculation with an ineffective strain depressed the nitrogen content of the beans by 12 % and 27 % at the two cuts.

The remaining four experiments in which the results were assessed on grain yield showed that beans grown in the acid Czechoslovakian soils at Lukavec responded slightly to inoculation and to nitrogen but little or not at all to other fertilisers. Wheat yields on the other hand, and rather surprisingly, were more increased by lime and phosphorus+potassium than by nitrogen fertiliser. The late-sown bean experiment on a fairly fertile UK soil, probably containing ample rhizobia, showed no treatment differences.

Discussion

The limitations of the IBP experiments

These rather complex experiments combining factorially three distinct sets of treatments at different levels, have been useful under diverse conditions in identifying the major microbiological, nutritional and climatic factors limiting fixation. The strip element in the design was satisfactory in preventing cross-infection of plots in the first year, and with two replicates there was adequate precision.

The results confirmed the importance of strain effectiveness and plant nutrition, and for the lucerne experiments have underlined the overriding importance of climatic factors, especially temperature and water supply, in determining the quantities of nitrogen that can be fixed.

In the Woburn and Rothamsted experiments the effective and ineffective inocula became rapidly established in the lucerne rhizosphere, especially when the indigenous populations of rhizobia were low, and formed most of the nodules in the first year. The inoculant strains persisted throughout the first season but later tended to be replaced by other strains. This information, though laborious to collect, was useful in interpreting the nitrogen uptake data.

An alternative agronomic approach to identify the factors that may be limiting fixation is to use very much simpler experiments, such as those of Brockwell (1971) who employed three treatments only (control v. inoculation v. nitrate) and then to investigate other factors in equally simple supplementary experiments. But this may neglect crucial interactions, and if treatments to measure available soil nitrogen are omitted, the basis for comparing one experiment with another is reduced or lost altogether. A case can indeed be made for extending the number of comparisons in a factorial design, using fewer or fractional replication, on the grounds of greater efficiency, as originally recommended by Yates (1937) and recently discussed by Boyd (1973) for fertiliser and varietal trials etc.

Nevertheless, our results showed that some treatments were more useful than others. It was, for example, more important to be able to measure the nitrogen yields in plots given no nitrogen fertiliser (N_0) than to determine exact response to nitrogen (usually it was enough to demonstrate whether or not there was such a response). Also, in our experiments the ryegrass given phosphorus, potassium etc. (GM) and usually the data from the ineffectively inoculated lucerne similarly treated (I, M) were little used, and when rhizobia were absent from the soil ineffectively inoculated plots were redundant. By omitting some of these factors the number of treatments could be reduced to 24, 18 or 10, but less information would be provided and the orthogonality of the design would be destroyed making plot arrangements and analysis difficult and less efficient.

More 32-treatment field and soil-core experiments are needed to examine further the correlation between yields in these two kinds of experiments, the effects of a range of nitrogen fertiliser dressings and what other criteria might be used to supplement nitrogen analysis. But where this is not feasible, a soil-core experiment could be combined with a reduced but fully factorial field experiment, employing 16 treatments (viz. O, I, E, $G \times U$, $M \times N_0$, N_1) with minimum replication. The average per cent standard errors per plot in the Rothamsted, Woburn and Stopsley experiments were 3.5 for the comparisons between uninoculated and effectively inoculated treatments (O v. E etc.) and 4.9 for the N_0 v. N_1 etc. comparisons. Reducing the experiment to 16 treatments and 32 plots increases the per cent standard errors to 5.0 and 7.0 respectively.

Since these experiments were proposed in 1967, the acetylene reduction method has come to be employed routinely to compare nitrogen-fixing activities in the field, and if used in our experiments would undoubtedly have given substantially parallel results. Its great advantages lie in being able to compare rates of fixation at different times and in helping to distinguish fixation from soil uptake, but it is not simple to integrate such data over a season, as discussed by other contributors. On the other hand, the analysis of the nitrogen in the crop integrates but does not distinguish causes, which must be deduced from treatment comparisons.

The use of ^{15}N-labelled fertiliser also helps to distinguish soil uptake from fixation (Walker, Adams & Orchiston, 1950; Allos & Bartholomew, 1955, 1959; Loginov, 1966; Ivanko, 1968; Mouchová & Apltaner, 1969; Trepachev & Kharbarova, 1966), and Döbereiner (1966) showed that the regression of total plant nitrogen on nodule weight gives an estimate of the uptake of nitrogen from the soil (and seed). These methods equally involve extrapolation and their validity depends on the extent to which the roots of the control plants explore the soil and on their efficiency in absorbing combined nitrogen compared with the better-grown, effectively nodulated or nitrate-fed plants. This difficulty was first recognised by Norman (1943) when comparing uninoculated soybeans and sudan grass with and without nitrogen fertiliser. Our results suggest that climatic conditions affect the estimates more than choice of control plant treatment.

Other estimates of nitrogen fixation

In the following account most of the estimates of nitrogen fixed by a legume are reported simply as the nitrogen content of the crop without reference to any control treatment. Like the IBP experiments this is likely to give a low estimate of the total amount fixed because uptake from the soil is usually less than the fixed nitrogen retained in roots and nodules and lost to the soil. Other estimates take account of contributions from or loss to the soil, or are based on how much nitrogen fertiliser is needed to increase growth.

Lucerne. Reports of amounts fixed annually range from 463 kg N/ha at Saxmundham, England (Williams & Cooke, 1971), through 184–333 kg N/ha in pure stands or 261–304 kg N/ha in rotations in the USA (Lyon & Bizzel, 1933, 1934; Bear, 1942; Washko & Marriott, 1960; Wedin, 1965; Dawson, 1970; Lee & Smith, 1972) to 56 kg N/ha in central Africa (Souza, 1969). First year *Medicago lupulina* is estimated to fix 144–204 kg N/ha in the UK (Radulovic & Nutman, 1962) and *Melilotus alba* 183 kg N/ha in the USA (Lyon & Bissel, 1934; Ashford & Bolton, 1961).

These estimates fall in the range reported in IBP experiments and support our finding that lucerne fixes more nitrogen in temperate than in hot climates. The average value for fixation by *Medicago* and

Melilotus species is199 kg N/ha (from thirty-five experiments excluding IBP work).

Clovers. Similar average and range values are reported for fixation by the perennial clovers (white, red, alsike, berseem and ladino clover), usually grown in swards with grass; viz. average 183 kg N/ha (from forty-five experiments), range 45–673 kg N/ha (Lyon & Bizzel, 1933, 1934; Melville & Sears, 1953; Sears, 1953; Butler & Bathurst, 1956; McAuliffe *et al.*, 1958; Røyset, 1958; Green & Cowling, 1960; Washko & Marriott, 1960; Rizk, 1962; Castle & Reid, 1963; Wedin, 1965; Kutuzova, 1966; Bland, 1967; Cowling & Lockyer, 1967; Raininko, 1968; Dawson, 1970; Munro, 1970; Sau, 1970; Williams & Cooke, 1971). Fixation is greatest in swards in moist temperate climates or with irrigation; one of the lowest values (80 kg N/ha) was for subtropical conditions in Rhodesia (Clatworthy, 1970). As with lucerne, fixation is much less in the first than in later years.

The annual clovers, such as subterranean clover, fix less, viz. 45–336 kg N/ha, than perennial species because of their shorter growing period. This range includes some analysis of soil enrichment of 24–55 kg N/ha per annum (Donald, 1960; Watson, 1969; Clarke, 1970).

Trigonella foenum-graecum (including roots) is reported to fix from 110 to 573 kg N/ha (Yankovitch, 1940; Essafi, 1960; Rizk, 1966), *Lespedeza* sp. 95 kg N/ha (Dawson, 1970), *Lotus corniculatus* 116 kg N/ha (Allos & Bartholomew, 1959; Washko & Marriott, 1960) and vetch 73–97 kg N/ha (Lyon & Bissel, 1934; Dawson, 1970).

Tropical and subtropical forage and browse legumes etc. Table 19.12 summarises reports on amounts fixed by this group of plants. Except for the very large fixation for *Desmodium* grown in Hawaii, these estimates come within the same range as those of temperate forage legumes, but with a larger average value, probably because such plants enjoy a longer growing period. Unlike lucerne or clovers their fixation appears not to be lowered by high temperatures; however, it should be pointed out that the very large fixation by *Desmodium* in Hawaii was achieved in an equable tropical climate under high illumination, and that in general forage crops do not grow satisfactorily in the tropics in regions with a pronounced dry season when soil temperatures are high.

Soyabean and other pulses. Among grain legumes soyabean production is pre-eminent and many estimates have been made of its nitrogen-fixing activity, some by using non-nodulating lines to assess the amount of combined nitrogen taken up from the soil (see Ham *et al.*, Chapter 20). Estimates of fixation range from 1 to 168 kg N/ha, the average of seventeen estimates being 103 kg N/ha (Lyon & Bizzel, 1934; Rizk, 1966; Weber, 1966*a*, *b*; Dawson, 1970; Vest, 1971; Johnson & Hume, 1972; Lagacherie & Obaton, 1973). Most of these take no account of the nitrogen in the haulm or root. This average value of 103 kg N/ha may be compared with the average yields of soyabean in 1971 of 53 kg N/ha (range 34–91), for the world's five largest producers (FAO, 1972). Other legumes show similar disparities between experimental and farm yields; this underlines the large improvements that should be attainable.

Table 19.12. *Nitrogen fixed by tropical and subtropical forage and browse plants, green manure and shade trees: (kg N/ha/yr)*

Plant	Average	Range	References
Centrosema pubescens (centro)	259	126–395	1, 2, 3, 4
Desmodium intortum and *D. canum* (tick clover)	897	–	5
Leucaena glauca	277	74–584	6
Lotonosis bainesii	62	–	7
Sesbania cannabina	542	–	8
Stylosanthes sp. (stylo)	124	34–220	3, 9, 10
Mixtures of centro and stylo	115	–	11
Phaseolus atropurpurea (siratro)	291	–	12
Mikanea cordata	120	–	13
Pueraria phaseoloides (kudzu)	99	–	14
Enterolobium saman	150	–	15

1, Akhurst (1940); 2, Moore (1960); 3, Odu *et al.* (1971); 4, Watson (1957); 5, Younge *et al.* (1964); 6, Hutton & Bonner (1970); 7. Clatworthy (1960); 8, Singh (1971); 9, Crack (1972); 10, Wetselaar (1968); 11, Hornell & Newhouse (1965); 12, Henzell (1970); 13, Watson *et al.* (1964); 14, Dawson (1970); 15, Jagoe (1949).

Estimates for other temperate and tropical pulse legumes are listed in Table 19.13 and do not differ much in their average value (152 kg N/ha) or range from soyabean fixation. The large value for pigeon pea may reflect its perennial habit.

The outstanding feature of these estimates is their very large range over and above any variability that could reasonably be attributed to differences in method of measurement. Many of the factors mentioned in the introduction to this chapter that are known to influence fixation were examined in this large body of work, but rarely more than two at a time, so that there is often doubt whether conditions were adequate

Table 19.13. *Nitrogen fixed by pulses (kg N/ha)*

Plant	Average	Range	References
Vicia faba	210	45–552	1, 2, 3, 4, 5
Pisum sativum	65	52–77	3
Lupinus spp.	176	145–208	3, 4
Phaseolus aureus (green gram)	202*	63–342	10
Phaseolus aureus (mung)	61	–	6
Cajanus cajan (pigeon pea)	224	168–280	7
Vigna sinensis (cowpea)	198	73–354*	8, 9, 10
Canavalia ensiformis	49	–	8
Cicer arietinum (chickpea)	103	–	4
Lens culinaris (lentil)	101	88–114	4
Arachis hypogaea (groundnut)	124	72–124	4, 9, 11
Cyamopsis tetragonolobus (guar)	130	41–220	9, 12
Calopogonium mucunoides (calapo)	202*	370–450	10

* Extrapolated from pot experiments.
1, McEwen (1970); 2, Yankovitch (1940); 3, Lyon & Bizzel (1934); 4, Rizk (1966); 5, Dawson (1970); 6, Singh & Chonbey (1971); 7, Sen (1958); 8, Gargantini & Wutke (1961); 9, Wetselaar (1968); 10, Agboola & Fayemi (1972); 11, Seeger (1961); 12, Sanderson (1974).

for the full expression even of those factors that were being examined. Without this knowledge it is idle to attempt to assign their relative importance, and more work on the interaction of factors would therefore seem to be warranted. Without such valid comparative data, legume improvement may continue to fall far behind expectation.

Thanks are due to the many collaborators who freely submitted their work for inclusion in this report, much of which it is hoped will be published by them more fully elsewhere, and to Mr Dunwoody of the Statistics Department, Rothamsted for analysing the data.

References

Agboola, A. A. & Fayemi, A. A. A. (1972). Fixation and excretion of nitrogen by tropical legumes. *Agronomy Journal*, **64**, 409–12.

Akhurst, C. G. (1940). *Report of the Rubber Research Institute (Soils Division) for 1939*, pp. 61–82. Kuala Lumpur, Malaya.

Allos, H. F. & Bartholomew, W. V. (1955). Effect of available nitrogen on symbiotic fixation. *Proceedings of the Soil Science Society of America*, **19**, 182–4.

Allos, H. F. & Bartholomew, W. V. (1959). Replacement of symbiotic nitrogen by available nitrogen. *Soil Science*, **87**, 61–6.

Ashford, R. & Bolton, J. L. (1961). Effects of sulphur and nitrogen fertilisation and inoculation with *Rhizobium meliloti* on the growth of sweet clover (*Melilotus alba* Dear.). *Canadian Journal of Plant Science*, **41**, 81–90.

Bear, F. E. (1942). Making the most of our nitrogen resources. *Proceedings of the Soil Science Society of America*, **7**, 294–8.

Bell, F. & Nutman, P. S (1971) Experiments on nitrogen fixation by nodulated lucerne. *Plant and Soil*, Special Volume, 231–64.

Bland, B. F. (1967). The effect of cutting frequency and root segregation on the yield from perennial ryegrass–white clover associations. *Journal of Agricultural Science*, **69**, 391–7.

Boyd, D. A. (1973). Development in field experimentation with fertilisers. *Phosphorus in Agriculture*, No. 61, 7–17.

Brockwell, J. (1971). An appraisal of an IBP experiment on nitrogen fixation by nodulated legumes. *Plant and Soil*, Special Volume 265–72.

Butler, G. W. & Bathurst, N. O. (1956). The underground transfer of nitrogen from clover to associated grass. *Proceedings of the International Grasslands Congress, Palmerston North, New Zealand*, pp. 168–78.

Castle, M. E. & Reid, D. (1963). Nitrogen and herbage production. *Journal of the British Grassland Society*, **18**, 1–6.

Clarke, A. L. (1970). Nitrogen accretion by an impoverished red-brown earth soil under short-term leys. *Proceedings of the XI International Grasslands Congress, Surfers' Paradise, Queensland* (ed. M. J. T. Norman), pp. 461–5. University of Queensland Press.

Clatworthy, J. N. (1970). A comparison of legume and fertilizer nitrogen in Rhodesia. *Proceedings of the XI International Grasslands Congress, Surfers' Paradise, Queensland* (ed. M. J. T. Norman), pp. 408–11. University of Queensland Press.

Cowling, D. W. & Lockyer, D. R. (1967). A comparison of the reaction of different grass species to fertiliser nitrogen and to growth in association with white clover. II. Yield of nitrogen. *Journal of the British Grassland Society*, **22**, 53–61.

Crack, B. J. (1972). Changes in soil nitrogen following different establishment procedures for Townsville stylo on a solodic soil in north eastern Queensland. *Australian Journal of Experimental Agriculture and Animal Husbandry*, **12**, 274–80.

Dawson, R. C. (1970). Potential for increasing protein production by legume inoculation. *Plant and Soil*, **32**, 655–73.

Döbereiner, J. (1966). Evaluation of nitrogen fixation in legumes by regression of total plant nitrogen with nodule weight. *Nature, London*, **210**, 850–2.

Donald, C. M. (1960). The impact of cheap nitrogen. *Journal of the Australian Institute of Agricultural Science*, **26**, 319–38.

Essafi, A. (1960). Remarque sur l'importance et sur l'estimation de la fixation d'azote atmosphérique par les légumineuses sous le climat de la Tunisie. *Comptes Rendus de l'Academie Agricole France*, **46**, 611–13.

FAO (1972). *Production Yearbook, 1972.* Food and Agricultural Organisation, Rome.

Gargantini, H. & Wutke, A. C. P. (1961). Fixação do nitrogênio do ar pelas bactérias que vivem associadas às raízes do feijão de porco e do feijão baiano. *Bragantia* (1960) **19**, 639–652; *Field Crop Abstracts*, **14**, 191.

Green, J. O. & Cowling, D. W. (1960). The nitrogen nutrition of grassland. *Proceedings of the VIII International Grasslands Congress*, Reading, UK, pp. 126–9. Alden Press, Oxford.

Henzell, E. F. (1968). Sources of nitrogen for Queensland pastures. *Tropical Grasslands*, **2,** 19–30.

Horrell, C. R. & Newhouse, P. W. (1965). *Proceedings of the IX International Grasslands Congress*, Sao Paulo, vol. 2, pp. 1133–6. Edições Limitada.

Hutton, E. M. & Bonner, I. A. (1960). Dry matter and protein yields in farm strains of *Leucaena glauca* Benth. *Journal of the Australian Institute of Agricultural Science*, **26,** 276.

Ivanko, S. (1968). The effects of mineral nitrogen on the fixation of atmospheric nitrogen by *Vicia faba* L. In: *Isotope studies on the nitrogen chain* (ed. M. Krippner) pp. 169–74. International Atomic Energy Authority, Vienna.

Jagoe, R. B. (1949). Beneficial effects of some leguminous shade trees on grassland in Malaya. *Malaysian Agricultural Journal*, **32,** 77–90.

Johnson, H. S. & Hume, D. J. (1972). Effects of nitrogen sources and organic matter on nitrogen fixation and yield of soybeans. *Canadian Journal of Plant Science*, **52,** 991–6.

Kutuzova, A. A. (1966). Utilization of nitrogen from legumes on cultivated pastures in the central regions of the forest zone of the USSR. *Proceedings of the X International Grasslands Congress*, Helsinki (ed. A. G. G. Hill), pp. 191–4. Valtioneuvoston Kirjapaino.

Lagacherie, B. & Obaton, M. (1973). L'inoculation du soja résultats d'essais et orientation du travail. *Académie d'Agriculture de France*, *Séance*, Jan., 67–79.

Lee, C. & Smith, D. (1972). Influence of nitrogen fertiliser on stands, yield of herbage and protein and nitrogenous fractions of field-grown alfalfa. *Agronomy Journal*, **64,** 527–30.

Loginov, Yu. M. (1966). Investigation of fixation by lupin of atmospheric nitrogen, using molecular nitrogen marked with N^{15}. *Agrokhimiya*, **11,** 21–8.

Lyon, T. L. & Bizzel, J. A. (1933). Nitrogen accumulation in soil as influenced by the cropping system. *Journal of the American Society of Agronomy*, **25,** 266–72.

Lyon, T. L. & Bizzel, J. A. (1934). A comparison of several legumes with respect to nitrogen accretion. *Journal of the American Society of Agronomy*, **26,** 657.

McAuliffe, C., Chamblee, D. S., Uribe-Arango, H. & Woodhouse, W. W. (1958). Influence of inorganic nitrogen on nitrogen fixation by legumes as revealed by N^{15}. *Agronomy Journal*, **50,** 334–7.

McEwan, J. (1970). Fertiliser nitrogen and growth regulators for field beans (*Vicia faba* L.). I. Effect of seed bed applications of large dressings of fertiliser nitrogen and residual effects on following winter wheat. *Journal of Agricultural Science, Cambridge*, **74,** 61–6.

Melville, J. & Sears, P. O. (1953). Pasture growth and soil fertility. II. The influence of red and white clovers, superphosphate, lime, and dung and urine on the chemical composition of pasture. *New Zealand Journal of Science and Technology*, Supplement **1,** 30–41.

Moore, A. W. (1960). Symbiotic nitrogen fixation in a grazed tropical grass–legume pasture. *Nature, London*, **185,** 638.

Mouchová, H. & Apltauer, J. (1969). Effect of mineral nitrogen on the fixation of atmospheric nitrogen and on the yield of horse bean. *Védecké Práce Výzkumnýck Uštavá Rostlinne Výroby Praga-Ruzyné*, **15,** 257–64.

Munro, J. M. M. (1970). The role of white clover in hill areas. *Journal of the British Grassland Society, White Clover Research, Occasional Symposium*, **6,** 259–66.

Norman, A. G. (1943). The nitrogen nutrition of soybeans. I. Effect of inoculation and nitrogen fertiliser on the yield and composition of beans on Marshall silt loam. *Proceedings of the Soil Science Society of America*, **8,** 226–38.

Odu, C. T. I., Fayemi, A. A. & Ogunwale, J. A. (1971). Effect of pH on the growth, nodulation and nitrogen fixation of *Centrosema pubescens* and *Stylosanthes gracilis*. *Journal of the Science of Food and Agriculture*, **22,** 57–9.

Radulovic, V. & Nutman, P. S. (1962). Field experiments on legume inoculation. Report of the Rothamsted Experimental Station for 1962, part 1, pp. 79–80.

Raininko, K. (1968). The effects of nitrogen fertilization, irrigation and number of harvestings upon leys established with various seed mixtures. *Suomen Maataloustieteellinen Seuran Julkaisuja*, **113,** 137 pp.

Rizk, S. G. (1962). Atmospheric nitrogen fixed by legumes under Egyptian conditions. I. Egyptian clover (*Trifolium alexandrinum*). *Journal of Soil Science of the United Arab Republic*, **2,** 253–70.

Rizk, S. G. (1966). Atmospheric nitrogen fixation by legumes under Egyptian conditions. *Journal of Microbiology of the United Arab Republic*, **1,** 33–45.

Røyset, S. (1958). Trials with nodule bacteria for white clover. *Meddelelser det Norske Myrselsk*, **56,** 114–22.

Sanderson, K. W. (1974). Guar (*Cyamopsis tetragonaloba*) in the Rhodesian low veld. *Rhodesia Agricultural Journal*, **71,** 17–21.

Sau, A. (1970). Legumes and fertilisers and sources of grassland nitrogen in temperate regions of the USSR. *Proceedings of the XI International Grasslands Congress*, Surfers' Paradise, Queensland (ed. M. J. T. Norman), pp. 416–20. University of Queensland Press.

Sears, P. D. (1953). Pasture growth and soil fertility. I. The influence of red and white clovers, superphosphate, lime and sheep grazing on pasture yields and botanical composition. *New Zealand Journal of Science and Technology*, A**35** Supplement 1, 1–29.

Seeger, J. R. (1961). Effects of nitrogen fertilizing on the nodulation and yield of groundnut. *Bulletin de l'Institute Agronomique de Gembloux*, **29,** 197–218.

Sen, A. N. (1958). Nitrogen economy of soil under rahar (*Cajanus cajan*). *Indian Society of Soil Science*, **6,** 171–6.

Singh, P. & Choubey, S. D. (1971). Inoculation – a cheap source of nitrogen to legumes. *Indian Farming*, **20,** 33–4.

Singh, R. G. (1971). Effects of phosphate and molybdenum on growth, nodulation and seed yield of dhaincha (*Sesbania cannabina* (Retz.) Pers.). *Indian Journal of Agricultural Science*, **41,** 231–8.

Souza, D. I. A. de (1969). Legume nodulation and nitrogen fixation studies in Kenya. *East African Agricultural and Forestry Journal*, **34,** 299–305.

Trepachev, E. P. & Kharbarova, A. I. (1966). Determination of actual nitrogen fixation by legumes. *Vestnik Sel'skok-hozyaistvennoi Nauki, Moscow*, **12,** 105–8.

Vest, G. (1971). Nitrogen increases in a non-nodulating soybean genotype grown with nodulating genotypes. *Agronomy Journal*, **63,** 356–9.

Virtanen, A. I. & Holmberg, A. M. (1958). The quantitative determination of molecular nitrogen fixed by pea plants in pot cultures and field experiments. *Suomen Kemistilehti* Ser. B. **31,** 98–102.

Walker, T. W., Adams, A. F. R. & Orchiston, H. D. (1950). Fate of labeled nitrate and ammonium nitrogen when applied to grass and clover grown separately and together. *Soil Science*, **81,** 339–51.

Washko, J. B. & Marriott, L. F. (1960). Yield and nutritive value of grass herbage as influenced by nitrogen fertilisation in the north eastern United States. *Proceedings of the VIII International Grassland Congress* Reading, UK, pp. 137–41. Alden Press, Oxford.

Watson, E. R. (1969). The influence of subterranean clover pastures on soil fertility. 3. The effect of applied phosphorus and sulphur. *Australian Journal of Agricultural Research*, **20,** 447–56.

Watson, G. A. (1957). Nitrogen fixation by *Centrosema pubescens. Journal of the Rubber Research Institute of Malaya*, **15,** 2–18.

Watson, G. A., Wong, P. W. & Narayan, R. (1964). Effect of cover plants on soil nutrient status and on growth of *Herea.* III. A comparison of leguminous creepers with grasses and *Mikania cordata. Journal of the Rubber Research Institute of Malaya*, **18,** 80–95.

Weber, C. R. (1966*a*). Nodulating and non-nodulating soybean isolines. I. Agronomic and chemical attributes. *Agronomy Journal*, **58,** 43–6.

Weber, C. R. (1966*b*). Nodulating and non-nodulating soybean isolines. II. Response to applied nitrogen and modified soil conditions. *Agronomy Journal*, **58,** 46–9.

Wedin, W. F. (1965). Legume and inorganic nitrogen for pasture swards in subhumid, microthermal climates of the United States. *Proceedings of the IX International Grasslands Congress*, Sao Paulo, vol. 2, pp. 1163–9. Edições Limitada.

Wetselaar, R. (1968). Estimation of nitrogen fixation by four legumes in a dry monsoonal area of north-western Australia. *Australian Journal of Experimental Agriculture and Animal Husbandry*, **7,** 518–22.

Williams, C. H. (1970). Pasture N in Australia. *Journal of the Australian Institute of Agricultural Science*, **36,** 199–205.

Williams, R. J. B. & Cooke, G. N. (1971). Experiments on herbage crops at Saxmundham, 1967–1971. Report of Rothamsted Experimental Station for 1971, Part 2, 95–121.

Yankovitch, L. (1940). Essai de la détermination de la fixation de l'azote atmospherique par les légumineuses. *Annales du Service botaniques (et agronomique) de Tunisie*, **16,** 303.

Yates, F. (1937). The design and analysis of factorial experiments. Imperial Bureau of Soil Science, Technical Communication No. 35.

Younge, O. R., Plucknett, D. L. & Rotar, P. P. (1964). Culture and yield performance of *Desmodium intortum* and *D. canum* in Hawaii. Technical Bulletin of the Hawaii Agricultural Experiment Station No. 59, pp. 1–28.

20. Influence of inoculation, nitrogen fertilizers and photosynthetic source–sink manipulations on field-grown soybeans

G. E. HAM, R. J. LAWN & W. A. BRUN

A well-nodulated legume is capable of obtaining a portion of its nitrogen from the atmosphere by symbiotic fixation when nodulated with effective strains of rhizobia, and because of this ability, nitrogen is not generally considered a limiting factor with respect to soybean seed yields. However, the large amount of nitrogen required in the short growth period of 3–4 months places a high demand on soil nitrogen and on nitrogen fixation, especially. With the development of higher-yielding varieties, seed yield and/or protein content may be increased with nitrogen fertilizer. This chapter summarizes studies designed to examine the effect of inoculation and fertilizer nitrogen on nitrogen fixation of the soybean plant under different conditions of photosynthate source–sink activity.

Field inoculation studies

Inoculation of soybeans (*Glycine max* L. Merr.) with effective strains of *Rhizobium japonicum* may increase seed yields and seed protein percentage when rhizobia are not already present (Abel & Erdman, 1964; Caldwell & Vest, 1970).

Whether inoculation is beneficial when seed is planted in soils where effectively nodulated plants have been produced previously is sometimes in doubt. This was examined using soybeans grown on soils containing a naturalized population of moderately effective *R. japonicum*.

In 1967, eight paired samples of different lots of soybean seed were obtained before and after commercial pre-inoculation with *R. japonicum* and stored at 5–7 °C until planting. In addition, a commercial humus inoculant was applied to samples of seed immediately before planting. The uninoculated checks were adequately nodulated by *R. japonicum* present in the soil from previous soybean crops. No differences in plant height, nodule scores or maturity were found between paired samples, and seed yields were not increased significantly by any inoculant treatment (Table 20.1). Table 20.1 also shows that some samples contained few or no rhizobia per seed and would not have been expected to increase seed yield (Ham, Cardwell & Johnson, 1971*a*). The recommended minimum number of rhizobia per seed for successful inoculation is 300 (Vincent, 1970).

Table 20.1. *Effect of inoculation of soybeans with commercial cultures of* R. japonicum *in 1967 when grown at two sites in a soil containing a naturalized population of rhizobia.* (After Ham *et al.*, 1971*a*)

Type of inoculant	Date inoculated	No. of rhizobia/seed*	Variety	Seed yield (kg/ha)	
				Lamberton	Waseca
None	—	—	Chippewa 64	1646	2735
Humus	May 6	3 400	Chippewa 64	1566	2408
None	—	—	Chippewa 64	1875	2567
Humus	April 9	690	Chippewa 64	1788	2392
None	—	—	Chippewa 64	2070	2329
Liquid	April 22	5.8	Chippewa 64	1935	2507
None	—	—	Chippewa 64	1680	2614
Liquid	April 22	0	Chippewa 64	1626	2439
None	—	—	Chippewa 64	1956	2426
Liquid	May 8	0	Chippewa 64	2043	2560
None	—	—	A-100	1761	2500
Humus	April 5	580	A-100	2016	2433
None	—	—	Chippewa 64	1942	2473
Humus	April 26	3 100	Chippewa 64	2150	2399
None	—	—	Harosoy 63	1976	2513
Humus	April 25	1 600	Harosoy 63	1875	2231
None	—	—	Chippewa 64	1707	2513
Humus†		7 000	Chippewa 64	1720	2500
Humus‡		18 000	Chippewa 64	1781	2325
			Uninoculated average	1846	2520
			Inoculated average	1850	2419

* Sampled on May 15.
† Inoculated at the recommended rate.
‡ Inoculated with maximum amount of inoculum which would adhere to seed. These samples were inoculated within 24 hours of planting and date inoculated depends on date planted at each location.

In 1968 more extensive field studies were conducted at four locations in co-operation with various inoculant companies (Ham *et al.*, 1971*a*) which were sent seed from a common lot to be inoculated by any method. One uninoculated control and two treated samples were returned from each company. All samples were delivered within 24 hours of inoculation by air freight or in person and stored at 5–7 °C until planting. In addition, two seed samples were treated with humus inoculants from each firm within 24 hours of planting. Some plots were also sown with nodulating and non-nodulating isolines developed from Chippewa and inoculated with the three most effective strains from studies on rhizobia-free soils. The numbers of *R. japonicum* per gram of soil at planting were 1000, 800, 2400 and 6200 at Lamberton, Morris, Rosemount and Waseca, respectively. Neither seed yield nor protein was significantly affected by pre-inoculation or inoculation immediately before planting

Table 20.2 *Effect of inoculation of soybeans with* R. japonicum *when grown at four sites in a soil containing a naturalized population of rhizobia.* (After Ham *et al.*, 1971*a*)

Inoculant	Seed yield (kg/ha)				Protein (%)	
	Morris	Rose-mount	Waseca	Lamber-ton	Morris	Waseca
Control	1673	1868	2903	2634	39.7	41.3
Strain 110	1693	1781	2890	2654	39.3	41.1
Strain 138	1761	1835	2869	2540	40.3	41.3
Strain 126	1828	1808	2843	2554	39.7	41.7
Non-nodulating isoline*	1189	1250	2251	1969	33.2	36.5
Nodulating isoline*	1895	2124	3011	3486	39.1	40.2

* Developed from Chippewa variety soybeans.

(Table 20.2). Comparison of seed yields of the non-nodulating isolines indicated that 20 % to 41 % of the seed yield of the nodulating isoline was attributed to nodulation. Serotyping the nodules from inoculants containing single strains indicated a substantial range in recovery of the applied strain (Table 20.3). None of the three strains was recovered from nodules of uninoculated plants. Strain 110 was the most successful competitor with the naturalized population in the soil at Rosemount but was only a mediocre competitor at the other locations. Strain 138 was a better competitor at Morris than at Rosemount. At Waseca and Lamberton, strains 110 and 138 were about equally competitive. Strain 126 was a poor competitor at all locations. No attempt was made to serotype the commercial inoculants since they usually contain a mixture of several strains which may or may not be distinguishable from each other or from the rhizobia present in the soil. Ham, Frederick &

Anderson (1971*b*) reported that individual serogroups of *R. japonicum* were related to at least two and as many as six soil properties, which may explain the differences in competitiveness of the various strains in different soils.

Table 20.3. *The incidence of inoculated strains of* R. japonicum *recovered from soybean nodules on plants grown at four sites containing a naturalized population of rhizobia.* (After Ham *et al.*, 1977*a*)

	Percentage recovery of serologically identifiable strains*			
Strain	Morris	Rosemount	Waseca	Lamberton
110	4–8	8–16	0–2	0–8
126	0–4	0–8	—	—
138	16–20	0–8	4–16	8–16

* Range is maximum and minimum among the four replicates (24 nodules serotyped per replicate).

Our failure to increase seed yield and/or protein content by introducing known strains of *R. japonicum* applied at the recommended rate immediately before planting agrees with the results of Caldwell & Vest (1970); moreover, Johnson, Means & Weber (1965) reported that inoculation at up to 800 times the normal rate did not form enough of the nodules to significantly increase seed yield.

Testing of strains of *R. japonicum* needs to be broadened, in conjunction with improved inoculation procedures. At present, most strains are screened only for effectiveness (nitrogen-fixing ability) with a given host plant. Competitiveness should also be evaluated, since added strains must be able to compete for nodule sites with strains already present in the soil. Persistence of introduced rhizobia in competition with the natural soil flora over long periods of time is virtually unknown, and information on this aspect is essential if successful inoculation is to be extended.

Field nitrogen fertilizer studies

Assessing the relative contributions of symbiotic nitrogen fixation and soil and fertilizer nitrogen to soybean yields is difficult under field conditions because the almost universal presence of rhizobia in our agricultural soils causes uninoculated plants to become well-nodulated.

Weber (1966) used nodulating and non-nodulating isolines of soybean to examine this question, after showing that these lines gave the same yield when grown with adequate fertilizer nitrogen.

An experiment was undertaken to examine the yield and nitrogen responses of Chippewa 64 and a nodulating and non-nodulating pair of isolines of Chippewa background to 224 kg N/ha of different forms of fertilizer nitrogen. Two slow-release forms (sulphur-coated urea (S-urea) and urea–formaldehyde) and three readily available forms of nitrogen (ammonium nitrate, urea, and urea with added sulphur) were applied before planting and mixed in the top 20 cm of soil. Table 20.4 shows the results obtained at one of three locations. The non-nodulating isoline responded well to all sources of fertilizer nitrogen with yields equal to or exceeding the control yield of the nodulated isoline (G. E. Ham, unpublished data).

Table 20.4. *Effect of nitrogen fertilizer sources on soybean seed yield, protein content and nodulation at Waseca, 1969*

Genotype	Nitrogen fertiliser	Seed yield*	Seed protein*		Nodule fresh wt/plant*	Nodule no./plant*
		(kg/ha)	(%)	(kg/ha)	(g)	
Non-nodulating isoline	Control	2157 a†	31.2	672	–	–
	Ammonium nitrate	136 bc	125	171	–	–
	Urea	140 c	125	175	–	–
	S-urea	127 b	118	150	–	–
	Urea + S	135 bc	125	169	–	–
	Urea-formaldehyde	126 b	109	138	–	–
Nodulating isoline	Control	2688 a	37.8	1016	1.3	83
	Ammonium nitrate	113 c	104	118	46	69
	Urea	111 bc	105	116	54	72
	S-urea	103 ab	102	105	92	86
	Urea + S	104 ab	104	109	46	73
	Urea-formaldehyde	106 abc	101	106	85	88
Chippewa 64	Control	2292 a	38.8	889	1.7	88
	Ammonium nitrate	118 bcd	104	122	24	53
	Urea	120 cd	105	126	24	45
	S-urea	108 ab	103	111	76	91
	Urea + S	123 d	104	128	18	49
	Urea-formaldehyde	111 bc	101	113	76	88

* Actual values are reported for controls and treatment values are expressed as a percentage of the control.
† Values followed by the same letter(s) are not significantly different at the 0.05 level.

The nodulating isoline and Chippewa 64 showed small, but in some cases significant, seed yield and protein content responses to the fertilizer nitrogen. The three readily available nitrogen fertilizer forms severely decreased both nodule weight and number per plant. The slow-release forms of nitrogen had relatively little effect on nodule number and weight but also had the least effect on seed yield and protein content. Similar results were obtained at two other locations.

The effects of adding fertilizer nitrogen to the soil at the end of flowering, at the approximate time when nodule activity was rapidly declining (as evidenced by the acetylene reduction assay), was examined using Chippewa 64 and Clay soybeans. These were planted on June 2 in 50 cm rows and thinned to 8 cm apart in the row. At the end of flowering (August 11 and 17 for Clay and Chippewa 64, respectively), ammonium nitrate was applied at the rates of 0, 224 and 448 kg N/ha. The nitrogen was placed 5 cm below the soil surface along either side of the center row. The plots were then watered using sprinkler irrigation.

Table 20.5. *Effect of ammonium nitrate applied at the end of flowering on nitrogenase activity and nodule fresh weight for two soybean varieties. Samples taken three weeks after applying fertilizer.* (After Lawn & Brun, 1974*b*)

Variety	Ammonium nitrate (kg/ha)	Nodule fresh weight (g/plant)	Total activity (μmoles C_2H_4/ plant/h)	Specific activity (μmoles C_2H_4/g nodule/h)
Chippewa 64	0	2.142	65.7	30.2
	224	2.096	58.5	29.2
	448	1.625*	17.0**	9.6
Clay	0	2.681	45.4	16.9
	224	2.184*	18.4**	6.9*
	448	1.529**	12.3**	7.0**

*, ** denotes significant difference from the respective controls at $P \leqslant 0.05$, $P \leqslant 0.01$, respectively.

Nodules were sampled on August 2 and 11, and September 1 to determine fresh weight and acetylene-reducing activity. Samples were taken at sunrise following a sunny day from 24 cm of row (3 plants) in each plot. Table 20.5 shows that the addition of fertilizer nitrogen at the end of flowering reduced nitrogenase activity for both varieties even further than the usual decline during podfilling (as shown in Fig. 20.1). The addition of nitrogen also caused premature decay of nodules, and a consequent reduction in nodule fresh weight per plant.

The effect of nitrogen on various parameters of yield is shown for both varieties in Table 20.6. For Clay, nitrogen fertilizer significantly increased both seed yield and seed protein content. The combined effect was an approximate 12 % increase in seed protein yield for the highest level of added nitrogen. Unlike Chippewa 64, Clay showed differences in the protein content of the vegetative material.

In contrast to the response of Clay, the seed yield of Chippewa 64 was not increased significantly with added nitrogen. Seed protein content was increased significantly with the addition of 448 kg N/ha. The protein content and protein yield of the Chippewa 64 vegetative material was also increased significantly by adding nitrogen fertilizer.

The reasons for the differential varietal response are not known.

Table 20.6. *Effect of nitrogen fertilizer application at the end of flowering on seed yield etc. of two soybean varieties.* (After Lawn & Brun, 1974*b*)

Variety	Supplemental nitrogen (kg/ha)	Seed yield (kg/ha)	Seed protein content (%)	Seed protein yield (kg/ha)	Seed size (g/100)	Vegetative protein content (%)	Vegetative protein yield (kg/ha)
Chippewa 64	0	3452	43.52	1504	16.48	6.35	260
	224	3376	43.71	1476	16.48	7.60**	324*
	448	3540	44.25*	1568	17.08	7.86**	332*
Clay	0	3848	41.06	1580	18.60	5.30	172
	224	4128*	41.94*	1732*	1923	5.53	184
	448	4188*	42.15**	1786**	19.36	5.46	184

*, ** denotes significant difference from the respective controls at $P \leqslant 0.05$, $P \leqslant 0.01$, respectively.

However, the absence of a seed yield response in Chippewa 64 indicates a limitation to yield imposed by factors other than nitrogen supply during podfilling. In particular, potential sink size may have already been determined in this variety by events prior to podfilling, e.g. a possibly lower capacity of this variety to support nitrogen fixation. Chippewa 64 responded to fertilizer nitrogen at planting time by producing more seed/ha.

Photosynthetic source–sink relationships

In our experiments, as in others (Pate, 1958; LaRue & Kurz, 1973), the senescence of nodules which commenced during the early podfilling stage was associated with a marked decline in nitrogen fixation.

A possible explanation of loss in nitrogen-fixing activity by nodules during early podfilling would be competition from the developing pods and seed for limited photosynthates. The importance of photosynthetic products for symbiotic nitrogen fixation has long been recognized (Lindstrom, Newton & Wilson, 1952; Virtanen, Moisio & Burns, 1955). Bergensen (1970), Wheeler (1971) and Mague & Burris (1972) have established marked diurnal variation in nitrogen fixation in several species, indicating that the process is quite sensitive to supply of photosynthetic assimilates.

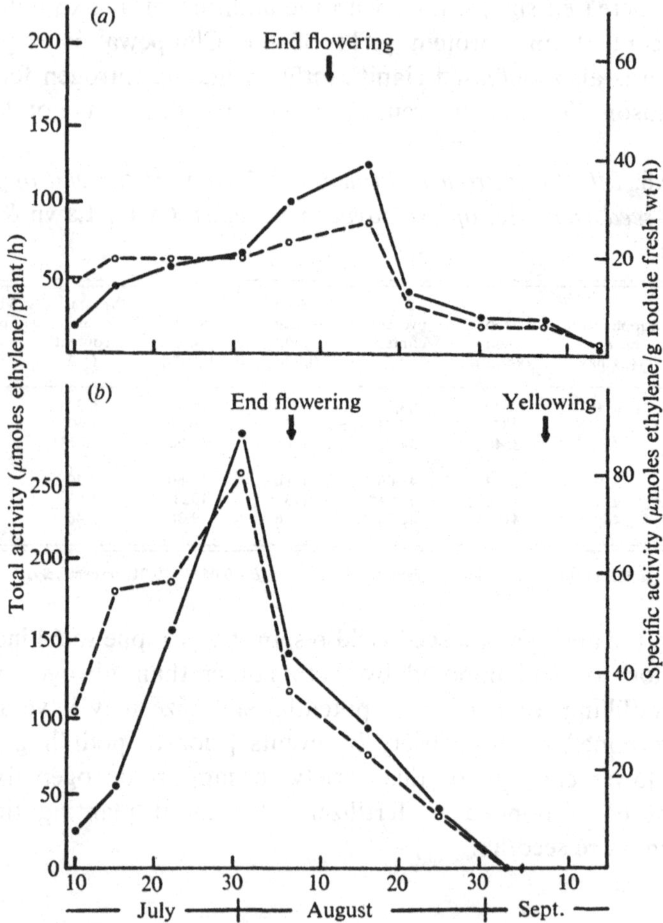

Fig. 20.1. Seasonal variation in acetylene reduction activity of nodules in two soybean varieties: (*a*), Chippewa 64; (*b*), Clay. Solid line represents total activity, dashed line specific activity.

The seasonal pattern of nitrogen fixation relative to physiological development of the plant was examined next using treatments designed to alter the supply of photosynthetic assimilates to the nodules during the podfilling stage.

Clay and Chippewa 64 soybeans were planted at St Paul on May 28, 1971. At the end of flowering (August 6 and 11 for Clay and Chippewa 64, respectively) the following treatments were applied: (1) supplementary light, (2) partial shade, (3) partial depodding, (4) partial defoliation, and (5) a control. Approximately 13×10^3 lux of supplementary light was provided to the lower leaves of the treated rows by placing four 75 watt cool-white fluorescent tubes under and to either side of the treated row. Lights came on 15 minutes after sunrise and went off 15 minutes prior to sunset. Partial shading was applied by suspending 50 % transmission Saran screening over the treated rows. Partial depodding involved the removal of all pods from each alternate node on the main stem and branches of each plant in the treated row, and partial defoliation the removal of the two basal leaflets of each trifoliolate leaf, leaving approximately 40 % of the leaf area. Irrigation was supplied as necessary to ensure good plant growth.

Samples were taken at 7–10 day intervals from July 10 (early flowering) to near maturity to determine plant dry matter accumulation, and number, fresh weight, and acetylene-reducing activity of nodules. Nodulated roots were dug immediately after the tops were cut. As far as possible, nodules were kept attached to the roots during sampling. Within 10 minutes of digging, the nodulated root systems were incubated at room temperature in 60 ml plastic syringes containing a helium–oxygen–acetylene mixture (72:18:10 by volume) and analyzed for ethylene after 30 minutes by gas chromotography following the general procedure of Hardy *et al.* (1968).

Total nitrogenase activity per plant (μ moles ethylene/plant/h) increased during the flowering period in each variety (Fig. 20.1). Peak activity in Clay was measured on July 31, shortly before the cessation of flowering, while peak activity in Chippewa 64 occurred August 16, shortly after flowering ceased. The increase in activity per plant during flowering was associated with an increase in the specific activity of the nodules (μmoles ethylene/g nodule fresh wt/h), and in nodule fresh weight per plant. The maximum nodule fresh weight and nodule number per plant were similar for the two varieties, but activities were higher for Clay.

Total activity per plant decreased substantially for both varieties during the post-flowering period, primarily due to a drop in specific activity which coincided with an increase in the numbers of green nodules. Nodule number and nodule fresh weight per plant declined during the early post-flowering period in Chippewa 64 but not in Clay.

In both varieties, the decay of senescent nodules per plant towards maturity resulted in a sharp drop in nodule fresh weight per plant.

Pod dry matter was approximately 10 and 30 % of the maximum for Clay and Chippewa 64, respectively at the time of the decline in nitrogenase activity, which thus occurred early in the podfilling stage in each variety, long before maximum pod dry matter was attained. The decline

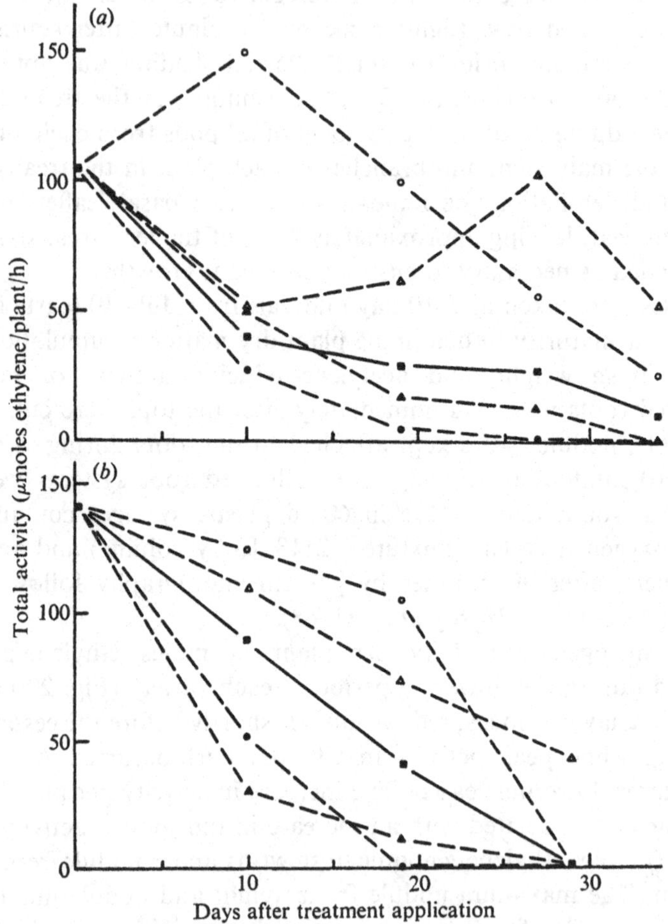

Fig. 20.2. Effect of altering the photosynthetic source–sink relationship on total acetylene reduction activity of nodules per plant for two soybean varieties: (*a*) Chippewa 64; (*b*), Clay. ■——■, control; ○---○, supplemental light; ●---●, partial shade; △---△, partial depodding; ▲---▲, partial defoliation.

occurred immediately prior to the stage at which the rate of pod growth was equal to the growth rate of the rest of the total tops in both varieties. These data are interpreted as evidence that the reduction of symbiotic activity of the nodules of these two varieties was associated with the development of the pods as a strong assimilate sink.

The response in nitrogenase activity to the treatments is shown in Fig. 20.2. The treatments designed to enhance the photosynthetic source–sink ratio for the nodules (supplementary light and depodding) stimulated activity to above that of the controls in both varieties. In contrast, reducing the source–sink ratio for the nodules by shading and defoliation reduced activity to levels below that of the controls.

The depodding treatment stimulated vigorous, renewed root growth and new nodule formation in Chippewa 64 and, to a lesser extent, in Clay. This response was reflected in the late recovery in nitrogenase activity for depodded Chippewa 64 plants.

Table 20.7. *Effect of changing the photosynthetic source–sink relationship on soybean nodule characteristics.* (After Lawn & Brun, 1974*a*)

Treatment	No. nodules/ plant*	Nodule fresh wt* (g/plant)	Nodule size* (g/100 nodules)	Specific activity* (μmoles C_2H_4/ g fresh wt/h)	Total activity* (μmoles C_2H_4/ plant/h)
Control	170 c†	3.90 b	2.14 a	12.5 bc	50.0 bc
Supplemental light	236 a	5.37 a	2.24 a	23.6 a	119.7 a
Partial depodding	206 b	4.13 b	1.98 ab	17.6 b	74.3 b
Partial shade	156 c	2.87 c	1.80 b	8.6 cd	27.3 cd
Partial defoliation	131 d	2.33 c	1.77 b	5.6 d	18.3 d

* Values are the average of samples taken 10 and 19 days after treatment application for two varieties Chippewa 64 and Clay.

† Means followed by the same letter are not significantly different ($P \leq 0.05$) using Duncan's New Multiple Range test.

Table 20.7 shows treatment means for nodule number per plant, nodule fresh weight per plant, nodule size and total and specific nitrogenase activities for the samples taken 10 and 19 days after treatment application. Total and specific activities and nodule number per plant, were significantly lower 19 days after treatment application than at 10 days. Plant fresh weight, nodule number per plant, and nitrogenase activities were all higher in the supplementary light or after depodding, and lower with shading and defoliation. Nodule size was significantly reduced by shading and defoliation.

The average number of nodules per plant immediately prior to the application of treatments was 251, so that the differences among treatment means appear to have been the result of differential rates of decay of the nodules. The formation of new nodules after treatment was evident only in the depodded plants, and then only after three to four weeks had elapsed.

Table 20.8. *Effect of changing photosynthetic source–sink relationships on seed, plant, and protein yields for two soybean varieties.* (After Lawn & Brun, 1974*a*)

Treatments	Seed yield (g/plant)		Seed protein content (%)		Seed protein yield (g/plant)		Total plant yield (g/plant)		Total protein yield (g/plant)	
	Chippewa 64	Clay	Chippewa 64	Clay	Chippewa 64	Clay	Chippewa 64	Clay	Chippewa 64	Clay
Control	12.32 b*	13.53 b	40.9 ab	39.4 a	5.05 b	5.35 b	25.57 b	23.67 ab	5.82 b	5.93 b
Supplemental light	15.84 a	16.08 a	38.4 b	38.2 a	6.09 a	6.15 a	32.31 a	26.61 a	6.90 a	6.71 a
Partial depodding	10.98 b	12.38 b	41.3 a	40.4 a	4.53 b	5.00 b	25.88 b	22.93 ab	6.01 b	6.04 b
Partial shade	9.51 c	10.76 c	39.9 ab	38.5 a	3.77 c	4.13 c	21.05 c	20.07 b	4.56 c	4.76 c
Partial defoliation	9.94 bc	9.81 c	38.7 ab	38.4 a	3.83 c	3.74 c	22.97 bc	19.72 b	4.51 c	4.26 c

* Means followed by the same letter within columns are not significantly different ($P \leqslant 0.05$) using Duncan's New Multiple Range test.

These data agree closely with the recent findings of Weil & Ohlrogge (1972) who reported that a reduction in interplant soybean competition by thinning at the beginning of the podfilling stage resulted in higher percentages of red nodules and a delay in nodule senescence by several weeks.

Total plant top yield at maturity (seed, pod and stem) was increased by supplementary lighting, decreased by shading and defoliation, and relatively unaffected by depodding (Table 20.8). The effect of shading was more severe than defoliation for Chippewa 64, while the reverse was true for Clay. Seed yield was also increased by supplementary light in both varieties. Pod number per plant, seeds per pod and seed size were all increased by this treatment.

Shading and defoliation reduced seed yield primarily by affecting number and seed size. Removal of 50 % of the pods at the end of flowering reduced yield by only 9–10 % because of a compensatory increase in seed size and the number of pods set at the non-depodded nodes. Seed protein content showed relatively small variation, with highest values for the depodded plants and lowest for plants given the supplementary light. Consequently, treatment differences in protein yield of the seed closely reflect those in seed yield. Similarly, treatment differences in total protein found at maturity in the seed, pod, and stem, in general reflected the differences in total plant dry matter.

These results indicate that the decline in activity was associated with the development of the pods as a competing assimilate sink, since the decline in nitrogenase activity in each variety coincided with the time when pod growth rate first exceeded total top growth rate; the latter being indicative of mobilization of previously assimilated material into the pods. The responses in nitrogenase activity and protein yield to treatments designed to alter the photosynthetic supply to the nodules were consistent with the hypothesis of competition for photosynthate between the developing pods and nodules.

Factors, both genotypic and environmental, which tend to lessen competitive effects by enhancing the photosynthetic source–sink ratio, may be expected to minimize such a decline in activity, and should be considered in the future development of higher yielding varieties. Agronomic and environmental factors (e.g. narrow row spacing, higher plant populations, or reduced solar radiation) which tend to enhance inter- and intra-plant competition during the podfilling stage may accentuate the terminal decline in symbiotic nitrogen fixation and lead to much lower yields.

Summary

Seed yield and protein percentage were not increased significantly by inoculating soybean seeds with *R. japonicum* at planting time when nodulated soybeans had been grown on the soil before. Serotyping of single strain inoculants indicated a substantial range of recovery (0–17 %) depending on strain and location.

Fertilizer nitrogen applied at planting increased the seed yield and protein percentage of Chippewa 64 soybeans and decreased nodule weight and number per plant. When applied at the end of flowering to Chippewa 64 and Clay strains it caused a decline in nitrogenase activity. Seed yield and seed protein percentage for Clay, and seed and vegetative protein yield for Chippewa 64 were all increased by nitrogen applications.

Acetylene-reducing activity per plant increased during flowering, reached a maximum near the end of flowering and declined sharply during early pod-filling. The decline occurred immediately before the growth rate of the pods (including seed) exceeded 50 % of that of the total tops in the respective varieties. Treatments to enhance the photosynthetic source–sink ratio (supplementary light and depodding), maintained nitrogenase activity well above the control, while shading and defoliation, which reduce the source–sink ratio, decreased it below the level of the control. These results are interpreted as evidence that nitrogen fixation declined during pod filling as a result of inadequate supply of assimilate to the nodules.

This research was supported in part by the Cooperative States Research Service of the United States Department of Agriculture and the National Soybean Processors Association.

References

Abel, G. H. & Erdman, L. W. (1964). Response of Lee soybeans to different strains of *Rhizobium japonicum*. *Agronomy Journal*, **56**, 423–4.

Bergensen, F. J. (1970). The quantitative relationship between nitrogen fixation and the acetylene-reduction assay. *Australian Journal of Biological Science*, **23**, 1015–25.

Caldwell, B. E. & Vest, G. (1970). Effects of *Rhizobium japonicum* on soybean yields. *Crop Science*, **10**, 19–21.

Ham, G. E., Cardwell, V. B. & Johnson, H. W. (1971*a*). Evaluation of *Rhizobium japonicum* inoculants in soils containing naturalized populations of rhizobia. *Agronomy Journal*, **63**, 301–3.

Ham, G. E., Frederick, L. R. & Anderson, I. C. (1971*b*). Serogroups of

Rhizobium japonicum in soybean nodules sampled in Iowa. *Agronomy Journal*, **63,** 69–72.

Hardy, R. W. F., Holsten, R. D., Jackson, E. K. & Burns, R. C. (1968). The acetylene–ethylene assay for N_2 fixation: laboratory and field evaluation. *Plant Physiology*, **43,** 1185–1207.

Johnson, H. W., Means, U. M. & Weber, C. R. (1965). Competition for nodule sites between strains of *Rhizobium japonicum* applied as inoculum and strains in the soil. *Agronomy Journal*, **57,** 179–85.

LaRue, T. A. G. & Kurz, W. G. W. (1973). Estimation of nitrogenase in intact legumes. *Canadian Journal of Microbiology*, **19,** 304–5.

Lawn, R. J. & Brun, W. A. (1974*a*). Symbiotic nitrogen fixation in soybeans. I. Effect of photosynthetic source–sink manipulations. *Crop Science*, **14,** 11–16.

Lawn, R. J. & Brun, W. A. (1974*b*). Symbiotic nitrogen fixation in soybeans. III. Effect of supplemental nitrogen and intervarietal grafting. *Crop Science*, **14,** 22–5.

Lindstrom, E. S., Newton, J. W. & Wilson, P. W. (1952). The relationship between photosynthesis and nitrogen fixation. *Proceedings of the National Academy of Sciences*, USA, **38,** 392–6.

Mague, T. H. & Burris, R. H. (1972). Reduction of acetylene and nitrogen by field-grown soybeans. *New Phytologist*, **71,** 275–86.

Pate, J. S. (1958). Nodulation studies in legumes. II. The influence of various environmental factors on symbiotic expression in the vetch (*Vicia sativa* L.) and other legumes. *Australian Journal of Biological Science*, **11,** 496–515.

Vincent, J. M. (1970). *Manual for the practical study of root-nodule bacteria.* IBP Handbook No. 15. Blackwell Scientific Publications, Oxford.

Virtanen, A. I., Moisio, T. & Burris, R. H. (1955). Fixation of nitrogen by nodules excised from illuminated and darkened pea plants. *Acta Chemica Scandinavica*, **9,** 184–6.

Weber, C. R. (1966). Nodulating and non-nodulating soybean isolines. II. Response to applied nitrogen and modified soil conditions. *Agronomy Journal*, **58,** 46–9.

Weil, R. R. & Ohlrogge, A. J. (1972). The seasonal development of and effect of inter-plant competition on soybean nodules. *Agronomy Abstracts*, p. 39.

Wheeler, C. T. (1971). The causation of the diurnal changes in nitrogen fixation in the nodules of *Alnus glutinosa*. *New Phytologist*, **70,** 487–95.

21. Field response of legumes in India to inoculation and fertiliser applications

N. S. SUBBA RAO

The lowest per capita daily intake of protein in the world (FAO, 1969) is in the Indian subcontinent (47.9 g) as opposed to New Zealand which has the highest daily intake (108.49 g). The bulk of this meagre protein in India is of vegetable origin coming from the edible seeds of leguminous annual crop plants, commonly known as 'pulses'. The pulses belong to diverse genera and species such as *Cicer arietinum* L. (gram, Bengal gram, chick pea or channa), *Cajanus cajan* (L.) Millsp. (pigeon pea, red gram, arhar or tur), *Phaseolus mungo* L. (black gram or urid), *P. aureus* Roxb. (green gram or mung bean), *P. aconitifolius* Jacq. (moth bean), *Lens culinaris* Medic. (lentil or masur), *Dolichos biflorus* L. (horse gram or kulthi), *Pisum arvense* L., *P. sativum* L. (peas or matar) and *Lathyrus sativus* L. (chickling vetch or khesari). Other minor pulses are *Vigna catjang* Walp. and *Vigna sinensis* (L.) Savi. (cowpea or lobia), *Cyamopsis tetragonoloba* (L.) Taub. (cluster bean or guar), *Dolichos lablab* L. (Indian bean, sem or avarai), *Phaseolus vulgaris* L. (French bean or fras bean) and *P. trilobus* Ait. (pillipersara). Of these Bengal gram and red gram contribute more than half of the edible leguminous seed production in India.

Two other important leguminous crops, *Arachis hypogea* L. (groundnut, peanut or moongphali) and *Glycine max* (L.) Merr. (soybean) are classed in India as oil-yielding crops rather than pulses. Excluding these approximately 22 million hectares are under cultivation of edible legumes (pulses) covering about $\frac{1}{6}$ of the total area under food grain crops. The annual pulse production in 1968–9 was 11 million tonnes of seed. Nearly 8 million hectares are under groundnut cultivation providing 6 million tonnes of pods annually. The cultivation of soybean was 60 000 hectares in 1972 with an annual production of 60 000 tonnes of seed.

While soybean is an irrigated crop, most of the other pulses are grown under rain-fed conditions. Limitation of water supply to plants is the primary hazard which the nodulating legume has to face, apart from high subsoil temperatures in summer months reaching 40 °C. Nearly 70 % of the cultivable area of the country is rain-fed of which 40 % is dry with an annual rainfall of less than 75 cm. Bengal gram (7.1 million hectares) and red gram (2.5 million hectares) are invariably grown in rain-fed areas and can tolerate drought by virtue of their deep root systems.

Field experiments

Field experiments were done at various locations in India (Fig. 21.1) having different NPK status (Table 21.1 shows average figures) with the following objectives: (1) to find out the relative merits of inorganic

Fig. 21.1. Map of India showing the various locations where field experiments were undertaken.

nitrogen application and *Rhizobium* inoculation in increasing grain yields so as to recommend rhizobial inoculations wherever possible and thereby save inorganic nitrogen which is in short supply in India and (2) to see how far *Rhizobium* inoculations would be effective in the presence of phosphorus and potassium fertilisers.

Table 21.1. *Available NPK status of soils at different locations in India**

Place	N	P	K
Delhi (Union Territory)	Low	Medium	Medium
Hissar (Haryana)	Low	Medium	Medium
Ludhiana (Punjab)	Medium	Medium	Medium
Jabalpur (Madhya Pradesh)	Medium	Medium	Medium
Sehore (Madhya Pradesh)	Low	Medium	High
Pantnagar (Uttar Pradesh)	Medium	High	Medium
Kanpur (Uttar Pradesh)	Low	Medium	Medium
Badnapur (Maharashtra)	Low	Low	Medium
Nagpur (Maharashtra)	Low	Low	Medium
Taloda (Maharashtra)	Medium	High	High
Akola (Maharashtra)	Low	Low	High
Jalgaon (Maharashtra)	Low	Medium	High
Sheikpura (Near Patna, Bihar)	Medium	Low	High
Kalyani (West Bengal)	Low	Medium	Medium
Dahod (Gujarat)	Low	Medium	Low
Junagadh (Gujarat)	Medium	Low	Low
Baroda (Gujarat)	Low	Medium	Low
Semiliguda (Andhra Pradesh)	Medium	Medium	Low
Jayamkundam (Andhra Pradesh)	Medium	Medium	Medium
Hyderabad (Andhra Pradesh)	Medium	Medium	Medium
Rajendranagar (Andhra Pradesh)	Low	Medium	Medium
Kudiyamalai (Tamil Nadu)	Low	Low	Low
Tindivanam (Tamil Nadu)	Low	Low	—
Bangalore (Mysore)	Medium	Low	Medium
Dharwar (Mysore)	Medium	Low	Medium
Raichur (Mysore)	Medium	Low	Medium

* Mostly based on Ramamoorthy & Bajaj (1969).
For N: low, 280 kg/ha; medium, 280–560 kg/ha. For P: low, 10 kg/ha; medium, 10–24.6 kg/ha; high, >24.6 kg/ha. For K: low, 108 kg/ha; medium, 108–280 kg/ha; high, >280 kg/ha.

The fertilisers used were ammonium sulphate or urea, single superphosphate ($Ca_3(PO_4)_2$, containing 16 % P_2O_5) and potassium sulphate (K_2SO_4). The different doses of fertilisers are given in terms of kilograms per hectare N, P_2O_5 and K_2O. Efficient strains of *Rhizobium* were prepared in the form of peat-based cultures and used to inoculate seeds before sowing. Strains were chosen after preliminary screening in pots with corresponding legumes in different soils. Grain yield was the sole criterion for assessing inoculation efficiency. Since varieties were not always specified, different local varieties were used at each location.

Field experiments with Cicer arietinum

The results of experiments conducted at various locations under rain-fed conditions are reported in Tables 21.2 and 21.3. Significant increases in yields (e.g. 31.6 % over the uninoculated control at Dahod) were

Table 21.2. *Field experiments with* Cicer arietinum, *winter 1967–8*

Place	Treatment	Grain yield (kg/ha)	Difference from control (%)
Uninoculated control		1437	
Delhi	*Rhizobium*	1364	− 5.1
Experiment 1	50 kg N/ha	1447	0.7
	100 kg N/ha	1240*	−13.7
Uninoculated control		1032	
Delhi	*Rhizobium*	1322*	28.1
Experiment 2	15 kg N/ha	838*	−18.8
	15 kg N/ha+*Rhizobium*	892	−13.5
	30 kg N/ha	937	− 9.2
	30 kg N/ha+*Rhizobium*	1107	7.3
Uninoculated control		993	
Kanpur	*Rhizobium*	881	−11.3
	15 kg N/ha	913	− 8.0
	15 kg N/ha+*Rhizobium*	746	−24.9
	30 kg N/ha	623	−37.3
	30 kg N/ha+*Rhizobium*	690	−30.5
Uninoculated control		774	
Jabalpur	*Rhizobium*	668	−13.7
	15 kg N/ha	694	−10.3
	15 kg N/ha+*Rhizobium*	746	− 3.6
	30 kg N/ha	681	−12.0
	30 kg N/ha+*Rhizobium*	731	− 5.5

* Significant difference from control at 5 % level.

obtained at many sites through *Rhizobium* inoculations. Four kinds of response were noticeable with regard to *Rhizobium* inoculation in relation to nitrogen. (1), In the presence of nitrogen (100 kg/ha) without *Rhizobium* inoculation, the yield of grain was reduced significantly (Delhi, Experiment 1; see Table 21.2); (2), *Rhizobium* inoculation either with or without nitrogen (25 kg/ha) increased yield significantly whereas nitrogen alone did not do so (Sheikpura; see Table 21.3);

Table 21.3. *Field experiments with* Cicer arietinum, *winter 1971–2*

Place	Treatment	Grain yield (kg/ha)	Difference from control (%)
Uninoculated control		1776	
Hissar	*Rhizobium*	1911	7.6
	25 kg/N/ha	1655	− 6.9
Uninoculated control		1240	
Badnapur	*Rhizobium*	1303	5.1
	25 kg N/ha	1330	7.2
Uninoculated control		186	
Akola	*Rhizobium*	191	2.7
	25 kg N/ha	144	−22.6
Uninoculated control		2081	
Sehore	*Rhizobium*	2025	− 2.7
	25 kg N/ha	2113	1.5
Uninoculated control		1708	
Rajendranagar	*Rhizobium*	1986	16.2
Experiment 1	25 kg N/ha	1876	9.8
Uninoculated control		1714	
Rajendranagar	*Rhizobium* S_1	1966*	14.7
Experiment 2	*Rhizobium* S_3	2047*	19.4
	Rhizobium A_1	1971*	15.0
	Mixture of S_1, S_3 and A_1	1817	6.0
Uninoculated control		1000	
Nagpur	*Rhizobium*	1260*	26.0
	25 kg N/ha	1243*	24.3
Uninoculated control		1041	
Sheikpura	*Rhizobium*	1252*	20.3
	25 kg N/ha	1054	1.3
	25 kg + *Rhizobium*	1650*	58.5
Uninoculated control		150	
Jabalpur	*Rhizobium* S_1	1732*	15.5
	Rhizobium S_3	1576*	5.1
	Rhizobium A_1	1621*	8.1
	Mixture of A_1, S_3 and S_1	1670*	11.3
Uninoculated control		955	
Dahod	*Rhizobium* S_1	1146*	20.0
	Rhizobium S_3	1257*	31.6
	Rhizobium A_1	1145*	19.9
Uninoculated control		2419	
Raichur, var. *Chaffa*	*Rhizobium* S_1	2967	22.7
Uninoculated control		1941	
var. *N*59	*Rhizobium* S_1	2113	8.9
	Rhizobium A_1	2108*	8.6

* Significant difference from control at 5 % level.

(3), application of nitrogen (25 kg/ha) or *Rhizobium* inoculation increased yield significantly (Nagpur; see Table 21.3); and (4), neither nitrogen (25 kg/ha) application nor *Rhizobium* inoculation increased yield significantly (Hissar, Badnapur, Akola, Sehore and Rajendranagar; see Table 21.3).

In assessing the results of these experiments with Bengal gram, it is to be borne in mind that this grain legume is predominantly cultivated under rain-fed conditions in India, receiving no water by way of organised irrigation. One of the objectives of field experiments under such dry land conditions was to find out the combined effects of paucity of moisture and high surface soil temperature (maximum 38 °C), as the dissolution and absorption of fertilisers and proper functioning of nodules are dependent on the frequency and intensity of rain. In spite of these complications in Bengal gram cultivation, the results show that simple rhizobial application can enhance yields at rain-fed areas such as Rajendranagar, Nagpur, Sheikpura, Jabalpur, Dahod and Raichur.

Field experiments with Cajanus cajan

In field experiments conducted in 1967–8 at Pantnagar and in 1972–3 at Kanpur, neither *Rhizobium* inoculation nor application of nitrogen (40–80 kg/ha), P_2O_5 (50–150 kg/ha) or K_2O (25–50 kg/ha) affected yield either individually or in combinations. The average grain yield at Pantnagar was 1800 kg/ha while at Kanpur it was of the order of 2000 kg/ha. In experiments done in 1972 under rain-fed conditions at

Table 21.4. *Field experiment with* Cajanus cajan *var. 21 at Pantnagar, winter 1972–3*

Treatment	Grain yield (kg/ha)	Difference from control (%)
Uninoculated control	1183	—
Rhizobium, strain SB16	1558	31.7
Rhizobium, strain GBPUAT	1549	30.9
Rhizobium SB16+talc-pelleting	1080	−8.7
Rhizobium GBPUAT+talc-pelleting	1471	24.3
Rhizobium SB16+charcoal-pelleting	1072	−9.4
Rhizobium GBPUAT+charcoal-pelleting	1655*	39.9
Rhizobium SB16+lime-pelleting	1762*	48.9
Rhizobium GBPUAT+lime-pelleting	1476	24.8

* Significant difference from control at 5 % level.

four other locations viz., Junagadh, Kudiyamalai, Baroda and Bangalore, there were general increases in yield from *Rhizobium* inoculation, the maximum being 68 % at Kudiyamalai. In these experiments different local varieties were used and hence yields were variable (200–1000 kg/ha). The variation in yield could also be attributed to factors met with under rain-fed cultivation of this crop. Red gram wilt caused by *Fusarium udum* was also reported at certain sites which may have contributed to low yields.

In 1972–3, field experiments were also conducted at Pantnagar, Hissar, Badnapur, Bangalore and Ludhiana to find out the efficacy of pelleting *Rhizobium*-inoculated seed with talc (magnesium silicate), wood charcoal and lime. In Pantnagar (soil pH 7.0), pelleting with charcoal or lime increased yield significantly, whereas conventional inoculation failed to do so (Table 21.4). Lime-pelleting significantly increased yields at Ludhiana where the pH of soil was also nearly neutral (Table 21.5).

Table 21.5. *Grain yield (kg/ha) of* Cajanus cajan *in field experiments at different sites, winter 1972–3*

	Locations			
Treatment	Hissar, soil pH 8.0 (var. *T*-21)	Badnapur, soil pH 8.0 (var. not reported)	Bangalore, soil pH 6.2 (var. *S*-5)	Ludhiana, soil pH 7.5 (var. *T*-21)
Uninoculated control	2405	786	536	1730
Rhizobium	2847	882	569	1720
Rhizobium+talc-pelleting	3030	861	639	1410*
Rhizobium+charcoal-pelleting	2638	852	549	1510
Rhizobium+lime-pelleting	2880	849	511	2010*

* Significant difference from control at 5 % level.

Field experiments with Phaseolus mungo *and* P. aureus

In an experiment at Pantnagar in 1967–8 with *P. aureus*, inoculation with *Rhizobium* increased yield up to 12.5 % over the uninoculated control, whereas application of nitrogen (up to 50 kg/ha) or P_2O_5 (up to 100 kg/ha) in the absence of *Rhizobium* was not beneficial and K_2O (50 kg/ha) depressed yield. The Pantnagar soil is rich in organic matter and phosphate, which may be one of the reasons for the lack of response to fertiliser applications.

In 1972, experiments were conducted at several locations (Ludhiana,

Baroda, Taloda, Junagadh, Jayamkundam) with different varieties of *P. mungo* and *P. aureus* on the effect of *Rhizobium* inoculation on the yield of grains. Although no significant effects on yield were obtained, increased yields were generally noticeable in the range of 4 to 53 %. Yields varied from 300 to 1000 kg/ha, depending on the varieties used and incidence of diseases and pests.

Field experiments on Lens culinaris

A comparison between fertiliser and *Rhizobium* applications was made at Pantnagar and Ludhiana in 1967–8. The level of N, P_2O_5 and K_2O were 0, 50 and 100 kg/ha in different combinations. At Pantnagar, the

Table 21.6. *Field experiments with* Lens culinaris, *winter 1971–2*

Place	Treatment	Grain yield (kg/ha)	Difference from control (%)
Uninoculated control		513	—
Rajendranagar	*Rhizobium* L_1	552	7.6
	Rhizobium L_4	587	14.4
	Rhizobium L_7	535	4.3
	Mixture of L_1, L_4 and L_7	528	3.0
Uninoculated control		354	—
Jabalpur	*Rhizobium* L_1	606*	70.9
	Rhizobium L_4	568*	60.4
	Rhizobium L_7	552*	55.7
	Mixture of L_1, L_4 and L_7	560*	57.9
Uninoculated control		1380	—
Ludhiana	*Rhizobium* L_4	1679*	21.7
Uninoculated control		1675	—
Hissar	*Rhizobium*	1256*	25.0
	25 kg N/ha	1578	−5.8
	50 kg P_2O_5/ha	2121*	26.6
Uninoculated control		808	—
Sehore	*Rhizobium*	957	18.4
	50 kg P_2O_5/ha	843	4.3
	50 kg P_2O_5/ha + *Rhizobium*	1137*	40.7
	50 kg P_2O_5/ha + 25 kg N/ha	953	17.9
Uninoculated control		1095	—
Sheikpura	50 kg P_2O_5/ha	1285*	17.3
	50 kg P_2O_5/ha + *Rhizobium*	1550*	41.5
	25 kg N/ha + 50 kg P_2O_5/ha	1407*	28.5
	25 kg N/ha + 50 kg P_2O_5/ha + *Rhizobium*	1600*	46.1

* Significant difference from control at 5 % level.

results were not significant. In general, increasing levels of N and K_2O decreased yield of grains and application of P_2O_5 or *Rhizobium* inoculation had no effect. At Ludhiana, on the other hand, P_2O_5 with or without *Rhizobium* inoculation generally increased yields significantly.

Table 21.6 shows results of field experiments repeated in 1971–2 on simple inoculation and at some locations on *Rhizobium* inoculation in the presence and absence of fertilisers. *Rhizobium* inoculation significantly increased yields at Ludhiana and Jabalpur (70.9 %) but not at Rajendranagar, and yield was significantly reduced at Hissar. Application of P_2O_5 with or without *Rhizobium* inoculation significantly increased yield at Sehore and Sheikpura.

Field experiments with Pisum sativum

At Pantangar and Delhi (1967–8) the effect of *Rhizobium* inoculation was compared with that of fertiliser applications. The results revealed that *Rhizobium* inoculation increased yields up to 12 % at Pantnagar and application of N, P_2O_5 or K_2O individually from 15 kg/ha to 100 kg/ha was not so effective. At Delhi, on the other hand, *Rhizobium* inoculation, either with or without nitrogen fertiliser (30 kg/ha) showed a general tendency to decrease yield. The grain yield varied from 1800 to 2200 kg/ha.

Field experiments with Arachis hypogea

Experiments were done during three consecutive years. In 1970, a promising strain of *Rhizobium* (from pot trials) was tried at three centres in the presence and absence of fertilisers (Table 21.7). Simple *Rhizobium* inoculation with or without nitrogen (from 10 to 20 kg/ha) tended to reduce yield at Dharwar and Ludhiana and had no effects at Tindivanam. At Ludhiana, a combination of 80 kg P_2O_5/ha and 20 kg N/ha showed a general tendency to improve yield of pods, which was, however, offset by *Rhizobium* inoculation. In 1971 experiments were repeated using three promising *Rhizobium* strains at five different locations (Table 21.8). Simple *Rhizobium* inoculation decreased yield at all places except Junagadh, where, however, the increase was not significant.

In 1972, several Indian isolates were compared in pot culture experiments, with *Rhizobium* isolates for groundnut obtained from abroad, and several promising strains were selected for field trials at four

Table 21.7. *Field experiments with* Arachis hypogea, *1970*

Place	Treatment	Pod yield (kg/ha)	Difference from control (%)
Uninoculated control		1267	—
Dharwar	*Rhizobium*†	1116*	−11.9
	10 kg N/ha	1169	− 7.7
	10 kg N/ha+*Rhizobium*	1097*	−13.4
	20 kg N/ha	1160	− 8.4
	20 kg N/ha+*Rhizobium*	1077*	−17.6
Uninoculated control		994	—
Tindivanam	*Rhizobium*	1048	10.5
	10 kg N/ha	907	− 8.7
	10 kg N/ha+*Rhizobium*	925	− 6.9
	20 kg N/ha	966	− 2.8
	20 kg N/ha+*Rhizobium*	1098	11.0
Uninoculated control		1609	—
Ludhiana	80 kg P_2O_5/ha	1678	4.3
	80 kg P_2O_5/ha+*Rhizobium*	1094*	−32.0
	80 kg P_2O_5/ha+10 kg N/ha	1684	4.7
	80 kg P_2O_5/ha+10 kg N/ha+ *Rhizobium*	1365	−15.2
	80 kg P_2O_5/ha+20 kg N/ha	1730	7.5
	80 kg P_2O_5/ha+20 kg N/ha+ *Rhizobium*	1447	−10.1

* Significant difference from control at 5 % level.
† *Rhizobium* strain 402-B was used in all these experiments.

different centres. An increase (not significant) in yield of pods was obtained at Bangalore with a South African *Rhizobium* strain from Pretoria. Other strains tended to reduce yields, especially at Tindivanam

Table 21.8. *Pod yield (kg/ha) of* Arachis hypogea *with different* Rhizobium *strains in field experiments in 1971*

Rhizobium strain	Location and variety used: Tindivanam (TMV-7)	Jalgaon (SB-XI)	Junagadh (Junagadh-II)	Ludhiana (M-145)	Hyderabad (TMV-2)
Uninoculated control	770 (0.0)	724 (0.0)	671 (0.0)	1104 (0.0)	395 (0.0)
402-B	740 (−3.9)	617 (−14.8)	744 (10.9)	1102 (−0.2)	302 (−23.5)
CO1	755 (−1.9)	570 (−21.3)	718 (7.0)	1156 (4.7)	352 (−10.9)
Pollachi isolate	747 (−3.0)	561 (−22.5)	690 (2.8)	1081 (−2.1)	368 (−6.8)

Figures in brackets give percentage from control. Data were not significant at the 5 % level.

Table 21.9. *Pod yield (kg/ha) of* Arachis hypogea *with different* Rhizobium *strains in field experiments in 1972*

Rhizobium strain	Location and variety used: Bangalore (TMV-2)	Jalgaon (SB–11)	Tindivanam (TMV-7)	Hyderabad (SB-11)
Uninoculated control	440 (0.0)	1545 (0.0)	1304 (0.0)	319 (0.0)
Gt. 21*	439 (−0.2)	1711 (10.7)	1197 (−8.2)	352 (10.3)
XBL6†	515 (17.0)	1637 (5.9)	1248 (−4.3)	338 (5.9)
R1283A‡	475 (7.9)	1675 (8.4)	1257 (−3.6)	284 (−11.0)
Composite of all strains	492 (11.8)	1694 (9.6)	1241 (−4.8)	324 (1.6)

Figures in brackets give percentage difference from uninoculated control. Data were not significant at the 5 % level.

* Isolated by plating 'Nitragin' peat culture.

† From Pretoria, S. Africa.

‡ From Schiffmann, Israel.

(Table 21.9). As the data show, large differences in the yield of pods were noticeable which can be due to different varieties used in different places. Secondly, all the varieties used were susceptible to *Cercospora* leaf spot disease, which may have influenced yields depending upon the intensity of symptoms.

Field experiments with Glycine max

One of the methods followed to obtain efficient cultures of *R. japonicum* to suit Indian conditions was to raise soybean plants from seeds inoculated with 'Nitragin' brand peat-based inoculant obtained from the USA. Well-formed nodules from such plants were then plated and

Table 21.10. *Yield of soybean* (Glycine max) *as influenced by* R. japonicum *(SB 16) inoculation at different locations in India in 1970.* (After Subba Rao & Balasundaram, 1971)

Location	Grain yield (kg/ha): Uninoculated	Grain yield (kg/ha): Inoculated	Difference from control (%)
Delhi	936	1583	69.1
Pantnagar	2307	2637	14.3
Jabalpur	1639	2920	78.1
Kalyani	1997	2561	28.2
Junagadh	573	1924	235.7

several *R. japonicum* strains isolated which proved highly effective in pots and later in field experiments in different parts of the country.

Significant increases in yield of grain were obtained by *Rhizobium* inoculation in trials from 1969 to 1972. The results obtained at five locations are shown in Table 21.10. The differences in yield between locations may have been due to the agro-climatic conditions characteristic of each location. For instance, Delhi and Junagadh are semi-arid in character (hence yields are low) compared with Pantnagar which is a terai area receiving high rainfall and whose soils are rich in organic matter and phosphates.

Conclusions

Four important points emerge from the results of field experiments presented in this chapter.

(1) Simple *Rhizobium* inoculation increased yields of certain leguminous crops in certain soils up to a maximum of 71 % over corresponding uninoculated controls. These included such important crops as *Cicer arietinum*, *Cajanus cajan* (after pelleting inoculated seed with wood charcoal or lime) *Lens culinaris* and *Glycine max*.

(2) Inoculations with *Rhizobium* alone sometimes decreased yields, especially in *Arachis hypogea*. This may be partially explained in terms of competition between the introduced strain of *Rhizobium* and the native strains already present in soil. Carroll (1934), Raju (1936), Allen & Allen (1940) and Gaur, Sen & Subba Rao (1973) observed that groundnut is nodulated not only by *Rhizobium* strains isolated from the genus *Arachis* but also from strains belonging to the 'cowpea miscellany' comprising isolates from fifty-one other species of legumes. Likewise, rhizobia isolated from nodules of *A. hypogea* were shown to nodulate roots of thirty-two other legume species. An extensive survey of nodulation of *A. hypogea* in different soils of India over the last five years shows that rhizobia capable of nodulating groundnut were generally prevalent, even in semi-arid desert soil samples from Jodhpur where neither cultivated nor wild species of *Arachis* occur. Dadarwal, Singh & Subba Rao (1973) studied the serology of forty-four isolates from nodules of *A. hypogea* (a cultivated species), *A. duranensis*, *A. prostrata*, *A. villosa*, *A. glaberata* and *A. marginata* (wild species). Twenty-seven serotypes were distinguished and all isolates nodulated all the species of *Arachis* studied. Such results underline the heterogeneity within *Arachis* rhizobia predominant in Indian soils and also

indicate the extent of competition to which the artifically applied *Rhizobium* strain is subjected.

(3) Application of inorganic phosphatic fertiliser generally improved yields of some legume crops either with or without *Rhizobium* inoculation. The importance of phosphate in the nutrition of nodulated legumes has been well-documented. It is, therefore, hardly necessary to emphasise the fact that adequate phosphate fertilisation of soil is a necessary prerequisite for the success of rhizobial inoculations.

(4) Application of inorganic nitrogen fertiliser to soil in the absence of *Rhizobium* inoculation generally decreased yield and in many instances even such small amounts as 10 to 25 kg N/ha reduced yields. The depressive effects of combined nitrogen on nodulation and nitrogen fixation are also well-known. However, in the absence of clear-cut experiments under controlled conditions to understand the different influences of inorganic nitrogenous fertilisers on the diverse phases of symbiosis, it is difficult to interpret the outcome of experiments reported here, except to point out the merits of inoculation practices in certain locations in India. In the present crisis relating to shortage of inorganic nitrogen fertilisers, it is necessary to define the optimum amounts of inorganic nitrogen required to improve the yields of different legume crops in conjunction with *Rhizobium* inoculation in different soils.

I thank Dr S. Ramanujam and Dr S. S. Rajan, the project co-ordinators in pulses and oil-seeds, respectively, and their staff for help in conducting field experiments and permission to present the findings here. Thanks are also due to the research staff of the Division of Microbiology, Indian Agricultural Research Institute, particularly Dr R. B. Rewari and Mr V. R. Balasundram, for evaluation of the efficiency of different rhizobia.

References

Allen, O. N. & Allen, E. K. (1940). Response of peanut plant to inoculation with rhizobia with special reference to morphological development of nodules. *Botanical Gazette*, **102,** 121–42.

Carroll, W. R. (1934). A study of *Rhizobium* species in relation to nodule formation on the roots of Florida legumes. *Soil Science*, **37,** 117–34.

Dadarwal, K. R., Singh, C. S. & Subba Rao, N. S. (1974). Nodulation and serological studies of rhizobia from six species of *Arachis. Plant and Soil*, **40,** 535–44.

FAO (1969). Nutritional studies, No. 19, p. 138. Food and Agricultural Organisation, Rome.

Gaur, Y. D., Sen, A. N. & Subba Rao, N. S. (1973). Promiscuity in groundnut *Rhizobium* association. *Zentralblatt für Bakteriologie*, II, **129,** 225–30.

Raju, M. S. (1936). Studies on the bacterial plant groups of cowpea, *Cicer* and Dhaincha. I. Classification. *Zentralblatt für Bakteriologie*, II, **94**, 249–62.
Ramamoorthy, B. & Bajaj, J. C. (1969). Available nitrogen, phosphorus and potassium status of Indian soils. *Fertiliser News*, (*India*), **14**, 1–12.
Subba Rao, N. S. & Balasundaram, V. R. (1971). *Rhizobium* inoculants for soybean. *Indian Farming*, **21**, 22–3.

22. Effect of inoculation and fertiliser application on the growth of soybeans in Rumania

C. HERA

With the increasing world population and world-wide nutritional deficiencies, additional sources of food proteins need to be developed, for example by extending the area and yield of pulse legumes.

Among the pulses, the culture of soybeans (*Glycine max*) warrants particular attention in Rumania because it is now the most widely grown of grain legumes, being followed by peas (*Pisum sativum*) and

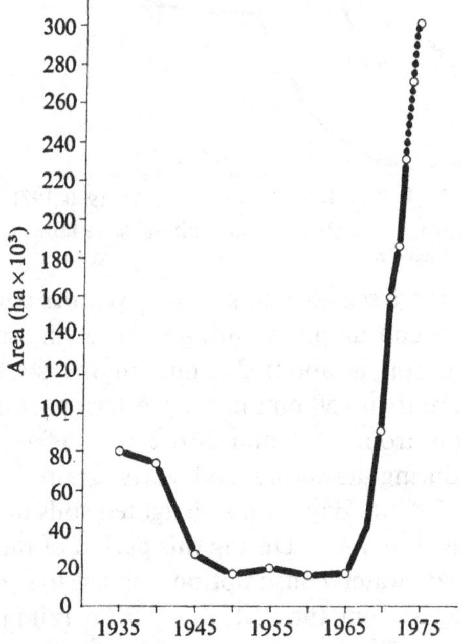

Fig. 22.1. Changes in the area planted with soybeans in Rumania.

beans (*Phaseolus vulgaris*), (Diaconescu & Miclea, 1971; Hera, 1971), although as recently as 1968–9 peas and beans were cultivated on a larger area than soybeans. This change was stimulated by the enhanced economic value of soybeans and their importance for animal feeding and in industry. Rumania is now the major producer of soybeans in

Europe, in terms of cultivated area (Diaconescu & Miclea, 1971); Fig. 22.1 shows changes in the area under soybeans since 1935.

Concomitant with this increase, improvements in the technologies of cultivation and release of better varieties has also increased the yield per unit area (Hera, 1971), as illustrated by data from state agricultural farms for 1963–72 (Fig. 22.2).

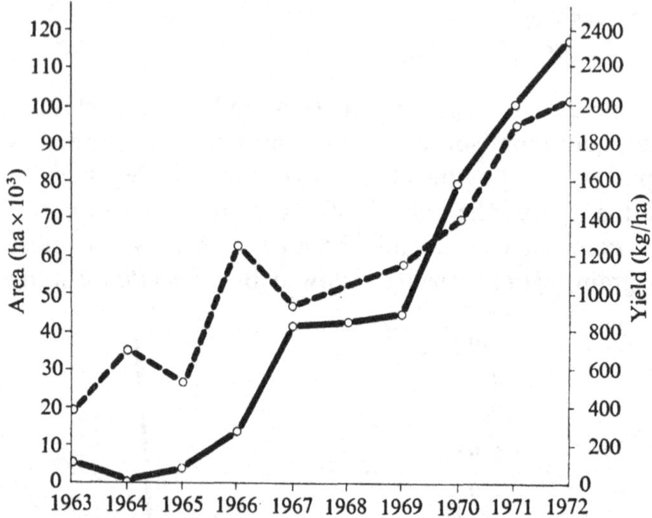

Fig. 22.2. Area planted with soybeans in agricultural state farms (solid lines) and average yields (dashed line) 1963–72.

Soybeans are more sensitive to soil and weather conditions than peas and beans. Water consumption during the growing season by soybeans on non-irrigated soil is about 340 mm in the southern part of the country and more than 450 mm in the western part of the country. On irrigated soils it is around 650 mm. Most water is needed in the months of June–July, during flowering and early grain formation, when it amounts to 3.4–3.8 mm/day on non-irrigated soils and 6.4–6.5 mm/day on irrigated soils (Fig. 22.3). During this period of rapid growth rainfall is much less than water consumption: under irrigation, 19 % of the consumed water is from the soil, 41 % from rainfall and 40 % from irrigation.

In Rumania there are two specific ecological zones that can be considered suitable for soybean. First, the Moldavia, Transylvania and Banat regions, where soils are fertile and the climate good, especially during flowering and the ripening of the pods. In this zone yields normally exceed 2000 kg/ha. Water supply here is reliable and comes from ground water, rainfall and irrigation. The second zone occupies the largest part of the Moldavia Highland, the Rumanian Plain and Crisana Highland as well as the adjacent zones of these areas. Here water availability and the low fertility are limiting factors, and to obtain high yields, irrigation and fertilisation are necessary. The rest of the country has no favourable conditions for growing soybeans economically.

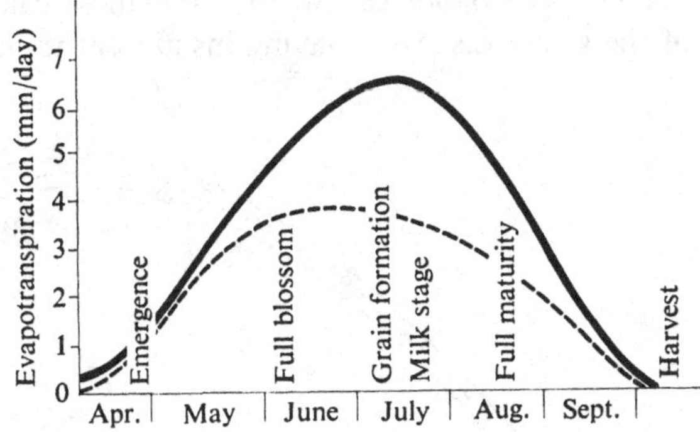

Fig. 22.3. Evapotranspiration of soybeans at Fundulea at different growth stages on irrigated (solid line) and non-irrigated (dashed line) soils.

Without irrigation soybeans are grown in the rotation soybeans – wheat – corn, and under irrigation in a four-crop rotation (soybeans – wheat + a second crop following wheat in the same year – corn) or a four year rotation with five crops (including sugarbeet). Because it fixes nitrogen and improves soil fertility, soybean is also a valuable crop for preceding wheat. Table 22.1 shows that this beneficial effect is relatively larger when the wheat is not fertilised.

Table 22.1. *The influence of soybeans on a following crop of wheat at ICCPT, Fundulea, 1966–70*

	Wheat yield (kg/ha)		
Previous crop	Unfertilised	Fertiliser 1*	Fertiliser 2†
Soybean	3310	4300	4390
Sunflower	2780	4270	4311
Corn	2390	4150	4022

* 120 kg N/ha + 60 kg P_2O_5/ha.
† 15 t/ha manure + 120 kg N/ha + 60 kg P_2O_5/ha.

A major problem in the use of fertilisers, especially nitrogenous ones, for soybeans is to determine the adequate and economic levels of application. High levels of nitrogen fertilisers are not thought to be necessary for inoculated and well-nodulated plants, but there is now some evidence that if applied at sowing they are beneficial. They are thought to act by allowing undisturbed early growth which prevents early nitrogen exhaustion and stimulates the onset of symbiosis. Nutman (1963) has shown that nitrogen fixation is under the influence of physical, chemical and biological factors, and these can affect the development of the symbiosis. Soils having insufficient rainfall for the

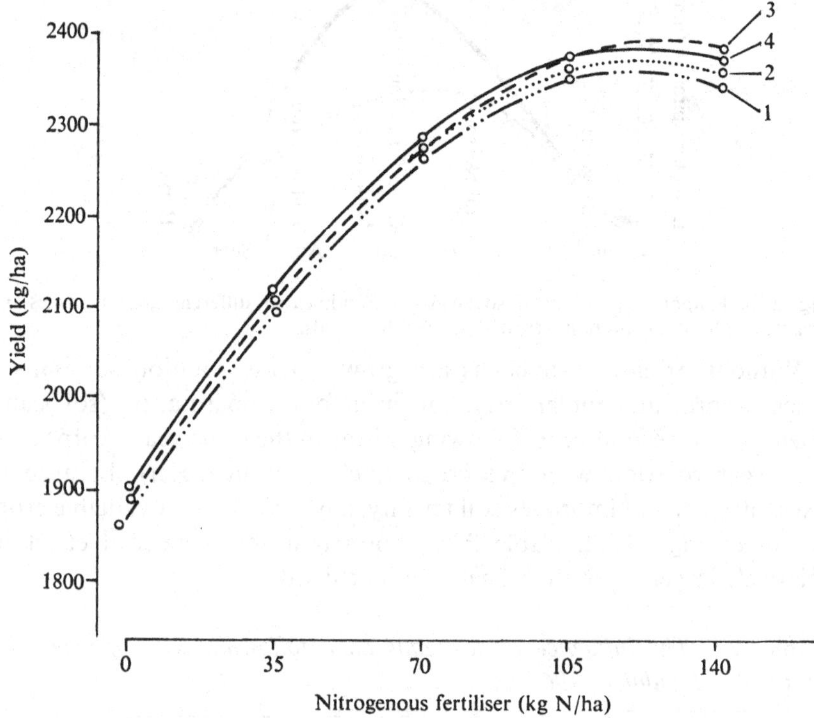

Fig. 22.4. Fertiliser efficiency in irrigated soybean crops. Mean values for eight sites over a four-year period. Regression lines for the graphs 1–4 are: 1, $y = 1866+7.82x-0.030x^2$; 2, $y = 1906+7.49x-0.029x^2$; 3, $y = 1897+6.97-0.023x^2$; 4, $y = 1903+6.75x-0.023x^2$.

Graph no.	Rate of and PK* (kg/ha)	Optimum N rate (kg/ha)	Yield (kg/ha)	Yield gain: kg/ha	Yield gain: kg/kg fertiliser
1	$P_0 K_0$	128	2370	500	3.90
2	$P_{40}K_0$	127	2360	510	3.05
3	$P_{60}K_0$	147	2410	540	2.60
4	$P_{80}K_{80}$	146	2400	530	1.76

* As P_2O_5 and K_2O.

normal development of the bacterial symbiosis process may respond to application of low rates of nitrogenous fertilisers. It is also known that the presence of the nitrate and ammonia nitrogen in the soil in larger quantities progressively reduces the effectiveness of a symbiosis (Weber, 1966; Nelson, 1971; Harper & Cooper, 1971).

We have found that when soybean is cultivated without being inoculated, or where inoculation is not successful for some reason, large amounts of nitrogenous fertilisers need to be applied, especially under irrigation. Thus, at eight sites, over a period of four years, the nitrogen level for maximum yield ranged from 127 to 147 kg N/ha. In these trials phosphorus fertilisation did not influence the level of yields (Fig. 22.4). Maximum net income ranged from 998 to 1092 lei/ha (12.75 lei being equivalent to £1.00), and was obtained at levels of 112 to 124 kg N/ha as ammonium nitrate. The use of phosphorus or phosphorus and potassium fertilisation diminished the net income (Hulpoi, Picu & Tianu, 1971).

Table 22.2. *Interaction of nitrogen and phosphorus levels on the yield (kg/ha) of uninoculated soybeans in a meadow chernozem at Lovrin in 1972*

Phosphorus level	Nitrogen level (kg N/ha)					
(kg P_2O_5/ha)	0	30	60	90	120	Mean
0	1580	2060	2380	2730	3040	2360
30	1700	2110	2920	3340	3510	2720
60	1600	2200	2730	3280	3530	2670
90	1640	2110	2680	3130	3390	2590
120	1630	2120	2660	3120	3160	2540
Mean	1630	2120	2670	3120	3320	

Significant differences at: 5 %, 110 kg/ha; 1 %, 150 kg/ha; 0.1 %, 200 kg/ha.

In dryland conditions, on a meadow chernozem, similar results were obtained, 90–120 kg N/ha being required for maximum yield (Table 22.2). In this soil, however, phosphorus mobility is reduced, and there was a response to phosphorus fertilisation at 30–60 kg P_2O_5/ha.

In recent years, symbiotically very active *Rhizobium* strains have been introduced and established (Balan *et al.*, 1968), among which the strain SO-69 should be mentioned. This strain produced an average yield gain of 470 kg/ha above the uninoculated control during four years on five soil types under irrigation and using 64 kg N/ha and 64 kg P_2O_5/ha fertiliser application to all plots. In two soils without irrigation the yield gain amounted to 300 kg/ha (see Table 22.3; also Balan, 1970).

Since 1970, experiments have examined further, under various soil conditions, the interaction between inoculation with strain SO-69 and nitrogen fertiliser application. Table 22.4 shows the average production,

Table 22.3. *Soybean yield (kg/ha) after inoculating with different strains of* Rhizobium japonicum *and a basal dressing of 64 kg N/ha+64 kg P_2O_5/ha (1966–9)*

Type of soil	Location	Yield of un-inoculated plots	*Rhizobium* strain				Significant differences at 5 %
			SO-69	SO-75	SO-78	SO-85	
		Irrigated					
Carbonated chernozem	Brăila	1710	2110	2322	2344	2190	307
Chestnut-coloured chernozem	Mărculeşti	1460	2311	2011	2190	2077	395
Leached chernozem	Caracal	2144	2522	2480	2566	2470	209
Forest reddish-brown soil	Simnic	2355	2910	2770	2770	2706	241
Exogleic podzol	Albota	2380	2560	2780	2322	2510	145
Mean		2011	2480	2480	2444	2390	141
		Non-irrigated					
Forest brown	Tg. Mureş	1677	1940	1810	1810	1880	146
Podzol and brown medium forest soil	Livada	2180	2511	2390	2333	2280	110
Mean		1913	2233	2110	2080	2080	83

over two and three years in four soil types, in response to inoculation and/or fertiliser. The highest yields were recorded in the inoculated treatments, inoculation alone providing gains ranging from 16 to 78 %. In fertile chernozem-like soils, e.g. those from Fundulea or Mărculeşti, the plants' requirement for nitrogen was met by inoculation only, whereas in less fertile soils, e.g. the exogleic podzol in Albota, inoculation did not eliminate the need for fertiliser (Balan *et al.*, 1973).

Based on the data in Table 22.4, regression equations for nitrogen fertilisation with or without inoculation (Fig. 22.5) were computed. Without inoculation, for all four soil types nitrogen fertilisers markedly

Table 22.4. *The effect of inoculation with* R. japonicum *SO-69 and nitrogen fertiliser application on soybean yield* (*kg/ha*)

	Nitrogen level (kg N/ha)							
	Non-inoculated				Inoculated			
Soil type and location	0	30	60	90	0	30	60	90
Exogleic podzol, Albota (1970–2)	1530	1970	1780	1960	1800	2250	2190	1940
Leached medium chernozem, Fundulea (1970–1)	2360	2550	2610	2440	2740	2680	2730	2500
Chestnut-coloured chernozem, Mărculeşti (1970–1)	1350	1640	1810	2090	2400	2290	2310	2220
Carbonated-chestnut-coloured chernozem, Valul Traian (1970–1)	2090	2200	2550	2720	2600	2570	2660	2760

increased production, the best economic yield being obtained with a fertilisation rate of 95 kg N/ha. With inoculation, the best response to nitrogen was at 23 kg N/ha. The utilisation of nitrogen applied at inappropriate rates might under some circumstances contribute to lower yields, by adversely influencing nodule formation and activity.

Thus, on the leached medium chernozem in Caracal, increasing the

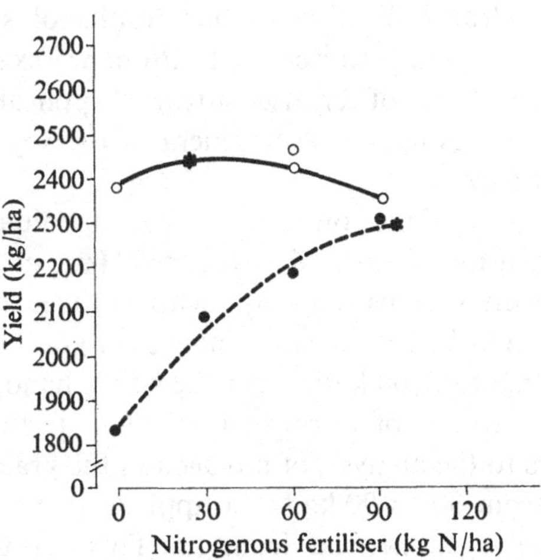

Fig. 22.5. Effect of nitrogen fertilisation on yield of irrigated soybeans. Inoculated plants (solid line) show maximum response (asterisk) at 23 kg N/ha, uninoculated plants (dashed line) at 95 kg N/ha. Regression lines for the two graphs are: $y = 2387+2.75x-0.0353x^2$ and $y = 1840+8.50x-0.0389x^2$ respectively.

nitrogen fertiliser rate both for the inoculated and uninoculated treatments, decreased the incidence of plants with nodules, the number of nodules per plant and average nodule volume and weight, as well as their content of nitrogen and iron (Table 22.5). In all cases the recorded yield was higher for inoculated plants than for uninoculated plants and was smaller for the 100 kg N/ha as compared with the 50 kg N/ha rate, particularly in inoculated treatments (Balan *et al.*, 1973).

Table 22.5. *The influence of nitrogen fertiliser and inoculation on the yield of soybean and on nodulation in the leached medium chernozem soil at Caracal* (1968–9)

Fertiliser (kg N/ha)	Inoculation	Yield (kg/ha)	% plants with nodules	No. of nodules: Main root	Lateral root	Total	Nodule volume (mm³)	Weight (mg): Fresh	Dry	Total nitrogen (mg)	Iron (mg)
0	Non-inoculated	2203	68.1	0.9	12.1	13.0	464	475	150	8.6	3.09
	Inoculated	2757	97.2	3.7	25.9	29.6	1130	1204	282	15.9	5.62
50	Non-inoculated	2587	33.3	0.9	5.8	6.7	384	365	89	5.04	2.05
	Inoculated	2817	70.8	1.6	16.5	18.1	854	884	223	12.3	4.70
100	Non-inoculated	2499	15.8	0.2	2.7	2.9	276	216	52	3.1	0.47
	Inoculated	2565	48.7	0.8	3.8	4.6	388	347	93	5.8	3.26

Because of a clear indication in our results of synergistic effects between inoculation and fertiliser application, we examined the interaction of rates and times of fertiliser nitrogen application and inoculation at different stages in the development of the soybean plant, using ^{15}N-labelled fertiliser.

Experiments were done on the leached medium chernozem at Fundulea, 1972, using inoculated soybeans (Hera *et al.*, 1972). Table 22.6 shows that the various rates, application times and techniques of nitrogen fertilisation had no positive effects on grain production, yield ranging from 3020 to 3360 kg/ha. On the other hand, clear-cut differences in the absorption of nitrogen applied as fertiliser were noted; Table 22.6 refers to the analysis of nitrogen in the grain.

An increase from 30 to 90 kg N/ha applied at planting, caused the percentage of nitrogen derived from the fertiliser to increase from 6 to 24.7 % and hence the amount of nitrogen absorbed from the fertiliser to increase from 9.5 to 39.5 kg N/ha. This produced an increase in the fertiliser utilisation rate from 32.0 % for 30 kg N/ha to 44.0 %

for 90 kg N/ha. Because grain production and total amount of absorbed nitrogen are practically constant over the range 0 to 90 kg N/ha, the higher rate of nitrogen absorption from fertiliser occurred at the expense of uptake of soil nitrogen and symbiotic nitrogen fixation.

When nitrogen fertiliser was applied within the growth period instead of at planting, the rate of utilisation was less whilst the amount of nitrogen derived from soil and symbiotic activity was at least equal to that obtaining without fertiliser. Fertilising during the growth period in soybean is thus more efficient than fertilisers applied at planting.

When fertilisers were applied during the growth period as well as at planting (30 or 60 kg N/ha), a decrease in the absorption of the nitrogen from soil and symbiotic activity was noted as well as an increase in the rate of fertiliser utilisation; this was accounted for by the detrimental effect of nitrogen fertiliser applied at planting time as a starter.

Table 22.6. *The influence of rate and time of application of nitrogen fertiliser on nitrogen uptake by nodulated soybeans*

Nitrogen levels (kg N/ha)				Nitrogen absorbed (kg/ha)				
At planting	During growth	Yield (kg/ha)	% nitrogen	From fertilisers	From soil and symbioses	Total	% nitrogen derived from fertiliser	Fertiliser utilisation rate (%)
—	—	3110	5.21	—	162.0	162	—	—
30	—	3020	5.25	9.5	149.5	159	6.0	32
60	—	3111	5.18	25.5	135.5	161	15.8	42
90	—	3022	5.28	39.5	120.5	160	24.7	44
—	—	3110	5.21	—	162.0	162	—	—
—	30	3360	5.63	6.1	182.9	189	3.2	20
—	60	3290	5.73	16.0	172.0	188	8.5	27
30	—	3020	5.25	9.5	149.5	159	6.0	32
30	30	3111	5.34	18.3	147.7	166	11.0	30
30	60	3260	5.30	27.3	145.7	173	15.8	30
60	—	3110	5.18	25.5	135.5	161	15.8	42
60	60	3030	5.31	35.7	125.3	161	22.2	30

Similar results were recorded with regard to nitrogen content in the stems (Table 22.7). With nitrogen applied during the growth period, fertiliser utilisation rate was less for the same amount of absorbed nitrogen. For the whole of the plant (grain plus stems), the total amount of absorbed nitrogen was greater when the fertiliser was applied during the growth period (220–230 kg N/ha) than when applied at planting (190–210 kg N/ha); see Table 22.8. The rate of utilisation of nitrogen fertiliser decreases in the following order: fertilisation at planting only; starter fertilisation at planting plus nitrogen application during the

Table 22.7. *The influence of rate, time and methods of application of nitrogen fertilisers on nitrogen content of the soybean stem*

Nitrogen level (kg N/ha)				Nitrogen absorbed (kg/ha)			% nitrogen derived from fertiliser	Fertiliser utilisation rate (%)
At planting	During growth	Yield (kg/ha)	% nitrogen	From fertiliser	From soil and symbioses	Total		
—	—	4580	0.67	—	30.7	30.7	—	—
30	—	4400	0.68	3.68	26.2	29.9	12.3	12.3
60	—	4870	0.84	9.16	31.7	40.9	22.4	15.3
90	—	4980	0.95	15.14	32.2	47.3	32.0	16.8
—	—	4580	0.67	—	30.7	30.7	—	—
—	30	4810	0.70	2.02	31.7	33.7	6.0	6.7
—	60	4800	0.80	4.0	34.2	38.4	11.0	7.0
30	—	4400	0.68	3.68	26.2	29.9	12.3	12.3
30	30	5160	0.95	7.64	41.4	49.0	15.6	12.7
30	60	5180	0.87	8.97	36.1	45.1	19.9	10.0
60	—	4870	0.84	9.16	31.7	40.9	22.4	15.3
60	60	4920	0.86	12.77	29.5	42.3	30.2	10.6

growth period; fertilisation during the growth period only. The amount of nitrogen derived from soil and symbiotic activity increased in the same order.

Our data thus show an important interaction between inoculation and nitrogen fertiliser application, by demonstrating a potentially detrimental effect of nitrogen applied at planting on symbiosis, whereas

Table 22.8. *Total nitrogen uptake by soybeans (grain and stem) depending on the rates and times of nitrogen fertilising*

Nitrogen level (kg N/ha)		Nitrogen absorbed (kg/ha)			Fertiliser utilisation rate (%)
At planting	During growth	From fertiliser	From soil and symbioses	Total	
—	—	—	192.7	192.7	—
30	—	13.2	175.3	188.5	44.0
60	—	34.6	167.4	202.0	57.7
90	—	54.5	152.3	206.8	60.5
—	—	—	192.7	192.7	—
—	30	8.1	214.7	222.9	27.0
—	60	20.2	206.7	226.9	33.4
30	—	13.2	175.3	188.5	44.0
30	30	25.9	189.2	215.1	43.2
30	60	36.3	181.6	217.9	40.3
60	—	34.6	167.4	202.0	57.7
60	60	48.5	154.7	203.2	40.4

application within the growing period is accommodated by the biological control mechanism of soybean plants that determines uptake and fixation rates. In this way fertiliser nitrogen increases the assimilation of nitrogen from soil and from symbiotic activity.

These data show the need for a thorough study of the interaction between inoculation and nitrogen fertilisation in order to ascertain the actual contributions of symbiotic activity and inhibitory factors and the point where their influence becomes deleterious, so as to obtain most benefit from biological nitrogen fixation.

References

Balan, N. (1970). Importanţa bacterizării la culturile de plante leguminoase. *Revista agricultura,* No. 13.

Balan, N., Dinca, D., Galbenu, E., Munteanu, O. & Tabaranu, T. (1968). Selecţia de tulpini bacteriene din populaţiile autohtone de *Rhizobium japonicum. Analele ICCPT, Fundulea,* Ser. B, **36,** 593–617.

Balan, N., Galbenu, El., Balan, C., Isfan, D., Armeanu, M., Enciu, M., Enescu, P., Pascu, A. M., Picu, I. & Tianu, A. (1973). Cercetări asupra relaţiei dintre bacterizare cu azot la soia. *Analele ICCPT, Fundulea,* Ser. B, **40,** in press.

Diaconescu, O. & Miclea, E. (1971). *Soia.* Editura Ceres, Bucharest.

Harper, J. E. & Cooper, R. L. (1971). Nodulation response of soybeans to application rate and placement of combined nitrogen. *Crop Science,* **11,** 438–40.

Hera, C. (1971). Response and current fertilization practice of grain legume crops in Romania. In: *Proceedings of a panel on the use of isotopes for study of fertilizer utilization by legume crops.* IAEA Publication No. 149, pp. 158–82. International Atomic Energy Agency, Vienna.

Hera, C., Suteu, G., Triboi, E., Mihaila, V., Bologa, M., Burlacu, G. & Stanciu, A. (1972). Cercetări cu ajutorul izotopilor ^{15}N şi ^{32}P privind sistemul de fertilizare la soia. *Analele ICCPT, Fundulea,* Ser. B, **39,** 309–14.

Hulpoi, N., Picu, I. & Tianu, A. (1971). Cercetări privind aplicarea îngrăşămintelor la culturile de cîmp irigate. *Probleme Agricole,* **8,** 48–50.

Nelson, W. L. (1971). Fertilization of soybeans. *Oleagineaux,* **26,** 101–6.

Nutman, P. S. (1963). Factor influencing the balance of mutual advantage in legume symbiosis. In *Symbiotic associations,* pp. 51–71. Cambridge University Press, London.

Weber, C. R. (1966). Nodulating and nonnodulating soybean isolines. *Agronomy Journal,* **58,** 43–9.

23. Inoculation and nitrogen fertiliser experiments on soybeans in Cuba

E. SISTACHS

The search for protein sources has become an important part of the livestock improvement programmes which have been developed in Cuba over the last few years. Soybeans (*Glycine max.* (L.) Merril) are of particular interest in this connection and the potential productivity and general agronomic properties of some introduced varieties are being examined (Zambrana, 1972; Sistachs, unpublished data). The fact that soybeans can use, in addition to nitrogen from the soil, atmospheric nitrogen made available by their symbiotic root nodule bacteria is especially important in developing countries where nitrogen is often the limiting factor in plant production.

The present work was conducted to study the symbiotic response of Pelikan variety of soybeans to inoculation with the recommended Australian strain of *Rhizobium japonicum* CB-1809. The effect of different amounts of nitrogen fertiliser applied at different stages of plant growth was also measured.

Materials and methods

The experiment was done during 1972 in a Truffin clay soil (Bennett & Allison, 1922) with a pH of 5.5. Berra & Preston (1967) have given details on general climatic characteristics of the region. A split-plot design replicated four times was used with inoculation or no inoculation in the main plots and different nitrogen fertiliser treatments as the subplots. Zero, 25, 50 or 75 kg N/ha (as urea) was applied at planting or when the cotyledon reserves had been exhausted. The subplots were 15 m long and 2.8 m wide with a net experimental area of 18.2 m^2. A direct drilling machine was used to sow the seeds on a soil which had lain fallow for two years. A basal phosphorus and potassium fertiliser at 30 kg/ha each was also applied with the drilling machine at sowing. Immediately after sowing 3 l/ha of Gramoxone was applied to control grass regrowth. Sowing was on 28 September 1972 (later than is normal for soybeans), in rows 70 cm apart with 7 cm sowing intervals. The urea was added as a side-dress application. Germination was by spray irrigation. *Rhizobium japonicum* strain CB-1809 was used for seed inoculation and a heavy suspension prepared from an agar culture applied to the seed immediately before sowing. The plots were sampled on 29 November 1972 at an advanced flowering stage to evaluate nodulation. Nodulation was examined in all the plants included in 1.0 m long strips chosen at random from the two outside rows of each plot. The crop was harvested on 27 December 1972 and the total yield and nitrogen content of the seed determined.

Results

An apparently highly effective nodulation occurred on all the plants sampled from the inoculated plots at flowering; plants from the uninoculated plots were without nodules. The direct drilling method used on a clay soil did not allow reliable measurement of nodule number or mass.

Table 23.1. *Yield responses of soybeans to inoculation with Strain CB-1809 and nitrogen fertiliser (kg/ha at 13 % humidity)*

Nitrogen treatment	Uninoculated	Inoculated
None	515.36[l]	863.62[defghijk]
At planting		
25 kg N/ha	944.03[cdefghi]	1068.73[bcdef]
50 kg N/ha	993.48[cdefgh]	1033.57[cdefg]
75 kg N/ha	1450.56[a]	1307.60[ab]
After cotyledon exhaustion		
25 kg N/ha	1098.01[cd]	960.05[cdefghi]
50 kg N/ha	1134.85[bc]	777.96[ghijk]
75 kg N/ha	1094.74[bcde]	1427.96[a]

[abcdefghijkl] values without letter in common differ at $P = 0.05$.

Table 23.1 shows yield responses to inoculation and nitrogen fertilisation at two different stages of plant growth. Strain CB-1809 proved to be effective for producing yields equivalent to those achieved with a nitrogen fertilisation up to 50 kg/ha; heavier application of fertilizer always resulted in higher grain production. Thus, 75 kg N/ha gave an increase in grain yield of 69 % over inoculation alone. The interaction between the factors studied was highly significant ($P = 0.01$). Additions of up to 50 kg N/ha to uninoculated plants produced significant yield increases regardless of the time of application. Increases of up to 120 % over the unfertilised non-nodulated plants were achieved *with*

50 kg N/ha. Responses of nodulated plants to lower nitrogen rates depended on the time of application. If the fertiliser was applied at planting, the yield improved, though not significantly, but when added after cotyledon exhaustion it had a depressive effect. Responses to heavier nitrogen fertilisation depended both on the time of application and occurrence of nodulation. Nodulated plants with up to 50 kg N/ha added at planting tended to give better yields than uninoculated plants

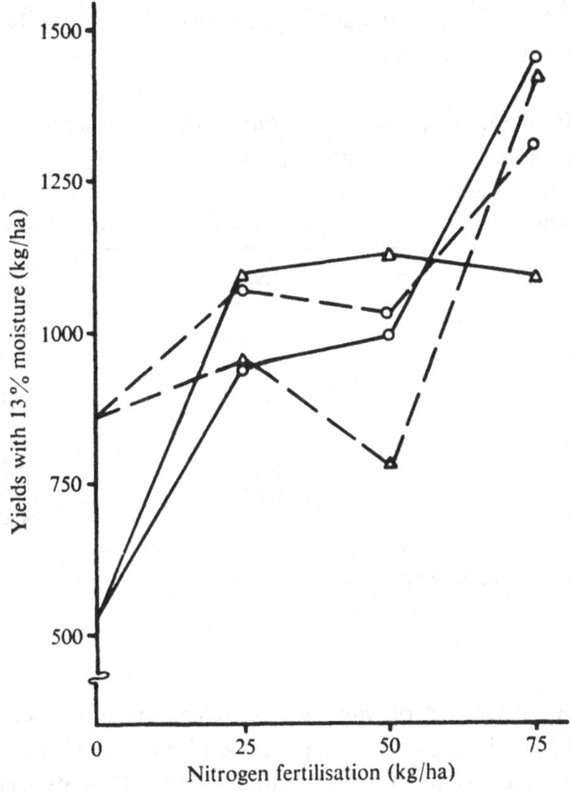

Fig. 23.1. Yield responses of soybeans to inoculation and nitrogen fertilisation. Uninoculated (——) and inoculated (---) plants were given nitrogen at planting (○) or after cotyledon exhaustion (△).

with the same fertiliser treatment (see Fig. 23.1). This tendency decreased gradually from high levels of significance ($P = 0.01$) when no nitrogen was added, to no difference at the two heavier nitrogen rates. Responses to delayed additions of nitrogen were contradictory. Twenty-five kg N/ha was apparently enough to inhibit nodule function and higher rates were rather depressive as previously noted. However, no yield responses were obtained to heavier application of nitrogen unless the plants were nodulated (see Fig. 23.1).

Table 23.2 shows the percentages of nitrogen in the seed. Although no statistical significance can be attached to the differences between treatments, it appears that even the highest rates of fertiliser nitrogen did not increase the nitrogen content of the seed to that of the inoculated plants given no fertiliser. In contrast, the nitrogen content of the seed of nodulated plants tended to be negatively related to fertiliser nitrogen at planting, while delayed nitrogen application had no effects on the nitrogen content of the seed.

Table 23.2. *Percentage of nitrogen in the seeds of soybeans as influenced by inoculation and nitrogen fertilisation*

Nitrogen treatment	Uninoculated	Inoculated
None	5.92	6.40
At planting		
25 kg N/ha	5.83	6.22
50 kg N/ha	5.32	5.94
75 kg N/ha	6.00	6.10
After cotyledon exhaustion		
25 kg N/ha	5.89	5.97
50 kg N/ha	6.20	5.94
75 kg N/ha	6.16	6.01

No significant differences were attributable to treatments.

Discussion

Photoperiodic responses of varieties of soybeans have been reviewed by Howell (1960). Pelikan variety must be planted in Cuba in late July or during August to attain maximum yields. This experiment was planted in late September, which resulted in an early flowering and accounted for the generally low grain yields obtained. Soybeans as well as the inoculum applied are sensitive to salt concentration, so the planting machine was arranged so that the basal fertiliser (phosphorus and potassium) was broadcast on the soil surface; this may have reduced the availability of these elements.

The higher yields obtained with nitrogen fertilisation support the widely accepted view that soybeans cannot attain maximum yields when wholly dependent on symbiotically fixed nitrogen (Weiss, 1949;

Ohlrogge, 1960; Ohlrogge & Kamprath, 1968). This apparent inadequacy of the symbiotic mechanism is supported by Allos & Bartholomew (1959), who concluded that only about one-half to three-quarters of the total nitrogen for maximum yields is generally supplied by the fixation process. However, in more recent studies with nodulating and non-nodulating soybean isolines, no better yields were obtained from nitrogen-fertilised non-nodulating lines than from unfertilised nodulated plants. The well-nodulated variety gave no significant yield response to nitrogen fertilisation (Wagner, 1962; Weber, 1966*a*, *b*). Beard & Hoover (1971) also report no benefit in yields from nitrogen fertiliser up to 112 kg/ha.

In our experiment strain CB-1809 of *R. japonicum* was used as an inoculant in Cuba for the first time. It will be necessary to screen the Pelikan variety against a wider variety of inoculants before any conclusion can be drawn from the results as it may be that there are strains with a higher specificity for the variety than CB-1809. Plant factors in symbiosis must be borne in mind (Nutman, 1969). Soybeans have been intensively bred by man, generally regardless of their complex nodulation response. Norris (1967) described a glasshouse trial with nine strains of *R. japonicum* and five cultivars of soybeans and found their performance differed greatly from variety to variety. Strain CB-1809 was selected for outstanding performance, but even its effectiveness varied according to the soybean variety, although only to a small extent. Döbereiner, Arruda & Penteado (1966) and Arruda, Döbereiner & Germer (1968) found highly significant differences in the responses of twenty-five soybean varieties to inoculation. In their tests Pelikan showed poor nodulating ability.

Stimulatory effects of low rates of combined nitrogen on symbiosis are indicated, though they are not significant, in Fig. 23.1. Starter doses of fertiliser nitrogen have been reported to stimulate fixation in *Vigna sinensis* (Ezedinma, 1964; Dart & Mercer, 1965*b*; Dart & Wildon, 1970), *Phaseolus vulgaris* (Burton, Allen & Berger, 1961) and *Glycine max* (Norman & Krampitz, 1946). On the other hand, Allos & Bartholomew (1959) reported decreases in the symbiotic contribution as the combined nitrogen supply increased. Comparing the two curves in Fig. 23.1 for nitrogen fertiliser applied at planting, the differences in yield of nodulated and non-nodulated plants may be attributed to nitrogen fixation. This difference decreased gradually as increasing nitrogen was applied. A similar effect was explained by Oghoghorie & Pate (1971) to be due to a competition between the nitrogen-fixing nodules and the assimilatory centres of the vegetative plants for supplies of reductant and carbon skeletons.

Depressive effects by delaying nitrogen fertilisation until the cotyledons were exhausted are difficult to explain. Drastic effects of combined nitrogen on nodules with their bacteroids already formed have been reported by Dart & Mercer (1965*b*). Different responses to delayed heavier application of nitrogen to nodulated and non-nodulated plants could suggest that by the time the plants had exhausted their cotyledon reserves the nodules were already fixing nitrogen.

Nitrogen fertilisation has been reported as significantly decreasing the nitrogen content in the seeds of soybeans (Kushizaki, Ishizuka & Akamatzu, 1964) and black beans (Sistachs, 1970). Döbereiner & Arruda (1967) found similar results but in soybean tops. Data presented in Table 23.2 suggest such an effect. Very little is known about the ways in which legumes partition their dry matter and nitrogen between grain and 'non-usable' parts.

Our results suggest that an acceptable potential production of medium yield can be obtained from currently available soybean cultivars and *Rhizobium* strains using a small dressing of nitrogen fertiliser or none; a fact that is relevant in areas where nitrogen is not readily available.

I am grateful to Messrs R. Frías, P. Rodríguez and L. Avila for technical assistance. Thanks are due to the Biometrics Division of the Institute of Animal Science for their statistical analyses and to Rothamsted Experimental Station for providing strain CB-1809.

References

Allos, H. F. & Bartholomew, W. V. (1959). Replacement of symbiotic fixation by available nitrogen. *Soil Science*, **87**, 61–6.

Arruda, N. B., Döbereiner, J. & Germer, C. M. (1968). Inoculacao, adubacao nitro-genada e revestimento calcário ém tres variedades de soja (*Glycine max* (L) Merril). *Pesquisa Agropecuaria Brasileira*, **3**, 201–5.

Beard, B. H. & Hoover, R. M. (1971). Effect of nitrogen on nodulation and yield of irrigated soybeans. *Agronomy Journal*, **63**, 815–16.

Bennett, H. H. & Allison, R. V. (1922). *The soils of Cuba.* Tropical Plant Research Foundation, Washington, DC.

Berra, E. & Preston, T. R. (1967). Effect of population and levels of fertilizer nitrogen on yield of grain sorghum. *Revista cubana de Ciencia agricola* (English edition), **1**, 61–70.

Burton, J. C., Allen, O. N. & Berger, K. C. (1961). Effects of certain mineral nutrients on growth and nitrogen fixation of inoculated bean plants, *Phaseolus vulgaris* L. *Agricultural and Food Chemistry*, **9**, 187–90.

Dart, P. J. & Mercer, F. V. (1965*a*). The influence of ammonium nitrate on the fine structure of nodules of *Medicago tribulcides* Desr and *Trifolium subterraneum*, L. *Archiv für Mikrobiologie*, **51,** 233–57.

Dart, P. J. & Mercer, F. V. (1965*b*). The effect of growth temperature, level of ammonium nitrate, and light intensity on the growth and nodulation of cowpea (*Vigna sinensis* Endl. ex Hassk'.). *Australian Journal of Agricultural Research*, **16,** 321–45.

Dart, P. J. & Wildon, D. C. (1970). Nodulation and nitrogen fixation by *Vigna sinensis* and *Vicia atropurpurea*: the influence of concentration, form and site of application of combined nitrogen. *Australian Journal of Agricultural Research*, **21,** 45–56.

Döbereiner, J. & Arruda, N. B. (1967). Interrelacoes entre variedades e nutricao na nodulacao e simbiose da soja (*Glycine max* (L) Merril). *Pesquisa Agropecuaria Brasileira*, **2,** 475–87.

Döbereiner, J., Arruda, N. B. & Penteado, A. P. (1966). Especificidade hospedera, em variedades de soja, na simbiose com *Rhizobium*. *Pesquisa Agropecuaria Brasileira*, **1,** 207–10.

Ezedinma, F. O. C. (1964). Effects of inoculation with local isolates of cowpea *Rhizobium* and application of nitrate nitrogen on the development of cowpeas. *Tropical Agriculture of Trinidad*, **41,** 243–9.

Howell, R. W. (1960). Physiology of the soybean. *Advances in Agronomy*, **12,** 265–310.

Kushizaki, M., Ishizuka, J. & Akamatzu, F. (1964). Physiological studies on the nutrition of soybean plants. I. Effects on growth yield and nitrogen content of soybean plants. *Journal of the Science of Soil Manure, Tokyo*, **35,** 323–7.

Norman, A. G. & Krampitz, L. O. (1946). The nitrogen nutrition of soybeans. II. Effect of available soil nitrogen on growth and nitrogen fixation. *Proceedings of the Soil Science Society of America*, **10,** 191–6.

Norris, D. C. (1967). The intelligent use of inoculants and lime pelleting for tropical legumes. *Tropical Grasslands*, **1,** 107–21.

Nutman, P. S. (1969). Genetics of symbiosis and nitrogen fixation in legumes. *Proceedings of the Royal Society*, Ser. B, **172,** 417–37.

Oghoghorie, C. G. O. & Pate, J. S. (1971). The nitrate stress syndrome of the nodulated field pea (*Pisum arvense* L.). *Plant and Soil*, Special Volume, 185–202.

Ohlrogge, A. J. (1960). Mineral nutrition of soybeans. *Advances in Agronomy*, **12,** 229–63.

Ohlrogge, A. J. & Kamprath, E. J. (1968). Fertilizer use on soybeans. In *Changing patterns in fertilizer use*, ed. L. B. Nelson, pp. 273–95. Soil Science Society of America, Madison, Wisconsin.

Sistachs, E. (1970). Effect of N fertilization and inoculation on yield and N content of black beans (*Phaseolus vulgaris*). *Revista cubana de Ciencia agricola*, (English edition), **4,** 227–30.

Wagner, G. H. (1962). Nitrogen fertilization of soybeans. Missouri Agricultural Experiment Station Bulletin No. 797.

Weber, C. R. (1966*a*). Nodulating and non-nodulating soybean isolines. 1. Agronomic and chemical attributes. *Agronomy Journal*, **58,** 43–6.

Weber, C. R. (1966*b*). Nodulating and non-nodulating soybean isolines. 2. Response to applied nitrogen and modified soil conditions. *Agronomy Journal*, **58**, 46–9.

Weiss, M. G. (1949). Soybeans. *Advances in Agronomy*, **1**, 90–7.

Zambrana, T. (1972). Instituto de Ciencia Animal: Report for 1971 and 1972. Instituto de Ciéncia Animal, Havana, Cuba.

24. Field and greenhouse experiments on the response of legumes in Egypt to inoculation and fertilisers

Y. A. HAMDI

Research on symbiotic nitrogen fixation is a recent development in Egypt. Selim (1947) presented the first Arabic review of root nodule bacteria and their importance in agriculture and Fahmy (1955) reviewed the role of these organisms in soil fertility. Selim (1947) prepared a liquid-type inoculant in the Department of Microbiology of the Faculty of Agriculture at Cairo University over the period 1931–46. Before this Forbes (1921) imported a few rhizobia cultures and inoculants for experimental purposes from the United States Department of Agriculture (USDA). Liquid-type cultures were also prepared by the section of Agricultural Chemistry at the Ministry of Agriculture from the early forties until 1954. Difficulties were encountered with this type of inoculant which led Loutfi & Fahmy (1958) to develop the solid base carrier 'Okadin'. With this development the production and application of legume inoculants progressed and research was extended to evaluate the response of different legumes to inoculation and their nitrogen-fixing capacities. This chapter presents a summary of the work.

Response to inoculation

Loutfi *et al.* (1966*a*) studied the response of broad beans (*Vicia faba*) to inoculation at four experimental farms in different parts of the country; at two sites there was a significant response to inoculation (Table 24.1). The average yield improvement from inoculation was 199.3 kg/ha and 287.8 kg/ha for the two significant values.

The field response of peanut to inoculation and to nitrogen and phosphorous fertilisers was investigated by Loutfi *et al.* (1966*b*). The results of four field experiments which were run in three sandy and one loamy soil at four different localities are shown in Table 24.2. As can be seen from the table, yield compared with unfertilised plants was only increased significantly by a fertiliser treatment of 233 kg P/ha+117 kg N/ha. The average increase in the yield of peanut was 212.5 kg/ha from inoculation without fertilisers. With inoculation and 233 kg superphosphate/ha, an increase of 312.4 kg/ha was obtained and with inoculation and 233 kg/ha superphosphate with 117 kg calcium nitrate, an increase of 444.5 kg/ha. However, with two applications of 117 kg N/ha, 303.5 kg/ha of peanut were obtained. Inoculation, therefore, increased the yield of peanut only with superphosphate and calcium nitrate at the levels of 233 and 117 kg/ha, respectively.

Table 24.1. *Effect of inoculation on horse bean yield in Delta and Upper Egypt*. (After Loutfi *et al.*, 1966*a*)

	No. of trials	Gain yield kg/ha	Gain yield %
Delta	1	221.4	12.2
Upper Egypt	3	173.5	5.6
Total	4	199.3	8.9

Similar studies were carried out by Taha, Mahmoud & Salem (1967*a*, *b*) to determine the response of broad bean (Table 24.3) and lentils (Table 24.4) to inoculation and the application of phosphorus and nitrogen fertilisers. Application of phosphorus to the soil without inoculation generally gave a marked increase in yield of both lentil and broad bean. Inoculation also increased the yield of both crops, but when combined with phosphorus gave significantly higher yield than

Table 24.2. *Increase in yield (kg/ha) of peanut as a result of inoculation and fertilising with superphosphate and calcium nitrate*. (After Loutfi *et al.*, 1966*b*)

Fertiliser (kg(ha)	Localities: Kafr-el-Gabal	Kafr-el-Gabal	Badr	Bilbess	Average
None	423.0	267.7	23.2	166.0	212.5
233 P+0 N	419.5	362.4	192.8	276.7	312.4
233 P+117 N	669.4 246.3	790.8 523.0	146.4	176.7	444.5
233 P+233 N	133.9 298.2	342.7 75.0	158.9	176.7	303.5
Average	410.5	440.9	130.3	199.9	294.5

inoculation alone. The best economic rate of superphosphate was 233 kg/ha for lentil and from 117 to 233 kg/ha for broad bean.

Taha *et al.* (1967*b*) showed that calcium nitrate increased the seed yield of lentils (Table 24.5) and broad beans (Table 24.6) in the uninoculated plots. Inoculation and application of calcium nitrate up to

Table 24.3. *Effect of inoculation in presence of superphosphate on broad bean yield and the total nitrogen content of the seeds.* (After Taha *et al.*, 1967*a*)

Superphosphate (kg/ha)	Mean yield (kg/ha)		Increase in yield (kg/ha)	Total nitrogen of seeds (mg/g)		Increase in nitrogen (mg/g)
	Uninoculated	Inoculated		Uninoculated	Inoculated	
0	1310	1550	240	62.9	65.4	2.5
117	1572	1716	144	64.3	68.5	4.2
233	1716	1856	140	64.9	72.4	7.5
350	1786	1856	70	65.2	70.7	5.5
466	1642	1893	251	65.4	73.4	8.0

117 kg/ha showed a significant increase in the yields of seeds of both broad bean and lentils. Further increase of nitrogen fertiliser reduced the yield of seeds but increased the dry weight of plants (straw) and deleteriously affected nodulation. Application of increasing amounts of nitrogen fertiliser progressively increased the total nitrogen of lentil seeds in both inoculated and uninoculated treatments. Inoculation alone increased the total nitrogen of both lentil and broad bean seeds.

Table 24.4. *Effect of inoculation in the presence of superphosphate on lentil yield and the total nitrogen content of the seeds.* (After Taha *et al.*, 1967*a*)

Superphosphate (kg/ha)	Mean yield (kg/ha)		Increase in yield (kg/ha)	Total nitrogen of seeds (mg/g)		Increase in nitrogen (mg/g)
	Uninoculated	Inoculated		Uninoculated	Inoculated	
0	1376	1638	262	40.5	50.2	9.72
117	1638	1733	95	40.6	51.2	10.60
233	1615	1873	258	41.4	52.2	10.85
350	1638	1873	236	42.5	57.0	14.46
466	1756	1972	217	44.8	58.3	13.52

For these crops the recommended rate of fertilisers with inoculation was 233 kg superphosphate and 117 kg calcium nitrate/ha.

Ashour *et al.* (1969) studied the effect of inoculation and fertilisers on the soybean variety Hampton cultivated in pots containing 40 kg of Nile silt and fertilised with 6 or 12 g/pot of calcium nitrate. Inoculation alone or fertilisation alone increased the dry weight and nitrogen content of the soybean plants. Addition of nitrogen to the inoculated

Table 24.5. *Effect of calcium nitrate and inoculation in the presence of superphosphate (233 kg/ha) on lentil yield and total nitrogen content of the seeds.* (After Taha *et al.*, 1967*b*)

Calcium nitrate (kg)	Mean yield of seeds (kg/plot)		Increase in yield (kg/plot)	Nitrogen content of seeds (mg/g)		Increase in nitrogen (mg/g)
	Un-inoculated	Inoculated		Un-inoculated	Inoculated	
58	12.4	15.6	3.2	41.40	45.90	5.50
117	13.4	17.0	3.6	44.76	48.34	3.58
175	13.6	19.2	5.6	53.00	56.86	3.86
233	14.6	17.6	3.0	57.33	62.56	5.23
291	16.0	16.2	0.2	59.03	61.26	2.23

LSD for inoculation at 5 %, 1.05 kg/plot; LSD for treatments at 5 %, 1.66 kg/plot.

plants increased the growth and the nitrogen content more than either inoculation or fertiliser alone. Inoculation increased the weight of seeds, pods and the number of seeds per plant (Table 24.7). Addition of nitrogen in combination with inoculation markedly increased several other yield components (Table 24.8). The percentage of crude protein in the seeds seemed to be more sensitive to nitrogen fertilisation than to inoculation. The percentage oil in the seeds was not affected by inoculation or by nitrogen fertiliser.

Table 24.6. *Effect of calcium nitrate and inoculation in the presence of superphosphate on broad bean yield.* (After Taha *et al.*, 1967*b*)

Calcium nitrate (kg)	Yield of seeds (kg/plot)	Increase in yield (kg/plot)	Yield of seeds (kg/ha)	Increase in yield (kg/ha)
58	21.2	–	2524	–
117	26.0	4.8	3096	572
175	24.0	2.8	2856	332
233	23.4	2.2	2786	262
291	24.0	2.8	2856	332

The response of the soybean varieties Hampton, Hill and Lee, cultivated in pots filled with Nile silt, to single and composite strains of *R. japonicum* grown together or grown separately and then combined before inoculation was evaluated by Hamdi *et al.* (1968). Host response (Table 24.9) to strain E41 was significantly different from the control. Inoculation with mixed strains gave inconsistent results.

Table 24.7. *Effect of inoculation and nitrogen fertiliser on the yield (g/plant) of soybean plants variety Hampton.* (After Ashour *et al.*, 1969)

Calcium nitrate (g/40 kg pot)	Inoculated			Uninoculated		
	No. of seeds	Weight of seeds	Weight of pods	No. of seeds	Weight of seeds	Weight of pods
0	91	18.2	24.4	75	12.0	17.2
6	124	20.2	28.5	91	13.2	21.6
12	147	25.6	37.1	76	14.1	23.4

Hamdi, Abd-el-Samea & Loutfi (1973) evaluated the response of soybean variety Clark to the amount of *R. japonicum* inoculated onto the seed. The seeds were inoculated at the rate of 28×10^6, 28×10^5 and 28×10^4 cells/seed. Inoculated seeds were planted in field plots or in pots containing the same silty loam soil and kept under greenhouse conditions at Giza Research Station. Nodulation was abundant within three weeks in the pot experiments where seeds were inoculated with 28×10^6 cells/seed. No nodules were observed after three weeks under field conditions or in plants in pots receiving 28×10^4 cells/seed.

After fifty-seven days in the field, nodulation was between 87 to 92 % when rhizobia were inoculated at the rate of 28×10^6/seed.

Table 24.8. *Effect of inoculation and nitrogen fertiliser on the percentage nitrogen, crude protein and oil in soybean seeds of variety Hampton.* (After Ashour *et al.*, 1959)

Calcium nitrate (g/40 kg pot)	Inoculated			Uninoculated		
	Nitrogen	Crude protein	Oil	Nitrogen	Crude protein	Oil
0	1.0	6.3	20.4	0.4	5.6	21.1
6	5.5	34.4	20.8	4.6	28.8	21.1
12	6.8	42.5	21.3	6.1	38.1	20.0

Nodulation dropped to 47 % with an inoculum of 28×10^4 cells/seed. Nodulation of 100 % was seen in plants receiving 28×10^6 cells/seed in pot cultures and decreased to 50 % in those inoculated with 28×10^4 cells/seed. After 102 days, nodulation frequency, fresh weight of nodules and dry weight of stems and pods were determined; it was found that they increased with increasing numbers of rhizobia in the inoculum (Table 24.10).

Table 24.9. *Nitrogen contents (g/100 plants) of soybean varieties inoculated with various strains of* R. japonicum. (After Hamdi *et al.*, 1968)

	Soybean varieties		
R. japonicum strain	Hampton	Hill	Lee
UAR-1	2.8	1.9	2.4
E40	3.0	6.1	3.7
E41	6.7	7.1	5.9
E45	4.6	5.6	4.2
511	3.3	5.4	4.2
524	4.0	3.3	3.4
UAR-1+E40*	6.2	5.3	4.8
UAR-1+E41*	4.2	4.6	4.6
UAR-1+E45*	4.8	5.6	4.6
UAR-1+511*	5.7	6.6	5.2
UAR-1+524*	5.6	4.9	4.0
UAR-1+E40†	6.0	5.2	2.2
UAR-1+E41†	3.8	3.9	2.8
UAR-1+E45†	4.6	3.7	3.2
UAR-1+511†	5.1	3.6	3.5
UAR-1+524†	4.6	5.0	5.2
Control	3.7	3.5	3.6

LSD between inocula and control, 1.6.
* Cultures of the two strains grown together.
† Cultures of the two strains grown separately and mixed just before planting.

In co-operation with a USDA project for developing more effective strains, Loutfi *et al.* (1968) reported that out of twenty foreign strains four were significantly more effective than local strains.

Response of inoculated cowpea to boron and molybdenum was studied by Hamdi, Moharram & Loutfi (1972). Seeds of cowpea were planted in earthenware pots, filled with Nile silt and inoculated with one of three rhizobial strains of the cowpea miscellany. Boron, molybdenum or a mixture of the two elements were added after one week, in solution, at the levels of 0, 4, 6, 10 and 12 ppm. Addition of molybdenum, boron or both usually significantly increased dry weight and nitrogen content. With strain 600, the nitrogen content of the stems was significantly increased by addition of 4, 6 or 10 ppm boron and molybdenum, but nitrogen content of pods was increased only by the 4 and 10 ppm additions. Twelve ppm of either of the elements caused no significant additional response with this strain.

Table 24.10. *Response of soybean var. Clark to the number of* R. japonicum *cells/seed used as inoculum after 102 days under field conditions*. (After Hamdi *et al.*, 1973)

R. japonicum (cells/seed)	Dry weight (g/plant)	Nodulation %	Nodulation Fresh weight (g/plant)	Pods per plant Number	Pods per plant (g dry wt)
Uninoculated control	9.8±1.0	0	0	33.0±2.63	8.7±1.8
28×10^6	12.1±1.2	93.3	1.30	46.5±6.41	17.7±3.0
28×10^5	10.5±1.9	81.0	0.70	28.0±5.9	14.7±3.7
28×10^4	9.7±1.4	57.0	0.30	31.1±9.1	8.5±2.2

The effect of seed inoculation and fertiliser application on the rhizosphere microflora of chickpea (*Cicer arietinum*) was evaluated by Badawy (1970). Inoculated plants harboured larger total microbial populations in their rhizospheres than uninoculated plants. Seed inoculation increased the number of methylene blue reducers as well as producers of gas and acid from glucose, and reduced the number of fungi, aerobic spore-formers and ammonifiers in the rhizosphere of unfertilised plants. The number of fungi in the rhizosphere of both inoculated and uninoculated plants was markedly increased by fertilising with nitrate and superphosphate. Fertilising, however, reduced the number of sporulating bacteria and had no consistent effect on the physiological groups of micro-organisms. In root-free soil, the counts of total microflora, fungi, sporulating bacteria etc. were lower than in the rhizosphere area and were not markedly affected by fertilisers.

The response of nine varieties of *Phaseolus vulgaris*, viz: Seminole, Contender, Giza 3, Triumph, Resistant Tender Green, Perla, Regalfin, Processor and Tender Long to the inoculation with *R. phaseoli* strains 402, 403, 404, 405 and D-400 was studied by El-Nadi *et al.* (1972). Plants were grown in sterile Nile silt soil for fifty days, after which examination of nodulation took place (Table 24.11). Strain 402 nodulated all varieties except the Tender Long. Strain 403 did not nodulate Seminole, Triumph, Resistant Tender Green, Tender Long or

Regalfin and 404 failed to nodulate Seminole and Tender Long. Strain 405 was non-infective on Seminole and Perla. These results indicated that the absence of the proper rhizobia may contribute to the failure of bean varieties to nodulate.

In a further study, Nasser *et al.* (1972) examined failure of nodulation in soils normally cropped to vegetables. Seeds of the bean variety Seminole were planted in pots containing sterile and non-sterile Nile silt and soils normally cropped with vegetables. The seeds were inoculated with rhizobia at 5×10^7 or 5×10^3 cells per pot at the time of

Table 24.11. *Average number of nodules per five plants of different varieties of* Phaseolus vulgaris *inoculated with different strains of* R. phaseoli. (After El-Nadi *et al.*, 1971)

	R. phaseoli strains					
Phaseolus variety	402	403	404	405	D-400	Mean
Contender	22.6	2.0	3.6	21.6	18.3	13.7
Seminole	2.6	–	0.3	0.6	3.3	1.4
Giza 3	8.3	2.3	4.6	3.3	0.6	3.9
Triumph	5.0	–	5.3	5.0	16.3	6.3
Resistant Tender Green	2.0	–	8.0	22.3	6.7	8.0
Processor	7.0	3.3	2.0	21.0	8.6	8.4
Tender Long	0.3	0.3	0.7	4.0	10.7	3.2
Perla	3.3	3.7	1.3	–	20.3	5.7
Regalfin	3.3	0.7	5.0	19.7	33.0	12.7
Mean	7.8	1.7	4.4	13.9	17.0	

Less than one nodule per 5 plants is considered in the text as failure to nodulate.
LSD at 5 % for variety, 3.3; LSD at 5 % for strains, 7.5.

planting or seven days later. Survival of rhizobia in the soils was determined by the plate count method. Because nodulation was absent or sparse in the vegetable-cropped soils, whether or not they were sterilised, the soils were examined for the presence of *Rhizobium* inhibitors. Soils were extracted with water, ethanol, acetone and chloroform. Filter disks were impregnated with these extracts and used for testing for the presence of the inhibitory substances. Nitrogen content and pH of the different soils were also determined. No inhibitory substances were detected in these soils either by extraction or by the survival tests; possibly, the combination of their high nitrogen level (230 ppm) and pH (8.3) contributed to nodulation failure.

Risk (1966) evaluated the nitrogen fixed by grain legumes under

Egyptian conditions. Grain legumes of lupin, horse bean, chick-pea, fenugreek, lentil, peanut and soybean fixed, respectively, 138.7, 135.7, 104.8, 97.6, 83.3, 78.5, and 40.5 kg/hectare.

Conclusions

The response of certain legumes, e.g. lentils, broad bean and peanut, to inoculation was modified by the application of nitrogen and phosphorus fertilisers. Soybean responded to nitrogen and inoculation. However, successful nodulation was correlated with a higher density of rhizobia/seed. Nodulation of *Phaseolus vulgaris* is generally sparse or lacking under field conditions. A degree of variety–strain specificity was evident among the varieties of *Phaseolus vulgaris* tested. Trace elements, e.g. molybdenum and boron, increased the nitrogen fixation in cowpea when applied at rates of 4 to 10 ppm. The amount of nitrogen-fixed by various grain legumes was different and ranged between 40.5 and 138.7 kg N/hectare.

References

Ashour, N. I., Moawad, A. A. & El-Sherif, A. F. (1969). The effect of inoculation by nodule bacteria and nitrogen fertilisation on growth, yield, and nitrogen content of soybean. 6th Arab Science Congress, Damascus, 1–7 Nov, Part 2, pp. 443–8.

Badawy, F. H. (1970). Effect of seed inoculation and P and N fertilisation of chick-pea (*Cicer arietinum*) on rhizosphere microflora. 2nd Congress of Microbiology, Cairo, 28–30 April 1970, p. 93. (Abstract.)

El-Nadi, M. A., Hamdi, Y. A., Loutfi, M., Nassar, S. H. & Faris, F. S. (1972). Response of different varieties of common beans to certain strains of *Rhizobium phaseoli*. *Agricultural Research Review*, **49**, 125–30.

Fahmy, M. (1955). Nitrogen fixation by root nodule bacteria. *Agricultural Science*, **8**, 1–26. (In Arabic.)

Forbes, R. H. (1921). Mokus lima beans. *Sult. Agric.* Bulletin No. 9 of the Sult. Agricultural Society, Cairo.

Hamdi, Y. A., Abd-el-Samea, M. E. & Loutfi, M. (1973). Nodulation of soybean under field conditions. *Zentralblatt für Bakteriologie*, in press.

Hamdi, Y. A., Gafar Zeinab, A., & Loutfi, M. (1968). Response of soybean varieties to single and composite strains of *Rhizobium japonicum*. *Agricultural Research Review, Arab Republic of Egypt*, **46**, 11–16.

Hamdi, Y. A., Moharram, A. A. & Loutfi, M. (1972). Response of inoculated cowpea to boron and molybdenum. *Agricultural Research Review, Arab Republic of Egypt*, **49**, 117–24.

Loutfi, M. & Fahmy, M. (1958). A new powdered type legume inoculant. *Agricultural Research Review, Arab Republic of Egypt*, **38**, 325.

Loutfi, M., El-Sherbini, M. E., Ibrahim, A. N. & Casdy, A. (1966*a*). Effect of inoculation of legumes by nodule bacteria on yield. I. Horse bean. *Journal of Microbiology of the United Arab Republic*, **1,** 161–8.

Loutfi, M., El-Sherbini, M. F., Ibrahim, A. N. & Moustafa, M. A. (1966*b*). Effect of inoculation of legumes by nodule bacteria on yield. II. Peanut. *Journal of Microbiology of the United Arab Republic*, **1,** 169–76.

Loutfi, M., El-Sherbini, M. E., El-Nadi, M., Moharram, A. M., Afifi, N. M., Badawy, E. H. & Abou-el-Fadl, M. (1968). Comparative efficiency of some local and foreign strains of rhizobia. *Agricultural Research Review, Arab Republic of Egypt*, **46,** 17–28.

Nassar, S., Hamdi, Y. A., El-Nadi, M. A., Loutfi, M. & Faris, F. S. (1972). Nodulation failure of *Phaseolus vulgaris* in the vegetable-cropped soils of Egypt. *Agricultural Research Review, Arab Republic of Egypt*, **50,** 95–101.

Risk, S. G. (1966). Atmospheric nitrogen fixation by legumes under Egyptian conditions. II. Grain legumes. *Journal of Microbiology of the United Arab Republic*, **7,** 33–45.

Selim, M. (1947). *Root nodule bacteria and their importance*. University of Cairo Press. (Monograph in Arabic.)

Taha, S. M., Mahmoud, S. Z. & Salem, S. H. (1967*a*). Effect of inoculation with rhizobia on some leguminous plants in UAR. 1-Phosphorous manuring. *Journal of Microbiology of the United Arab Republic*, **2,** 17–29.

Taha, S. M., Mahmoud, S. Z. & Salem, S. H. (1967*b*). Effect of inoculation with rhizobia on some leguminous plants in UAR. II. Nitrogen fertilization. *Journal of Microbiology of the United Arab Republic*, **2,** 31–41.

25. Application of the acetylene reduction technique to the study of nitrogen fixation by white clover in the field

C. L. MASTERSON & P. M. MURPHY

The acetylene reduction technique (Dilworth, 1966; Schöllhorn & Burris, 1966, 1967; Hardy *et al.*, 1968) is widely used as an index of nitrogen fixation, in both laboratory and field studies. The method is rapid and sensitive, but is affected by technical and environmental factors, making care necessary both in procedure and in interpreting results.

Comparatively little work has been done using this method on white clover (*Trifolium repens* L.) under field conditions. Moustafa, Ball & Field (1969) have described the effect of nitrogen fertiliser and defoliation on pure white clover swards, Sinclair (1973) has investigated the same species grown in soil in pots under glasshouse conditions and Engin & Sprent (1973) have studied the effects of water stress on growth and nitrogen fixation by white clover. The present report describes the use of the acetylene reduction method to determine the pattern of nitrogen fixation by white clover in grazed swards under Irish conditions and the effects of environmental and other factors on the process and on the assay procedures.

Materials and methods

Established pastures were selected at five sites in Co. Wexford, characteristics of which have been published (Gardiner & Ryan, 1964) and are summarised in Table 25.1.

On the Johnstown Castle site, four plots each measuring 18×5.5 m were sampled at two week intervals through 1971–2. In January 1972, twelve additional plots were marked out, four of which received 250 kg calcium ammonium nitrate per ha, on either 1 or 25 February or 15 March. On each of the other sites, six similar plots were marked out in January 1972. All plots received in spring 500 kg/ha of compound fertiliser containing 10 % phosphorus and 20 % potassium. No nitrogenous fertiliser was applied, except as noted above. All plots were rotationally grazed by cattle throughout the experimental period, except on the Screen site where cattle were maintained from February onwards. Cultivar S100 white clover was grown in pots in a growth chamber (i) in soil of very low nitrogen supplying power, using 65 mm diameter and 50 mm deep plastic containers, the sides and base of which were perforated to facilitate drainage and gas exchange, or (ii) in acid-washed quartz sand in pots 125 mm diameter and 120 mm deep. One plant was grown per pot. All plants were inoculated with *Rhizobium trifolii* (strain J708) effective with S100, and received water and Crone's nitrogen-free nutrient solution (Allen, 1957) as required. The day length in the growth chamber was 16 hours and day/night temperatures were 19 °C and 14 °C respectively.

Table 25.1. *Soil characteristics of five sites in Co. Wexford*

Soil	Description
Johnstown Castle	Moderately well-drained heavy loam, moderately well-developed structure
Clonroche	Well-drained loam with well-developed structure throughout the rooting zone
Rathangan	Imperfectly drained, with poor structure in lower horizons
Macamore	Poorly drained, with high clay content and weak structure
Screen	Highly permeable, very light sandy soil, liable to drought

Single S100 white clover plants were also grown outdoors in 125 mm diameter pots containing the soil described above. The plants were inoculated and received water and nitrogen-free nutrient solution as required.

The plots were sampled at two week intervals. Eight 100 × 50 × 50 mm subplots were sampled at random. All samples were taken between 9 and 10.30 a.m. When desired, the clover was washed free of soil and grass using water at near-ambient temperature.

In 1971, assays of acetylene reduction at the Johnstown Castle site were carried out on clover roots washed free from soil with water and incubated in 29 ml McCartney bottles. In 1972, assays were carried out on undisturbed sods incubated in 1600 ml Kilner jars.

Acetylene reduction was determined by incubating sods, plants, roots or nodules in air in suitably sized glass containers for the required time at the desired temperature. The 1600 ml Kilner jars used for sod incubation had perforated screw cap metal lids fitted with rubber liners to facilitate gas sampling by syringe. Plants grown in plastic pots were assayed in smaller (825 ml) Kilner jars. Root and nodule samples were incubated in 29 ml or 7 ml McCartney bottles in a waterbath in the laboratory, and in the field by maintaining the containers at ambient soil temperature. The normal incubation period was one hour using an acetylene partial pressure ($P_{C_2H_2}$) of 0.1 atm. When Kilner jars were used, some of the gas phase was transferred to an evacuated McCartney bottle at the end of incubation and assayed as soon as possible.

Acetylene and ethylene were detected by gas chromatography using a Perkin Elmer F11 gas chromatograph fitted with a hydrogen-flame ionisation detector. For each assay 200 μl of gas sample were injected onto the column using a 500 μl Hamilton glass gas-tight syringe. The gases were separated on a 1.2 m $\times$ 2 mm column of Porapak N at 50 °C using nitrogen as a carrier gas. Retention times were 68 s for ethylene and 124 s for acetylene. A standard curve for ethylene concentrations over the range 9.3×10^{-7} moles–4.4×10^{-11} moles C_2H_4 was made and used to determine ethylene concentrations in the analytical samples.

Analyses with $^{15}N_2$ were done on washed-out clover roots from the field which were placed in McCartney bottles, using a gas phase of 78 % argon, 21.6 % oxygen and 0.4 % carbon dioxide. $^{15}N_2$ gas (93 atoms % $^{15}N_2$) was then incorporated at 0.2 atm. After 80 min the reaction was stopped by injecting 10 ml sulphuric acid. The roots and inactivating acid were then digested in Kjeldahl flasks containing 3 ml conc. sulphuric acid, 1.5 g potassium sulphate and 1.5 ml mercury catalyst. Total nitrogen was measured by distillation and Nesslerisation. These extracts, in a total of 50 ml, were boiled down to 6 ml and the $^{15}N_2$ content determined by measuring masses 28, 29 and 30 using a Consolidated Electrodynamics Corporation (CEC) model 21 – 130 mass spectrometer.

Results and discussion

Nitrogen fixation in the field

Pattern of acetylene reduction. The annual patterns of acetylene reduction by white clover in a grazed sward at Johnstown Castle for the years 1971–2 and at the four other sites (1972) are shown in Fig. 25.1.

Acetylene reduction activity at Johnstown Castle (1971) remained low throughout the 1970–1 winter and started to rise in late February, increasing rapidly until the end of May. The slight reduction in activity in April coincided with the period immediately following the first grazing. A marked reduction in activity occurred during June and the first half of July, coinciding with the flowering of the clover. After flowering, activity rapidly increased to a maximum value for the year and then decreased until the low winter level was reached in November.

In 1972 the pattern for this site was similar in the early part of the year. However, during April and May activity did not continue to increase but remained relatively constant. This is attributed, at least to a considerable extent, to lower temperatures in 1972. The 25 mm soil and air temperatures (means of 6 day periods ending on sampling dates), were lower by 3.5 deg C and 1 deg C respectively during this period in 1972. As in 1971, a reduction in activity coincided with flowering, but the effect was less marked in 1972, probably because of the lower activity during the pre-flowering period.

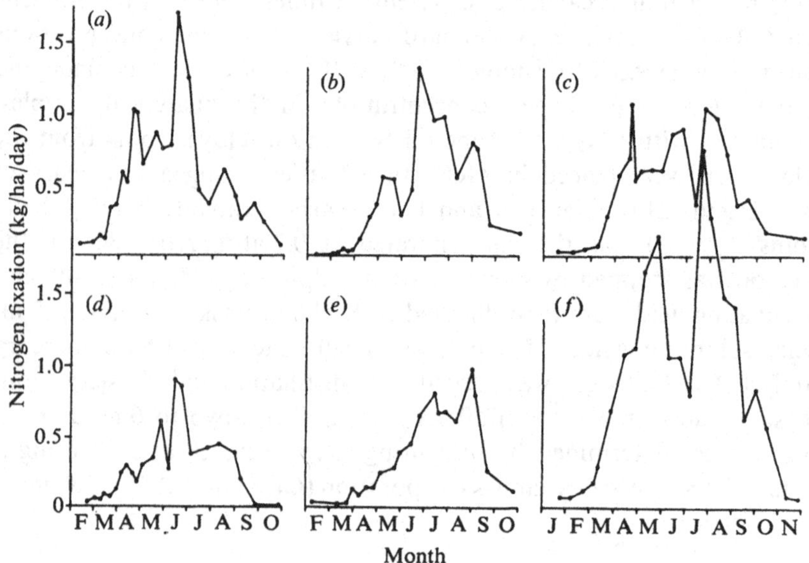

Fig. 25.1. Patterns of nitrogen fixation by white clover in grazed grass – clover swards 1971–2. (*a*), Clonroche 1972; (*b*) Rathangan 1972; (*c*) Johnstown Castle 1972; (*d*) Screen 1972; (*e*) Macamore 1972; (*f*) Johnstown Castle 1971.

The acetylene reduction patterns for four other Co. Wexford sites are also shown in Fig. 25.1. Most activity was found in the Clonroche site, a four-year-old sward on a well-drained highly productive soil. The Macamore pattern was characterised by a slow increase in activity in spring, possibly influenced by the extremely wet condition of this poorly drained soil at this time. The low activity observed at the Screen site throughout the year may have been because the site was heavily stocked with cattle from mid-February onwards. Acetylene reduction rates during April and May from areas on this site protected by cages were 40 % higher than on the unprotected area. The Rathangan site, like the Johnstown and Clonroche sites, showed a lower increase in activity during April and May related to grazing. The effect of flowering was discernible but not marked at all sites in 1972.

The variation from soil to soil in early spring was correlated with soil wetness. Activity commenced about three weeks earlier in the drier than in the wetter soils, although at this period the soil temperatures in the drier areas were lower than in the wetter ones, suggesting that soil wetness inhibits nitrogen fixation at this period more than soil temperature.

Using the theoretical $C_2H_2:N_2$ ratio factor of 3, the amounts of nitrogen fixed (kg/ha) are as follows: Johnstown Castle (1971) 296, Johnstown Castle (1972) 183, Clonroche 163, Rathangan 141, Macamore 111 and Screen 83. The Johnstown Castle (1971) figure was calculated from assays on samples washed free from soil with water. These were multiplied by 2.64 (see p. 307) to allow for decreased activity compared with undisturbed sods and by 1.17 to allow for decreased activity with air instead of the $Ar:O_2:CO_2$ as gas phase in the incubation vessels. All the 1972 figures shown were calculated using the 1.17 factor.

Although large differences were found between soils, these estimates of fixation are within the range of nitrogen fixation usually quoted for clovers (Nutman, 1971).

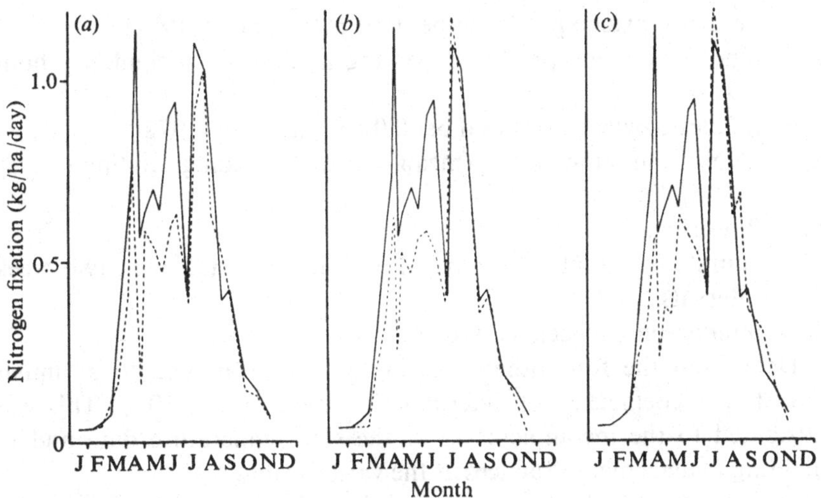

Fig. 25.2. Effect of applying 250 kg calcium ammonium nitrate/ha on (*a*) 1 February, (*b*) 25 February and (*c*) 15 March on acetylene reduction in a grazed grass–clover sward. Dashed line represents fertiliser treatment, solid line unfertilised control.

Effect of nitrogen fertiliser on acetylene reduction. The effect of calcium ammonium nitrate on acetylene reduction by white clover in the field was studied in 1972. The fertiliser was applied at the rate of 250 kg/ha on one of three dates in February–March. Each application caused a reduction in activity which was discernible after a few days and lasted until after flowering (Fig. 25.2). The earliest application (1 Feb) caused a smaller reduction (17.7 %) in activity over the year than application on 25 February (27.6 %) or 15 March (27.4 %).

Effect of field environmental factors on acetylene reduction. The relationship between acetylene reduction and certain environmental criteria measured at the Johnstown site are shown in Table 25.2.

Of the factors examined, the mean soil temperature at 25 mm for six days ending on sampling day was most highly correlated with activity. A similar trend was found with respect to air temperature and hours of sunshine. Rainfall was not correlated with acetylene reduction.

The following regression equations were fitted to the data:

$$y_1 = -1590 + \underset{(50.6)}{400x_1} - \underset{(221.3)}{1362x_2} + \underset{(1.7)}{5.2x_3} - \underset{(0.6)}{1.08x_4} \qquad R^2 = 93\,\%,$$

$$y_2 = -124.1 + \underset{(4.5)}{33.2x_1} - \underset{(19.6)}{79.2x_2} + \underset{(0.23)}{1.3x_3} - \underset{(0.03)}{0.15x_4} \qquad R^2 = 85\,\%,$$

$$y_3 = -2451 + \underset{(94.7)}{656x_1} - \underset{(524.8)}{732x_2} + \underset{(2.4)}{7.0x_3} - \underset{(0.5)}{1.86x_4} \qquad R_2 = 80\,\%;$$

y_1 = nmoles ethylene produced per 0.01 m^2 per hour (1971),
y_2 = nmoles ethylene produced per mg nodule dry weight per hour (1971),
y_3 = nmoles ethylene produced per 0.01 m^2 per hour (1972),
x_1 = 25 mm soil temperature (mean of 6 daily readings ending on day of sampling),
x_2 = flowering,
x_3 = cumulative number of days after 1 January, taken at two week intervals,
x_4 = interaction between x_1 and x_3.

Data from the four other sites in 1972 were applied to a similar model. The coefficients of determination were low, *c.* 50 %. This was attributed to the incompleteness of the soil temperature data and to differing management practices at the various sites.

In the equations shown above a number of parameters were tested as the x_3 variable, viz. cumulative air temperature, cumulative 12.5 mm soil temperature and cumulative hours of light before sampling time.

The coefficients of variation of the resulting equations did not differ significantly. The equations given above are the simplest, involving the least number of measurements. The x_4 variable denotes a changing relationship between ethylene production and soil temperature with time. The reason for this is probably the different rates of change in relative growth of nodules in the first and second halves of the year and which will be considered later (Fig. 25.9). Attempts to combine data from both years to give a single or parallel equation were unsuccessful.

Table 25.2. *Correlation coefficients of environmental factors with acetylene reduction*

	1971		1972
Environmental factor	Activity/unit area df16	Activity/g nodule dry wt df16	Activity/unit area df21
25 mm soil temperature (mean of 6 days ending on sampling date)	0.82***	0.61**	0.69***
25 mm soil temperature (on sampling date)	0.77***	0.53*	0.64**
Air temperature (mean of 6 days ending on sampling date)	0.66**	0.50*	0.60**
Hours sunshine (mean of 6 days ending day prior to sampling)	0.75***	0.56*	0.48*
Hours sunshine (on day before sampling)	0.49*	0.45	0.17
Grazing	0.69**	0.63**	0.47*
Flowering	0.25	0.03	0.21
Rainfall	−0.01	0.08	−0.06

* $P = 0.05$; ** $P = 0.01$; ***$P = 0.001$.

Total nodule weight was highly correlated with ethylene production per unit area ($r = 0.93$***) and with soil temperature ($r = 0.89$***). Percentage number of new nodules was also correlated with ethylene production per unit area ($r = 0.77$**). These variables were omitted from the above equations to permit predictions to be made in terms of environmental factors only.

The three equations show that soil temperature was the environmental factor having the greatest single influence on clover nitrogen fixation and growth. This finding is in accord with those of Brown (1939) and Soper & Mitchell (1956) working with various pasture species.

The importance of temperature on nitrogen fixation is further illustrated in Fig. 25.3, which shows acetylene reduction during 1971 on field samples from the Johnstown Castle site assayed at ambient 25 mm soil temperature and at 15 °C. Up to 9 June, and after 30 August, activity at 15 °C was greater than at ambient. In the June–August period, greater activity was obtained from the ambient temperature incubations. Since soil temperatures were above 15 °C in the June–August period and below this level throughout the remainder of the year, these data illustrate the close relationship which exists between temperature of incubation and nodule activity. It is also apparent, particularly in the early part of the year, that nodules possess a potential for higher activity which is immediately realised when temperature is raised.

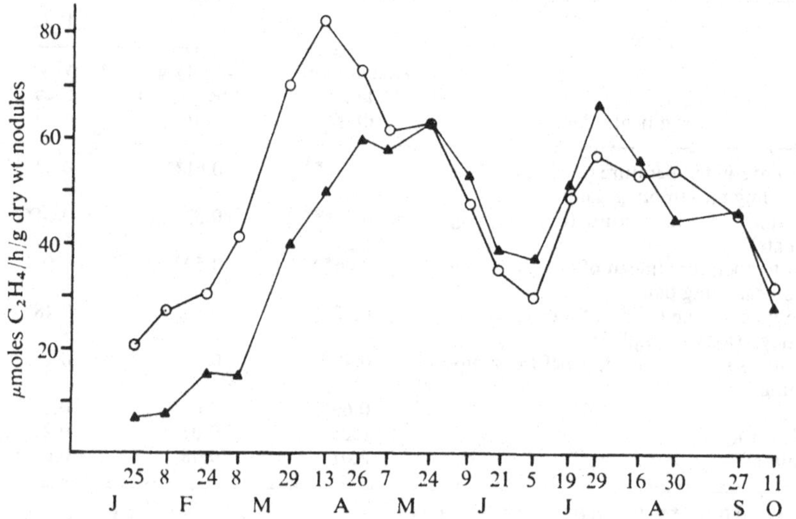

Fig. 25.3. Effect of incubation temperature on ethylene production by white clover samples taken from the field. ▲——▲, ambient temperature; ○——○, 15 °C.

The changing nature of the 15 °C values in Fig. 25.3 may also be affected by the varying 'quality' of the nodules. Thus as spring progressed, the number of new nodules greatly increased, resulting in greater specific activity. The reduced activities observed during flowering and towards the end of the season were due to increased proportions of senescent nodules at these periods.

Examination of some experimental and environmental variables

Effect of sampling method on nodule activity and $C_2H_2 : N_2$ ***ratio.*** Acetylene reducing activity of clovers in the undisturbed sod is greater than in corresponding plant material washed free from the sod and assayed in

McCartney bottles; undisturbed samples were on average 2.6 times higher in activity. On removing clover plants from the sod some nodule material is lost or damaged, a factor which will contribute to lower activities. It is usually necessary to wet the sod to facilitate removal of the clover and this procedure also exposes the nodular material to the risk of water saturation, a condition which is known to lead to inhibition of nitrogenase activity (Sprent, 1971) and affect the ratio of acetylene reduction to actual amount of nitrogen fixed.

Several experiments were conducted to determine this ratio, details of which are shown in Table 25.3 and in Figs. 25.4 and 25.5. In experiment 1, in which $^{15}N_2$ enrichment was compared with acetylene reduction, a

Table 25.3. *Experiments to determine* C_2H_2: N_2 *ratio*

No.	Location	Species	Growth medium	Preparation	No. of samples	C_2H_2:N_2 ratio	
						Kjeldahl	$^{15}N_2$
1	Field	*Trifolium repens*	Soil	Roots washed with water	11	—	4.6 (0.77) *
2	Pot (growth chamber)	*Trifolium repens*	Soil	Undisturbed	52	2.8 (0.24)	—
	Pot (growth chamber)	*Trifolium repens*	Soil	Roots washed with water	52	1.5 (0.21)	—
3	Pot (outdoor)	*Trifolium repens*	Soil	Roots washed with water	76	2.1 (0.27)	—
4	Pot (growth chamber)	*Trifolium repens*	Sand	Roots shaken free from sand – not washed	26	2.6 (0.24)	—
5	Pot (outdoor)	*Medicago sativa*	Soil	Roots washed with water	80	1.5 (0.20)	—
6	Pot (outdoor)	*Pisum sativum*	Soil	Roots washed with water	76	2.9 (0.26)	—
7	Pot (outdoor)	*Vicia faba*	Soil	Roots washed with water	80	2.9 (0.25)	—

* Figures given in brackets are the standard errors.

ratio of 4.6 was obtained, indicating that acetylene reduction overestimated nitrogen fixation in these samples. In experiment 2 (Fig. 25.4) when Kjeldahl analysis and acetylene reduction were compared, undisturbed plants (*a*) gave a ratio of 2.8 while the comparable figure for plants washed before assay was 1.5 (*b*). These differences may be attributed to damage to, and loss of, nodules, and their saturation with water. Such factors would seriously reduce acetylene reduction activity, but would have comparatively little effect on Kjeldahl nitrogen determination. The ratios obtained in experiment 2 may be artificially low because of some soil nitrogen uptake by the plants during growth. The plants in experiment 3 were grown outdoors in pots using soil. Again, the ratio was lower than that for the undisturbed plants in experiment 2 and consistent with an effect of washing. The reason for the difference

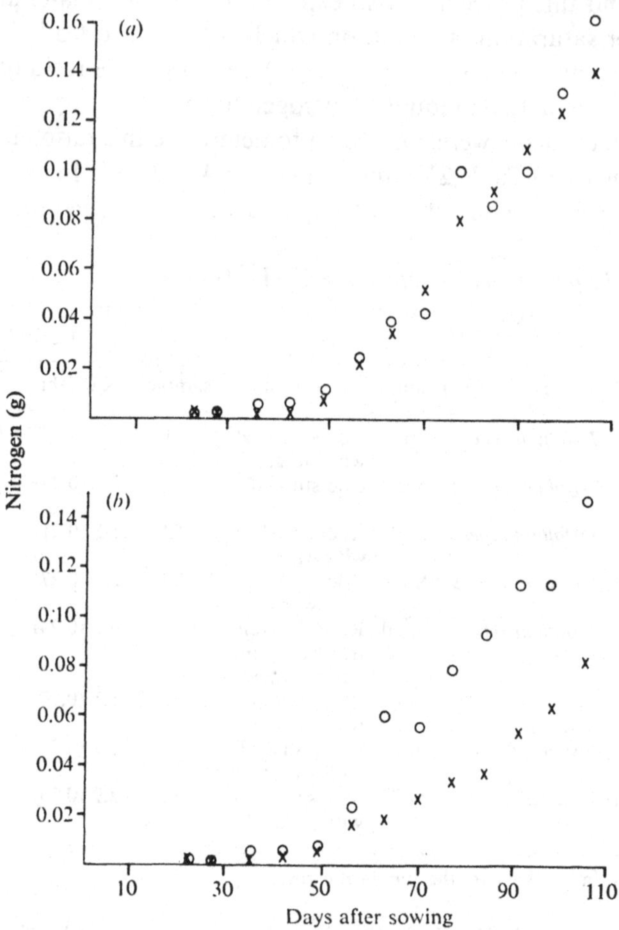

Fig. 25.4. $C_2H_2:N_2$ ratios (×) obtained by assaying white clover in (*a*) undisturbed sods and (*b*) after washing the plants free of soil. Total plant nitrogen (○) was determined by Kjeldahl analysis. Equivalent nitrogen values were obtained using the theoretical conversion value of 3.

between the ratios 2.1 and 1.5 is less clear. However, some mineralisation of soil nitrogen would have occurred in the soil and this may have been less in experiment 3 due to the lower temperatures outdoors than in the growth chamber and the possible loss of some soil nitrogen by leaching. In experiment 4, a ratio similar to that in experiment 2 was obtained; these plants would have suffered little damage during preparation but nevertheless some slight effect was to be expected.

The ratios obtained in these experiments, while within the range reported in the literature, nevertheless show considerable variation, and were clearly affected by the experimental conditions including the kind of sample treatment before assay. Further work on this aspect is desirable.

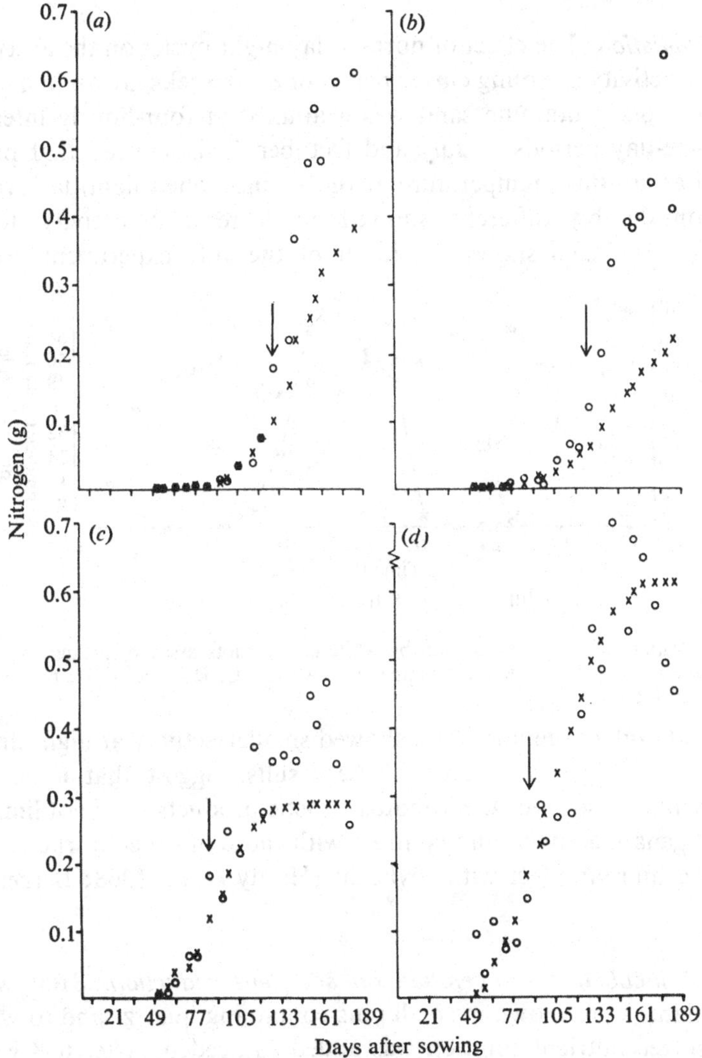

Fig. 25. 5. Comparison of total nitrogen fixed by four species grown in soil as measured by acetylene reduction (×) and Kjeldahl (○) assays. (*a*), white clover; (*b*), lucerne; (*c*) pea; and (*d*), broad bean. Arrows mark start of flowering

Fig. 25.5 shows the nitrogen fixation profiles of four legumes using Kjeldahl and acetylene reduction assays for estimating fixation. The patterns observed are generally similar. The correlation between acetylene reduction and Kjeldahl analyses is highest until the start of flowering. Thereafter the acetylene reduction estimate becomes progressively lower than the Kjeldahl. Considerable difference between species is apparent as shown by the ratios in Table 25.3.

Diurnal variation. The effect of normal day/night cycles on the acetylene reducing activity of young clover plants of 8–10 weeks grown in a glasshouse in pots containing sand was examined at four-hourly intervals over three-day periods in July and October. This showed that plants assayed at constant temperature at these times when light/dark ratios were considerably different, showed no decrease in activity during darkness. Fig. 25.6 shows the result of the July experiment. Plants

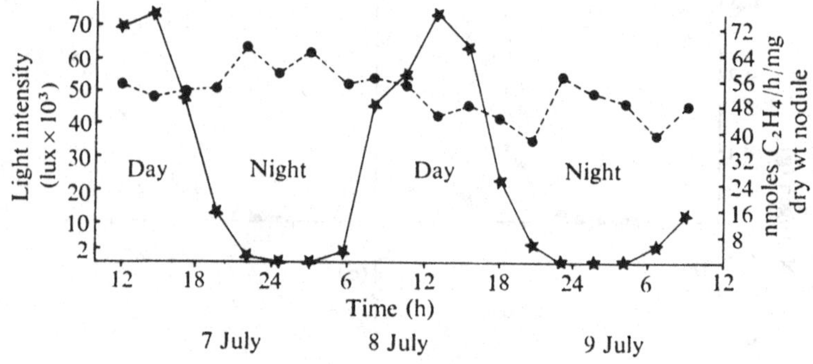

Fig. 25.6. Pattern of C_2H_4 production by white clover roots assayed through two consecutive day/night cycles. Incubation temperature was 21 °C. ●---●, C_2H_4 production; *——*, light intensity.

assayed at ambient temperature showed smaller activity at night due to the lower night temperatures. These results suggest that under the experimental conditions used photosynthetic products were not limiting for nitrogenase activity, and contrast with those of other workers, who showed a diurnal effect with soybeans (Hardy *et al.*, 1968; Bergersen, 1970).

Effect of incubation temperature on acetylene reduction. S100 white clover plants were grown in small pots containing quartz sand to which nitrogen-free nutrient solution was added as needed. After 6–8 weeks the plants were washed free from sand, placed in McCartney bottles, acetylene added, and incubated in a water bath at the required temperature. Fig. 25.7 shows that the increase in activity was approximately linear as temperature increased to 21 °C, where a peak occurred. A slightly higher optimum temperature for *T. pratense* has been reported by Dart & Day (1971).

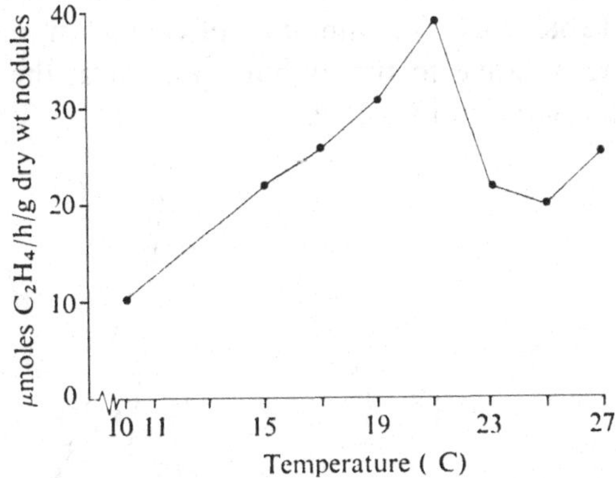

Fig. 25.7. Effect of incubation temperature on C_2H_4 production by white clover roots incubated for one hour at various temperatures.

Effect of oxygen concentration (P_{O_2}). The effect of P_{O_2} on acetylene reduction was determined using twelve-week-old white clover plants grown in sand. Washed roots were incubated in 29 ml McCartney bottles containing argon and oxygen at various concentrations. A $P_{C_2H_2}$ of 0.1 atm was used. Results in Fig. 25.8 show that optimum activity was reached at a P_{O_2} of 0.2 atm and that levels up to 0.4 atm were not inhibitory. These findings show that the P_{O_2} relationships for

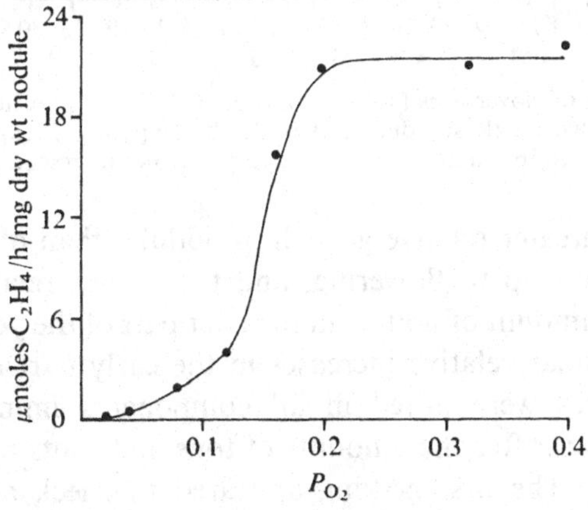

Fig. 25.8. Effect of oxygen concentration on white clover nodule nitrogenase.

white clover and other legume nodule nitrogenases are very similar (Koch & Evans, 1966; Bergersen, 1970; Dart & Day, 1971).

Seasonal characteristics of clover growth. Clover tops, roots and nodules at the Johnstown Castle site were weighed during 1971 and results are shown in Table 25.4. The amounts of each component, recorded at each sampling, relative to the amount present at the first sampling (25 January), are shown in Fig. 25.9.

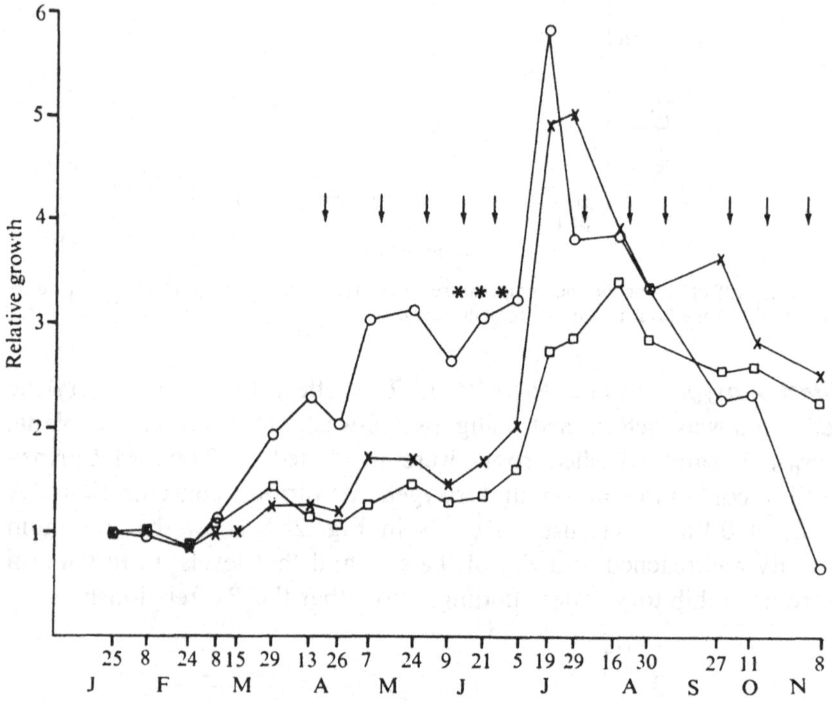

Fig. 25.9. Amounts of clover tops (×——×), roots (□——□) and nodules (○——○) recorded at Johnstown Castle site during 1971. Each component is expressed relative to the amount recorded at first sampling on 25 January. Arrows represent grazing dates and asterisks flowering.

There was a greater relative growth of nodules than of roots or tops during the season up to flowering, and the largest relative reduction occurred in the amount of nodule in the later part of the year. Roots and tops showed similar relative increases in the early part of the season. Marked increases were noted in all components immediately after flowering and thereafter the amounts of tops and roots much exceeded that of nodules. The first grazing appeared to check all growth but particularly that of nodules. A more marked reduction coincided with

Table 25.4. *Amounts of clover, and of ethylene, produced from October 1970–December 1971 at Johnstown Castle*

	Dry matter (g/m^2)				µmoles C_2H_4/h/g dry material							
					At ambient temperature				At 15 °C			
Date	Nodules	Roots	Tops	Total	Nodules	Roots	Tops	Total	Nodules	Roots	Tops	Total
22. Oct. 70	2.17	15.7	70.2	88.1	31.9	4.5	1.0	0.8	60.9	8.4	1.9	1.5
4. Nov. 70	1.44	11.6	46.7	59.8	12.4	1.5	0.4	0.3	20.3	2.7	0.7	0.5
18. Nov. 70	1.86	10.5	49.9	62.3	6.1	1.1	0.2	0.1	17.6	3.3	0.7	0.5
2. Dec 70	0.96	7.0	32.7	40.6	14.5	2.2	0.5	0.4	22.7	2.9	0.6	0.5
14. Dec. 70	1.24	10.5	52.6	64.3	8.7	1.0	0.2	0.1	18.6	2.2	0.4	0.3
25. Jan. 71	1.03	7.3	35.3	43.6	7.2	1.1	0.2	0.1	20.7	3.1	0.6	0.5
8. Feb. 71	0.99	7.6	37.2	45.8	8.0	1.0	0.2	0.1	27.4	3.7	0.8	0.6
24. Feb. 71	0.86	6.4	29.3	36.5	15.5	2.2	0.5	0.4	30.9	4.1	0.9	0.7
8. Mar. 71	1.23	8.2	35.1	44.5	15.3	2.3	0.5	0.4	41.3	6.1	1.4	1.1
15. Mar. 71	ND	10.9*	35.6	46.6	ND	2.9*	0.9	0.7	ND	5.7*	1.8	1.3
29. Mar. 71	1.99	10.5	44.1	56.7	40.9	7.4	1.7	1.3	70.1	14.6	3.3	2.6
13. Apr. 71	2.38	8.4	44.9	55.6	50.4	15.2	2.7	2.2	82.6	24.9	4.5	3.6
26. Apr. 71	2.12	7.9	42.4	52.4	60.5	15.8	2.9	2.3	73.3	20.7	3.7	3.0
7. May 71	3.17	9.4	61.0	73.5	58.5	19.5	2.9	2.4	61.8	22.5	3.3	2.8
24. May 71	3.26	10.9	60.9	75.0	63.2	20.3	3.5	2.8	63.2	20.3	3.5	2.8
9. June 71	2.75	9.6	52.8	65.2	53.5	13.9	2.6	2.1	48.0	13.8	2.5	2.0
21. June 71	3.18	10.2	60.3	73.7	39.3	12.5	2.1	1.7	35.5	11.9	2.0	1.7
5. July 71	3.36	11.8	72.9	88.0	37.7	10.8	1.6	1.3	30.3	9.4	1.4	1.2
19 July 71	6.07	20.3	177.2	203.6	51.5	15.8	1.8	1.6	49.3	14.0	1.6	1.4
29. July 71	3.96	21.0	181.6	206.6	67.1	13.4	1.5	1.3	57.1	10.1	1.2	1.0
16. Aug. 71	3.99	25.2	141.0	170.2	56.8	8.4	1.5	1.3	53.3	8.5	1.5	1.3
30. Aug. 71	3.49	21.0	121.1	145.6	45.6	7.6	1.2	1.0	54.9	8.5	1.4	1.2
13. Sept. 71	ND	22.9*	95.1	118.0	ND	3.8*	0.9	0.7	ND	3.8*	0.8	0.7
27. Sept. 71	2.39	18.8	131.3	152.5	46.8	6.2	0.9	0.7	46.4	5.6	0.8	0.7
11. Oct. 71	2.43	19.2	101.7	123.3	28.7	3.9	0.7	0.6	32.3	3.4	0.6	0.5
8. Nov. 71	0.69	16.7	89.4	106.7	21.2	2.2	1.6	2.0				
23. Nov. 71	1.35	14.4	85.7	101.5	23.7	2.2	1.6	1.9				

ND, not determined.
* Includes nodules.

the onset of flowering, again with greatest effect on the nodules. While flowering is known to be associated with a degeneration of existing nodules and to inhibit new nodule and root formation (Pate, 1958), in the present work the effect is complicated by grazing which occurred on and after 25 May.

Most growth of all components occurred in the post-flowering period with the peak for nodule production occurring before those for tops or roots.

Nodule characteristics were observed throughout 1971. In January and early February, only old nodules were found, brown in colour, many obviously degenerate and damaged by soil animals. Some had small amounts of leghaemoglobin visible in their centres and very low acetylene reduction rates were recorded. New nodules and new growth on old nodules were first noticed on 24 February, characterised by the appearance of new white tissue at the distal ends of the nodules. The percentage of nodules showing new growth increased from 20 % on 24 February to 25 % on 8 March, and 55 % on 29 March, and closely paralleled acetylene reduction activity. During this period the acetylene reduction per unit weight of nodule also increased.

During February–March the formation of new growth on old nodules preceded the appearance of new root or shoot growth by some weeks. It is not clear why new nodule growth should have occurred so early in the year when temperatures were in the region of 4–5 °C and other conditions for growth were minimal.

Conclusions

Our results confirm that acetylene reduction is a valuable means of studying nitrogen fixation by white clover under field conditions. The considerable variation in fixation observed between years and between soils was not unexpected, considering the number of environmental and management factors affecting the process. Where these factors can be adequately measured, their effect on fixation can be assessed and expressed in the form of a regression, as illustrated by the Johnstown Castle data.

Nodule quantity was the greatest determinant of nitrogen fixing activity and soil temperature was the most important environmental factor. The appearance of new nodules, and of new growth on old nodules in early spring was the first visible sign of plant growth and occurred when soil temperatures were in the region of 4–5 °C and when other conditions for plant growth were minimal. The plant mechanism triggering this sudden nodule meristematic activity is not evident.

Flowering of the clover caused a reduction in fixation of *c.* 15 % in 1971, but this effect was much less in 1972. It is clear, however, that total annual fixation could be significantly increased if flowering could be either prevented, postponed or controlled. A similar increase would accrue if it became possible to retard the rapid drop in fixation activity which occurs after the post-flowering peak.

Fixation was reduced by applying nitrogen fertiliser in spring. The effect lasted until after flowering and was more pronounced after later fertiliser application. There was some evidence that soil wetness delays the onset of fixation in spring.

The estimation of total nitrogen fixed depends on the C_2H_2:N_2 conversion factor. We have shown this to vary from 1.5 to 4.6, underlining the susceptibility of the acetylene reduction technique to experimental conditions, and emphasising the caution necessary in interpreting results when quantified in terms of kilograms nitrogen fixed per hectare. Our estimate of this factor, though wide, compares well with the values obtained by other workers as reviewed by Hardy Burns & Holsten (1973).

The authors wish to acknowledge the following: financial assistance from the Irish National IBP Committee; Mrs Marie T. Sherwood for her constant interest and for many useful discussions, especially during the preparation of this paper; Dr R. D. Hauck, TVA, Muscle Shoals, Alabama, USA for $^{15}N_2$ analyses; Messrs S. Turner, S. Reynolds, M. Murphy, J. Dwyer and E. McDonald for skilled technical assistance; Mr M. O'Keeffe for the statistical analyses and helpful discussions and Miss Olive Shudall for preparing diagrams.

References

Allen, O. N. (1957). *Experiments in soil bacteriology*, pp. 93–4. Burgess Publishing Co., Minneapolis.

Bergersen, F. J. (1970). The quantitative relationship between nitrogen fixation and the acetylene reduction assay. *Australian Journal of Biological Science*, **23,** 1015–25.

Brown, M. E. (1939). Some effects of temperature on the growth and chemical composition of certain pasture grasses. Mo. Agricultural Experiment Station Research Bulletin, No. 299.

Dart, P. J. & Day, J. M. (1971). Effects of incubation temperature and oxygen tension on nitrogenase activity of legume root nodules. *Plant and Soil*, Special volume, 167–84.

Dilworth, M. (1966). Acetylene reduction by nitrogen-fixing preparations from *Clostridium pasteurianum. Biochimica et Biophysica Acta*, **127,** 285–94.

Engin, M. & Sprent, J. (1973). Effects of water stress on growth and nitrogen-fixing activity of *Trifolium repens*. *New Phytologist*, **72**, 117–26.

Gardiner, M. J. & Ryan, P. (1964). *Soils of Co. Wexford*. An Foras Taluntais, Dublin.

Hardy, R., Holsten, R., Jackson, E. & Burns, R. (1968). The acetylene–ethylene assay for N_2 fixation: laboratory and field evaluation. *Plant Physiology*, **43**, 1185–1207.

Hardy, R., Burns, R. & Holsten, R. (1973). Applications of the acetylene–ethylene assay for measurement of nitrogen fixation. *Soil Biology and Biochemistry*, **5**, 47–81.

Koch, B. & Evans, H. J. (1966). Reduction of acetylene to ethylene by soybean root nodules. *Plant Physiology*, **41**, 1748–50.

Moustafa, E., Ball, R. & Field, T. (1969). The use of acetylene reduction to study the effect of nitrogen fertiliser and defoliation on nitrogen fixation by field grown white clover. *New Zealand Journal of Agricultural Research*, **12**, 691–6.

Nutman, P. S. (1971). Perspectives in biological nitrogen fixation. *Science Progress*, **59**, 55–74.

Pate, J. S. (1958). Nodulation studies in legumes. 11. The influence of various environmental factors on symbiotic expression in the vetch (*Vicia sativa* L.) and other legumes. *Australian Journal of Biological Science*, **2**, 496–515.

Schollhorn, R. & Burris, R. H. (1966). Study of intermediates in nitrogen fixation. *Federation Proceedings*, **24**, 710.

Schollhorn, R. & Burris, R. H. (1967). Acetylene as a competitive inhibitor of N_2 fixation. *Proceedings of the National Academy of Sciences, USA*, **58**, 213–16.

Sinclair, A. G. (1973). Non-destructive acetylene reduction assay of nitrogen fixation applied to white clover plants growing in soil. *New Zealand Journal of Agricultural Research*, **16**, 263–70.

Soper, K. & Mitchell, K. J. (1956). The developmental anatomy of perennial ryegrass. *New Zealand Journal of Science and Technology*, A**37**, 484–504.

Sprent, J. I. (1971). Effects of water stress on nitrogen fixation in root nodules. *Plant and Soil*, Special volume, 225–8.

PART IV

Legume nitrogen fixation and the environment

26. Symbiotic specialisation in pea plants: some environmental effects on nodulation and nitrogen fixation

T. A. LIE, D. HILLE, RIS LAMBERS & A. HOUWERS

Of the plants used for experimental work, the pea (*Pisum sativum*) is particularly good for physiological and genetic studies. Ever since the classical experiments of Mendel, these plants have been much used for genetic studies so that a vast body of data is now available enabling gene maps to be constructed (see Blixt, 1972). Due to the self-fertilisation of the species it is easy to obtain inbred, homozygous plants of remarkably uniform phenotypic expression (Went, 1957). Moreover, seeds can be obtained from a plant within three to four months after sowing and will germinate promptly even when immature (Went, 1957; Lie, unpublished observations), thus enabling several generations to be grown within one year. Also, reproducible growth can be achieved under defined climatic conditions in artificial light, and in these situations nodulation is excellent and uniform whether in sand, perlite or in nutrient solution. Under such conditions, where roots are allowed to develop in a nitrogen-free nutrient solution under aseptic conditions (Virtanen & Miettinen, 1963) nodule formation is remarkably uniform and takes place mostly within a three- to five-day period following inoculation of the culture (Lie, unpublished observations).

The size of the plant might be considered a disadvantage but this is offset by its monopodial habit which permits it to be grown in relatively cramped conditions and yet obtain sufficient nodular material for physiological studies. Because of its rapid growth, differences due to treatment can sometimes be observed within four to six weeks after germination. A final advantage is that small plantlets resembling the intact plant in physiological performance can be obtained by removing cotyledons from the seedlings at an early stage. This feature is made use of in the experiments described here.

Genotypes or cultivars of peas are very widely grown, but since growth of the species is restricted by high phototemperatures (Went, 1957), it is generally unsuited to tropical climates, and in regions with a hot summer period the crop is grown only during the winter or early spring. In this article attention will be directed to a group of pea genotypes, collected from Central Asia and the Middle East. These are characterised by very small, speckled seeds and can be regarded as more or less primitive cultivars of the modern garden pea, to which they are nevertheless closely related as shown by the ease with which the two groups can be crossed (see later). However, as shown earlier (Lie, 1971*b*,*c*) and in this article, nodulation of certain of these primitive pea cultivars seems to be basically different from that of the cultivated types, at least as far as nodulation by the *Rhizobium* strains used in our work is concerned.

A survey of the symbiosis of pea cultivars

In a preliminary experiment forty-two pea cultivars, collected from different parts of the world, were tested for nodulation and symbiotic nitrogen fixation with *Rhizobium* strain PRE. The temperature of the greenhouse used for culture varied between 20 °C and 24 °C. The plants were grown in perlite, and fed with a nitrogen-free nutrient solution. The results obtained from representative cultivars are shown in Table 26.1. All plants became nodulated except for pea cv. Afghanistan. Pea cv. Iran (57.426) produced only a few nodules and the amount of nitrogen fixed was very small. At the other extreme the cultivar Mexico

Table 26.1. *Nodulation and nitrogen fixation of different pea cultivars, inoculated with* Rhizobium *strain PRE*

	Nodulation		mg N per plant		
Pea cultivar	No. per plant	Fresh weight (mg/plant)	Nodules	Shoot	% N in the shoot
57.425 Afghanistan	0	0	0	0.91	1.45
57.426 Iran	13	60	0.42	2.63	1.70
57.591 India	85	144	0.50	10.31	3.20
57.423 Manitoba	40	190	1.03	12.10	4.11
57.584 India	66	191	1.24	12.40	3.44
57.629 Ethiopia	39	211	1.00	12.90	3.52
57.500 Turkey	42	349	1.02	14.40	3.92
57.420 Austria	74	233	1.90	15.09	3.67
57.447 Argentine	43	236	0.87	16.73	3.55
57.441 Iran	38	217	1.68	17.18	3.22
57.403 Brazil	51	218	1.87	17.48	3.54
57.435 Mexico	46	249	1.28	18.50	3.78
57.724 Sweden	35	272	1.69	20.68	4.43
57.438 Mexico	40	302	1.98	29.80	4.10

The plants were grown for six weeks in perlite under greenhouse conditions.

(57.438) seemed to be the best nitrogen fixer. There was some variation in the amount of nitrogen fixed and the amount of nodule tissue formed by the various cultivars but further experiments under more controlled environmental conditions are needed before definite conclusions can be drawn regarding the intermediate responses of some of the cultivars.

Nodulation-resistance in pea plants

In earlier studies (Lie, 1971*b*,*c*), using controlled conditions, we showed that the cultivar Afghanistan did not form root nodules in solution culture with any of the twenty *Rhizobium* strains tested and at any of the temperatures used. Occasionally some plants of this cultivar when grown in perlite formed one or two nodules which fixed nitrogen as revealed by acetylene reduction. Similarly, in soil cultures containing a natural population of *Rhizobium*, some plants of this cultivar formed a few nodules, but the *Rhizobium* strains isolated from these nodules did not appear to be any better adapted to the cultivar Afghanistan than did the other *Rhizobium* strains from the collection at Wageningen (Lie, unpublished data).

Table 26.2. *The effect of temperature on root nodule formation of six pea cultivars in symbiosis with four* Rhizobium *strains*

	Nodulation at 20 °C with *Rhizobium* strain				Nodulation at 26 °C with *Rhizobium* strain			
Pea cultivar	PRE	PF2	310a	313	PRE	PF2	310a	313
Rondo	+++	+++	+++	+++	+++	+++	+++	+++
Che-Un-To	++	+++	+++	+++	+	++	+	++
Iran	−	+	+++	+	++	+++	+++	+++
Kabul	±	++	±	±	+	++	−	++
Isfahan	−	+	−	±	±	++	−	++
Afghanistan	−	−	−	−	−	±	−	−

−, no nodules; ±, a few plants with one nodule; +, a small number ++, an intermediate number and +++, a large number of nodules per plant.

Nodulation of pea cv. Iran is markedly temperature-dependent, i.e. at 20 °C nodulation is either absent or poor with nearly all the *Rhizobium* strains used, but nodulation will occur normally at a slightly higher temperature (26 °C). One *Rhizobium* strain (310a) was selected, that formed nodules on pea cv. Iran at either of these temperatures (Lie, 1971*a*,*c*).

In 1971 two additional pea cultivars were collected from the markets in Afghanistan (cv. Kabul) and Iran (cv. Isfahan). These, together with some other pea cultivars were then tested for root nodule formation with four *Rhizobium* strains at two temperatures. The results obtained from experiments in water culture and in perlite, are summarised in Table 26.2. The Dutch cultivar Rondo nodulated profusely with all *Rhizobium* strains tested at all temperatures, whereas at the other extreme the cultivar Afghanistan was almost completely devoid of root nodules under all conditions. The cultivar Che-Un-To, a pea of unknown origin but presumably coming from China, nodulated poorly at the higher temperature but was well-nodulated at 20 °C, whilst the reverse situation was observed for Iran. In contrast to Rondo, where almost no difference was observed in nodulation response between the four *Rhizobium* strains, Che-Un-To formed relatively fewer nodules with *Rhizobium* strain PRE than with the other strains. Nodulation of the cultivars Kabul and Isfahan was consistently poor, the best results being obtained with *Rhizobium* strain PF2 followed by strain 313. In contrast with results obtained for Iran, *Rhizobium* strain 310a is a poor strain for nodulation of the two cultivars Kabul and Isfahan.

From these results it is obvious that nodulating ability of the four *Rhizobium* strains is a highly specific characteristic determined primarily by the host cultivar.

Inheritance of nodulation-resistance in pea plants

It has been known for some time that nodulation is genetically determined both in the host plant and in the rhizobial symbiont and the comprehensive review by Nutman (1969) on this subject deals particularly with work on clover (*Trifolium* spp.). A survey of the work carried out with soybean in the USA has been given by Weber *et al.*, (1971), whilst the study of Gelin & Blixt (1964) suggests that sparse nodulation in pea plants may be due to the action of two recessive genes. However, since this work was conducted in soil-grown plants, presumably exposed to a mixture of *Rhizobium* strains, the results must be viewed with caution. As shown here and in earlier papers (Lie, 1971*b*,*c*), it is important to carry out experiments of this nature under controlled conditions using single strains of *Rhizobium*, since influences due to bacteria, temperature conditions and growth medium can affect the response of a plant to nodulation.

In our studies at Wageningen genetic analysis has been carried out under controlled climatic conditions using aseptic nutrient solutions. Crossings were effected between Rondo (R) and Iran (I), between Rondo and Afghanistan (A) and between Iran and Afghanistan. In each case reciprocal crossings were made, and in addition back crosses were carried out between the F_1 plants and their parents.

Rondo × Iran (R×I or I×R)

The F_1 plants derived from crossings of R×I or I×R resemble the Iran parental plants in the nodulation status, i.e. at 20 °C they are resistant to *Rhizobium* strain PRE but not to *Rhizobium* strain 310a. At 26 °C they are susceptible to all *Rhizobium* strains used. By growing the F_2 plants at 20 °C with *Rhizobium* strain PRE as a test strain for nodulation, segregation into non-nodulating and nodulating plants occurred in a ratio close to 3:1. These results indicate that the 'temperature-dependent resistance' factor of Iran is determined by one major dominant gene.

Rondo × Afghanistan (R×A or A×R)

When Rondo was crossed with Afghanistan basic differences in nodulating capacity were observed between the F_1 plants from the reciprocal crosses. When Afghanistan is the mother plant (A×R) a complete resistance, similar to that of Afghanistan parents, was observed in the F_1 plants. With the reciprocal cross (R×A) no nodules were formed with *Rhizobium* strains PRE and 310a, but intermediate or low numbers of nodules were formed with *Rhizobium* strains PF2 and 313 respectively, on the F_1 plants growing at 26 °C. At 20 °C the F_1 plants were completely resistant. However, reciprocal comparison of populations revealed no differences, thus ruling out a cytoplasmic factor for nodulation in this case.

The F_2 analysis and back crosses again showed that resistance to nodulation in Afghanistan is due to one major gene. When *Rhizobium* strain PRE was used as a test strain, the gene conferring resistance appeared to be fully dominant, but not with *Rhizobium* strain PF2.

ran × Afghanistan (I×A or A×I)

Crosses of Iran and Afghanistan produce offspring bearing either no nodules or only very few nodules when exposed to *Rhizobium* strains 310a or PF2 at a temperature of 26 °C. No nodules developed at all when these associations were grown at 20 °C. When the F_2 plants were inoculated with *Rhizobium* strain 310a and kept at 20 °C, the plants segregated into non-nodulating and nodulating plants in a ratio close to 3:1, indicating that the resistance conferred by Afghanistan is of a dominant type. However, when this test was performed with the same *Rhizobium* strain at 26 °C the results again suggested that one major gene was involved and probably of an intermediate type.

The breeding experiments clearly show that the resistance in Afghanistan and Iran is due to major genes, probably a different gene being involved in each of these cultivars. The degree of dominance appears to be strongly affected by temperature and by the *Rhizobium* strain used.

The effect of temperature on pea cv. Iran

Pea cv. Iran was selected for further study on the effect of temperature. The possibilities of controlling nodulation through temperature and *Rhizobium* strain interactions (Lie, 1971*b*,*c*) offer an ideal system for physiological studies. And, as shown above, temperature-dependent nodulation is heritable and open to genetic analysis.

The effect on root nodule formation

The effect of temperature on nodule formation is known to be strictly localised. When only the roots or part of the roots were kept at a temperature (26 °C) permitting nodulation, nodules appeared only on that part of the roots exposed to the conducive temperature. Within

Table 26.3. *The effect of temperature on root nodule formation of pea cv. Iran inoculated with* Rhizobium *strain 313 or 310a*

Plants growing at 20 °C			Plants growing at 26 °C		
	No. of nodules per plant			No. of nodules per plant	
Days at 26 °C	313	310a	Days at 20 °C	313	310a
None	0	4.5	None	16.1	22.3
0–1	0	5.6	0–1	18.3	19.4
0–2	12.3	10.2	0–2	4.4	16.2
0–3	13.5	16.4	0–3	1.1	10.3
1–2	6.6	14.0	1–2	4.0	13.2
2–3	1.8	16.0	2–3	11.0	14.4

The plants were either kept at 20 °C or at 26 °C, except for the periods indicated in columns 1 and 4 when they were exposed to other temperatures. Day 0 is the time of inoculation.

limits shoot temperatures were found to have no effect on nodulation and it seemed immaterial whether the shoots were kept in the light or dark during the temperature treatment (Lie, unpublished observations).

By growing the plants (or keeping the roots) at a temperature (20 °C) restricting nodulation and then exposing them for short periods (24 h) to a temperature of 26 °C which permitted nodulation, it was shown that it was only those processes taking place during the second and/or third day after inoculation which required the higher temperature (Lie, 1971*c*). This was confirmed by conducting the opposite experiment, i.e. growing plants at a temperature permitting nodulation except for short exposures to a temperature known to restrict nodulation. As shown in Table 26.3 a pronounced inhibition of nodulation was observed when the roots received the adverse temperature treatment during the second and/or third day after infection but not if this temperature treatment occurred at the other times in the infection and nodulation sequence.

Table 26.4. *The effect of temperature on root nodule formation of pea cv. Iran, inoculated with* Rhizobium *strain 313*

Period in days		Nodule no. per plant	% nodulated plants
0–2	2–3		
25 °C	25 °C	4.3	90
25 °C	30 °C	4.1	90
25 °C	35 °C	0.8	50

The plants were kept at 20 °C except for the periods indicated in columns 1 and 2. Day 0 is the time of inoculation.

Reducing the time of exposure to conducive temperatures (26 °C) to five or ten hours, did not elicit nodulation even when this temperature was increased to 30 °C or 35 °C. These results suggest that it is unlikely that a single trigger mechanism is involved. The temperature range permitting nodulation of Iran is a rather narrow one. For example, plants raised at 20 °C and exposed during the second day to 26 °C will form nodules satisfactorily but not when exposed on this day to 30 °C or 35 °C (Lie, unpublished results).

To understand the nature of induction by higher temperatures, some experiments were conducted to attempt to counteract the stimulatory effect of the 26 °C temperature on nodulation. Plants which had been induced for two days at 26 °C, were subsequently exposed for 24 hours to higher temperatures. Table 26.4 shows that the processes of induction at 26 °C were eliminated by subsequent exposure to 35 °C but not 30 °C.

It was found that processes taking place during the first day following inoculation were not affected seriously by temperature since microscopic examinations revealed that rhizobial growth on the roots was quite normal. However, the inoculated roots at 20 °C were thickened and covered by swellings and these swellings were absent in roots kept at 26 °C. Many root hairs formed at 20 °C and, although less dense than at 26 °C, distorted root hairs were still detected at 20 °C. In a few cases infection threads were observed penetrating the cortex at 20 °C.

The outgrowth of the nodule is not inhibited at 20 °C, once nodules have been initiated at 26 °C. Indeed, in such experiments the size of the individual nodule and the total quantity of nodules formed, may be larger in plants transferred from 26 °C to 20 °C than in those maintained at 26 °C.

The effect on nitrogen fixation

To compare the nitrogen-fixing activity of the root nodules at the two temperatures, pot experiments were carried out with the cultivar Iran growing in perlite. To study the functioning of root nodules on plants at 20 °C, the plants were kept for one week at 26 °C to nodulate and then returned to 20 °C to fix nitrogen. Rondo was included in the experiments for purposes of comparison.

Table 26.5 shows that both cultivars grew equally well at the two temperatures when combined nitrogen was provided. When inoculated with *Rhizobium* strain PRE or PF2, almost the same amount of nitrogen was fixed by Rondo at both temperatures, but with Iran less was fixed at a higher temperature. The most interesting result was found for *Rhizobium* strain 310a, which is the only *Rhizobium* strain capable of forming nodules on Iran at 20 °C. In both cultivars nitrogen-fixing activity with 310a was good to moderate at 20 °C but at 26 °C was considerably reduced, especially in Iran. In a number of experiments it was observed that nodules induced by *Rhizobium* strain 310a very quickly lost their red colour at 26 °C giving rise to green or partly green nodules. *Rhizobium* strain 313 held an intermediate position between 310a and PRE in its fixation performance on the two varieties at the two temperatures.

To study the effect of temperature on nitrogen fixation more directly a parallel experiment investigated the acetylene-reducing capacities of nodulated roots, derived from peas, grown for four weeks at different temperatures. Roots were incubated first for two hours at 20 °C under acetylene and then re-gassed before being incubated for a further two hours at 26 °C. Table 26.6 lists only the results of the 20 °C incubation

Table 26.5. *The effect of temperature on nitrogen fixation and nitrogen uptake of pea cv. Rondo and cv. Iran, inoculated with the ineffective* Rhizobium *strain P8 or the effective* Rhizobium *strains PRE, PF2, 310a and 313, or supplied with ammonium nitrate*

	Pea cv. Rondo (mg N/plant)			Pea cv. Iran (mg N/plant)		
Treatment	20 °C	26 °C	% decrease/ increase at 26 °C	20 °C	26 °C	% decrease/ increase at 26 °C
Uninoculated	2.1	2.5	–	0.5	0.9	–
P8	2.3	2.6	–	0.5	0.9	–
PRE	27.8	24.3	−13	22.3	11.2	−49
PF2	26.0	27.5	+ 6	23.4	17.5	−25
310a	22.0	5.8	−74	13.5	1.8	−86
313	37.8	18.8	−50	21.2	6.5	−69
Ammonium nitrate	32.8	35.9	+ 8	28.7	35.3	+22

since those obtained from the later incubation at 26 °C were considered less reliable, presumably due to a decline in the activity of the nodules after some hours. The data show that the acetylene-reducing activity of root nodules formed by strain PRE was almost identical at the two temperatures, whereas the nodules formed by strain 310a at 26 °C had a

Table 26.6. *The effect of temperature on nitrogen fixation of pea cv. Iran inoculated with* Rhizobium *strains PRE, 313 or 310a*

Rhizobium strain	μmoles C_2H_4 produced/g fresh wt nodules/h by plants growing at		% decrease 26 °C
	20 °C	26 °C	
PRE	5.0	4.5	10
313	5.0	3.0	40
310a	5.8	1.75	65

The detached nodulated roots of plants, grown either at 20 °C or at 26 °C, were incubated in air enriched with 10% C_2H_2 at 20 °C.

considerably lower activity than those formed at 20 °C. Again *Rhizobium* strain 313 formed nodules with activity intermediate between that of PRE and 310a. There were no indications whatsoever that the activity of a nodule was consistently higher if assayed at the temperature at which it had developed.

The use of 'embryo cultures'

For microscopic studies the large root system of pea plants is very inconvenient and would be less so if plant size could be reduced by removing the cotyledons at an early stage. In an attempt to do this embryos were first isolated from dry or water-soaked seeds, but the subsequent growth of these embryos on inorganic agar medium was found to be slow and erratic. However, if the cotyledons were removed somewhat later, when the radicle was about 1 cm (seedling about 2 days old), satisfactory growth in an inorganic nutrient medium was achieved even without the addition of organic substances to the medium. In this manner diminutive plantlets of 1–2 cm height could be obtained after only one month's growth. At first the embryos were grown on agar slopes completely enclosed within cotton-stoppered tubes, but under these conditions nodulation was found to be erratic and if nodules formed at all they appeared only after considerable delay. However, by growing the plantlets with only their roots within the tubes of aseptic nutrient solution, normal and more regular nodulation was achieved within about one week after inoculation.

The effect of light intensity

When the plantlets were kept at the same light intensity as intact plants, only some of the plants became nodulated and very few nodules were formed per plant. Also, the plantlets turned yellow, especially at high temperatures. A considerable improvement in nodulation and growth could be achieved, however, by reducing the light intensity of the plantlets to 25% of that used for intact plants. The inhibition by high light intensities was not found to be specific for nodulation but rather a more general effect on plant growth, since plantlets provided with combined nitrogen showed inhibited growth at high light intensities. In this respect these plantlets resemble detached rooted leaves which are also sensitive to high light intensity (Mothes & Englebrecht, 1956; Lie, 1971*a*).

The effect of combined nitrogen

Good growth was obtained when ammonium nitrate was supplied to the plantlets and much bigger plants resulted from such treatment than when nodules alone provided nitrogen. A strong inhibition (90%) of nodulation was observed at 40 ppm N (as ammonium nitrate) while at 80 ppm N and above nodulation of plantlets was completely prevented.

The effect of Rhizobium *strain and temperature*

Good nodulation was observed in plantlets derived from Rondo inoculated with *Rhizobium* strains PRE, PF2, 310a and 313. However, in contrast to the results obtained with intact plants where these four *Rhizobium* strains behaved almost similarly, strain PF2 proved to be far superior to the others in nodulation and nitrogen fixation when associated with the plantlets.

The results using plantlets of Iran are in general agreement with those obtained with intact plants. At 20 °C no nodules were formed with strain PRE but consistent nodulation was obtained with 310a. Strains PF2 and 313 nodulated plantlets at 20 °C occasionally. At 26 °C good nodulation was achieved by strains PF2, 313 and 310a but in contrast with the intact plants PRE proved to be incapable of forming nodules on plantlets at 26 °C. At the latter temperature PF2 proved to be the best strain for the Iran plantlets.

The above-mentioned results appear to justify the use of plantlets derived from pea seedlings the cotyledons of which had been removed at an early stage as material representative of host performance in nodulation. A remarkable observation was that nodules on plantlets became almost as large as those developing on intact sister plants, so that despite the diminutive size of the plantlet shoot, quite large amounts of nodule tissue could still be formed on its roots. Nitrogen fixation in the nodulated plantlets was detected using Kjeldahl analysis. Further study is needed to explain why nodules of such large size can develop on so small a plantlet.

Discussion

When considering the results of these studies it must be kept in mind that only a small collection of pea genotypes was examined and that only *Rhizobium* strains collected from temperate regions were included in the experiments. Moreover, no *Rhizobium* strains indigenous to soils where these primitive pea genotypes occur were available for study. Nevertheless, within these limits, it is still of interest to note that of the forty-two pea cultivars tested, only two genotypes, one from Afghanistan and one from Iran, turned out to be more or less resistant to the *Rhizobium* strains used. In addition, only those two genotypes collected by us directly from market places in Iran and Afghanistan, proved to be quite distinct in respect of their symbiotic affinities. The two concerned (Kabul and Isfahan) seemed also to differ basically from the genotypes from these regions used earlier. H. Ohlendorf in Sweden (personal communication) has also found that in a collection of 120 pea introductions five genotypes proved to be resistant to nodulation, three of these coming from Afghanistan, one from Iran and one from China. These results strongly support the hypothesis that the Iranian/Afghanistan region, which is considered to be one of the main centres of origin of peas (Zohari, 1970), harbours pea genotypes which are resistant to *Rhizobium* collected from more advanced cultivars of the species. It should be noted that this accords with the general rule that gene centres of cultivated plants are usually the best places to locate plants resistant to certain diseases (cf. Leppik, 1970).

So far, however, nothing is known of the nodulation status of these plants in these regions and it seems imperative to explore whether they are nodulated, and, if so, whether a strong specificity exists between these pea cultivars and their associated root nodule bacteria. In this connection the remarkable symbiotic specificity of the African clovers should be mentioned. As Norris & 't Mannetje (1964) have shown, symbiotic specialisation in these clovers occurs not only between the different clover species or varieties but even between the different plant introductions of the same species and their associated root nodule bacteria. On the other hand it is, nevertheless, remarkable that so many pea genotypes will symbiose effectively with the one *Rhizobium* strain, a promiscuity which suggests one common genetical trait in relation to nodulation, introduced long ago into the genetic constitution of ancestors of the cultivated pea and perpetuated during later selections. It may be assumed that in the early days of agriculture, when nitrogen fertilisers were not widely used, the most vigorous plants might well have been those carrying effective nodules and that these plants, yielding better than others, would have been selected by the farmer.

Our results with pea cv. Iran and cv. Afghanistan are clearly in accordance with the results from other leguminous plants (see Nutman, 1969) that nodulation is host-controlled and heritable. One major gene appears to be responsible for failure to form root nodules in these pea cultivars. However, the general nodulating propensity of the host can also clearly be modified by environmental factors like temperature and *Rhizobium* strain interactions which were not fully appreciated in earlier investigations.

Variation in the symbiotic ability of *Rhizobium* can be assessed satisfactorily only in association with its host. This situation is similar to that for plant–parasite relationships in which it has been suggested that for each gene conditioning susceptibility of the host plant, a corresponding gene conditioning aggressiveness must be present in the associated parasite (Flor, 1971). Pea cultivars described which behave so differently in respect of nodulation with different *Rhizobium* strains may enable us to test that this gene-for-gene hypothesis holds for the symbiotic situation. Furthermore, the location of the different genes for the different *Rhizobium* strains in chromosomes of the host plant might be explored by using widely different strains of *Rhizobium*, such as 310a and PF2, as microbial partners.

From the physiological point of view, pea cv. Iran is of special interest. The possibility of controlling its nodulation by simply varying the temperature for only a short period of time, and the availability of *Rhizobium* strains which differ in their ability to nodulate it at 20 °C and 26 °C, makes this cultivar ideal for definitive studies on nodulation. From the results obtained it is evident that only the processes during the second or third day are temperature-dependent and our aim now is to determine which part of the nodulation process is involved. An interesting case was reported by Syono & Furuya (1971) of a mutant tobacco callus tissue which was able to grow rapidly in a medium without kinetin when the temperature of incubation was 26 °C, but not able to grow at all at 16 °C unless kinetin was present. The obvious conclusion is that synthesis of cytokinin in this tissue (Einset & Skoog, 1973), is blocked at low temperatures, and it seems of interest to know whether a similar explanation might hold for nodule formation in the pea cv. Iran. However, so far we have not been able to effect nodulation of the cultivar at 20 °C by supplying kinetin to its roots (Lie, unpublished observations). The nodulation of Iran bears a striking resemblance to that described by Antonelli & Daly (1966) for wheat infected by *Puccinia graminis*. A wheat line was found to contain a temperature-sensitive gene which apparently conferred resistance to *Puccinia* at 20 °C, but not at 26 °C. By transferring plants from one temperature to the other at different periods of time, the conclusion was drawn that the reactions which took place during the first three days after infection are essentially temperature-insensitive, but that this is then followed by a temperature-sensitive period. It would be of interest to conduct parallel studies on these two systems in further investigations.

Summary

A survey of the *Rhizobium* symbiosis of pea cultivars collected from different parts of the world has been given and a number of cultivars from Iran and Afghanistan described which fail to nodulate under certain environmental conditions. Special attention was paid to cv. Iran, which requires a short period at a higher temperature for nodulation. From breeding experiments it has been shown that the characteristic 'temperature-dependent nodulation' is heritable and determined by one major gene of the host plant. We analysed which part of the symbiotic process is temperature-sensitive and the differential effects of temperature on nodulation and nitrogen fixation using both intact plants and plantlets derived from seedlings the cotyledons of which had been removed shortly after germination.

References

Antonelli, E. & Daly, J. M. (1966). Decarboxylation of indoleacetic acid by near-isogenic lines of wheat resistant or susceptible to *Puccinia graminis* f. sp. *tritici*. *Phytopathology*, **56**, 610–18.

Blixt, S. (1972). Mutation genetics in *Pisum*. *Agri Hortique Genetica*, **30**, 1–293.

Einset, J. W. & Skoog, F. (1973). Biosynthesis of cytokinins in cytokinin-autotrophic tobacco callus. *Proceedings of the National Academy of Sciences, USA*, **70**, 658–60.

Flor, H. H. (1971). Current status of the gene-for-gene concept. *Annual Review of Phytopathology*, **9**, 275–96.

Gelin, O. & Blixt, S. (1964). Root nodulation in peas. *Agri Hortique Genetica*, **22**, 149–59.

Leppik, E. E. (1970). Gene centers of plants as sources of disease resistance. *Annual Review of Phytopathology*, **8**, 323–44.

Lie, T. A. (1971*a*). Nodulation of rooted leaves in leguminous plants. *Plant and Soil*, **34**, 663–73.

Lie, T. A. (1971*b*). Temperature-dependent root-nodule formation in pea cv. Iran. *Plant and Soil*, **34**, 751–2.

Lie, T. A. (1971*c*). Symbiotic nitrogen fixation under stress conditions. In: *Biological nitrogen fixation in natural and agricultural habitats* (ed. T. A. Lie & E. G. Mulder), *Plant and Soil Special Volume*, pp. 127–37.

Mothes, K. & Englebrecht, L. (1956). Über den Stickstoffumsatz in Blattstecklingen. *Flora, Jena*, **143,** 428–72.

Norris, D. O. & 't Mannetje, L. (1964). The symbiotic specialization of African *Trifolium* spp. in relation to their taxonomic and their agronomic use. *East African Agricultural and Forestry Journal*, **29,** 214–35.

Nutman, P. S. (1969). Genetics of symbiosis and nitrogen fixation in legumes. *Proceedings of the Royal Society of London*, Ser. B, **172,** 417–37.

Syono, K. & Furuya, T. (1971). Effects of temperature on the cytokinin requirement of tobacco calluses. *Plant and Cell Physiology*, **12,** 61–71.

Virtanen, A. I. & Miettinen, J. K. (1963). Biological nitrogen fixation. In: *Plant physiology: a treatise* (ed. F. C. Stewart), vol. 3, pp. 539–668. Academic Press, New York & London.

Weber, D. F., Caldwell, B. E., Sloger, C. & Vest, H. G. (1971). Some USDA studies on the soybean-*Rhizobium* symbiosis. In: *Biological nitrogen fixation in natural and agricultural habitats* (ed. T. A. Lie & E. G. Mulder), *Plant and Soil Special Volume*, pp. 293–304.

Went, F. W. (1957). The pea. In: *The experimental control of plant growth* (ed. F. W. Went), pp. 115–23. Chronica Botanica Company, Waltham, Massachusetts.

Zohari, D. (1970). Centers of diversities and centers of origin. In: *Genetic resources in plants – their exploration and conservation* (ed. O. H. Frankel & E. Bennett), pp. 33–42. Blackwell Scientific Publications, Oxford & Edinburgh.

27. Physiology of the reaction of nodulated legumes to environment

J. S. PATE

The normal practice of reviewers dealing with the effect of environment on legume symbiosis has been to deal in turn with the various factors known to influence nodulation and nitrogen fixation, listing for each the beneficial or deleterious effects on symbiosis reported within the range of legumes studied. This approach has its merits, if only for its comprehensiveness and ability to serve as a basis for making generalised statements regarding symbiotic response, but since several such exhaustive treatments have recently appeared (e.g. Gibson, 1971; Lonergan, 1972; Nutman, 1972; Pate, 1974) this chapter will employ a somewhat different treatment.

Arguing that it is only by describing in detail the normal symbiosis that one can make correct deductions regarding what is likely to happen in a potentially hostile environment, the first section concentrates especially on the basic elements of mature nodule functioning in conditions close to the optimum for fixation. Emphasis is placed on the nutritional economy of the nodule and host plant, as seen in terms of the acquisition and utilisation of the essential raw materials consumed. The likely manner in which environmental stresses might affect these processes is discussed, and, in order to achieve continuity between sections dealing with various commodities of functioning, most of the data considered relates to symbiosis of one genus, *Pisum*, a legume extensively used for physiological study within our laboratory.

A smaller section considers adaptations of host and bacterium to environmental stress, and the practical issue of improving legume yield in difficult agricultural environments.

The basic elements of symbiotic functioning

Apart from its special ability to fix atmospheric nitrogen, the nodulated legume probably functions in a basically similar way to any other green higher plant, so that dissimilarities between legume and non-legume in the requirement for and handling of raw materials from the environment are likely to be relative and to stem largely from the nutritional and functional interdependencies of the nodule–plant system rather than from any inherent properties of the host plant. The economy of the symbiotic system will be discussed in relation to four nutritional themes:

(1) The fixation of carbon in photosynthesis and its utilisation in nourishment of the nodules.

(2) The formation of organic compounds of nitrogen in nodular fixation and their release to and assimilation by the host.

(3) The uptake and use of water in plant growth, transpiration and priming the export system of the nodules.

(4) The exchange of gases with the root environment involving carbon dioxide and oxygen in root and nodule respiration, nitrogen gas in nodule fixation, and trace amounts of other gases with possible inhibitory effects on symbiosis.

The essential chemical nutritional elements provided by the rooting medium, are considered to fall outside the scope of this chapter, except as regards combined nitrogen, which is specially relevant because of its potentially competitive influence on the fixation process.

The carbon economy of host plant and nodules

Ultrastructural observations on nodule tissues and their contained bacteroids, and chemical assays of nodule contents (e.g. see Schlegel, 1962; Wong & Evans, 1971; Minchin & Pate, 1974) suggest that, nodules in general maintain meagre reserves of readily utilisable carbohydrate relative to their requirements for fixation, so that they probably rely heavily for their growth and functioning on photosynthetic products currently translocated from the leaves, or on reserves of carbohydrate mobilised from other regions of the plant. Experimental evidence is fully consistent with this view, especially concerning the very close relationship that has been observed between photosynthesis and nitrogen fixation. Firstly there are the many reports in which reduced photosynthesis, by reducing light or after defoliation, causes drastic and immediate reductions in fixation (see Jones, Davies & Waite, 1967; Hardy *et al.*, 1968; Gibson, Chapter 29; Sprent, Chapter 30). Secondly, there is the group of findings in which improvement of photosynthetic capacity either by increasing day length and light intensity (e.g. Day, 1972), or by increasing the ambient concentration of carbon dioxide around the photosynthesising shoot (Hardy & Havelka,

1973, Chapter 31) or by grafting on to the plant a second shoot system (Streeter, 1973) stimulates fixation. Finally, there are reports of artificial reduction or removal of demands of shoot apices or fruits for assimilates causing diversion of extra assimilates to nodules and a corresponding increase in their fixation activity (Carr & Pate, 1967; Roponen & Virtanen, 1968; Halliday, unpublished data). The effects observed cannot always be explained solely in terms of competition between nodules and other organs for a fixed amount of photosynthate. For example, in some annual legumes highest rates and efficiencies of nodular fixation are encountered during fruiting (Pate, 1958; Day, 1972), despite the extra demands of assimilates for seed filling. The presence of flowers or fruits in some way stimulates extra photosynthetic activity, thereby allowing more, not less, assimilate to flow to the nodules. A similar sink-controlled raising of assimilation rate may also operate in some non-leguminous species (Wardlaw, 1968). Because the period of fruit filling is often when the annual legume accumulates the major part of its nitrogen (Hardy *et al.*, 1971; Pate & Flinn, 1973), we need urgently to know more about the physiology of the sharing of assimilates by fruits and nodules.

Central to an understanding of host–microsymbiont inter-relationships is the estimation of the proportion of the total photosynthetic product required by the nodules, and how this is influenced by each of the consuming processes within the nodules. An early study by Bond (1941), measured dry matter and nitrogen accumulation and root and nodule respiration, and compared growth rates of symbiotic plants with those of plants utilising equivalent amounts of combined nitrogen, from which he estimated that the nodules of soybean (*Glycine max*) consumed 19 mg carbohydrate (7.6 mg C) per mg nitrogen fixed. This corresponds to a diversion to the nodules of 16 % of the total carbohydrate formed by the host plant in photosynthesis. In *Trifolium subterraneum* comparisons made by Gibson (1966), of dry matter accumulation in symbiotically-fed or nitrate-fed plants suggested that 1.2–4 mg carbon per mg nitrogen fixed were required during the early establishment of symbiosis, but considerably less (0.32 mg C per mg N fixed) once nodules were fully developed. For *Pisum sativum*, Minchin & Pate (1973) constructed a balance sheet for the utilisation of photosynthetically fixed carbon for the period 21–30 days after sowing – the period of most rapid root growth and nodular activity preceding flowering. Nodules were found to command 32 % of the net gain of carbon by the shoot, compared with 42 % by roots, and of the nodules' share 16 % was used in growth, 37 % in respiration and the remaining 47 % returned to the shoot combined with amino acids produced in fixation. The respiration of underground organs at this early stage of the life cycle dissipated 47 out of each 100 units net gain of carbon by the shoot, a remarkably large share of the carbon budget of the plant. Selection for economy in root respiration might be a sensible way to improve legume yield.

No attempt has yet been made to measure photorespiratory losses from intact nodulated legumes to determine their importance in the plant's economy. Legumes studied so far (e.g. Hellmuth, 1971) appear for the most part to possess the C_3 pathway of photosynthesis, and the recent demonstration by Quebedeaux & Hardy (1973) of greatly improved vegetative growth of soybeans (*Glycine max*) in subatmospheric levels of oxygen suggests that photorespiration may effectively lower the general efficiency of the plant. A search for legumes exhibiting economy in this connection might also be of value.

The concept of carbohydrate supply to the nodule from the shoot as the natural pace regulator of fixation, extends also to observations on diurnal variations in nodular activity. It has long been known that nodules detached from plants at night fix nitrogen poorly compared with those harvested in the day (Virtanen, Moisio & Burris, 1955), and the diurnal changes in acetylene-reducing activity constructed for whole nodulated roots, or from samples of nodules freshly excised from roots at different times in the daily cycle, usually show higher rates of fixation during the daylight hours than at night (Hardy *et al.*, 1968; Bergersen, 1970*a*; Mague & Burris, 1972; Day, 1972). Since fixation by detached nodules is consistently less than that of equivalent weights of attached nodules (Mague & Burris, 1972), and since this difference is relatively greater during the day than at night (Bergersen, 1970*a*), nodules appear to be very dependent on current translocate and therefore likely to perform most actively when the phloem is being loaded with new supplies of photosynthate. It comes as no surprise, therefore, to find that the level of soluble carbohydrate in nodules falls and their rate of fixation quickly declines once the photosynthesising shoot has been removed (Minchin & Pate, 1974).

Differences in fixation rate between day and night in natural environments are usually recorded as being of the order of 1.5–3:1, suggesting a day:night difference of the same order of magnitude in supply of translocate to nodules. However, this need not necessarily apply to plants raised in controlled environments, as a recent study on *Pisum sativum* has illustrated (Minchin & Pate, 1974). The effects of two environments were examined, each employing a 12 hour (27 000 lux) day but one with a constant temperature day and night (18 °C), the other with a day:night regime of 18 °C:12 °C. In both environments fixation rates (by acetylene-reduction assay) and starch and sugar levels in the nodules increased during the photoperiod, suggesting that sufficient translocate arrived to satisfy not only the demands of the nitrogenase but also to replenish pools of carbohydrate in the nodule tissues. In the night these trends became reversed, and at its end minimal fixation rates and nodule sugar levels were recorded.

In the constant-temperature environment fixation during the photoperiod was 1.2 times higher than at night, but in the 18 °:12 °C environment the unusual situation prevailed in which slightly less nitrogen (1.41 mg N) was fixed during the photoperiod than in the following night (1.60 mg N). So it would appear that periods of darkness

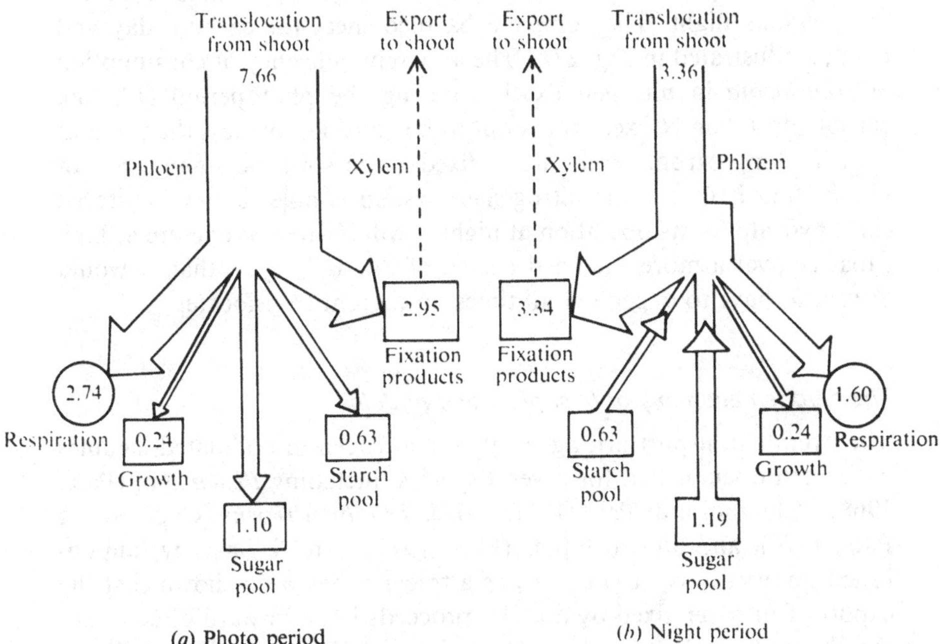

Fig. 27.1. Balance sheets for carbon utilisation by root nodules of *Pisum sativum* for (*a*) the 12-hour photoperiod at 18 °C and (*b*) the 12-hour night period at 12 °C of the 28th day after sowing. Data after Minchin & Pate (1974). All values mg C/plant.

can be times of high nodule activity, provided that sufficient carbohydrate has built up in the nodule during the previous photoperiod, and that the night temperature is optimum for efficient use of carbohydrate for fixation. In the 18 °C night as much stored carbohydrate was available and utilised by the nodules as in the 12 °C night, but in the latter the carbohydrate appeared to be used more effectively for nitrogen fixation. Others have commented on the unfavourable influence of high night temperature on the fixation of temperate legumes (Roponen, Valle & Ettala, 1970; Gibson, 1971), and the general relationship between nitrogen fixation and root respiration may, as suggested above, provide the basis on which temperature affects symbiosis.

An important corollary of the results for the study of the 18 °:12 °C environment was that although slightly less nitrogen was fixed in the photoperiod than during the night, 2.3 times more carbohydrate was estimated to have arrived as translocate in the nodules during the photoperiod than at night. The two large items of expenditure of carbon during the day were the relatively high rate of nodule respiration, and the daytime replenishment of pools of carbohydrate exhausted during the previous night. The complete balance sheet for carbon, day and night, is illustrated in Fig. 27.1. The apparent efficiency of consumption of *translocate* in nitrogen fixation during the photoperiod (13.6 mg carbohydrate/mg N fixed) turns out to be considerably less than that at night (5.3 mg carbohydrate/mg N fixed). The strategic significance of this finding is that if the nitrogenase system is able to rely on stored carbohydrate for its operation at night it will be able to operate at high capacity over a more extended period of the daily cycle than it would were it to have to depend at all times on current translocate.

The nitrogen economy of host plant and nodules

The overall transport arrangements for nitrogen in nodulated legumes have been documented for several species, including *Pisum* spp. (Pate, 1968; Oghoghorie & Pate, 1971, 1972), *Trifolium repens* (Copeland & Pate, 1970), and annual lupins (Pate, 1973; Pate & Sharkey, unpublished observations). Using $^{15}N_2$ as a tracer it has been shown that the export of nitrogen fixed by nodules proceeds in an upward direction to the shoot through the xylem (Oghoghorie & Pate, 1972), and that it involves the secretion to the xylem of specific amino compounds at concentrations several times higher than in the bacterial tissues (Pate, Gunning & Briarty, 1969; Wong & Evans, 1971). The pericycle cells of the nodule vascular bundles are believed to implement export to the xylem and in certain legumes these cells become modified as 'transfer cells' through the formation of elaborate finger-like ingrowths of their walls. Two recent articles discuss at length the efficiency of functioning of these transfer cells in the nodule (Pate & Gunning, 1972; Gunning *et al.*, 1974). The nitrogenous products exported from nodules in the xylem consist principally of nitrogen-rich molecules such as amides or ureides and these obviously permit a considerable economy to be made in relation to the amount of carbon which has to be cycled through the nodules in association with the export of fixed nitrogen, (Minchin & Pate, 1973; Pate, 1974). On arrival in the shoot, these amino compounds either serve as substrates for protein synthesis, or become diverted to pools of soluble nitrogen which may be used later in the life cycle as substrates for seed nutrition (Pate, 1971, 1973).

Much of the nitrogen received by the shoot from the nodules later returns to the root attached to amino compounds in the phloem and amino acid analyses of phloem sap (Pate & Sharkey, unpublished data) suggest that the phloem carries not only the nitrogenous solutes typical of root metabolism, but also those amino compounds generated in photosynthesis. This recycling of nitrogen through the shoot is held to be important quantitatively in the nutrition of outlying parts of the root, especially when the root does not have access to any form of nitrogen other than that from its nodules, (Oghoghorie & Pate, 1972; Pate, 1974). Moreover, it is likely that shoot-derived translocate might have a special qualitative significance to root growth since it contains certain photosynthetically-produced amino acids likely to be formed only sparingly in the root and nodule.

The scheme of functioning outlined above helps the understanding of effects which combined nitrogen may have on the physiology of an already nodulated legume. In the first place, if combined nitrogen is available directly to the root from the surrounding medium it is likely that the growing root's dependence on nitrogen cycled via the shoot will be partially circumvented, if not abolished. Secondly, if an unreduced source of nitrogen such as nitrate be given to a root, some of this nitrogen will be reduced below ground thereby causing a substantial drain on the carbon reserves of the root. In pea roots, for instance, an active nitrate reductase system is present (Wallace & Pate, 1965) and it is estimated that in association with its activity 6.2 mg carbon is consumed by the root for every milligram of nitrate-nitrogen which is assimilated (Minchin & Pate, 1973). Finally, if some of the combined nitrogen escapes unchanged from root to shoot in the xylem – as is known to occur in pea if either urea or nitrate is fed to its roots (Pate &

Wallace, 1964) – photosynthetically-associated assimilation of this nitrogen may lead to a considerable reduction in the total amount of translocate available to the roots. In this situation it has been shown by means of $^{14}CO_2$ feeding studies that it is the nodules and not the roots which suffer most from the shortage of translocated photosynthate (Small & Leonard, 1969; Oghoghorie, 1971).

Whatever the mechanisms involved in suppression of nodular fixation by combined nitrogen, it is now apparent that the nodule's performance can be altered quite rapidly by changing the nitrate supply in the rooting medium. Oghoghorie (1971), working with field pea, showed that exposure of roots of nodulated plants raised in minus-nitrogen medium to a culture solution containing 315 ppm nitrate-nitrogen caused, within 48 hours, a drastic curtailment of nitrogenase activity as assayed by acetylene reduction. This time is somewhat longer than that required to induce active reductase systems in root and shoot but long enough for effects mediated through translocation to impose restriction on nodule activity. Conversely, when plants whose fixation had been suppressed by long exposure to 315 ppm nitrate-nitrogen have their nitrate supply to the rooting medium removed, it takes about the same time (50–60 hours) before nitrogenase activity is significantly restored. Again, the time scale of events suggested regulation by feedback control operating via the translocatory system rather than by any immediate effect operating, for example, through nodule metabolism.

There is convincing evidence in agriculturally important legumes that small supplements of combined nitrogen, if correctly timed in their application, can stimulate nodulation and nitrogen fixation (e.g. see reviews by Loneragan, 1972; Olsen, 1972; Pate, 1974; Gibson, Chapter 29), and this effect relates to the fact that many cultivated legumes, especially the high-yielding ones, possess a growth potential which can not be fully satisfied by even the most efficient types of symbiosis. Benefit of symbiosis is then likely to be indirect and to come from the general stimulus which the added nitrogen may give to plant growth and dry matter production. Stages in the growth cycle most likely to benefit from fertiliser nitrogen are either during the early establishment of the seedling before fixation has commenced (Pate & Dart, 1961), or during fruit filling when there is a peak demand for nitrogen in synthesis of seed protein, and when symptoms of starvation and senescence may be manifest in the nodule population.

Unfortunately we are still largely in ignorance concerning the effects of mineral nitrogen in the soil on the symbiosis of legumes of natural vegetation. As a general rule these act as pioneer accumulators of nitrogen in nitrogen-poor habitats or as maintainers of nitrogen capital in stable ecosystems such as grassland or forest, and in either of these situations there may be good evidence of nitrogen being the limiting factor in productivity of the biomass (Hardy & Holsten, 1972). The legume component of natural vegetation may gain virtually all of its nitrogen from the atmosphere, if only because the energy-consuming root nodules will be on roots less well-equipped than competing non-legume roots for scavenging traces of combined nitrogen from the soil. Indeed, the shoot of the legume host may have become so conditioned to subsisting on the purely organic diet of amides, ureides and amino acids which it receives from its nodules, that both it and the supporting root may have lost the capacity to deal effectively with inorganic forms of nitrogen available in the surrounding medium. These speculations remain to be tested since virtually no work has been done on the response of naturally occurring species of legumes to added nitrogen in their native habitats.

The transfer of nitrogen to developing seeds is of special importance in the nitrogen economy of the pulse legumes as peanut (*Arachis hypogaea*), soybean (*Glycine max*), and pea (*Pisum*). In *Pisum* timed applications of ^{15}N at different times in the life cycle have shown that nitrogen is transferred to seeds with an efficiency of approximately 50 % if assimilated before flowering, and with even greater efficiency (70–6 %) if applied late in the life cycle when fruits are starting to form (Schilling & Schalldach, 1966; Pate & Finn (1974). Using this information and data for accumulation of total nitrogen in plant parts, it can be estimated that in the tall, sequentially flowering *Pisum arvense*, approximately one-fifth of the total nitrogen acquired by the seeds is taken up prior to flowering, with the remainder coming from the rooting medium after flowering has occurred (Pate & Flinn, 1974). Evidence was obtained that early- or late-fed ^{15}N was mobilised to the seeds rather slowly, so that all fifteen reproductive nodes were ultimately able to derive benefit; the younger ones as much, if not more, than those developing earlier. This was attributed to the fact that nitrogen entering the plant is first processed in vegetative structures and may not become incorporated into the seeds until after it has passed through several age series and types of vegetative organs. It is therefore difficult to judge whether selection to improve protein levels or yields in pulses should emphasise capacity to build up of large, mobilisable reserves of nitrogen in vegetative structures or ability to transfer rapidly and efficiently the recently absorbed nitrogen to the seeds. The danger in selecting for efficient and complete mobilisation of nitrogen from vegetative organs is that this may encourage premature senescence of photosynthetic surfaces essential for the carbon nutrition of the fruits. Carbon which is assimilated early in the life cycle, unlike nitrogen, becomes bound irretrievably into the vegetative framework of the plant, and as such will have little direct relevance to seed nutrition (Pate & Flinn, 1974). Indeed, in *Pisum arvense*, for example, the fruits developing at each reproductive node depend heavily (up to 60–70 %) for their carbon on current photosynthesis at organs subtended at their parent node (Flinn & Pate, 1970), so that if conditions become unsuitable for assimilation, seed nutrition at the relevant nodes is likely to be adversely affected. With great emphasis now being placed on grain legumes as sources of protein for humans or domestic animals, concerted efforts by plant physiologists are required in this general area. Only then can full advantage be taken of genetic studies now being carried out with a view to increasing protein levels in pulse legumes (Panton, Coke & Pierre, 1973; Müller & Gottschalk, 1973).

The water economy of host plant and nodules

The legume nodule requires water for the maintenance of turgidity of its tissues and for the export in the xylem of the products of fixation. It is not surprising therefore, to find ample evidence of nodular activity being acutely suppressed when whole plants or detached nodules are subject to water stress. The general topic of drought damage to symbiosis is discussed elsewhere by Sprent (1971, 1972*a*,*b*,*c*). Here it will be sufficient to summarise as follows:

(1) Loss of fixation activity on desiccation occurs in nodules with either determinate growth, e.g. *Glycine max* (Sprent, 1971), or indeterminate growth, e.g. *Trifolium repens* (Engin & Sprent, 1973), *Vicia faba* (Sprent, 1972*c*), *Pisum sativum* (Minchin & Pate, unpublished data).

(2) Effects of water stress are reversible, provided that water loss from a nodule does not amount to more than 20 % of its maximum fresh weight (Sprent, 1971).

(3) Irreversible structural damage occurs with severe drought damage; seen as collapse of cells in the nodule cortex and damage to plasmodesmatal connections in the nodule tissues (Sprent, 1972*a*).

(4) In the field fixation is found to be highest at or near field capacity but is severely suppressed (in *Vicia faba*) once flagging of the lower leaves has commenced (Sprent, 1972*c*, Chapter 30).

(5) In times of stress a deficiency of water associated specifically with nodules of the upper zone of the root may be made good by lateral transfer to these nodules of water retrieved by roots in lower horizons of the soil profile (see ^3H_2O studies on *Glycine max* by Sprent, 1972*c*).

(6) Osmotic damage to fixation may occur through the concentration of ions near the nodules (Sprent, 1972*b*). This is not necessarily associated with water deficiency, and can be seen in water cultures containing unusually high concentrations of nutrients, particularly if the nodule zone is above the liquid so that salts concentrate on the nodule surfaces. (Minchin & Pate, unpublished observations).

Minchin & Pate (1973) and Minchin (1973) have drawn up a water budget for the pea plant and its nodules. As in earlier discussion of the carbon economy of this species, attention was focussed on the period of rapid vegetative growth 21–30 days after sowing, using the 18 °:12 °C environment in growth cabinets known to be near optimal for development. During this period 27.3 mg of fixed nitrogen was exported from the nodule, and assuming that the concentration of nitrogenous solutes achieved in nodule xylem sap (1.5 % w/v) represents the maximum efficiency of operation of the export system, the water requirement for export during this 9 day period would have been at least 9.7 ml. In addition, 0.3 ml water was consumed in nodule growth during the same period, giving a total water requirement by the nodules of 10 ml. Three sources of water intake could contribute to this requirement: surface absorption by the nodule, mass flow into the nodule through the phloem, and the retrieval by the nodule of water from adjacent non-vascular tissues of the root. Data for rates of liquid output from the xylem of detached nodules suggest that if given unlimited access to water, the nodules might have taken up at least 1.3 ml by surface absorption during the 9 day period (Minchin, 1973), whilst in respect of inflow of water via the phloem recent measurements of the composition of phloem sap in legumes (Pate & Sharkey, unpublished observations) suggest that sugars flow into the nodule as a 14 % w/v solution so that approximately 2 ml of water would have been delivered to the nodule through the phloem in supplying the 112 mg carbon required by the nodules during the 9 day period (Minchin & Pate, 1973). Assuming that the extra 6.7 ml of water needed to meet the nodules' 10 ml water requirement came from the parent root, the total intake would be 20 % from the phloem, 13 % as surface uptake, and the remaining 67 % abstracted laterally from the adjacent root. It would appear, therefore, that the host root is relatively more important as a source of water to the nodule than the adjacent rooting medium, a finding which agrees well with the observations that nodules can survive and function satisfactorily in upper, dry strata of the soil, provided that the parent root has access to adequate supplies of water, and that it is only when quite severe drought symptoms are visible in the host shoot that nodules are permanently damaged (Sprent, 1972*c*). Of course if flagging of the lower leaves takes place, photosynthesis is likely to be arrested, and since these leaves are likely to be the main providers of carbon to the nodules, it is possible that the first reduction of nitrogen fixation during droughting will be caused by reduction in the assimilate supply. This suggestion is fully consistent with the reversibility of the effects of mild droughting damage and does not conflict with the ideas expressed above of other permanent structural manifestations of damage within the nodule once severe desiccation of host and nodule tissues has taken place (Sprent, 1972*a*). Whether moderate or severe, the permanency of symbiotic damage due to drought is often uncertain for, as shown by Wilson (1942) and Masefield (1961), shedding of one generation of droughted nodules may lead to a replacement set being formed. Also, in certain species (e.g. *Trifolium repens*) it has been shown that nodules whose older tissues have been permanently damaged by droughting, may, on relief from water stress, recommence apical growth and develop new zones of fixing tissue (Engin & Sprent, 1973).

Despite the uncertainty that exists regarding the precise water budget of nodules, it is clear that under normal circumstances the amount of water absorbed by or cycled through the nodules is relatively small compared with that absorbed by the root and passed to the atmosphere from the shoot in transpiration. In the example of pea mentioned above, 140 ml of water were transpired during the 9 day interval of study, more than fourteen times the amount of water calculated to have passed through the xylem of the nodules during the same period. Of course, the two xylem streams of nodule and root must be fully interconnected, so that during active transpiration a general condition of tension within the xylem may give rise to a more efficient displacement of exported products from the nodule xylem. Indeed, this is borne out by the demonstration that products of fixation tend to accumulate in nodules at night when water loss from shoots is reduced to a minimum, and that a reduction in transpiration during the day by raising the humidity around shoots also leads to a measurable build up of soluble nitrogen in nodules (Minchin & Pate, 1974). Nevertheless, it remains unlikely that accumulation of fixation products in the nodule exercises any direct inhibitory effect on fixation rate. Indeed, the cool, humid nights of the 18 °:12 °C environment caused a massive build up of soluble nitrogen in the nodules, but this did not prevent fixation at night from occurring less effectively than in the day when transpiration facilitated a much more rapid clearance of fixed nitrogen from the nodules (see Minchin & Pate, 1974).

The gaseous environment of the nodulated root and symbiotic functioning

In natural habitats, it is commonly observed that legumes are generally intolerant of waterlogged or poorly aerated soils, and that nodules which do develop in such conditions occupy positions above the water table or in the better aerated pockets of the soil profile. To study the phenomena more precisely, recourse is made to defined laboratory conditions using such techniques as bubble aeration with various gas mixtures or varying the water level in liquid or sand cultures. The following generalisations emerge from such studies:

(1) Fixation in nodulated plants in water culture is at a maximum when cultures are aerated with air (i.e. 20 % oxygen: Virtanen & von Hausen, 1936; van Schreven, 1958). Lowered oxygen level in the aerating gas stream causes a great suppression of fixation (Bond, 1950; Ferguson & Bond, 1954).

(2) Detrimental effects of waterlogging tend to be greater in plants relying on nodules for their nitrogen than in comparable plants having access to nitrate (Ferguson & Bond, 1954; Minchin, 1973).

(3) A cline of decreasing efficiency in nodular fixation down a root can be observed in a waterlogged environment, greatest mass of nodules and efficiency of fixation occurring closest to the surface of the rooting medium, i.e. closest to the source of oxygen (see Table 27.1).

(4) In a sand or soil medium fixation is greatest at field capacity (Engin & Sprent, 1973; Minchin, 1973).

(5) Nodule initiation appears to be more tolerant of low oxygen partial pressure than is subsequent nodule functioning; this may be of special adaptive significance for species nodulating in wet intractable soils in the spring.

(6) Respiration and nitrogen fixation studies on detached nodules exposed to different oxygen tensions show that the efficiency of carbohydrate consumption in fixation is highest at near-atmospheric levels of oxygen, and is much decreased with lowering of oxygen supply (Bergersen, 1971). The low fixation output and efficiency of deeply buried nodules recorded in Table 27.1, is therefore likely to result from misuse of, rather than restricted supply of, carbohydrate.

Table 27.1. *Effects of prolonged waterlogging on development and nitrogen fixation of nodules in different zones of sand-cultured roots of* Pisum sativum. (Unpublished data of Minchin & Pate)

Successive zone down root from cotyledonary node	Nodule fresh wt (g/plant)	N_2 (C_2H_2) fixation† (μg N/plant/h)	Nodule fixation efficiency (μg N/g fresh wt nodule/h)
Plants waterlogged for* 16 *days*			
0–1.5 cm	0.14±0.00	24.4±1.6	174±11
1.5–3.0 cm	0.16±0.06	21.1±4.4	132±57
3.0–4.5 cm	0.05±0.00	5.7±0.2	106± 4
Control plants with free drainage*			
0–1.5 cm	0.27±0.08	121.3± 2.7	458±134
1.5–3.0 cm	0.30±0.07	114.3±10.6	381± 96
3.0–4.5 cm	0.22±0.06	52.5± 2.3	238± 66

* Plants 30 days old at harvest.
† A factor of 2.4 was used to convert moles acetylene reduced to moles N_2 fixed. Six zones of 1.5 cm length per assay.

Although the special sensitivity of nodulated as opposed to nitrate-grown plants to waterlogging suggests some effect specific to symbiotic functioning, other more general effects of waterlogging on host plant performance cannot be discounted. Root growth and uptake of certain nutrients are known to be reduced by waterlogging (see Devitt & Francis, 1972). Leakage of metabolites can also occur (see review by Pate, 1974), as can the induction of abnormal metabolic pathways (e.g. McManmon & Crawford, 1971), and a lowering of output by roots of growth substances (Burrows & Carr, 1969). Moreover, although oxygen deficiency is generally regarded as the major damaging influence in the stagnant root environment, the presence of various hydrocarbons and excessively high levels of carbon dioxide can also exercise effects. Acetylene is strongly inhibitory to fixation through its competitive influence on nitrogenase, whilst its reduction product ethylene is highly damaging to nodule initiation (Grobbelaar, Clarke & Hough, 1971).

The effects of elevated levels of carbon dioxide in the root environment are ambiguous; concentrations in excess of 1 % carbon dioxide are reported by Mulder & Veen (1960) as stimulating fixation in certain legumes, whereas Stolwijk & Thimann (1957) have found this level to be strongly inhibitory to root growth of peas.

Studies on respiration of attached and detached nodules suggest that the symbiotic apparatus does not have any specially great requirement for oxygen, nodules not being noticeably more active in respiration than a comparable mass of young root tissue (Allison *et al.*, 1940; Bond, 1941). However, the overall requirement for oxygen by a root system engaging in nodular fixation might be noticeably greater than that of a comparable non-nodulated root consuming nitrate, since the nitrate ion might furnish a substantial proportion of the oxygen requirement of the root. This might well explain why nitrate-grown plants grow almost normally when waterlogged, and why addition of nitrate to waterlogged nodulated plants immediately improves their growth and alleviates symptoms of nitrogen deficiency (Minchin, 1973).

Under natural conditions neither waterlogging nor drought are likely to be permanent. Experiments on sand-cultured peas simulated fluctuating water regimes by employing timed intervals of flooding or droughting alternating regularly with periods when roots were given normal access to water and allowed to drain normally. The treatments all occurred over an 11 day period 19–30 days after sowing with plants raised to the 18 day stage under optimal water and nutrient supply which procured good nodulation and nitrogen fixation. Table 27.2 shows that conditions in the root medium optimal for fixation were achieved by watering on alternate days with free drainage at all times. Increasing the extent and duration of waterlogging or droughting beyond this optimum condition proportionally decreased the amounts of nitrogen fixed, the most severe flooding treatment having a greater inhibitory effect on nitrogen fixation than that induced by severest droughting. For equivalent reductions in total fixation, nitrogen percentages in dry matter were consistently lower in waterlogged plants than in droughted plants, so that the inhibitory effect of excess water had a more specific and dramatic effect on nitrogen fixation than was caused by water deficiency (see Table 27.2).

On the day of final harvest in the above experiment, acetylene reduction was measured on nodulated roots, and the sand from around the roots of the plants of the various treatments assayed for water content (see columns 4 and 5 of Table 27.2). The sand water contents, from 2 % of sand dry weight to the permanent wilting point of 1.2 % and from 15 % up to the fully waterlogged condition of 17 %, led to drastic reductions in fixation performance. But within the range 2–15 %

Table 27.2. *Effect of waterlogging or drought stresses on nitrogen fixation in nodules, and plant nitrogen content of sand-cultured* Pisum sativum. (*Unpublished data of Minchin & Pate*)

Stress	N_2 fixation in period 19–30 d after sowing (mg N/plant)	% N in plant dry wt (30 d harvest)	N_2 (C_2H_2) fixation on day 30 (μg N/plant/h)	Water content of sand on day 30 (% of dry wt; w/w)
5 d waterlogged*: 1 d recovery†	6.87	2.33	41.8	16.9
3 d waterlogged: 1 d recovery	10.54	2.74	61.1	16.0
1 d waterlogged: 1 d recovery	17.95	3.60	124.3	15.5
watered daily free drainage	22.74	4.39	130.4	15.6
1 d watered: 1 d drying out	25.68	4.49	145.5	7.7
1 d watered: 3 d drying out	18.27	4.06	88.3	2.0
1 d watered: 5 d drying out	16.10	3.89	12.7	1.2

* By capillary rise of water placed in dish under basal drainage holes of pot.
† By removal of dish to allow free drainage.

water content had little effect on fixation. It would be interesting to see whether this broad tolerance range of symbiosis to the moisture:gas content of the rooting medium holds generally for legumes grown under field conditions. Studies on other species, e.g. *Trifolium repens* (Engin & Sprent, 1973) suggest that symbiotic tolerance to water stress, and ability to recover from quite severe and prolonged stress is much greater than had been supposed.

Examples of adaptation of host and bacterium to physiological stress

Symbiotic response to physiological stress is a highly variable quantity in which tolerance or susceptibility appears to result from clearly heritable factors in both host and *Rhizobium* (see Table 27.3). Differences between genera and species of host are usually the largest and

Table 27.3. *Examples of adaptation in host and bacterium to physiological stress*

Presumed environmental variable to which tolerance or susceptibility is shown	Source of adaptation in host and bacterium to physiological stress			
	In symbiosis			
	Host legume		*Rhizobium* strain	
	Intergeneric and interspecific levels	Intraspecific level		Free-living *Rhizobium*
Chemical nutrients in excess of deficiency, extreme of pH	Andrew & Milligan (1954) Andrew & Norris (1961) Nutman (1956) Odu, Fayemi & Ogunwale (1971) Parle (1958) Robson & Loneragan (1970)	Epstein (1972) Snaydon (1962)	Jones & Rees (1970) Lynch & Sears (1950 Masterson & Sherwood (1970) Roberts & Olsen (1942)	Holding & King (1963) Jones *et al.* (1964) Lie (1971) Masterson & Sherwood (1970) Robson & Loneragan (1970b)
Grazing	Butler, Greenwood & Soper (1959) Whiteman & Lulham (1970)			
Drought	Sprent (1972*a*)			
Waterlogging	McManmon & Crawford (1971)			
Temperature	Allen, Allen & Klebesadel (1963) Dart & Day (1971) Gibson (1971) Joffe, Weyer & Saubert (1961) Mes (1959) Meyer & Anderson (1959)	Gibson (1961, 1971) Lie (1971) Roponen, Valle & Ettala (1970) Roughley & Dart (1969, 1970)	Ek-Jander & Fahraeus (1971) Gibson (1961) Lie (1971) Pate (1961) Petrosyan (1959) Roughley & Dart (1969) Roughley (1970) Vartiovaara (1937)	Marshall (1964) Wilkins (1967)
Combined nitrogen	Allos & Bartholomew (1955, 1959) Dart & Wildon (1970) Gibson (Chapter 29) Pate & Dart (1961) Small & Leonard (1969)	Richardson, Jordan & Garrard (1957)	Gibson (Chapter 29) Pate & Dart (1961)	

most easy to detect (column 1), and indeed it is these which underly many established practices of legume selection for climate, soil type or agricultural management. However, a second order of variation is manifest at the level of host cultivar and *Rhizobium* strain (columns 2 and 3) sufficient to justify selection for improved symbiotic yield, e.g. in acid soils (see Masterson & Sherwood, 1970; Lie 1971) in soils rich in combined nitrogen (Copeland & Pate, 1970; Gibson, Chapter 29), or in exacting temperature conditions (Gibson, 1971; Lie, 1971 Dart *et al.*, Chapter 28). The final source of variation concerns survival of the free-living *Rhizobium* in soil (column 4). Again, differences may be quite large – especially between *Rhizobium* of different 'host groupings' – but the introduction and establishment of more desirable *Rhizobium* strains in soil environments already containing infective rhizobia often proves difficult (Bergersen, 1970*b*).

A symbiotic association is exposed to a continually changing pattern of stresses, each potentially limiting, and it is no easy task to discover which may have had greater impact, nor indeed to ascertain which were responsible for permitting one particular association to perform better than others in a given set of exacting conditions. In annual arable crops, in which factors such as sowing time, inoculation practice, soil texture and nutrient and water status are likely to be arranged close to the optimum, ability to withstand environmental stress may confer little, if any, advantage. Yield is then likely to be determined largely by internal factors, including general physiological parameters of the host genotype, and the ability of its nodules to provide nitrogen early enough, fast enough and long enough to match the growth demands of the host. With great advances now being made in breeding legumes for high protein yield, it becomes especially important to select for matching improvements in symbiotic effectivity.

Pasture legumes may depend on somewhat different attributes, for example, ability to compete with grasses or other non-legume components, to survive persistent grazing or mowing, and to withstand wide fluctuations in water and nutrient levels. In perennial species, ability to overwinter, to tolerate dry hot summers, or to nodulate in cold, waterlogged soils in spring, may be among the more important adaptive features, whilst in the annual or ephemeral pasture species effective seed set and dispersal, rapidity of seedling establishment and nodulation, and survival of appropriate *Rhizobium* strains between successive generations of the host, may be especially desirable characteristics.

Finally, in the highly complex and little-explored situation of legumes in their native habitats, one can only guess at the key factors for the survival of the species. Properties mentioned above for pasture species will doubtless be important, but both bacterial and host partners must prove able to tolerate the wide extremes of climate and edaphic conditions encountered during the life span of the plant community, including, possibly, the ravages of pests, frost, drought, fire and flood. Here resources of a different kind must be deployed; for example the ability to accumulate reserves in resistant vegetative structures, to produce hard-coated seeds of great longevity, and, as has so often occurred during the evolution of legumes, to acquire toxins affording protection against predators. In such a situation qualities rated of importance to man in agriculture may take second place. Rate and efficiency of fixation of nitrogen may be much less important, say, than the nodules' ability to withstand extremes of soil temperature and moisture, and the growth rate of the legume may be less crucial than its ability to scavenge effectively for a limiting soil nutrient. Whatever the basis for superior performance, whether of symbiotic origin or not, long-term survival of species within particular ecological situations has undoubtedly encouraged evolution of associations superlatively equipped for operating in their home environment. This is particularly well-documented in respect of host variation in ecotypes of *Trifolium repens* (Snaydon, 1962) and *Rhizobium* strain adaptation in quite a variety of situations (Vartiovaara, 1937; Roberts & Olsen, 1942; Nutman & Read, 1952; Petrosyan, 1959; Ek-Jander & Fahraeus, 1971). Much is likely to be gained from genetic analyses, and auto-ecological studies of these types of variation, and exciting scope exists for the plant physiologist fully conversant with the intricacies of symbiotic functioning in the field.

References

Allen, E. K., Allen, O. N. & Klebesadel, L. J. (1963). An insight into symbiotic nitrogen-fixing plant associations in Alaska. In: *Proceedings of the 14th Alaskan Science Congress, Anchorage, Alaska*, pp. 54–62.

Allison, F. E., Ludwig, C. A., Minor, F. W. & Hoover, S. R. (1940). Biochemical nitrogen fixation studies. II. Comparative respiration of nodules and roots, including non-legume roots. *Botanical Gazette*, **101**, 534–49.

Allos, H. F. & Bartholomew, W. V. (1955). Effect of available nitrogen on symbiotic fixation. *Proceedings of the Soil Science Society of America*, **19**, 182–4.

Allos, H. F. & Bartholomew, W. V. (1959). Replacement of symbiotic fixation by available nitrogen. *Soil Science*, **87,** 61–6.

Andrew, W. D. & Milligan, R. T. (1954). Different molybdenum requirements of medics and subterranean clover on a red brown soil at Wagga, New South Wales. *Journal of the Australian Institute of Agricultural Research*, **20,** 123–4.

Andrew, C. S. & Norris, D. O. (1961). Comparative responses to calcium of five tropical and four temperate pasture legume species. *Australian Journal of Agricultural Research*, **12,** 40–55.

Bergersen, F. J. (1970*a*). The quantitative relationship between nitrogen fixation and the acetylene reduction assay. *Australian Journal of Biological Sciences*, **23,** 1015–25.

Bergersen, F. J. (1970*b*). Some Australian studies relating to the long-term effects of the inoculation of legume seeds. *Plant and Soil*, **32,** 727–36.

Bergerson, F. J. (1971). The biochemistry of symbiotic nitrogen fixation in legumes. *Annual Review of Plant Physiology*, **22,** 121–40.

Bond, G. (1941). Symbiosis of leguminous plants and nodule bacteria. *Annals of Botany*, **5,** 313–37.

Bond, G. (1950). Symbiosis of leguminous plants and nodule bacteria. 4. The importance of the oxygen factor in nodule formation and function. *Annals of Botany*, **15,** 95–108.

Burrows, W. J. & Carr, D. J. (1969). Effects of flooding the root system of sunflower plants on the cytokinin content in the xylem sap. *Physiologia Plantarum*, **22,** 1105–12.

Butler, G. W., Greenwood, R. M. & Soper, K. (1959). Effects of shading and defoliation on the turnover of root and nodule tissue of plants of *Trifolium repens*, *Trifolium pratense* and *Lotus uliginosus*. *New Zealand Journal of Agricultural Research*, **2,** 415–26.

Carr, D. J. & Pate, J. S. (1967). Ageing in the whole plant. *Symposia of the Society for Experimental Biology*, **21,** 559–600.

Copeland, R. & Pate, J. S. (1970). Nitrogen metabolism of nodulated white clover in the presence and absence of nitrate nitrogen. *Occasional Symposium No. 6 of the British Grassland Society*, pp. 71–7. British Grassland Society, Hurley, Berkshire.

Dart, P. J. & Day, J. M. (1971). Effects of incubation temperature and oxygen tension on nitrogenase activity of legume root nodules. *Plant and Soil*, Special Volume, 167–84.

Dart, P. J. & Wildon, D. C. (1970). Nodulation and nitrogen fixation by *Vigna sinensis* and *Vicia atropurpurea*: the influence of concentration, form and site of application of combined nitrogen. *Australian Journal of Agricultural Research*, **21,** 45–55.

Day, J. M. (1972). Studies of the role of light in legume symbiosis. PhD Thesis, University of London.

Devitt, A. C. & Francis, C. M. (1972). The effect of waterlogging on the mineral nutrient content of *Trifolium subterraneum*. *Australian Journal of Experimental Agriculture and Animal Husbandry*, **12,** 614–17.

Ek-Jander, J. & Fahraeus, G. (1971). Adaptation of *Rhizobium* to subarctic environments in Scandinavia. *Plant and Soil*, Special Volume, 129–38.

Engin, M. & Sprent, J. I. (1973). Effects of water stress on growth and nitrogen-fixing activity of *Trifolium repens. New Phytologist*, **72,** 117–26.

Epstein, E. (1972). *Mineral nutrition of plants: principles and perspectives.* Wiley, New York.

Ferguson, T. P. & Bond, G. (1954). Symbiosis of leguminous plants and nodule bacteria. V. The growth of red clover at different oxygen tensions. *Annals of Botany*, **18,** 385–96.

Flinn, A. M. & Pate, J. S. (1970). A quantitative study of carbon transfer from pod and subtending leaf to the ripening seeds of the field pea (*Pisum arvense* L.) *Journal of Experimental Botany*, **21,** 71–82.

Gibson, A. H. (1961). Root temperature and symbiotic nitrogen fixation. *Nature, London.* **191,** 1081.

Gibson, A. H. (1966). The carbohydrate requirements for symbiotic nitrogen fixation. A 'whole plant' growth analysis approach. *Australian Journal of Biological Sciences*, **19,** 499–515.

Gibson, A. H. (1971). Factors in the physical and biological environment affecting nodulation and nitrogen fixation in legumes. *Plant and Soil*, Special Volume, 139–52.

Grobbelaar, N., Clarke, B. & Hough, M. C. (1971). The nodulation and nitrogen fixation of isolated roots of *Phaseolus vulgaris* L. III. Effects of carbon dioxide and ethylene. *Plant and Soil*, Special Volume, 216–23.

Gunning, B. E. S., Pate, J. S., Minchin, F. R. & Marks, I. (1974). Quantitative aspects of transfer cell structure and function. *Symposia of the Society for Experimental Biology*, **28,** 87–126.

Hardy, R. W. F., Burns, R. C., Herbert, R. R., Holsten, R. D. & Jackson, E. K. (1971). Biological nitrogen fixation: a key to world protein. *Plant and Soil*, Special Volume, 561–90.

Hardy, R. W. F. & Havelka, V. D. (1973). Symbiotic N_2 fixation: multifold enhancement of CO_2-enrichment of field-grown soybean. *Plant Physiology*, Supplement, p. 35.

Hardy, R. W. F. & Holsten, R. D. (1972). Global nitrogen cycling: pools, evolution, transformations, transfers, quantitation and research needs. In: *Environmental protection agency symposium.* Washington, DC, USA.

Hardy, R. W. F., Holsten, R. D. Jackson, E. K. & Burns, R. C. (1968). The acetylene–ethylene assay for N_2 fixation: laboratory and field evaluation, *Plant Physiology*, **43,** 1185–1207.

Hellmuth, E. O. (1971). The effect of varying air–CO_2 level, leaf temperature and illuminance on the CO_2 exchange of the dwarf pea, *Pisum sativum* var. Meteor. *Photosynthetica*, **5,** 190–4.

Holding, A. J. & King, J. (1963). The effectiveness of indigenous populations of *Rhizobium trifolii* in relation to soil factors. *Plant and Soil*, **18,** 191–8.

Joffe, A., Weyer, F. & Saubert, S. (1961). The role of root temperature in symbiotic nitrogen fixation. *South African Journal of Science*, **57,** 278–80.

Jones, D. G., Munro, J. M. M., Hughes, R. & Davies, W. E. (1964). The contribution of white clover to a mixed upland sward. I. The effect of

Rhizobium inoculation on the early development of white clover. *Plant and Soil,* **21,** 63–9.

Jones, D. G. & Rees, L. (1970). Suitability of *Rhizobium trifolii* for use as inoculum in acid soils. *Occasional Symposium No. 6 of the British Grassland Society,* pp. 61–70. British Grassland Society, Hurley, Berkshire.

Jones, R. J., Davies, J. G. & Waite, R. B. (1967). The contribution of some tropical legumes to pasture yield of dry matter and nitrogen at Sandford, SE Queensland *Australian Journal of Experimental Agriculture and Animal Husbandry,* **7,** 57–65.

Lie, T. A. (1971). Symbiotic nitrogen fixation under stress conditions. *Plant and Soil,* Special Volume, 118–27.

Loneragan, J. F. (1972). The soil chemical environment in relation to symbiotic nitrogen fixation. In: *Use of isotopes for study of fertilizer utilization by legume crops.* IAEA Publication No. 149, pp. 17–54. International Atomic Energy Agency, Vienna.

Lynch, D. L. & Sears, O. H. (1950). Differential response of strains of *Lotus* nodule bacteria to soil treatment practices. *Proceedings of the Soil Science Society of America,* **15,** 176–80.

McManmon, M. & Crawford, R. M. M. (1971). A metabolic theory of flooding tolerance; the significance of enzyme distribution and behaviour. *New Phytologist,* **70,** 299–306.

Mague, T. H. & Burris, R. H. (1972). Reduction of acetylene and nitrogen by field-grown soybeans. *New Phytologist,* **71,** 275–86.

Marshall, K. C. (1964). Survival of root nodule bacteria in dry soils exposed to high temperatures. *Australian Journal of Agricultural Research,* **15,** 273–81.

Masefield, G. B. (1961). The effect of irrigation on nodulation of some leguminous crops. *Empire Journal of Experimental Agriculture,* **29,** 51–60.

Masterson, C. L. & Sherwood, M. T. (1970). White clover/*Rhizobium* symbiosis. *Occasional Symposium No. 6, of the British Grassland Society,* pp. 11–39. British Grassland Society, Hurley, Berkshire.

Mes, M. G. (1959). Influence of temperature on the symbiotic nitrogen fixation of legumes. *Nature, London,* **184,** 2032–3.

Meyer, D. R. & Anderson, A. J. (1959). Temperature and symbiotic nitrogen fixation. *Nature, London,* **183,** 61.

Minchin, F. R. (1973). Physiological functioning of the plant–nodule symbiotic system of garden pea (*Pisum sativum*) cv. Meteor. PhD Thesis, Queen's University, Belfast.

Minchin, F. R. & Pate, J. S. (1973). The carbon balance of a legume and the functional economy of its root nodules. *Journal of Experimental Botany,* **24,** 259–71.

Minchin, F. R. & Pate, J. S. (1974). Diurnal functioning of the legume root nodule. *Journal of Experimental Botany,* **25,** 295–308.

Mulder, E. G. & Veen, W. L. van. (1960). The influence of carbon dioxide on symbiotic nitrogen fixation. *Plant and Soil,* **13,** 265–78.

Müller, H. & Gottschalk, W. (1973). Quantitative and qualitative situation of seed proteins in mutants and recombinants of *Pisum sativum*. In: *Nuclear techniques for seed protein improvement*. International Atomic Energy Agency, Vienna.

Nutman, P. S. (1956). The influence of the legume in root nodule symbiosis. A comparative study of host determinants and functions. *Biological Reviews*, **31**, 109–52.

Nutman, P. S. (1972). The influence of physical environmental factors on the activity of *Rhizobium* in soil and symbiosis. In: *Use of isotopes for study of fertilizer utilization by legume crops*. IAEA Publication No. 149, pp. 55–84. International Atomic Energy Agency, Vienna.

Nutman, P. S. & Read, M. P. (1952). Symbiotic adaptation in local strains of red clover and nodule bacteria. *Plant and Soil*, **4**, 57–75.

Odu, C. J. I., Fayemi, A. A. & Ogunwale, J. A. (1971). Effect of pH on the growth, nodulation and nitrogen fixation of *Centrosema pubescens* and *Stylosanthes gracilis*. *Journal of the Science of Food and Agriculture*, **22**, 57–9.

Oghoghorie, C. G. O. (1971). The physiology of the field pea – *Rhizobium* symbiosis in the presence and absence of nitrate. PhD Thesis. Queen's University, Belfast.

Oghoghorie, C. G. O. & Pate, J. S. (1971). The nitrate stress syndrome of the nodulated field pea (*Pisum arvense* L.). Techniques for measurement and evaluation in physiological terms. *Plant and Soil*, Special Volume, 185–202.

Oghoghorie, C. G. O. & Pate, J. S. (1972). Exploration of the nitrogen transport system of a legume using ^{15}N. *Planta, Berlin*, **104**, 35–49.

Olsen, R. A. (1972). Fertilizing the soybean crop (*Glycine max* (L) Merill). In: *Use of isotopes for study of fertilizer utilization by legume crops*. IAEA Publication No. 149, pp. 101–10. International Atomic Energy Agency, Vienna.

Panton, C. A., Coke, L. B. & Pierre, R. E. (1973). Seed protein improvement in certain legumes through induced mutations. In: *Nuclear techniques for seed protein improvement*. International Atomic Energy Agency, Vienna.

Parle, J. (1958). Field observations of copper deficiency in legumes. In: *Nutrition of the legumes*, pp. 280–3. Butterworth & Co., London.

Pate, J. S. (1958). Nodulation studies in legumes. I. The synchronization of host and symbiotic development in the field pea *Pisum arvense* L. *Australian Journal of Biological Science*, **11**, 366–81.

Pate, J. S. (1961). Temperature characteristics of bacterial variation in legume symbiosis. *Nature, London*, **192**, 637–9.

Pate, J. S. (1968). *Physiological aspects of inorganic and intermediate nitrogen metabolism*, pp. 219–40. Academic Press, London & New York.

Pate, J. S. (1971). Movement of nitrogenous solutes in plants. In: *Nitrogen*^{15} *in soil–plant studies*. International Atomic Energy Agency, Vienna.

Pate, J. S. (1973). Uptake, assimilation and transport of nitrogen compounds by plants. *Soil Biology and Biochemistry*, **5**, 109–19.

Pate, J. S. (1974). The biology of dinitrogen fixation by legumes. In: *Dinitrogen fixation* (ed. R. W. F. Hardy). Wiley, New York. (In press.)

Pate, J. S. & Dart, P. J. (1961). Nodulation studies in legumes. IV. The influence of inoculum strain and time of application of ammonium nitrate on symbiotic response. *Plant and Soil*, **15,** 329–46.

Pate, J. S. & Flinn, A. M. (1973). Carbon and nitrogen transfer from vegetative organs to ripening seeds of field pea (*Pisum arvense* L.) *Journal of Experimental Botany*, **24,** 1090–9.

Pate, J. S. & Gunning, B. E. S. (1972). Transfer cells. *Annual Review of Plant Physiology*, **23,** 173–96.

Pate, J. S., Gunning, B. E. S. & Briarty, L. G. (1969). Ultrastructure and functioning of the transport system of the leguminous root nodule. *Planta, Berlin*, **85,** 11–34.

Pate, J. S. & Wallace, W. (1964). Movement of assimilated nitrogen from the root system of the field pea (*Pisum arvense* L.). *Annals of Botany*, **28,** 80–99.

Petrosyan, A. P. (1959). Ecological features of nodule bacteria in the Armenian SSR. *Izd-vo Ministerstva sel'skogo klozyaistva Arm SSR.*

Quebedeaux, B. & Hardy, R. W. F. (1973). Oxygen as a new factor controlling reproductive growth. *Nature, London*, **243**, 477–9.

Richardson, D. A., Jordan, D. C. & Garrard, E. H. (1957). The influence of combined nitrogen on nodulation and nitrogen fixation by *Rhizobium meliloti* Dangeard. *Canadian Journal of Plant Science*, **37,** 205–14.

Roberts, J. L. & Olsen, R. F. (1942). The relative efficiency of strains of *Rhizobium trifolii* as influenced by soil fertility. *Science*, **95,** 413–14.

Robson, A. D. & Loneragan, J. F. (1970*a*). Sensitivity of annual *Medicago* species to manganese toxicity as affected by calcium and pH. *Australian Journal of Agricultural Research*, **21,** 223–32.

Robson, A. D. & Loneragan, J. F. (1907*b*). Nodulation and growth of *Medicago truncatula* in acid soils. II. Colonization of acid soils by *Rhizobium meliloti. Australian Journal of Agricultural Research*, **21,** 435–45.

Roponen, I. E., Valle, E. & Ettala, T. (1970). Effect of temperature of the culture medium on growth and nitrogen fixation of inoculated legumes and rhizobia. *Physiologia Plantarum*, **23,** 1198–1205.

Roponen, I. E. & Virtanen, A. I. (1968). The effect of the prevention of flowering on the vegetative regrowth of inoculated pea plants. *Physiologia Plantarum*, **21,** 655–7.

Roughley, R. J. (1970). The influence of root temperature, *Rhizobium* strain and host selection on the structure and nitrogen fixing efficiency of the root nodules of *Trifolium subterraneum. Annals of Botany*, **34,** 631–46.

Roughley, R. J. & Dart, P. J. (1969). Reductions of acetylene by nodules of *Trifolium subterraneum* as affected by root temperature, *Rhizobium* strain and host cultivar. *Archiv für Mikrobiologie*, **69,** 171–9.

Roughley, R. J. & Dart, P. J. (1970). Growth of *Trifolium subterraneum* L. selected for sparse and abundant nodulation as affected by root temperature and *Rhizobium* strain. *Journal of Experimental Biology*, **21,** 776–86.

Schilling, G. & Schalldach, I. (1966). Translocation, incorporation and utilisation of late applied N in *Pisum sativum*. *Albrecht–Thaer–Archiv*, **10,** 895–907.

Schlegel, H. G. (1962). Die Isolierung von Poly-β-hydroxy-buttersaure aus Wurzelknollchen von Leguminosen. *Flora*, **152,** 236–40.

Schreven, D. A., van (1958). Some factors affecting the uptake of nitrogen by legumes. In: *Nutrition of the legumes*, pp. 137–63. Butterworth & Co. London.

Small, J. G. C. & Leonard, D. A. (1969). Translocation of ^{14}C-labelled photosynthate in nodulated legumes as influenced by nitrate nitrogen. *American Journal of Botany*, **56,** 187–96.

Snaydon, R. W. (1962). Micro-distribution of *Trifolium repens* L. and its relation to soil factors. *Journal of Ecology*, **50,** 133–43.

Sprent, J. I. (1971). The effects of water stress on nitrogen-fixing root nodules. I. Effects on the physiology of detached soybean nodules. *New Phytologist*, **70,** 9–17.

Sprent, J. I. (1972*a*). The effects of water stress on nitrogen-fixing root nodules. II. Effects on the fine structure of detached soybean nodules. *New Phytologist*, **71,** 443–50.

Sprent, J. I. (1972*b*). The effects of water stress on nitrogen-fixing root nodules. III. Effects of osmotically applied stress. *New Phytologist*, **71,** 451–60.

Sprent, J. I. (1972*c*). The effects of water stress on nitrogen-fixing root nodules. IV. Effects on whole plants of *Vicia faba* and *Glycine max*. *New Phytologist*, **71,** 603–11.

Stolwijk, J. A. J. & Thimann, K. V. (1957). On the uptake of carbon dioxide and bicarbonate by roots and its influence on growth. *Plant Physiology*, **32,** 513–20.

Streeter, J. G. (1973). Growth of two shoots on a single root as a technique for studying physiological factors limiting the rate of nitrogen fixation by nodulated legumes. *Plant Physiology*, Supplement, p. 35.

Vartiovaara, U. (1937). Investigation on the root nodule bacteria of leguminous plants. XXI. The growth of the root nodule organisms and inoculated peas at low temperature. *Journal of Agricultural Science*, **27,** 626–37.

Virtanen, A. I. & Hausen, S. von. (1936). Investigations on the root nodule bacteria of leguminous plants. XVIII. Continued investigations on the effect of air content of the medium on the development and function of the nodule. *Journal of Agricultural Science*, **26,** 281–7.

Virtanen, A. I., Moisio, T. & Burris, R. H. (1955). Fixation of nitrogen by nodules excised from illuminated and darkened pea plants. *Acta chemica Scandinavica*, **9,** 184–6.

Wallace, W. & Pate, J. S. (1965). Nitrate reductase in the field pea (*Pisum arvense* L.). *Annals of Botany*, **29,** 655–71.

Wardlaw, I. F. (1968). The control and pattern of movement of carbohydrates in plants. *Botanical Reviews*, **34,** 79–105.

Whiteman, P. C. & Lulham, A. (1970). Seasonal changes in growth and nodulation of perennial tropical pasture legumes in the field. I. The influence of planting date and grazing and cutting on *Desmodium uncinatum* and *Phaseolus atropurpureus*. *Australian Journal of Agricultural Research*, **21**, 195–206.

Wilkins, J. (1967). The effects of high temperatures on certain root nodule bacteria. *Australian Journal of Agricultural Research*, **18**, 299–304.

Wilson, J. K. (1942). The loss of nodules from legume roots and its significance. *Journal of the American Society of Agronomy*, **34**, 460–4.

Wong, P. P. & Evans, H. J. (1971). Poly-β-hydroxybutyrate utilisation by soybean (*Glycine max* Merr.) nodules and assessment of its role in maintenance of nitrogenase activity. *Plant Physiology*, **47**, 750–5.

28. Symbiosis in tropical grain legumes: some effects of temperature and the composition of the rooting medium

P. DART, J. DAY, R. ISLAM & JOHANNA DÖBEREINER

Grain legume production in most agricultural systems is dependent on symbiotic nitrogen fixation, which is affected by *Rhizobium* strain and host cultivar and influenced by environmental factors, especially temperature (e.g. Dart & Mercer, 1965; Dart & Day, 1971; Gibson, 1971), light intensity and daylength (Day & Dart, 1969, 1970), the soil nutrient status, particularly inorganic combined nitrogen (e.g. for cowpea, Pate & Dart, 1961; Ezedinma, 1964; Dart & Wildon, 1970; and soybean, Wilson, 1940; Allos & Bartholomew, 1959; Weber, 1966; Hanway & Weber, 1971; Ham, Lawn & Brun, Chapter 20), and soil moisture tension (e.g. Hume *et al.*, 1974; Sprent, Chapter 30). Grain yields vary greatly between countries and this may reflect the effect of different environments on nitrogen fixation. Soil temperatures, often very high in unmulched, tropical soils before the leaf canopy closes, and nitrate levels, also often high at the start of the wet season (e.g. Greenland, 1958) may restrict the establishment of the symbiosis. Organic matter is rapidly mineralised under tropical/subtropical conditions and soil levels are often low. The effect of this on growth and nodulation of legumes is not known.

We report some effects of soil temperature on the symbiosis of soybean *Glycine max* and chickpea *Cicer arietinum*, and the effects of organic matter and inorganic combined nitrogen on that of green gram *Vigna radiata* and black gram *Vigna mungo*.

Materials and methods

The soybeans, varieties Chippewa, Merit, Grant, Kent, and L2006, L652 from Brazil were grown from uniform size seed in a sterile 1:1

washed quartz sand: quartz grit mixture, usually two seedlings per 15 cm diameter plastic pot in Saxcil controlled-environment cabinets. Seeds, surface sterilised with mercuric chloride, were inoculated at sowing from broth cultures of the effective *Rhizobium japonicum* strains CB1809 (CSIRO Division of Tropical Agronomy, Brisbane, Australia), Sm1b (isolated from a forage soybean in Brazil) or CC705 (CSIRO Division of Plant Industry, Canberra – ex Wisconsin 505). Plants were watered with nitrogen-free nutrient solution (Dart & Pate, 1959) and grown in a 16 hour 26 000 lux photoperiod provided by warm-white fluorescent tubes. Day temperatures were 21, 27 or 33 °C reduced by 3 deg C at night. Four replicate pots for each treatment were harvested at intervals from 12–47 days after sowing.

Cicer arietinum Kabuli and Deshi types (seed purchased locally but believed to originate from India) and another variety from Iran, were grown similarly (four seedlings per pot). The pots were placed in water baths maintaining controlled root temperatures ranging from 15–33 °C. The air temperature for all plants fluctuated between 24–7 °C in the day and 18–21 °C at night. The 12 hour 16 000 lux photoperiod was provided by warm-white fluorescent tubes. Seeds were inoculated at sowing from broth cultures of *Rhizobium* strains CB1189 (CSIRO Division of Tropical Agronomy, Brisbane, Australia) 27A2 and 27A9 (Nitragin Co., Milwaukee, Wisconsin, USA), Ca-1 and Ca-2 (Division of Microbiology, Indian Agricultural Research Institute, New Delhi, India). Plants were watered daily to constant weight with nitrogen-free nutrient solution.

In one experiment, Kabuli plants inoculated with strain CA-1 were grown at 23 °C root temperature for 42 days before being transferred to baths at 33 and 36 °C giving root temperatures of 32.5 and 35 °C. Eight plants selected from three replicate pots were harvested 6, 12, 24, 48 and 72 hours after being transferred, and their nitrogenase activities assayed at the treatment root temperature. Control plants kept at 23 °C were also assayed.

Some plants grown at 23 °C for 45 days were also subjected to daily cycles in which the bath temperature either rose over 1 hour to 33 °C and was then maintained for a further 5 hours, or rose over 3 hours to 36 °C then maintained for 3 hours, before cooling gradually to 23 °C. The sand temperature lagged only slightly below these temperatures and reached 32.5 and 35 °C. Two replicate pots each containing four plants were harvested after three and after seven cycles and their nitrogenase activity assayed at 33 and 36 °C.

The response of nitrogenase in *Cicer* nodules to incubation temperatures from 6 to 40 °C was determined on excised roots of 42–5 day old Kabuli plants inoculated with strain Ca-1, grown in Saxcil cabinets at 23 °C day and 19 °C night. Roots trimmed to remove non-nodulated portions were placed singly in 60 ml bottles, capped loosely and placed in temperature-controlled water baths for temperatures of 15 °C or above or in rooms at 6 and 11 °C, for either 15 or 30 minutes before the caps were tightened and acetylene added. The air temperature within the vials equilibrated with the outside temperature within 7 minutes.

Vigna mungo and *V. radiata* seed was purchased locally but believed to be of Indian origin. Attempts to grow these legumes in sand:grit mixtures with nitrogen-free nutrient solution were unsuccessful, but plants grew when the quartz sand:grit was amended with 10 % by volume of Kettering clay-loam (0.29 % N), a soil used in potting composts. To find a suitable growth medium, plants were grown in glasshouses with day temperatures ranging from 25–35 °C and night temperatures of 25 °C, or in the Saxcil controlled-environment cabinets. The effect of adding ammonium nitrate to the culture solution at sowing was examined on plants inoculated with CB756, grown in sand+10 % Kettering loam in Saxcil cabinets, at 30 °C day temperature, 27 °C night with a 12 hour day of 26 000 lux. The 10 % Kettering loam provided 55–60 mg/N (nitrate and ammonium) per pot, and to this was added from 17.5 to 175 mg N per pot as ammonium nitrate equivalent to 10–100 kg N/ha calculated on a surface area basis with a pot area of 75 cm^2. Four replicate pots each containing two plants were harvested 30 and 45 days after sowing.

Residual nitrate in 100 g subsamples of the wet root medium at harvest was determined with a Corning–EEL nitrate ion specific electrode (Nair & Talibudeen, 1973) using potassium nitrate and ammonium nitrate standards.

Nitrogenase activities were determined by acetylene reduction (Dart, Day & Harris, 1972). Bottles of 60, 120 or 380 ml were used depending on root size, and the 10 % acetylene added within 10 minutes of returning the bottles to the temperature at which the roots had been growing. Ethylene production after 30 minutes incubation was determined on a Perkin–Elmer F11 gas chromatograph using a Poropak N column and a flame ionisation detector. Plant nitrogen contents were determined after Kjeldahl digestion by an autoanalyser method using an indophenol colour reaction (after Varley, 1966).

Results

Soybean (Glycine max)

Nitrogen fixation. Nodule formation, growth, and nitrogen fixation was most rapid at 27 °C day temperature; nodules formed with a slight delay at 33 °C. At 21 °C the slow nodulation and low initial nitrogenase activity induced a pronounced 'nitrogen hunger' state in which the leaves remained yellow and nitrogen deficient until 29–30 days from sowing when they greened up almost overnight. At 27 °C there was a 'nitrogen hunger' state between 18 and 21 days from sowing, with none at 33 °C. Nitrogen fixation was measurable by Kjeldahl analysis by 19 days at 27 °C, 22 days at 33 °C and 26 days at 21 °C. Fig. 28.1

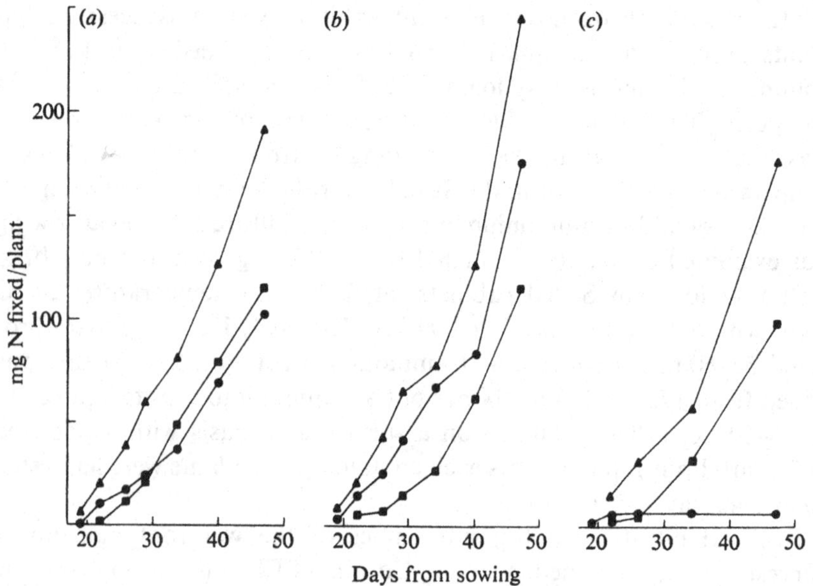

Fig. 28.1. Nitrogen fixation by Chippewa soybeans inoculated with *Rhizobium* strains (*a*) CB1809, (*b*) Sm1b or (*c*) CC705, and grown at day temperatures of 21 °C (■——■), 27 °C (▲——▲) or 33 °C (●——●).

shows marked differences between strains in their rates of nitrogen fixation at the three temperatures. At 21 °C, CB1809 fixed slightly more than Sm1b and CC705, but at 33 °C Sm1b was much more effective, particularly from 26 days after sowing; between 40 and 47 days this strain fixed almost twice as much as CB1809. At 27 °C there was little difference between strains before 40 days; after this Sm1b then fixed much more than the other strains. CC705 was only slightly less effective than the other two strains at 21 and 27 °C, and was quite ineffective at 33 °C.

Plant growth was poor, the leaves remained nitrogen deficient and senesced early. Nodules formed at this temperature and there was little difference in nodule weight per plant between strains until 34 days. The rapid nitrogen fixation by strain Sm1b at 33 °C after 40 days was associated with rapid individual nodule growth (Fig. 28.2).

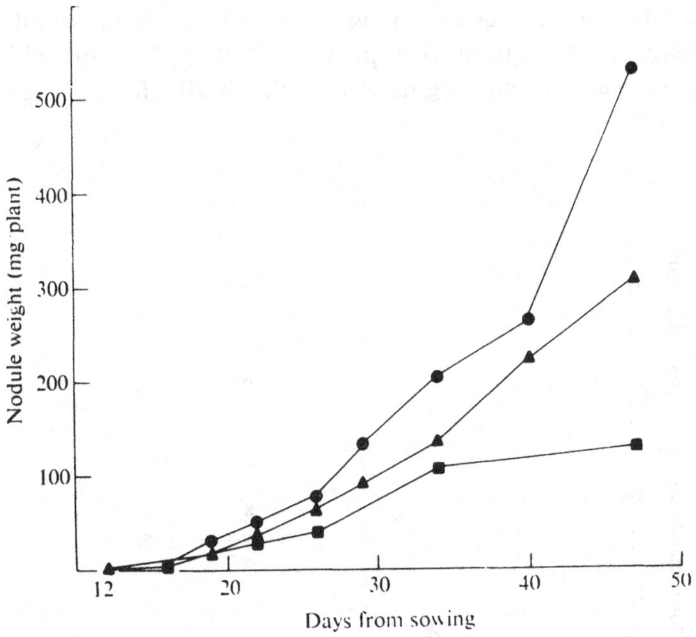

Fig. 28.2. Nodule production by Chippewa soybeans inoculated with *Rhizobium* strains Sm1b (●——●), CB1809 (▲——▲) or CC705 (■——■), and grown at 33 °C day temperature.

With the other soybean cultivars viz. Merit, Grant, Kent, and L2006 and L652 from Brazil, which are productive under warm tropical conditions, Sm1b was shown to be superior at the warmer temperatures and CC705 ineffective at 33 °C. With Grant CB1809 was markedly more effective at 21 °C than the other two strains.

Nitrogenase activity. At 21 °C little nitrogenase activity was detected before 19 days, after which it increased rapidly (Fig. 28.3). There was slightly more activity per plant for strain CB1809 than Sm1b and CC705. Activity was present at the first harvest 12 days after sowing at 27 °C and 33 °C, and showed most rapid increase thereafter at 27 °C. At 33 °C, nitrogenase activity per plant similarly increased with CB1809 and Sm1b until 34 days, but with CC705 activity increased only slightly to a maximum at 26 days and then declined.

Fig. 28.4 shows that nodule efficiency, i.e. nitrogenase activity per unit weight of nodule, rapidly increased at all temperatures to a maximum which was reached between 16 and 26 days from sowing, and then declined. The maximum rate of ethylene production was between 60 and 80 μmoles per gram dry weight nodule per hour for all strain–temperature combinations except CC705 at 33 °C, for which the maximum was 49 μmoles/g/h. Although CC705 fixed virtually no

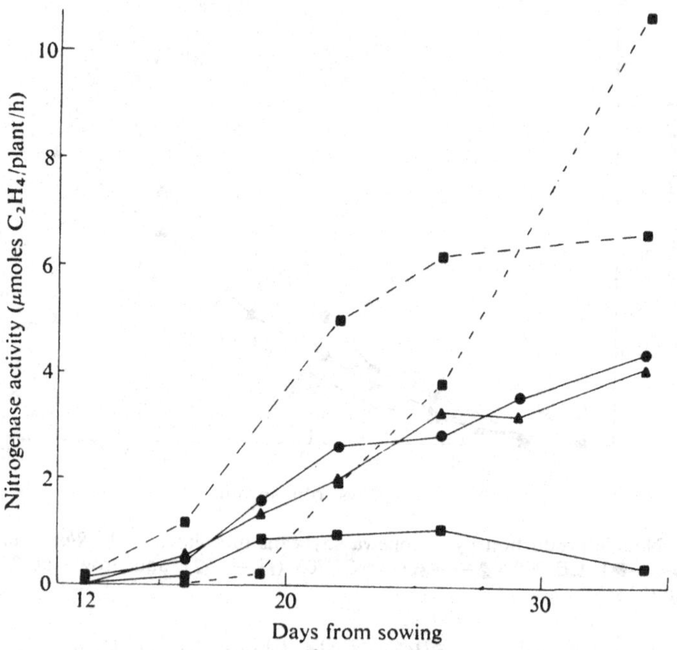

Fig. 28.3. Nitrogenase activity of nodulated roots of Chippewa soybeans inoculated with *Rhizobium* strain Sm1b (●), CB1809 (▲) or CC705 (■) and grown at 33 °C day temperature (——). Also shown for comparison are the activities of strain CC705 at 27 (---) and 21 °C (-·-).

nitrogen at 33 °C (Figs. 28.1, 28.3 and 28.4) acetylene reduction was appreciable, and using a conversion ratio of 3 moles acetylene equivalent to 1 mole nitrogen, the nitrogenase activity was much in excess of that needed for the amount of nitrogen fixed. For the other two strains the rates of acetylene reduction slightly underestimated the actual nitrogen fixation.

Leghaemoglobin (Lb) was present in the CC705 nodules at 33 °C at a concentration (Lb/g fresh weight nodules) greater than for CB1809 at 33 °C until 34 days from sowing. Acetylene reduction per gram fresh weight nodule actually decreased for both strains while the Lb concentration increased, with CC705 nodules having only 40 % of the activity of CB1809 at 34 days (Maskall & Dart, 1973).

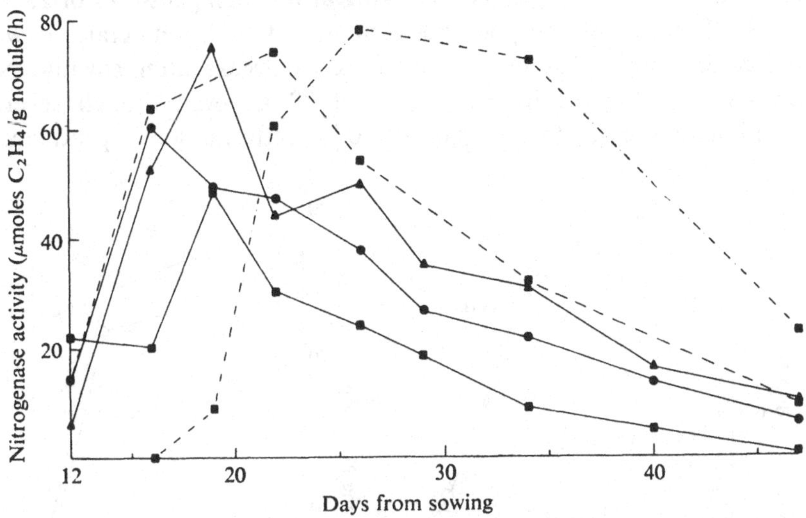

Fig. 28.4. Nitrogenase activity per gram dry weight nodule for nodulated roots of Chippewa soybeans grown at 33 °C (——) day temperature and inoculated with strains Sm1b (●), CB1809 (▲) and CC705 (■) and compared with activity of CC705 nodules formed at 27 °C (---) and 21 °C (-·-).

The ineffectiveness of CC705 at 33 °C could be attributed to a combination of slightly less nodule growth and nitrogenase activity per unit nodule weight, coupled with some defect (possibly the supply of carbon skeletons or the enzymes involved in amino acid synthesis) preventing the nitrogenase present from functioning in nitrogen fixation.

Chickpea (Cicer arietinum)

In a preliminary experiment the Deshi variety was grown at root temperatures of 15, 20, 25, 30 and 33 °C, inoculated with strains CB1189 or 27A2 and harvested, 3, 4, 5, 7 and 9 weeks after sowing. Nodules were present and active at 3 weeks at 20 °C and 25 °C and at 4 weeks at 15 °C, but not until 7 weeks at 30 °C. No nodules were formed by either strain at 33 °C. At 5 weeks dry matter production was slightly less at 30 and 33 °C, but by 9 weeks there were marked differences in growth with most at 25 °C and very little at 30 and 33 °C (Fig. 28.5).

Nitrogen fixation (total plant nitrogen – seed nitrogen) differed little between 15 and 25 °C but there was none at 30 °C. Strain CB1189 was slightly more effective than 27A2.

Five different *Rhizobium* strains were used to inoculate Deshi, Kabuli and the Iranian variety, growing at root temperatures of 23, 30 and 33 °C to see if the poor performance at high temperatures was affected by strain. Plants were harvested 6 weeks after sowing. No nodules were formed by any strain at 33 °C, even though several amendments such as the addition of 10 % Kettering loam, peat, small

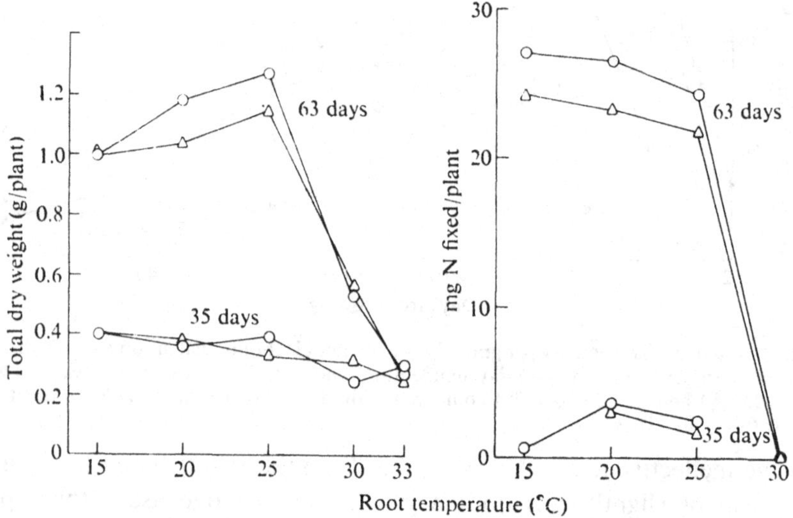

Fig. 28.5. Dry weight production and nitrogen fixation by 35 and 63 day-old *Cicer arietinum* plants inoculated with strains CB1189 (○) and 27A2 (△) and grown at root temperatures from 15 to 33 °C.

amounts of ammonium nitrate, or growth in soil were tried. Plants grew well if combined nitrogen was available. Table 28.1 shows that the strains differed slightly in the amount of nitrogen fixed at 23 °C, with slightly different ranking between varieties. The strains also differed in nodule weight produced, nitrogenase activity per plant and nitrogenase activity per gram nodule tissue.

At 30 °C, less nitrogen was fixed and the differences between strains were much greater. Strain Ca-2 was more effective than the others, fixing more than 60 % as much nitrogen as at 23 °C with all cultivars. Strain Ca-1 was moderately effective at this temperature but the others were quite ineffective. The effect of the higher temperature was to

Table 28.1. *Symbiotic performance of five* Rhizobium *strains with three* Cicer arietinum *varieties at two root temperatures*

	CB1189		27A2		27A9		Ca-1		Ca-2	
Temperature (°C)	23	30	23	30	23	30	23	30	23	30
Deshi										
Dry wt nodules/plant (mg)	89	56	79	55	84	82	82	64	87	76
Nitrogenase/plant	9.0*	3.1	4.5	2.1	6.3	5.1	8.3	4.8	7.6	5.5
Nitrogenase/g nodule	105*	51	59	36	75	63	105	76	92	69
Nitrogen fixed (mg/plant)	26	8	21	6	27	13	28	12	28	17
Kabuli										
Dry wt nodules/plant (mg)	98	96	93	65	108	90	78	93	103	85
Nitrogenase/plant	6.5*	3.9	6.9	1.32	11.3	3.1	10.8	5.7	7.5	6.5
Nitrogenase/g nodule	68*	41	70	20	104	34	142	65	74	60
Nitrogen fixed (mg/plant)	30	10	27	3	35	8	27	17	33	21
Iranian										
Dry wt nodules/plant (mg)	100	86	86	82	96	70	104	81	96	104
Nitrogenase/plant	6.1*	2.8	5.5	2.1	9.6	2.4	14.8	3.6	8.3	3.8
Nitrogenase/g nodule	61*	32	63	23	102	36	139	43	86	37
Nitrogen fixed (mg/plant)	33	11	29	8	32	9	35	12	21	22

Plants harvested at 42 days
* Measured as μmoles ethylene produced/hour.

reduce nitrogenase activity per gram nodule tissue, possibly by accelerating basal nodule senescence.

Strains Ca-2 and 27A2, which differed most in their response at 30 °C, were examined further, using Kabuli plants grown at 23 and 30 °C root temperatures and harvested at weekly intervals from 3 to 10 weeks after sowing. Fig. 28.6. shows that there was little difference in growth between treatments up to 5 weeks. Thereafter plants inoculated with strain 27A2 grew and fixed little at 30 °C, but those given strain Ca-2 continued growing and at 10 weeks had fixed c. 60 % as much nitrogen at 30 °C as at 23 °C, as in the previous experiment. Strain Ca-2 was again slightly more effective than strain 27A2 at 23 °C.

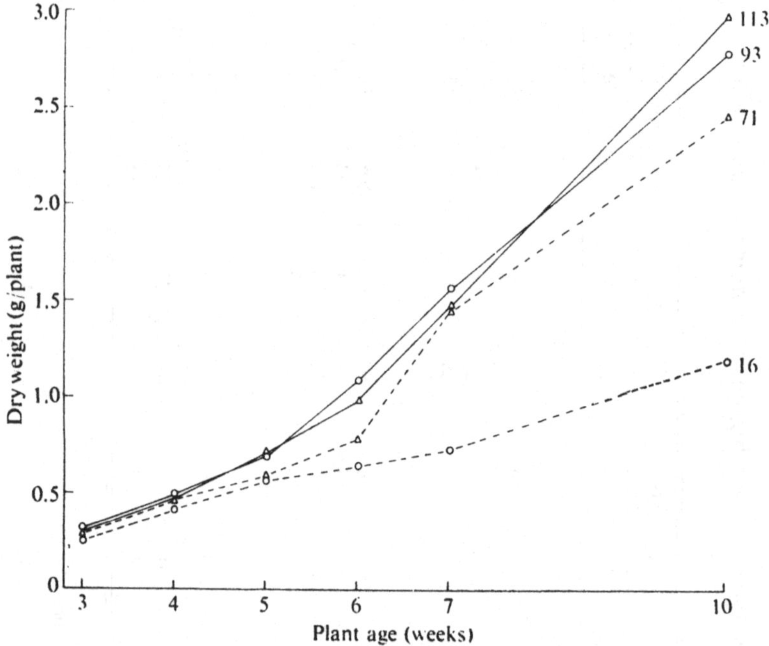

Fig. 28.6. Dry weight production by *Cicer arietinum* inoculated with strain Ca-2 (△) or 27A2 (○) and grown at root temperatures of 23 °C (—) or 30 °C (---). The amounts of nitrogen fixed (mg) over the 10-week period are also shown.

Fig. 28.7 shows that nodule growth continued similarly for both strains throughout the 10-week period except for 27A2 plants at 30 °C which produced less weight of nodule, although this reduction was not as marked as was plant growth and nitrogen fixation. This suggests that the differences in nitrogen fixation were primarily related to differences in efficiency (nitrogenase activity per gram nodule weight), rather than to total nodule weight. Fig. 28.8 shows that nitrogenase activities per

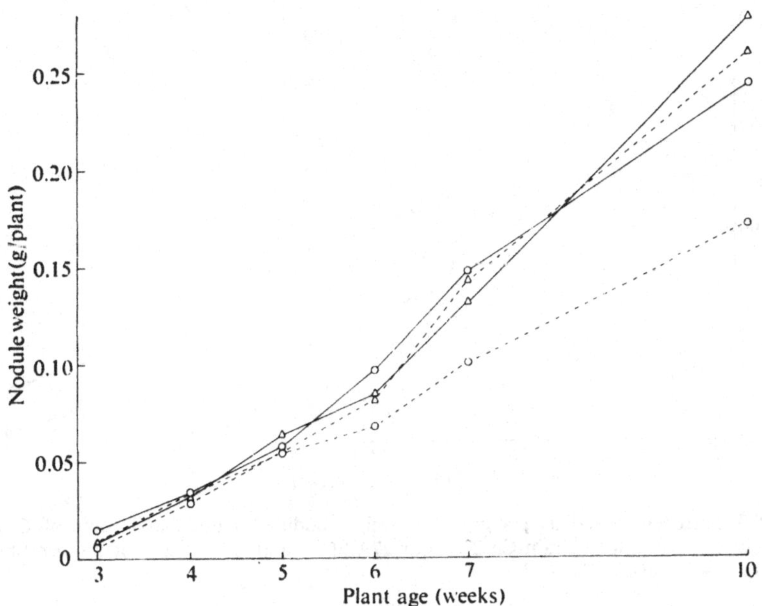

Fig. 28.7. Nodule production by *Cicer arietinum* plants inoculated with strains Ca-2 (△) or 27A2 (○) and grown at root temperatures of 23 °C (——) and 30 °C (---).

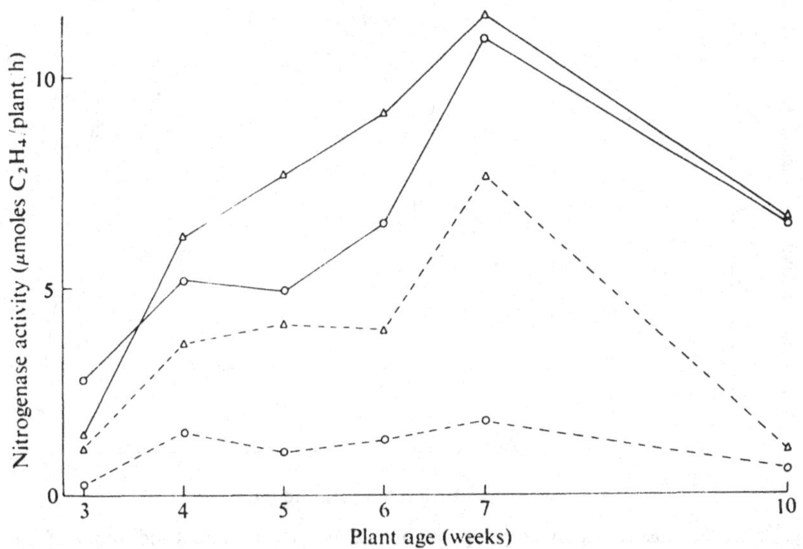

Fig. 28.8. Nitrogenase activity per plant for nodulated roots of *Cicer arietinum* inoculated with strain Ca-2 (△) or 27A2 (○) and grown at root temperatures of 23 °C (——) and 30 °C (---).

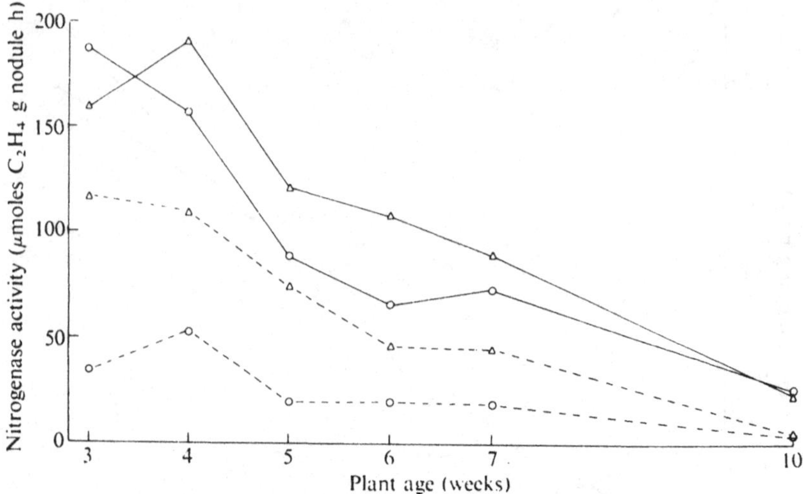

Fig. 28.9. Nitrogenase activity per gram dry weight nodule for nodulated roots of *Cicer arietinum* inoculated with strain Ca-2 (△) or 27A2 (○) and grown at root temperatures of 23 °C (——) and 30 °C (---).

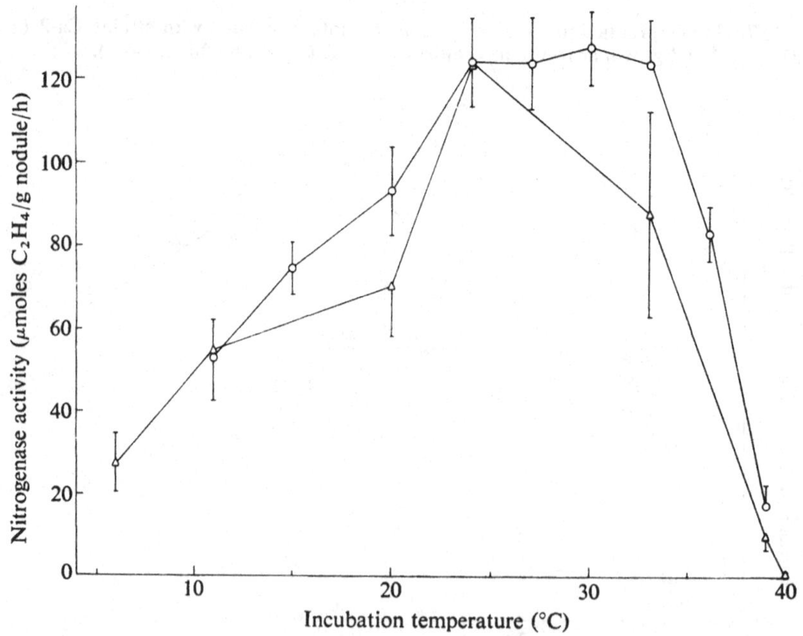

Fig. 28.10. Nitrogenase activity per gram nodule weight for nodulated roots of *Cicer arietinum*. Plants inoculated with strain Ca-1 incubated during assay at temperatures ranging from 6 °C to 40 °C. Bottles containing the roots were equilibrated at the assay temperatures for 10 min (○——○) or 30 min (△——△) before the acetylene was added. Bars represent standard errors of the means for the eight replicate roots.

plant increased until the seventh week for all combinations and then declined, although for 27A2 at 30 °C activity increased very little after the fourth week. Nitrogenase activities per plant correlated well with the amounts of N_2 fixed. The decline in nitrogenase activity was not associated with flowering which began c. 35 days from sowing. By the tenth week pod fill was well advanced, and plants were senescing. Fig. 28.9 shows that nodule efficiencies declined for all combinations after 4 weeks' plant growth, when nodules were less than 14 days old. Fig. 28.10 shows *Cicer* nitrogenase activity in nodules formed at 23 °C, near the optimum temperature for growth and nitrogen fixation, and

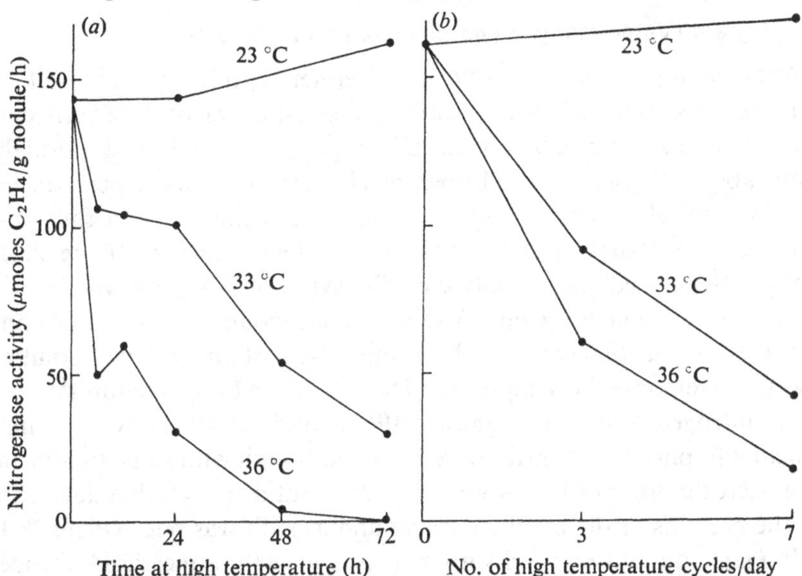

Fig. 28.11. Effect of temperature stress on *Cicer arietinum* nitrogenase. (*a*) Effect of transfer from a continuous root temperature of 23 °C to 33 °C or 36 °C on nitrogenase activity per gram dry weight nodule of *Cicer arietinum* – CA1 plants. Nodulated roots were assayed at the transfer temperatures. (*b*) Effect of daily increases of temperature (cycling) from 23 °C to 33 °C or 26 °C. Nitrogenase activity of nodulated roots of *Cicer arietinum*- Ca1 plants grown at 23 °C and assayed at the transfer temperature.

incubated during the assay at a range of temperatures from 6 °C to 40 °C. Acetylene reduction was found over the whole temperature range with maximum activity between 24 and 33 °C, and a rapid decline at higher temperatures. Thus the lower efficiencies of nodules on plants grown at 30 °C in the previous experiments were not due to an effect of temperature on the functioning of the nitrogenase enzyme but rather to the amount of enzyme present.

Plants nodulated and grown continuously at 23 °C were then transferred either to continuous or to intermittent root temperatures of 33 or 36 °C for periods up to 7 days. Fig. 28.11 shows that nitrogenase activity rapidly declined under continuous high temperature being zero after 48 hours at 36 °C, and only 17 % of the controls at 23 °C after 72 hours at 33 °C. When the high temperature treatment was experienced for 3–5 hours during the day, a situation perhaps nearer to that in field soils, nitrogenase activity also declined rapidly. Soil temperatures of 33 and 36 °C are not unusual in the subtropics. Three daily cycles of high temperature were sufficient to halve nitrogenase activities, with further decline after seven cycles.

Black gram (Vigna mungo) *and green gram* (V. radiata)

Effects of root medium on growth. Neither species grows well in nitrogen-free sand or water culture in glasshouses or in Saxcil controlled-environment cabinets. Seedlings germinate and grow normally until about 18 days, when brown-black necrotic spots appear on the seedling unifoliate leaves. These increase and within three or four days the leaves are shed, just as the trifoliate leaves appear (Plate 28.1*b* and *c*). Such seedlings usually die. The symptoms appear whether the plants are inoculated or not. Adding small amounts of inorganic combined nitrogen, or changing the composition of the nutrient solution did not overcome the symptoms – Hoaglands or Long Ashton low and high nitrogen solutions (Hewitt, 1966), and solutions with a high calcium to phosphorus ratio were tried – although with some treatments the necrotic primary leaves were not shed but a leaf crinkle developed at the margins of the trifoliate leaves and growth was poor (Plate 28.1*a*–*d*). Plants did not respond with increased growth to increased nitrogen supply. The severity of the symptoms appeared to vary with the season and was less in 1973 than in 1972 and 1971. Symptoms appeared at root temperatures from 20 to 33 °C.

Additions of aqueous soil extract, or charcoal, activated carbon, and bentonite, and growth in milled neutral sedge peat and sphagnum moss, John Innes and Levington commercial potting composts, vermiculite or perlite did not alleviate the symptoms. Growth was normal in Rothamsted silty loam (Batcombe series developed on clay-with-flints), Kettering loam, a neutral fen peat soil from Lincolnshire, and a fen silty-loam soil. Adding 10 to 30 % by volume of Kettering loam to the sand:grit mixture improved growth and nodulation, 10 % giving better growth than 30 % or 100 % (Plate 28.2*a*). The lack of nodulation in the 100 % loam may account for the poor growth. When an amount of Kettering loam soil equivalent to the 10 % v/v was ignited at 450 °C for 4 hours to remove the organic matter components of the soil, then mixed with sand:grit and minus-nitrogen culture solution added, growth of *V. radiata* and *V. mungo* was poor and plants eventually died.

Addition of 10 % by volume of Kettering loam to the sand:grit mixture produced a marked increase in growth of Prima cowpeas, *Cajanus cajan* and *Phaseolus vulgaris* var. Seafarer (Navy beans), greater than that obtained by adding nitrate (50 mg N/pot) equivalent to that available in the 10 % Kettering loam (Plate 28.2*b* and *c*). For *C. cajan* 10 % loam also gave greater growth and nodulation than 30 % loam or loam alone. There was no difference between these loam treatments with cowpeas.

The effect of different organic matter levels in soils on the growth of *V. mungo* and *V. radiata* was examined in Woburn sandy soil (Cottenham series developed on lower green sand) taken from twelve different long-term experimental treatments. Soil pH (0.01 M $CaCl_2$ extract) ranged from 5.4 to 6.4. The soils were passed through a 3 mm sieve

Table 28.2. *Manganese levels in* Vigna mungo *under different cultural conditions*

Treatment	Sample	ppm Mn in dry matter*
Sand:grit only	Leaf Root	50 40
Sand:grit + nutrient solution	Leaf Root	49 15
Sand:grit + 10 % Kettering loam + nutrient solution	Leaf Root	30 17
Nutrient solution only		0.08

* Plants harvested at 27 days; inoculated with *Rhizobium* strain CB756 at sowing.

and three volumes of soil mixed with one volume of 6 mm quartz grit. Soils were watered to 90 % field capacity every 2 days with de-ionised water. Surface sterilised seeds were inoculated at sowing with *Rhizobium* strain CB756. Generally growth was good in soils with high total percentage nitrogen with no symptoms appearing in the two soils having more than 0.17 % N. One soil with 0.14 % N also produced no symptoms. In seven other soils severe symptoms appeared on *V. mungo*, but not *V. radiata* (e.g. Plate 28.1*e* and *f*), with some variability between replicate pots, and in two soils symptoms appeared on both species. In some soil plants grew out of the symptoms after a further 2–3 weeks growth, but by 35 days the plant dry weight in the poorest soil was only 34 % of that in the best soil for *V. radiata* and 24 % for *V. mungo.* Growth in the 10 % Kettering loam was 50–60 % as much as in the best soil. Plants nodulated in all soils at varying levels. The results suggested that the form of the organic matter in the soil has some effect on the growth.

Symptoms developed in soils regardless of whether they were initially watered with minus-nitrogen culture solution, and similar results were obtained with nine other agronomically promising varieties obtained from the Indian Agricultural Research Institute, New Delhi, and the International Institute of Tropical Agriculture, Ibadan, Nigeria.

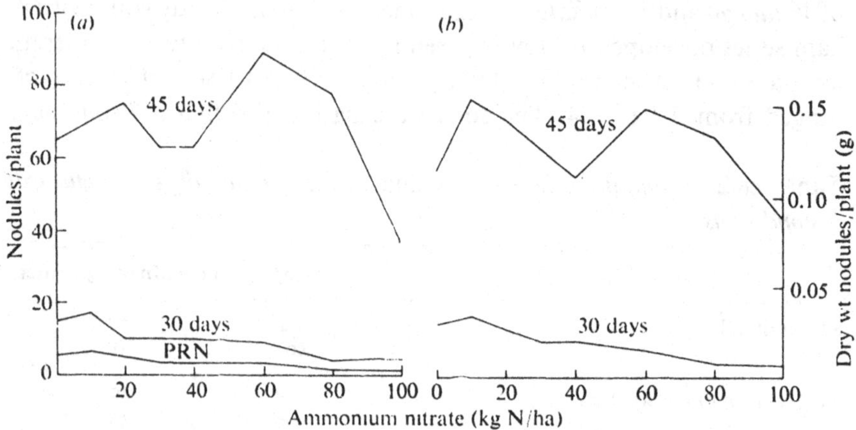

Fig. 28.12. Total and primary root (PRN) nodule number (*a*) and dry weight of nodules per plant (*b*) at 30 and 45 days from sowing for *Vigna mungo* plants grown in 10 % Kettering loam and supplied the equivalent of 0–100 kg N/ha as ammonium nitrate at sowing.

The necrotic spots on the seedling leaves did not contain any virus or fungus. The leaf crinkle symptoms superficially resembled manganese toxicity, but analysis of plants with and without symptoms showed that manganese levels were of a similar low order for both (Table 28.2).

The effect of inorganic combined nitrogen. *V. mungo* and *V. radiata* growth was examined in a further experiment with ammonium nitrate added at sowing to the 10 % Kettering loam sand:grit mixture. Primary root nodulation (PRN) on both species was slightly stimulated by 10 kg N/ha, but thereafter PRN decreased with increasing combined nitrogen levels although two nodules per plant formed even with 100 kg N/ha (Figs. 28.12 and 28.13). Total nodule number and weight at 30 days reflected this pattern. By 45 days many secondary root nodules had developed on all treatments with some stimulation of numbers and total nodule weight in both species with added nitrogen. *V. mungo* nodulation was more restricted by the 100 kg N/ha level than *V. radiata.*

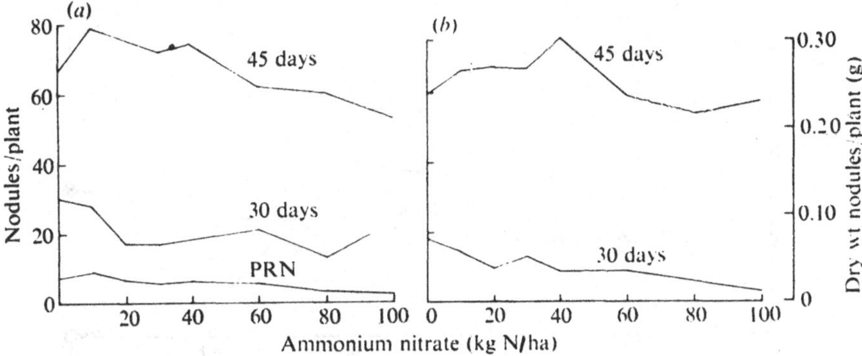

Fig. 28.13. Experimental details as for Fig. 28.12, but using *V. radiata*.

The nitrogen content of the plants at 30 days increased slightly with increasing nitrogen addition, but by 45 days this pattern was considerably altered (Figs. 28.14 and 28.15). For *V. mungo* the nitrogen content reflected the pattern for nodule weight, with plants supplied 10 and 60 kg N/ha containing more than those supplied none. Nitrogen fixation was estimated as total plant nitrogen plus residual nitrate in rooting medium minus (seed nitrogen+nitrogen supplied); this slightly overestimates fixation by the amount of residual ammonium-nitrogen. Fixation was stimulated slightly by 10 kg N/ha. For *V. radiata*, plant nitrogen content was stimulated by additions of up to 60 kg N/ha over plants supplied none, with nitrogen fixation stimulated by additions of up to 30 kg N/ha. Nitrogen fixation for both species was reduced by a greater supply of combined nitrogen, with virtually no fixation at 80 and 100 kg N/ha. By 45 days the residual nitrate levels in the root medium were similar and low for all treatments.

Discussion

Soil temperature has a large effect on nodulation and nitrogen fixation of soybean and *C. arietinum* inoculated with different strains of *Rhizobium*. Strain Sm1b for soybean and strain Ca-2 for *Cicer* were much more effective than others tested at higher temperatures and were also highly effective at more moderate temperatures.

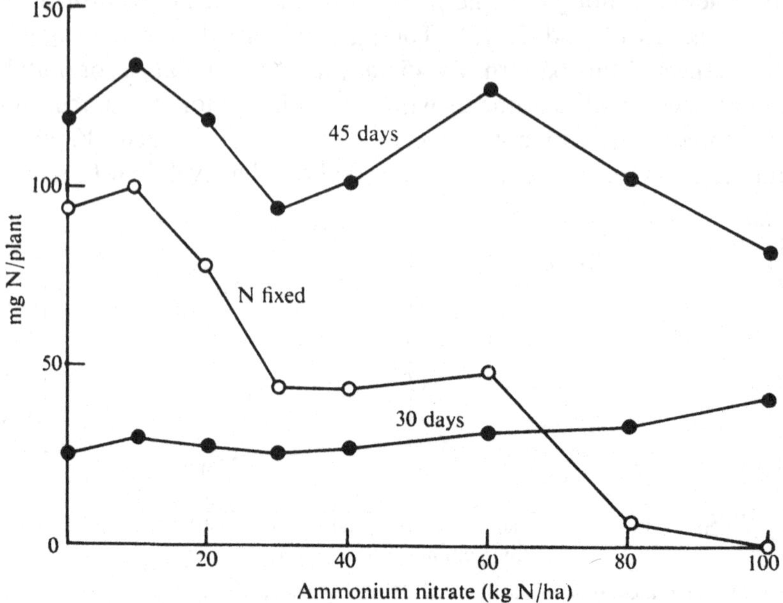

Fig. 28.14. Nitrogen content at 30 and 45 days and the amount of nitrogen fixed by 45 days for *Vigna mungo* plants grown in 10 % Kettering loam and supplied the equivalent of 0–100 kg N/ha as ammonium nitrate at sowing.

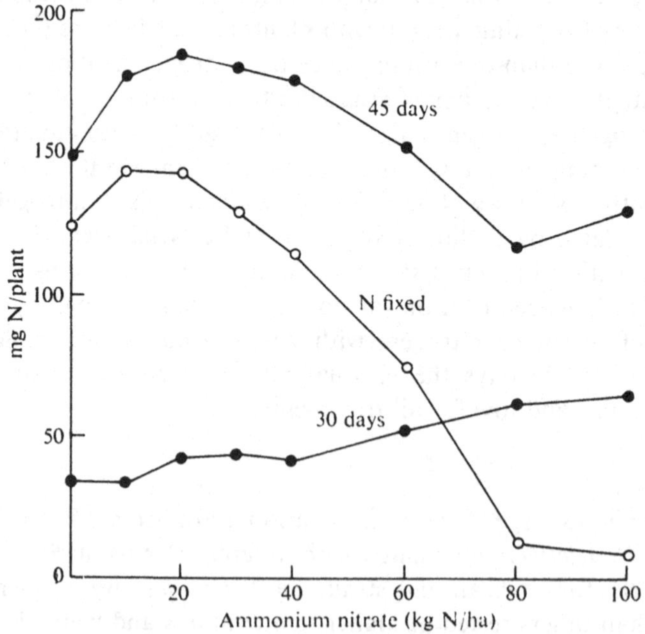

Fig. 28.15. Experimental details as for Fig. 28.14, but using *V. radiata*.

The increased effectiveness of strain Sm1b with soybean at high temperature apparently resulted from increased nodule tissue production, whereas for strain Ca-2 with *Cicer*, it resulted from increased nitrogenase activity per unit nodule weight, perhaps because nodule senescence was delayed. Although nodule formation was influenced by temperature, this was not a major factor in the differences between strains. It is not known whether the lack of nodulation of *Cicer* above 32 °C results from a restriction of infection or of nodule development. Nodulation of various tropical plants is very sensitive to 'low' root temperatures, being poor or absent at 18–21 °C but good at 24 °C (Dart & Mercer, 1965; Gibson, 1971). The leghaemoglobin content of nodules usually correlates well with nitrogen fixation (e.g. Graham & Parker, 1961; Johnson & Hume, 1973) but there was no such relationship for strain CC705 with soybean at 33 °C, where leghaemoglobin (Lb) concentration was high but little nitrogen was fixed. Nitrogenase activity per unit of Lb in the nodules varies with time, so that the nitrogen fixation does not appear to be closely controlled by Lb content. A mutant of *Pisum sativum* also produces ineffective nodules containing Lb (Holl, 1973). This suggests caution when using nodule colour as a guide to effectiveness.

Strains of *R. trifolii* differ greatly in nitrogen fixation at higher root temperatures; the response is also dependent on the *Trifolium subterraneum* cultivar (e.g. Gibson, 1963, 1965, 1971). Strain NA30, effective at lower root temperatures, was ineffective at 30 °C, because bacteroid development was restricted and abnormal, and tissue degeneration more rapid than for effective strains. Transferring NA30 nodulated plants from 22 to 30 °C induced a rapid decline in nitrogenase activity and extensive bacteroid degeneration (Pankhurst & Gibson, 1973). A similar degeneration is perhaps induced in the *Cicer* nodules transferred from 23 °C to 33 °C and 36 °C. The soybean varieties also interacted with temperature in their ability to fix nitrogen. Seven out of eight *Cicer* cultivars from diverse locations, uninoculated and supplied combined nitrogen, grew better at 22.5 °C than at 30 °C (Sandhu & Hedges, 1971). Thus both host variety and *Rhizobium* strain can be selected for growth at high soil temperatures.

It is tempting to suggest that strains adapted to high temperatures are selected for in high soil temperatures. Strain Sm1b came from hot tropical conditions in Brazil and CC705, ineffective at high temperature, is believed to have been isolated near Wisconsin where soil temperatures are cooler. The *Cicer* strains performing best at high temperatures were isolated near New Delhi where *Cicer* growth extends into the summer season. A similar relationship with cold tolerance of *R. trifolii* has been suggested (Roughley & Dart, 1970).

These results emphasise the need for strain selection at the soil temperatures likely to be encountered during growth and preferably in the soils where the plants are to be grown. The poor performance of *Cicer* above 30 °C indicates that extending the growth range of this plant into warmer soils will require selection of cultivars and strains which will nodulate in these conditions.

The increased growth of the *Vigna* spp., *P. vulgaris* and *Cajanus cajan* in root media high in organic matter is difficult to explain. Perhaps some subtle change in the availability of nutrients is affected, possibly the organic matter either supplies a plant hormone or adsorbs excess seedling-produced hormone. In many tropical soils, organic matter levels decline rapidly, and these experiments suggest that poor legume growth may then result. Some *Vigna* spp. varieties seem less sensitive than others to the leaf crinkle symptom so that plants can perhaps be bred for better growth in soils low in organic matter.

Tropical legumes may encounter high soil nitrate levels when sown at the beginning of the wet season. *Vigna* spp. nodulated well in the presence of the nitrate in the 10% Kettering loam addition. For *V. radiata*, and to a lesser extent *V. mungo*, nodulation was initially inhibited by high levels of combined nitrogen but as the concentration in the root medium was reduced during growth, nodule formation and growth increased rapidly so that nodule weight per plant in some treatments became greater than that of control plants. Perhaps this is a response similar to that found with delayed inoculation (Dart & Pate, 1959); nodules formed further away from the crown and lower at the root system with increasing nitrogen levels. The biological advantage of this may be that nodules will be placed away from the soil surface where high temperatures can disrupt the symbiosis.

The stimulation of nitrogen fixation of *Vigna* spp. by combined nitrogen present at sowing suggests that the symbiosis in these plants may be adapted for high fertility soils. This may be a *Rhizobium* strain response (e.g. Pate & Dart, 1961). Nitrogen fixation by soybeans over 10 weeks growth was considerably stimulated by weekly additions of ammonium nitrogen (Allos & Bartholomew, 1959). Whether stimulation persists and results in increased yields is not known; slight increases in yield of inoculated soybeans can sometimes be obtained with additions of nitrogenous fertiliser (see Ham *et al.*, Chapter 20; Hera,

Chapter 22). Soybeans also give high yields relying on nitrogen fixation, which are not stimulated by nitrogen fertiliser addition (e.g. Hanway & Weber, 1971; Chesney, Khan & Bisessar, 1973; E. E. Hartwig, personal communication).

One should, therefore aim to find *Rhizobium* strain-host variety combinations which yield well in the particular soil and environment without nitrogen fertiliser, thus using the full potential of the symbiotic system.

Summary

Chippewa soybeans inoculated with *Rhizobium* strains CB1809, Sm1b or CC705, were grown at 21, 27 or 33 °C day temperature in controlled-environment cabinets. There was little difference in nitrogen fixation between strains at 21 °C, but at 27 °C and especially 33 °C, Sm1b was the most effective. CC705, as effective as CB1809 at 27 °C, at 33 °C formed red nodules with some nitrogenase activity, but which fixed little nitrogen. *Cicer arietinum* grew and fixed similar amounts of nitrogen with all strains at root temperatures from 15–25 °C, but nodulation and nitrogen fixation at 30 °C was dependent on *Rhizobium* strain, the best strain Ca-2 fixing more than 60 % as much as at 23 °C. No nodules were formed by any strain at root temperature greater than 32 °C. Nitrogenase activity per plant increased until about 7 weeks from sowing and then declined, but nitrogenase activity per gram nodule weight declined from the fourth week. *Cicer* nodule nitrogenase functions over the temperature range from 6–40 °C, with most activity between 24 and 33 °C and little at higher temperatures. The nitrogenase activity of plants transferred from root temperatures of 23 °C to 33 or 36 °C declined rapidly.

Vigna mungo and *V. radiata* plants grew poorly in quartz sand: grit with several nutrient solutions with and without nitrogen, but growth was good if this was amended with 10 % of a clay loam.

R. Islam and J. Döbereiner thank the British Council for support during this work. J. Döbereiner was on leave from the Instituto Pesquisas e Experimentacao Agropecuarias do Centro-Sul, Brazil. We are grateful to S. Maskall and P. Quilt for use of their data, to A. E. Johnston and G. Mattingly for soil samples from their experiments, and to R. A. Date, N. S. Subba Rao and J. Burton for the *Rhizobium* strains. D. Jenkinson gave us much helpful advice.

References

Allos, H. F. & Bartholomew, M. W. (1959). Replacement of symbiotic nitrogen fixation by available nitrogen. *Soil Science*, **87**, 61–6.

Chesney, H. A. D., Khan, M. A. & Bisessar, S. (1973). Performance of soybeans in Guyana as affected by inoculum (*Rhizobium japonicum*) and nitrogen. *Turrialba*, **23,** 91–6.

Dart, P. J. & Day, J. M. (1971). Effects of incubation temperature and oxygen tension on nitrogenase activity of legume root nodules. *Plant and Soil*, Special Volume, 167–84.

Dart, P. J., Day, J. M. & Harris, D. (1972). Assay of nitrogenase activity by acetylene reduction. In: *Use of isotopes for study of fertilizer utilization by legume crops*, pp. 85–100. International Atomic Energy Agency/Food and Agricultural Organisation, Vienna.

Dart, P. J., & Mercer, F. V. (1965). The effect of growth temperature, level of ammonium nitrate, and light intensity on the growth and nodulation of cowpea (*Vigna sinensis* Endl ex Hassk). *Australian Journal of Agricultural Research*, **16,** 321–45.

Dart, P. J. & Pate, J. S. (1959). Nodulation studies in legumes. III. The effects of delaying inoculation on the seedling symbiosis of barrel medic, *Medicago tribuloides* Desr. *Australian Journal of Biological Sciences*, **12,** 427–44.

Dart, P. J. & Wildon, D. C. (1970). Nodulation and nitrogen fixation by *Vigna sinensis* and *Vicia atropurpurea*: the influence of concentration, form, and site of application of combined nitrogen. *Australian Journal of Agricultural Research*, **21,** 45–56.

Day, J. M. & Dart, P. J. (1969). Light intensity, nodulation and nitrogen fixation. Report of the Rothamsted Experimental Station, Harpenden, England. Part 1, p. 105.

Day, J. M. & Dart, P. J. (1970). Factors affecting nitrogenase activity of legume root nodules. Diurnal fluctuations. Flowering. Report of the Rothamsted Experimental Station, Part 1, pp. 85–6. Harpenden, England.

Ezedinma, F. O. C. (1964). Effects of inoculation with local isolates of cowpea *Rhizobium* and application of nitrate nitrogen on the development of cowpeas. *Tropical Agriculture, Trinidad*, **41,** 243–9.

Gibson, A. H. (1963). Physical environment and symbiotic nitrogen fixation. I. The effect of root temperature on recently nodulated *Trifolium subterraneum* L. plants. *Australian Journal of Biological Sciences*, **16,** 28–42.

Gibson, A. H. (1965). Physical environment and symbiotic nitrogen fixation. II. Root temperature effects on the relative nitrogen assimilation rate. *Australian Journal of Biological Sciences*, **18,** 295–310.

Gibson, A. H. (1971). Factors in the physical and biological environment affecting nodulation and nitrogen fixation by legumes. *Plant and Soil*, Special Volume, 139–52.

Graham, P. H. & Parker, C. A. (1961). Leghaemoglobin and symbiotic nitrogen fixation. *Australian Journal of Science*, **23,** 231–2.

Greenland, D. J. (1958). Nitrate fluctuations in tropical soils. *Journal of Agricultural Science*, **50,** 82–92.

Hanway, J. J. & Weber, C. R. (1971). Accumulation of N, P and K by soybean (*Glycine max* (L.) Merrill) plants. *Agronomy Journal*, **63,** 406–8.

Hewitt, E. J. (1966). *Sand and water culture methods used in the study of plant nutrition.* Commonwealth Agricultural Bureau, Farnham Royal, England.

Holl, F. B. (1973). A nodulating strain of *Pisum* unable to fix nitrogen. *Plant Physiology*, Lancaster, **51,** (Supplement) 35.

Hume, D. J., Johnson, H. S., Criswell, J. G. & Stevenson, K. R. (1974). Effects of moisture stress on nitrogen fixation in soybeans. *Canadian Journal of Plant Science*, in press.

Johnson, H. S. & Hume, D. J. (1973). Comparisons of nitrogen fixation estimates in soybeans by nodule weight, leghaemoglobin content and acetylene reduction. *Canadian Journal of Microbiology*, **19,** 1165–8.

Maskall, S. M. & Dart, P. J. (1973). Leghaemoglobin. Report of Rothamsted Experimental Station for 1972, Part 1, pp. 86–7. Harpenden, England.

Nair, P. K. R. & Talibudeen, O. (1973). Dynamics of K & NO_3 concentrations in the root zone of winter wheat at Broadbalk using specific ion electrodes. *Journal of Agricultural Science*, **81,** 327–37.

Pankhurst, C. E. & Gibson, A. H. (1973). *Rhizobium* strain influence on disruption of clover nodule development at high root temperature. *Journal of General Microbiology*, **74,** 219–31.

Pate, J. S. & Dart, P. J. (1961). Nodulation studies in legumes. IV. The influence of inoculum strain and time of application of ammonium nitrate on symbiotic response. *Plant and Soil*, **15,** 329–46.

Roughley, R. R. & Dart, P. J. (1970). Growth of *Trifolium subterraneum* L. selected for sparse and abundant nodulation as affected by root temperature and *Rhizobium* strain. *Journal of Experimental Botany*, **21,** 776–86.

Sandhu, S. S. & Hedges, H. F. (1971). Effects of photo-period, light intensity, and temperature on vegetative growth, flowering, and seed production in *Cicer arietinum* L. *Agronomy Journal*, **63,** 913–14.

Varley, J. A. (1966). Automatic methods for the determination of nitrogen, phosphorus and potassium in plant material. *The Analyst*, **91,** 119–26.

Weber, C. R. (1966). Nodulating and non-nodulating soybean isolines. II. Response to applied nitrogen and modified soil conditions. *Agronomy Journal*, **58,** 43–8.

Wilson, P. W. (1940). *The biochemistry of symbiotic nitrogen fixation.* University of Wisconsin Press, Madison.

Explanation of plates

Plate 28.1. All plants were inoculated with *Rhizobium* strain CB756. Plants in (*a*)–(*d*) were 28 days old. (*a*) *Vigna mungo* grown in quartz sand:grit mixture, initially watered with Hoaglands' complete nutrient solution containing 210 ppm N. Some seedling unifoliate leaves have dropped off, and some of the trifoliate leaves have crinkled margins. (*b*) *V. radiata* grown in sand:grit, watered with Long Ashton minus-nitrogen solution. Seedling leaves are covered with brown necrotic spots. (*c*) Normal and crinkled trifoliate leaves and necrotic unifoliate leaves below. (*d*) *V. mungo* grown in sand:grit mixture, with different nutrient solutions: DP−N, Dart and Pate minus-nitrogen; LA−N, Long Ashton minus-nitrogen; LA+N, Long Ashton complete nutrient solution (190 ppm N); H+N, Hoagland's complete nutrient solution. Plants were also grown in milled sedge peat watered with DP−N solution, or in Rothamsted silty loam soil. Growth was only symptom-free in the soil. (*e*) *V. mungo* grown in Woburn sandy soil mixed with 25 % v/v quartz grit, symptom-free from Road Piece (RP, 0.14 % N) but with severe leaf crinkle from Ley arable (LAr, 0.11 % N) site. (*f*) *V. radiata* grown in soil from the Ley arable was symptom-free.

Plate 28.2. All plants were inoculated with *Rhizobium* strain CB756. (*a*) *V. radiata*, 28 days from sowing in sand:grit alone (S) or amended with 10 % v/v Kettering loam (10 % K) watered with minus-nitrogen nutrient solution. Plants were also grown in different soils: W, Woburn sandy soil; K, Kettering clay-loam; F, a neutral fen peat soil from Lincolnshire; R, Rothamsted silty loam. Note that growth in 10 % Kettering loam is better than in the complete soil. (*b*) Prima cowpea, 35 days after sowing in sand:grit alone (C) in sand-grit plus 50 mg N per pot as ammonium nitrate, in sand:grit amended with 10 % v/v Rothamsted silty loam (10 % R) or Kettering loam (10 % K) or in Kettering clay-loam alone (K). All pots watered initially with minus-nitrogen culture solution. The 10 % Kettering loam provided about 50 mg available nitrate, so that its improved growth over the 50 mg N treatment is not due to nitrate alone. (*c*) *Cajanus cajan*, 40 days old, grown in sand:grit alone (C), sand:grit amended with 10 % (10 % K) or 30 % (30 % K) v/v Kettering loam, or in Kettering clay-loam alone (K). All pots except the K treatment were watered with minus-nitrogen culture solution. As for *V. radiata*, growth in 10 % K is better than in loam alone.

Plate 28.1 (*Facing p.* 384.)

Plate 28.2

29. Recovery and compensation by nodulated legumes to environmental stress

A. H. GIBSON

Periodic environmental stresses, of varying duration and kind, are a common feature during the growth of any crop or pasture legume. In order to derive the maximum benefit from the symbiotic system, it is essential to appreciate how these stresses will affect the system, and to know where corrective or ameliorative approaches should be adopted to minimise their various effects.

Nodulation stress is defined as that stress which adversely affects the formation and function of nodules such that there is a greater difference in nitrogen assimilation between inoculated plants and plants supplied combined nitrogen under stress conditions than there is under optimal conditions. The stresses may involve temperature, light, moisture, lowered oxygen concentration, adverse soil pH, nutrient deficiencies specifically affecting symbiotic behaviour, defoliation, or insufficient numbers of rhizobia to effect prompt, adequate nodulation. A specific form of biological stress concerns incomplete compatibility between the symbionts such that sufficient nitrogen is not provided to maintain plant growth at a level comparable to that achieved by combined nitrogen controls.

Two types of response to stress are considered: compensation and recovery. In the first the symbiotic system responds to a stress affecting the level of activity by increasing the capacity of the system for nitrogen fixation. The second type of response involves recovery of the symbiotic system following the removal of a stress which had previously affected nodule development or nitrogenase activity.

Temperature

Temperature stresses may be applied to the nodulated plant through either the roots or the shoots, and they may be of short- or long-term duration. Several lines of evidence indicate that plants respond to lower root temperatures by producing a greater volume of nodule tissue than would be otherwise anticipated, and that this may be regarded as a form of compensation against lower nitrogen-fixing activity per unit of tissue. With *Trifolium subterraneum*, proportionately more of the increase in total dry weight and total nitrogen, was found in the nodules when the plants were growing at 8 °C relative to those at 15 °C, and at 15 °C relative to those at 22 °C (Gibson, 1969*a*). Similarly the higher concentration of leghaemoglobin in the nodules of plants transferred from 22 to 12 °C, relative to those left at 22 °C, was considered as a compensatory effect, either through the production of higher levels of leghaemoglobin in individual nodules or through the maintenance of active tissue that might otherwise have degenerated (and been replaced) at the higher temperature (Davidson, Gibson & Birch, 1970). Roughley (1970) drew a similar conclusion from studies in which he showed a greater volume of bacteroid tissue, and a lower rate of nitrogen fixation per unit of bacteroid tissue, on plants grown at 11 °C root temperature, relative to those at 15 °C.

Observations indicating that the symbiotic system is able to respond to the adverse effects of lower temperatures are not confined to *T. subterraneum*. Both *Lupinus angustifolius* (lupin) and *Glycine max*

Table 29.1. *Dry weight (mg/plant) and total nitrogen (mg/plant) in nodules formed by two strains of* Rhizobium japonicum *on Lincoln soybeans*

Strain:	CB1809		CC705	
Temperature (°C):	18	25	18	25
Nodule dry weight				
Day 26	130		110	
Day 33	210	210	200	250
Day 40	360	290	370	370
Nodule nitrogen				
Day 26	6.8		4.5	
Day 33	9.7	9.5	9.8	11.1
Day 40	18.7	15.6	17.7	19.2
Distribution index (%)				
Dry weight	8.2	3.2	10.2	6.3
Nitrogen	14.8	6.8	22.9	13.9
Nitrogen fixation rate				
Days 26–33	0.57	0.91	0.39	0.77
Days 33–40	0.48	0.88	0.42	0.60

Plants grown in perlite/vermiculite under a 27/22 °C temperature regime (14 h light/day, 32 300 lux) then transferred to 18 or 25 °C 26 days after sowing. Distribution index shows the increase in nodule weight and nitrogen, over the period day 26 to day 40, as a proportion of the increase in total plant dry weight and nitrogen. The nitrogen fixation rate (mg N_2/mg nodule N/day) is based on mean nodule nitrogen values over each 7-day period.

(soybean) produced a higher proportion of their total dry weight and nitrogen in nodule tissue when grown at suboptimal than at optimal temperatures. In some instances, total nodule tissue exceeded that on plants grown under optimal temperature conditions (Table 29.1). In other experiments, nodulated soybean roots incubated at temperatures between 10 °C and 40 °C showed a marked effect of temperature on nitrogen-fixing activity (Fig. 29.1). The form of the response was

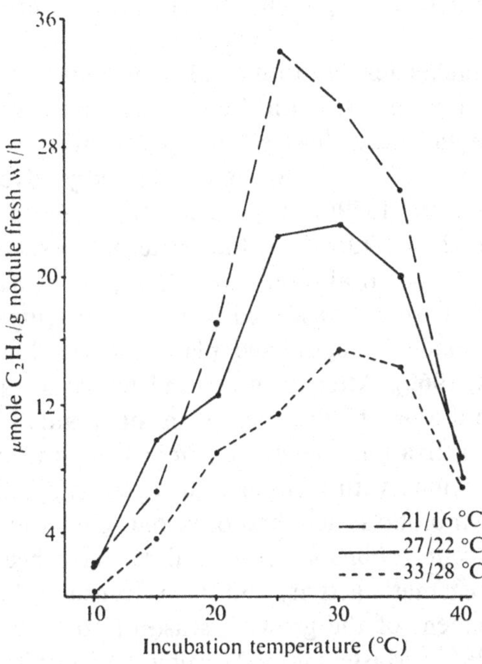

Fig. 29.1 Effect of incubation temperature on nitrogen-fixing activity (as determined by C_2H_2 reduction) of Lincoln soybeans inoculated with *Rhizobium* strain CB1809 and grown under three temperature regimes (other conditions: 14 h light/day, 32 300 lux). Optimum condition for growth and nitrogen fixation was the 27/22 °C regime.

dependent on the temperature regime under which the plants were grown, and the temperature for maximum activity was higher for the plants from warmer conditions. The highest specific activity (μmoles C_2H_4/g nodule fresh wt/hour) was obtained with the nodules from 21/16 °C when incubated at 25 °C, indicating that these nodules had potentially greater capacity for nitrogen fixation than the nodules from plants grown at higher temperatures. This could be interpreted as a compensatory response to the lower temperature of the growth conditions. Similar observations have been made with lupins (Gibson, 1973). These results indicate the need for caution when extrapolating from short assays to estimate nitrogen fixation over long periods during which effects due to diurnal and daily variation in temperature could be overlooked. The fact that the response to incubation temperature described here (Fig. 29.1) differs from some previously published results (Gibson, 1971; Dart & Day, 1971) indicates that species differences, plant age, and other features of the environmental conditions under which the plants are grown, influence the nature of the results obtained.

Responses to higher temperature conditions depend on the strain of rhizobia and the host plant. With *T. subterraneum*, exposure to 30 °C root temperature had little effect on nitrogen fixation when the plants were nodulated by strain TA1, but had a markedly adverse effect with other strains (Gibson, 1969*b*). This effect appeared to be associated with an accelerated breakdown of the bacteroid tissue (Pankhurst & Gibson, 1973). In a pot trial, nodulated *Medicago sativa* plants made very poor regrowth when transferred to high root temperatures, compared with that made by nodulated plants provided with combined nitrogen (Rogers, 1969). After recutting and return to favourable root temperature conditions, further regrowth of plants dependent on symbiotic nitrogen fixation was greatly improved, and comparable to that of plants provided with combined nitrogen, suggesting that rapid recovery of the symbiotic system had occurred. Under low temperature conditions, senescent *Trifolium repens* nodules have been observed to recommence meristematic activity and form (transient) bacteroid tissue at the commencement of the growing season (Pate, 1958; Bergersen, Hely & Costin, 1963; Masterson & Murphy, Chapter 25).

Light

Eaton (1931), Sironval, Bonnier & Verlinden (1957), and Day (1972) have shown marked effects of daylength on nodulation of soybeans, while the data of Lie (1969) with *Pisum sativum* indicate that phytochrome is probably involved in nodule development. Direct lighting of the roots of *Phaseolus vulgaris* adversely affects nodulation (Grobbelaar, Clarke & Hough, 1971), possibly through a high-energy photoreactive system, whilst with *T. subterraneum*, the effect of light on the roots is dependent on the strain of rhizobia forming the nodules (Gibson, 1968).

In other studies with *T. subterraneum* (Gibson, unpublished observations), marked effects of daylength on the rate of nitrogen fixation were observed, and these effects showed a strong interaction with root temperature. Light intensity also affected nitrogen fixation and growth, but as with the effects of photoperiod, the response by plants nodulated

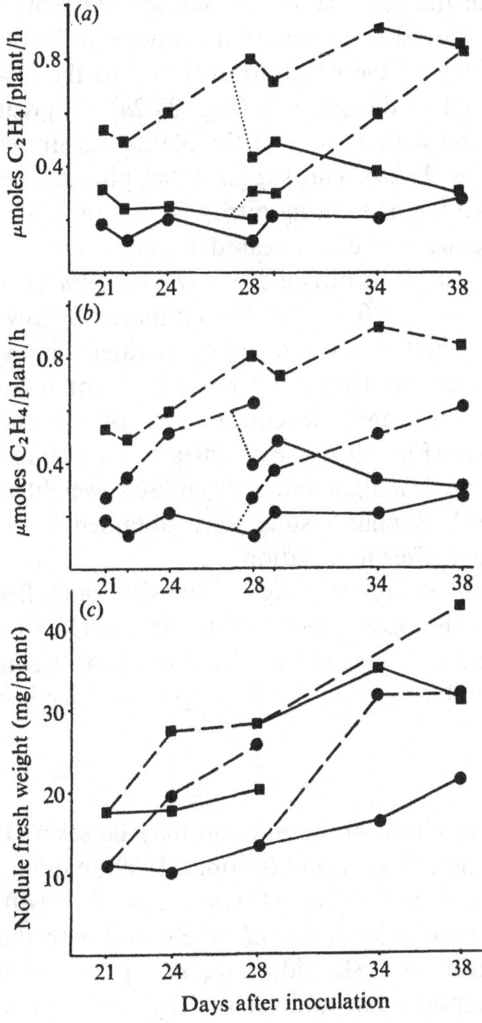

Fig. 29.2. Effect of light intensity on the nitrogen-fixing activity of *Trifolium subterraneum* plants, inoculated with strain TA1, and grown in test-tubes in a controlled-environment cabinet (Gibson, 1965) with 22 °C root temperature, and a 22/15 °C shoot temperature regime based on a 14 h daily light period: (*a*) shows effects of transferring plants from a high to low light intensity; (*b*) from a low to high light intensity; and (*c*) shows changes in nodule fresh weight for the principal treatments. (——, 8600 lux; ---, 32 300 lux; plants grown for first 21 days at low (●) or high (■) light intensity).

with the fully effective strain, TA1, was usually similar to that of plants provided with adequate mineral nitrogen.

In other studies of the effect of light intensity on nodulation and nitrogen fixation by *T. subterraneum*, plants were grown at low (8600 lux) and high (32 300 lux) light intensities for 20 days after inoculation, and at the end of the dark period preceding the twenty-first day, some plants were transferred to the alternate condition. Within 5 hours, the nitrogenase activity of the plants transferred to the low intensity was 40 % less than that of the controls (Fig. 29.2*a*). Of greater significance was the 50 % increase in activity of the plants transferred to the higher light intensity (Fig. 29.2*b*). This suggests that photosynthate supply had been the principal factor limiting nitrogen fixation, although the possibility that nitrogenase levels increased rapidly cannot be disregarded. Whatever the cause, it is obvious that *T. subterraneum* is capable of extremely rapid recovery from the stress imposed by low light intensity. Furthermore the higher level of photosynthate supply also affects nodule development, as shown by the significant increase in nodule weight within 3 days and the continuance of the effect during the subsequent 4 days (Fig. 29.2*c*). By contrast, for plants transferred to, and maintained at, low light intensity, nodule weight remained static at the initial level. Similar results were obtained when the transfers were made 28 days after inoculation.

That short-term changes in light intensity can affect nitrogenase activity supports the observations of Bergersen (1970) and Mague & Burris (1972), and indicates the need for caution in comparing values obtained under differing conditions of light intensity in the field.

Defoliation

The effects of defoliation on nodulation may be severe (Wilson, 1942; Bowen, 1958; Butler, Greenwood & Soper, 1959), moderate (Whiteman, 1970), or have little or no effect (Davidson *et al.*, 1970). They appear to vary with the species (Butler *et al.*, 1959) and with the intensity and frequency of defoliation (Davidson *et al.*, 1970; Whiteman, 1970). Kamata (1963) reports that nodules on *Trifolium repens* that had lost leghaemoglobin and bacteroid tissue recovered their activity by regeneration of new bacteroid tissue in the degenerating nodules, although this awaits confirmation.

Following defoliation of white and subterranean clovers, nitrogenase activity rapidly declines (Moustafa, Ball & Field, 1969; Gibson, unpublished observations) but recovery is well advanced within 7–10 days. In a study using *T. subterraneum* and three *R. trifolii* strains of different levels of symbiotic effectiveness, the response pattern to weekly defoliation was similar, regardless of the strain, over a four-week period (Gibson, unpublished observations).

Combined nitrogen

Despite an extensive literature on the effects of combined nitrogen on various aspects of the symbiosis between legumes and rhizobia, the prediction of the response to the application of combined nitrogen remains a hazardous procedure. Inhibitory effects of combined nitrogen on root hair infection (e.g. Munns, 1968), nodule initiation (e.g. Gibson & Nutman, 1960; Gibson, 1973), and nodule development and function (e.g. Oghoghorie & Pate, 1971) are well documented so that consideration of combined nitrogen as a form of stress is justified. Stimulatory effects of combined nitrogen are known (e.g. Pate & Dart, 1961) but are less clearly understood. In considering the effect of combined nitrogen fertilisers, little attention has been given to available soil nitrogen although this will influence the overall effect obtained. Despite this, laboratory and glasshouse experiments are usually done with nitrogen-free media, for experimental convenience when measuring nitrogen fixation and to allow better control of experimental conditions.

Concern about the overall effects of combined nitrogen on the ultimate expression of symbiotic effectiveness, especially following the demonstration that strains may vary in their symbiotic response to combined nitrogen (Pate & Dart, 1961), prompted a series of experiments with *T. subterraneum.* In one experiment, two cultivars, Yarloop and Bacchus Marsh, were each inoculated with six strains of *R. trifolii* and nitrate supplied (1 mg N/tube) to half the plants inoculated with each strain. With Yarloop, considerable variation existed between the strains in the degree to which nitrogen fixation was stimulated, the stimulation ranging from 0 to 100 % over the controls (Fig. 29.3*a*). Furthermore, the ranking of strains was markedly different in the '−N' and '+N' treatments. With Bacchus Marsh, the degree of stimulation of overall nitrogen fixation was similar for five of the six strains (c. 30 % more than the controls) and the ranking of the strains in the '−N' and '+N' treatments was essentially the same (Fig. 29.3*b*). The time to initial nodulation was retarded 0.5 days with all strains on cv. Yarloop, and by 0–1.5 days with cv. Bacchus Marsh.

Examination of the responses of five cultivars, each inoculated with strain TA1 and provided with a range of combined nitrogen applications (0–1.5 mg N/tube) showed considerable variation between cultivars and the rate of nitrogen application on nitrogen fixation. Only Yarloop showed increased fixation at all levels of nitrogen applied, whilst two

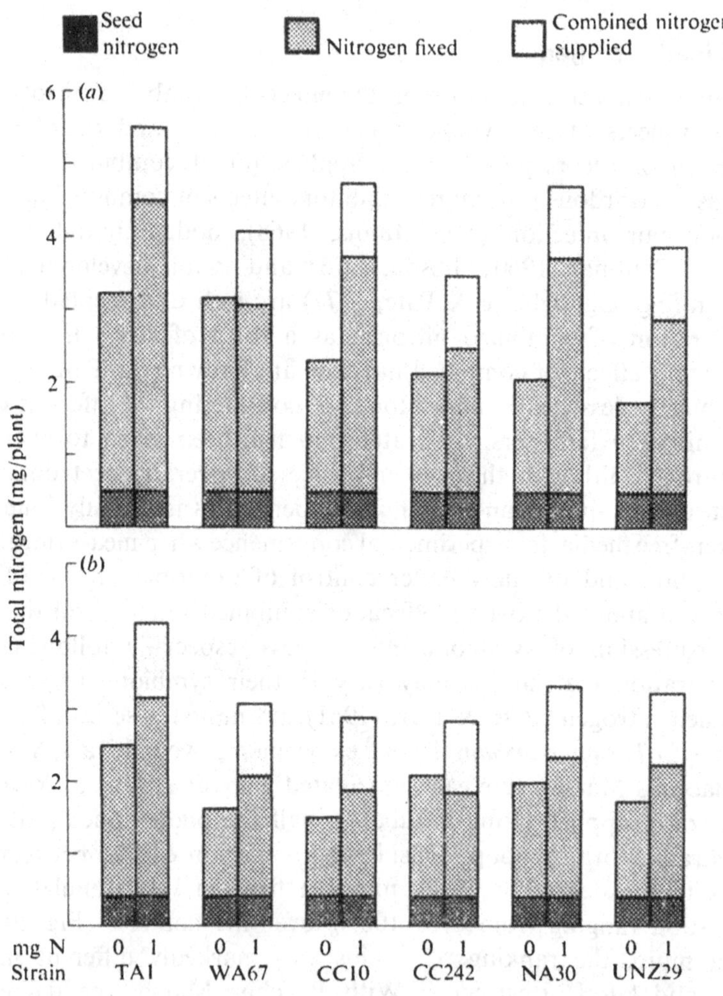

Fig. 29.3. Influence of supplemental combined nitrogen as nitrate (1 mg N/plant) on symbiotic nitrogen fixation by the *Trifolium subterraneum* cultivars Yarloop (*a*) and Bacchus Marsh (*b*) when each was inoculated with six strains of *Rhizobium trifolii*. Plants grown with 22 °C root temperature, a 22/15 °C shoot temperature regime based on a 14 h daily photoperiod, and harvested 27 days after inoculation. The estimate of nitrogen fixed was made by subtracting seed nitrogen and amount of nitrogen supplied from total plant nitrogen.

others showed a response at one level only (1.0 mg N/tube). With the other cultivars, Mount Barker and Dwalganup, estimated nitrogen fixation was depressed, or not affected, by the addition of combined nitrogen.

The nature of the stimulatory response was studied by making sequential harvests on cv. Tallarook plants inoculated with an effective strain, TA1, or less effective strain, CC10. Differences between the nitrogen treatments were evident 12 days after inoculation, and thereafter all treatments showed an exponential increase in total nitrogen (Fig. 29.4). The relative nitrogen assimilation rates (R_n)† for all treatments were similar, but because of better growth and nitrogen assimilation in the initial period, final differences in total plant nitrogen were considerable. Based on total plant nitrogen, strain CC10 was 70 % as effective as strain TA1 without added nitrogen, but 87 % as effective with 1 mg N supplementation. The practical significance of these observations was indicated in a glasshouse study using soils of different nitrogen levels; the largest differences in the growth of plants inoculated with these strains was found in the soil with the lowest level of soil nitrogen (Simpson & Gibson, 1970).

With *Medicago sativa*, a bimodal response to the application of combined nitrogen at sowing has been attributed to the tolerance of nodulation and nitrogen fixation to lower levels of nitrogen, and to the suppression of nodulation at higher levels (Hogland, 1973). Cultivars of *M. sativa* differed in their response to combined nitrogen, whilst the temperature conditions affected the yield relationship between supplemented plants and those completely dependent on symbiotic nitrogen fixation.

Both the author and J. S. Pate consider, from their experience with bacterial strain and host cultivar selection trials, that there is a need for a greater awareness of the influence of soil and/or fertiliser nitrogen on symbiotic responses. With levels of available nitrogen ranging from low to complete adequacy for plant growth, caution should be exercised in extrapolating freely from nitrogen-free culture studies to expected results in the field, and attention should be given in selection trials to simulating likely practical situations.

Marked stimulation of nitrogen fixation by early supplementation with combined nitrogen has also been observed with two grain legumes.

†$R_n = (\log N_2 - \log N_1)/(t_2 - t_1)$ where N_2 and N_1 are total plant nitrogen values at harvest 2 (t_2) and harvest 1 (t_1) respectively, and R_n is given as mg N assimilated/mg plant N/day (Gibson, 1965).

With soybeans, 7 mM nitrate-nitrogen was included in the nutrient solution for 14 days after sowing, and the pots then eluted with nitrogen-free nutrient. Nitrogen assimilation by these plants was comparable to that of inoculated plants supplied with 7 mM nitrate-nitrogen throughout, and nitrogen fixation was three times better than that of control plants not supplemented (Fig. 29.5*a*). The increase in plant nitrogen was exponential in all treatments, but the R_n values were larger following nitrogen supplementation. The threefold difference in total

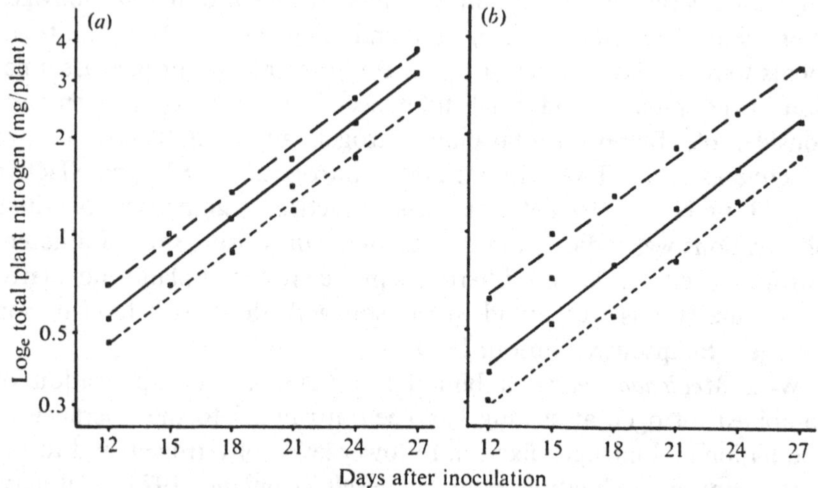

Fig. 29.4. Increase in total plant nitrogen for *Trifolium subterraneum* cv. Tallarook inoculated with *Rhizobium trifolii* strains TA1 (*a*) and CC10 (*b*) and supplied with 0 (-----), 0.3 (——), or 1.0 (– – –) mg nitrate-nitrogen at inoculation. Plants grown in test tubes at 22 °C root temperature with a 22/15 °C shoot temperature regime based on a 14 h daily photoperiod (Gibson, 1965).

nitrogen at 36 days was attributed to the better early growth of the supplemented plants. A similar situation was found with the lupins (Fig. 29.5*b*) although the failure to achieve exponential increase in plant nitrogen meant that the comparison of the R_n values was not appropriate. However the twofold difference in plant nitrogen observed at day 18 was maintained throughout the period of observation. A further difference between the treatments was in the lateral root development; supplementary nitrogen considerably increased number and length of lateral roots.

Of five tropical and subtropical legumes examined, only *Glycine wightii* showed a stimulatory response to combined nitrogen. This occurred only at 30 °C root temperature, and was modified by the shoot temperature (Table 29.2). The addition of combined nitrogen 12 days

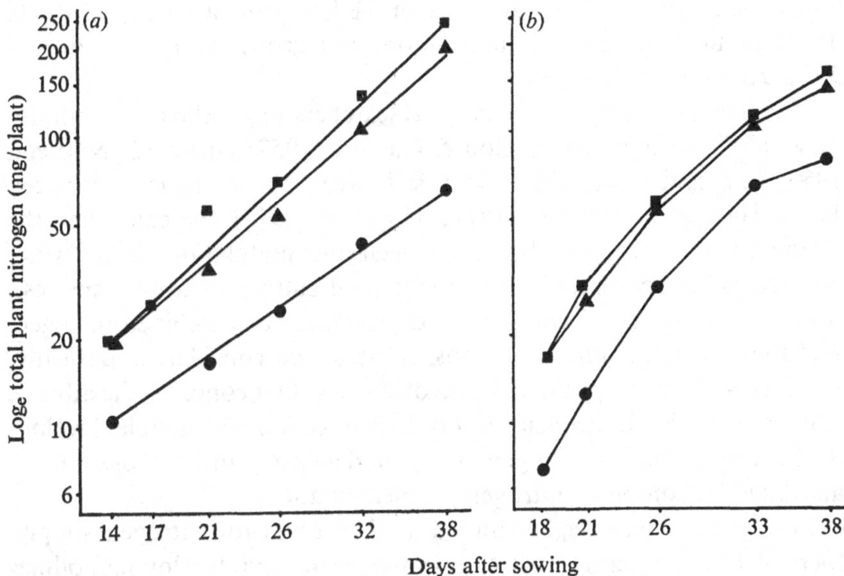

Fig. 29.5. Increase in total plant nitrogen for (*a*) soybeans (cv. Lincoln) inoculated with CB1809 and grown with a 27/22 °C temperature regime, and (*b*) lupins (cv. Uniharvest) inoculated with strain 425 and grown with a 21/16 °C temperature regime. The plants were grown in a vermiculite/perlite medium without supplemental nitrogen (●) or with 7 mM KNO_3 for 14 days after sowing (▲) or throughout the experiment (■). Daily light period 14 h and 32 300 lux.

after inoculation also promoted nitrogen fixation, but to a lesser extent than when given at inoculation. The other species examined were *Macroptilium atropurpureum*, *Stylosanthes humilis*, *Desmodium intortum*

Table 29.2. *Dry matter yield of* Glycine wightii *cv. Cooper when inoculated with CB756 or provided with adequate combined nitrogen (as ammonium nitrate)*

	Dry weight			Nitrogen fixed		
Treatment	25/15	30/20	35/25	25/15	30/20	35/25
Inoculated with CB756	82	79	82	1.69	1.66	1.39
+N at inoculation	157	213	191	3.28	4.64	3.96
+N at day 12	121	162	123	3.28	2.90	2.33
Nitrogen control	215	257	186	–	–	–

Some inoculated plants received 0.5 mg N (NH_4Cl) per tube at inoculation, or 12 days later, and this and seed nitrogen (0.3 mg/plant) is substracted from total plant nitrogen in calculating nitrogen fixed (mg/plant) after 35 days. Plants grown in test tubes (Gibson, 1965) with 30 °C root temperature under each of three shoot temperature regimes (25/15, 30/20, 35/25 °C) based on a 14 h daily light period 23 700 lux (with D. E. Byth).

and *D. uncinatum*, all inoculated with CB756, provided with 0.5 mg N as ammonium chloride at inoculation, and grown at 18, 24, 30 and 36 °C root temperature.

These results and those of other researchers (e.g. Allos & Bartholomew, 1955; Richardson, Jordon & Garrard, 1957; Gibson & Nutman, 1960; Pate and Dart, 1961; Dart & Mercer, 1965*a*; Dart & Wildon, 1970; Harper & Cooper, 1971; Hoglund, 1973) indicate that the relationship between combined nitrogen and nodulation (in its wider context) is very complex. Host species (and cultivars within a species), bacterial strains, the form, level and placement of combined nitrogen, and the environmental conditions, must all be considered. Basically, two forms of stress appear to be involved. The first concerns the adverse effect of combined nitrogen on root hair infection and nodule development. The second is nitrogen stress or deficiency in the host, and is alleviated by combined nitrogen supplementation.

In a young plant dependent on seed reserves for nitrogen supply, there will be competition between shoots, roots and developing nodules for substrates such as free amino acids, with the strength of the competition being dependent on the environmental conditions. The level of free amino acids in the root will decline as the seedling develops (Hubbell & Elkan, 1967), and competition for these substrates will place overall plant development, including nodule development, under a degree of stress. Supplementation with combined nitrogen should alleviate this nitrogen stress and at appropriate levels of supplementation, concomitant increases in the level of photosynthesis should raise the level of soluble sugars, the other vital substrate needed for nodule development (Orcutt & Wilson, 1935). As Pate & Dart (1961) suggest, the timing and amount of such supplementation will be critical for various stages of nodulation. However, in field situations, the only operational variables are the host cultivar, the bacterial strain and the amount, form, and placement of nitrogenous fertiliser. Until the interaction of these variables is better understood, and the effects of soil nitrogen levels, soil pH, and nutrient deficiencies are better known, the nitrogen fertilising of legumes will remain a hit-or-miss procedure.

Another aspect of combined nitrogen stress is the effect on nitrogen fixation *per se*. Dart & Mercer (1965*b*) described nodule breakdown due to ammonium nitrate application (10.7 mM) to *T. subterraneum* and *M. tribuloides*, with degeneration commencing within 2–3 days and being complete 6–8 days later. Small & Leonard (1969) showed that the level of ^{14}C translocated to nodules on *T. subterraneum* and *P.*

sativum plants provided with sodium nitrate for 5 days was reduced by 60–75 % relative to nitrogen-free controls. In studies with soybeans in this laboratory, nitrogenase activity was found to decline within 48 hours of supplying 7 mM nitrate-nitrogen in the nutrient medium, and further nodule development to cease. After six days, during which the level of nitrate was maintained, elution of the pots with nitrogen-free medium was followed, within 4 days, by complete recovery of nitrogenase specific activity and a resumption of nodule development. A similar recovery was observed (Fig. 29.6) following 10 days treatment. During this period, leghaemoglobin levels (Gibson, 1969*a*) were not affected. Nor was there any indication in the breakdown of the fine structure of the nodules, as examined by electron microscopy, four days after treatment commenced (D. J. Goodchild & A. H. Gibson, unpublished results).

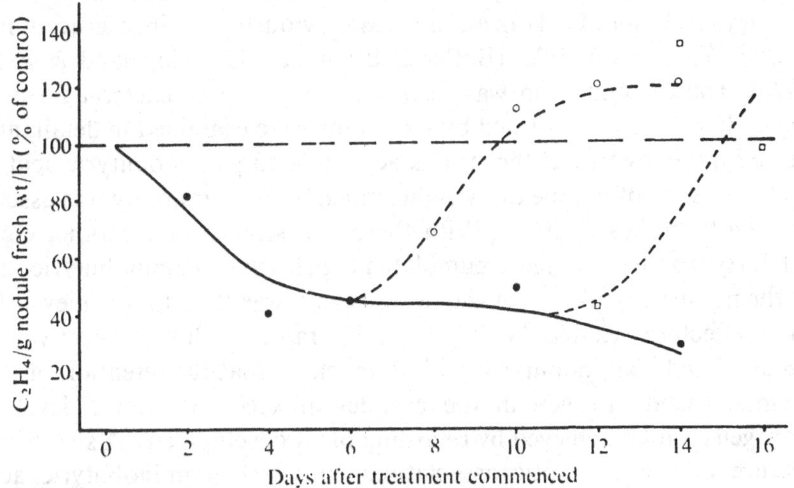

Fig. 29.6. Recovery of nitrogenase activity in soybean nodules (cv. Lincoln) exposed to 7 mM KNO_3 for 6 (○---○), or 10 (□---□), days from 24 days after sowing. Grown with 27/22 °C temperature regime and 14 h light/day (32 300 lux). Also shown (●——●) is the effect of exposure to 7 mM KNO_3 throughout the experiment.

In studies by Dr K. Hashimoto and the author (unpublished data), a decline in the rate of ^{14}C translocation to soybean nodules was observed 24 hours after supplying nitrate (7 mM) in the medium; the proportion of labelled photosynthate translocated from the fed leaf and found in the nodules decreased in proportion to the length of nitrate treatment. Whilst this is consistent with the hypothesis that the decline in nitrogen fixation is due to photosynthate deprivation, resulting from competition for photosynthate from the many sites of nitrate assimilation in the roots and possibly the shoots (Oghoghorie & Pate, 1971), it does not preclude the possibility of a more direct effect of nitrate within the nodules. Such an effect could reduce the sink strength of the nodules for photosynthate. In the same studies, ammonium ions were less inhibitory to nitrogen fixation than nitrate ions at the same applied concentration, and the inhibition of nitrogen fixation was less when the plants were grown at 15 °C than at 25 °C. This again indicates the importance of the form of combined nitrogen, and the environmental conditions, in influencing the response of the symbiotic system to combined nitrogen.

Accumulation of γ-aminobutyric acid in nodules

Non-protein nitrogen accumulating in the nodules of *T. subterraneum* (Gibson, 1969*a*) is primarily in the form of 'bound' γ-aminobutyric acid (Freney & Gibson, 1974) an amino acid previously described as accumulating in *T. repens* nodules (Butler & Bathurst, 1958; Copeland & Pate, 1970). The accumulation was related to the strain of bacteria forming the nodules. Nodules formed by six strains were examined in detail, and those formed by two of the strains accumulated γ-aminobutyric acid at up to 10–22 % of nodule dry weight; much of this variability was associated with the host cultivar. With these two strains, and another eight (of thirty examined) which accumulated high levels of γ-aminobutyric acid in the nodules, symbiotic effectiveness was lower than that achieved by fully effective strains. While not all strains of lower effectiveness accumulated γ-aminobutyric acid, it appeared that the retention of this form of fixed nitrogen in the nodules affected the overall level of nitrogen fixation achieved by retarding plant development. A significant feature of the symbiotic associations in which γ-aminobutyric acid accumulated in the nodules was that nodule weight increased by 20–40 % over that formed by non-accumulating strains. Whilst a proportion of this increased weight was due to γ-aminobutyric acid, the remainder may be considered as a compensatory response in an attempt to produce more fixed nitrogen for the host.

Accumulation of γ-aminobutyric acid was more rapid at 22 °C than at 12 °C root temperature. Plants grown at low light intensity (8600 lux) had 34 % of their nodule dry weight as γ-aminobutyric acid, compared with 14 % of nodule dry weight at high light intensity (32 300 lux). Defoliation of plants did not result in a decline in the level of γ-aminobutyric acid as may be expected if this was a form of nitrogen storage. Enzyme studies indicated that nodules accumulating γ-aminobutyric acid possessed a lower level of γ-aminobutyric acid transaminase activity than 'non-accumulating' nodules.

Other forms of stress

Recovery of stressed nodules has been described for two situations. Wilson (1970) found that saline conditions depressed nitrogen fixation, but on removal of the stress, the nodules regained their pigmentation and nitrogen fixation recommenced. Soybean nodules will withstand desiccation to 85 % of full turgor without suffering permanent damage (Sprent, 1972). The ability of nodule meristems to recommence growth, with a subsequent development of bacteroid tissue, may give this type of symbiotic system a greater ability to recover quickly from drought conditions than those with spherical, non-meristematic nodules (Engin & Sprent, 1973).

Conclusions

The studies reported in this chapter indicate that the symbiotic system can make a rapid recovery from various stresses, such as those involving temperature, light, defoliation, and nitrate (as it affects nitrogen fixation). In many cases, the direct cause of the stress appears to be associated with a reduction in the supply of photosynthate to the nodules, and recovery is dependent on an improvement to the overall level of photosynthesis by the host. Although more difficult to prove, there are strong indications that symbiotic systems can adapt to some adverse conditions which lower the rate of nitrogen fixation; this appears to be achieved through increasing the level of nitrogenase, increasing the amount of nodule tissue, or increased longevity of existing active tissue. The marked and rapid response to increased levels of atmospheric carbon dioxide, both in terms of nodule production and nitrogen fixation (Wilson, Fred & Salmon, 1933; Hardy & Havelka, 1973) also indicates the adaptability of the symbiotic system under changing conditions. Perhaps the most intriguing situation is that involving the effects of combined nitrogen which can either be inhibitory or stimulatory. The unravelling of these effects, involving both symbionts, the environmental conditions, and the form and level of fertiliser and soil nitrogen could pave the way for highly significant increases in symbiotic nitrogen fixation under field conditions.

References

Allos, H. F. & Bartholomew, W. V. (1955). Effect of available nitrogen on symbiotic fixation. *Proceedings of the Soil Science Society of America*, **19,** 182–4.

Bergersen, F. J. (1970). The quantitative relationship between nitrogen fixation and the acetylene-reduction assay. *Australian Journal of Biological Sciences*, **23,** 1015–25.

Bergersen, F. J., Hely, F. W. & Costin, A. B. (1963). Overwintering of clover nodules in alpine conditions. *Australian Journal of Biological Sciences*, **16,** 920–1.

Bowen, G. D. (1959). Field studies on nodulation and growth of *Centrosema pubescens* Benth. *Queensland Journal Agricultural Science*, **16,** 253–66.

Butler, G. W. & Bathurst, N. O. (1958). Free and bound amino acids in legume root nodules: bound γ-aminobutyric acid in the genus *Trifolium*. *Australian Journal of Biological Sciences*, **11,** 529–37.

Butler, G. W., Greenwood, R. M. & Soper, K. (1959). Effects of shading and defoliation on the turnover of root nodule tissue of plants of *Trifolium repens*, *Trifolium pratense*, and *Lotus uliginosus*. *New Zealand Journal of Agricultural Research*, **2,** 415–26.

Copeland, R. & Pate, J. S. (1970). Nitrogen metabolism of nodulated white clover in the presence and absence of nitrate nitrogen. In: *White clover research* (ed. J. Lowe), Occasional Symposium No. 6 of the British Grassland Society. British Grassland Society, Hurley, Berkshire.

Dart, P. J. & Day, J. M. (1971). Effects of incubation temperature and oxygen tension on nitrogenase activity of legume root nodules. *Plant and Soil*, Special Volume, 167–84.

Dart, P. J. & Mercer, F. V. (1965*a*). The effect of growth temperature, level of ammonium nitrate, and light intensity on the growth and nodulation of cowpea (*Vigna sinensis* Endl. ex Hassk.) *Australian Journal of Agriculture Research*, **16,** 321–45.

Dart, P. J. & Mercer, F. V. (1965*b*). The influence of ammonium nitrate on the fine structure of nodules of *Medicago tribuloides* Desr. and *Trifolium subterraneum* L. *Archiv für Mikrobiologie*, **51,** 233–57.

Dart, P. J. & Wildon, D. C. (1970). Nodulation and nitrogen fixation by *Vigna sinensis* and *Vicia atropurpurea*: the influence of concentration, form, and site of application of combined nitrogen. *Australian Journal of Agricultural Research*, **21,** 45–56.

Davidson, J. L., Gibson, A. H. & Birch, J. W. (1970). Effects of temperature and defoliation on growth and nitrogen fixation in subterranean clover. *Proceedings of the XI International Grassland Congress* (Surfer's Paradise Australia), pp. 542–5. University of Queensland Press.

Day, J. M. (1972). Studies on the role of light in legume symbiosis. PhD thesis, University of London.

Eaton, S. V. (1931). Effects of variation in daylength and clipping of plants on nodule development and growth of soy bean. *Botanical Gazette*, **91,** 113–43.

Engin, M. & Sprent, J. I. (1973). Effects of water stress on growth and nitrogen-fixing activity of *Trifolium repens. New Phytologist*, **72,** 117–26.

Freney, J. R. & Gibson, A. H. (1974). Non-protein nitrogen accumulation in legume nodules. In: *Mechanisms of regulation of plant growth* (ed. R. L. Bielski, A. R. Ferguson & M. M. Cresswell), pp. 37–40. Bulletin 12, The Royal Society of New Zealand, Wellington.

Gibson, A. H. (1965). Physical environment and symbiotic nitrogen fixation. II. Root temperature effects on the relative nitrogen assimilation rate. *Australian Journal of Biological Sciences*, **18,** 295–310.

Gibson, A. H. (1968). Nodulation failure in *Trifolium subterraneum* L. cv. Woogenellup (syn. Marrar). *Australian Journal of Agricultural Research*, **19,** 907–18.

Gibson, A. H. (1969*a*). Physical environment and symbiotic nitrogen fixation. VI. Nitrogen retention within the nodules of *Trifolium subterraneum* L. *Australian Journal of Biological Sciences*, **22,** 829–38.

Gibson, A. H. (1969*b*). Physical environment and symbiotic nitrogen fixation. VII. Effect of fluctuating root temperature on nitrogen fixation. *Australian Journal of Biological Sciences*, **22,** 839–46.

Gibson, A. H. (1971). Factors in the physical and biological environment affecting nodulation and nitrogen fixation by legumes. *Plant and Soil*, Special Volume, 139–52.

Gibson, A. H. (1974). Consideration of the growing legume as a symbiotic association. *Proceedings of the Indian National Science Academy*, **40B,** in press.

Gibson, A. H. & Nutman, P. S. (1960). Studies on the physiology of nodule formation. VII. A reappraisal of the effect of preplanting. *Annals of Botany*, **24,** 420–33.

Grobbelaar, N., Clarke, B. & Hough, M. C. (1971). The nodulation and nitrogen fixation of isolated roots of *Phaseolus vulgaris* L. II. The influence of light on nodulation. *Plant and Soil*, Special Volume, 203–14.

Hardy, R. W. F. & Havelka, U. D. (1973). Symbiotic N_2 fixation: multifold enhancement by CO_2-enrichment of field-grown soybeans. *Plant Physiology*, **51** (Supplement), 35.

Harper, J. E. & Cooper, R. L. (1971). Nodulation response of soybeans (*Glycine max* L. Merr.) to application rate and placement of combined nitrogen. *Crop Science*, **11,** 438–40.

Hoglund, J. H. (1973). Bimodal response by nodulated legumes to combined nitrogen. *Plant and Soil*, **39,** in press.

Hubbell, D. H. & Elkan, G. H. (1967). Host physiology as related to nodulation of soybean by rhizobia. *Phytochemistry*, **6,** 321–8.

Kamata, E. (1963). Morphological and physiological studies on nodule formation in leguminous crops. IX. Effects of the leaves clipping on nodule formation in Ladino clover. (*Trifolium repens* L.). *Proceedings of the Crop Science Society of Japan*, **31,** 245–8.

Lie, T. A. (1969). Non-photosynthetic effects of red and far red light on root nodule formation by leguminous plants. *Plant and Soil*, **30,** 391–404.

Mague, T. H. & Burris, R. H. (1972). Reduction of acetylene and nitrogen by field-grown soybeans. *New Phytologist*, **71,** 275–86.

Moustafa, E., Ball, R. & Field, T. R. O. (1969). The use of acetylene reduction to study the effect of nitrogen fertiliser and defoliation of nitrogen fixation by field-grown white clover. *New Zealand Journal of Agricultural Research*, **12,** 691–6.

Munns, D. N. (1968). Nodulation of *Medicago sativa* in solution culture. II. Compensating effects of nitrate and of prior nodulation. *Plant and Soil*, **28,** 246–57.

Oghoghorie, C. G. O. & Pate, J. S. (1971). The nitrate stress syndrome of the nodulated field pea (*Pisum arvense* L.) *Plant and Soil*, Special Volume, 185–202.

Orcutt, F. S. & Wilson, P. W. (1935). The effect of nitrate-nitrogen on the carbohydrate metabolism of inoculated soybeans. *Soil Science*, **39,** 289–96.

Pankhurst, C. E. & Gibson, A. H. (1973). *Rhizobium* strain influence on disruption of clover nodule development at high root temperature. *Journal of General Microbiology*, **74,** 219–31.

Pate, J. S. (1958). Nodulation studies in legumes. II. The influence of various environmental factors on symbiotic expression in the vetch (*Vicia sativa* L.) and other legumes. *Australian Journal of Biological Science*, **11,** 496–515.

Pate, J. S. & Dart, P. J. (1961). Nodulation studies in legumes. IV. The influence of inoculum strain and time of application of ammonium nitrate on symbiotic response. *Plant and Soil*, **15,** 329–46.

Richardson, D. A., Jordan, D. C. & Garrard, E. H. (1957). The influence of combined nitrogen on nodulation and nitrogen fixation by *Rhizobium meliloti* Dangeard. *Canadian Journal of Plant Science*, **37,** 205–14.

Rogers, V. E. (1969). Depression of nitrogen uptake and growth of lucerne at high soil temperatures. *Field Station Record, Division of Plant Industry, CSIRO* (*Australia*), **8,** 37–44.

Roughley, R. J. (1970). The influence of root temperature, *Rhizobium* strain and host selection on the structure and nitrogen-fixing efficiency of the root nodules of *Trifolium subterraneum. Annals of Botany*, **34,** 631–46.

Simpson, J. R. & Gibson, A. H. (1970). A comparison of the effectiveness of two strains of *Rhizobium trifolii* with *Trifolium subterraneum* in agar and three soils. *Soil Biology and Biochemistry*, **2,** 295–305.

Sironval, C., Bonnier, Ch. & Verlinden, J. P. (1957). Action of daylength on nodule formation and chlorophyll content in soybean. *Physiologia Plantarum*, **10,** 697–707.

Small, J. G. C. & Leonard, O. A. (1969). Translocation of C^{14}-labelled photosynthate in nodulated legumes as influenced by nitrate-nitrogen. *American Journal of Botany*, **56,** 187–94.

Sprent, J. I. (1972). The effects of water stress on nitrogen-fixing root nodules. II. Effects on the fine structure of detached soybean nodules. *New Phytologist*, **71,** 443–50.

Whiteman, P. C. (1970). Seasonal changes in growth and nodulation of perennial tropical pasture legumes in the field. II. Effects of controlled defoliation levels on nodulation of *Desmodium intortum* and *Phaseolus atropurpureus*. *Australian Journal of Agricultural Research*, **21**, 207–14.

Wilson, J. K. (1942). The loss of nodules from legume roots and its significance. *Journal of the American Society of Agronomy*, **34**, 460–71.

Wilson, J. R. (1970). Response to salinity in *Glycine*. VI. Some effects of a range of short-term salt stresses on the growth, nodulation, and nitrogen fixation of *Glycine wightii* (formerly *javanica*). *Australian Journal of Agricultural Research*, **21**, 571–82.

Wilson, P. W., Fred, E. B. & Salmon, M. R. (1933). Relation between carbon dioxide and elemental nitrogen assimilation in leguminous plants. *Soil Science*, **35**, 145–65.

30. Nitrogen fixation by legumes subjected to water and light stresses

JANET I. SPRENT

Light and soil moisture are two factors known to influence profoundly the growth of higher plants, yet the effect of these factors on the legume/*Rhizobium* symbiosis has received surprisingly little attention. The purpose of this article is to show that water stress, waterlogging and shading have adverse effects on nitrogen fixation in a wide range of legumes under laboratory and field conditions. No attempt is made to explain the mechanisms of these effects, the aim being to show how a consideration of these factors may help to improve yields in leguminous crops. The acetylene-reduction assay for nitrogen fixation will be used to make comparisons between treatments involving different degrees of environmental stress, but the data will not be converted into estimates of nitrogen fixed, although it has been shown both for water (Engin & Sprent, 1973) and light stress (i.e. shading, Sprent, 1973) that apart from small variations, such estimates are valid, when used on whole plants or on detached root systems. The acetylene-reduction assay has been modified to suit the species and type of stress under test. Where possible, whole detached root systems have been used, but in the clover pasture and the laboratory experiments with *Phaseolus vulgaris* nodulated root pieces and detached nodules were used respectively. These differences in method in no way invalidate comparisons made within experiments, but they preclude conversions of the acetylene-reduction assays into total nitrogen fixed.

Water stress

This is known to affect growth, yield and quality of various legume crops (e.g. Kilmer *et al.*, 1960; Bourget & Carson, 1962; Mériaux, 1972; Mack, 1973), but until recently, little information was available about its effects on nitrogen fixation. We have now examined the effects of water stress on nitrogen fixation in a range of species including herbage legumes such as *Trifolium repens* (Engin & Sprent, 1973) and grain legumes with both the spherical type of nodule as in *Glycine max* (Sprent, 1971*a*, 1972*b*) and the meristematic type of nodule as in *Vicia* *faba* (Sprent, 1972*b*) and *Lupinus arboreus* (Sprent, 1973). The pattern of response to stress is similar in all species, as shown in Fig. 30.1. Within the linear range of the curve, nodule activity is restored on watering, full activity being resumed in one or more hours (Sprent 1972*b*). Plants with meristematic nodules can recover from the more damaging kinds of stress by regrowth of existing nodules; this occurs

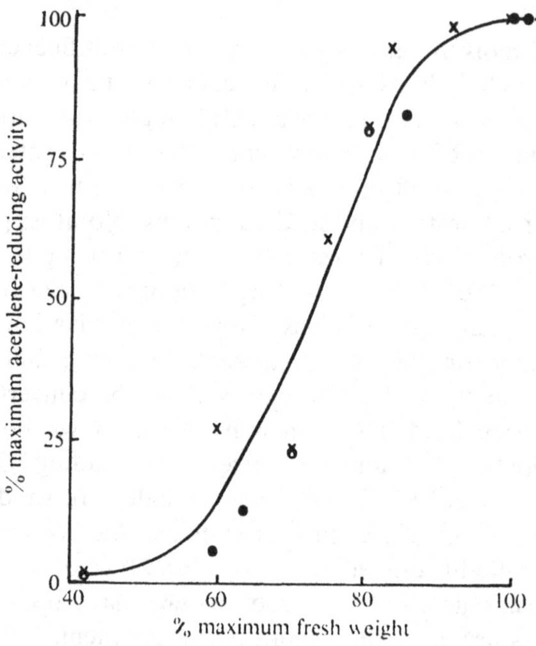

Fig. 30.1. Relationship between water content of nodules and maximum acetylene-reducing activity. Open and closed circles represent *Lupinus arboreus* (meristematic-type nodules) grown at high and low light intensity respectively, and the crosses soybean (spherical-type nodules).

two to three days after watering (Engin & Sprent, 1973). In plants with spherical nodules, severe stress causes nodule shedding (Wilson, 1931, 1942) and recovery is likely to be slower and to involve the formation of new nodules.

As well as reducing the activity of existing nodules, water stress affects the growth of young nodules and also formation of further nodules. Typical data are shown in Table 30.1. *Phaseolus vulgaris* cv. Glamis plants were germinated in sand at uniform water content until the first foliage leaves were expanding and then half were stressed by being given water only when at the point of wilting. They were grown on a 16 hour photoperiod and a day/night temperature of 22/12 °C.

Two strains of *Rhizobium* were used and these gave nodules with different specific acetylene-reducing activities. For both strains activity was reduced by about 90 % in stressed plants. Nodule number and size were also depressed by stress. We do not yet know whether the reduction in nodule number resulted from fewer rhizobia being available for infection (Foulds, 1971) or from effects on the infection process, or both.

Table 30.1 *Effect of water stress on nodule number, size and acetylene-reducing activity of* P. vulgaris *plants (44 days old) inoculated with two strains of* R. phaseoli

	Rhizobium strain			
	3601		3605	
	Control	Stressed	Control	Stressed
pmoles C_2H_4/mg/min	16.45	1.75	37.45	3.15
Nodule number	28.4	8.3	18.50	4.5
Average nodule wt (mg)	1.44	0.95	1.72	1.16
Water content of sand (%DW)	6.54	0.71	6.34	0.81

Under natural conditions, water stress usually occurs gradually and progressively as moisture is lost from the soil. Under such conditions, wilting in the lower leaves is generally a good indication that nodules are functioning at suboptimal rates (Sprent, 1972*b*), but it has been shown that, provided the soil remains moist, shoot systems of *Glycine max* and *Phaseolus vulgaris* under simulated wind conditions wilt, without there being a noticeable effect on acetylene reduction in the nodules (Sprent & Gallacher, unpublished observations). This is consistent with the observation that nodules are able to obtain tritiated water *en route* from roots to leaves (Sprent, 1972*b*). Stress in nodules occurs when the root system cannot supply sufficient water both to export materials from nodules in the xylem sap (see Pate p. 344 for discussion of this point) and also to compensate for losses from the nodule surface in drying soil. Because of the importance of water loss from nodule surfaces under stress conditions, we have examined their surface architecture under the scanning electron microscope (SEM). Since normal methods involving freeze-drying as part of the preparative procedure are clearly inappropriate, a replica method was used. Fresh nodules (stressed or otherwise) are pressed into viscous silicone rubber, previously mixed with a rapid hardener. A skin forms in contact with the nodule surface and the rubber is left to harden overnight, by which time nodules have shrunk and are readily removed. The negative image of the nodule surface is filled with araldite resin, left to harden and then coated with a 40 nm layer of gold–palladium for viewing in the SEM. Plate 30.1*a* shows a control (unstressed) soybean nodule slice and indicates how the replica method can also be used for cut surfaces. Note the highly localised activity of the cork cambium (Bieberdorf, 1938). Cells developed from this cambium form spherical to elongate outgrowths on the nodule surface whereas the areas between these lenticel-like ridges are composed of closely packed parenchyma-type cells (Plate 30.1*b* and *c*). Intercellular spaces amongst the ridge cells could be continuous with the internal intercellular space system, although the cork cambium would exert a barrier to strictly radial movement, but there is little sign of spaces between cells of the inter-ridge areas. Bergersen & Goodchild (1973) reported continuous air channels from centre to exterior of soybean root nodules that may be the major route for gaseous exchange. However, gaseous exchange through the cell surfaces could also be important, at least for the nodule cortical cells. These cells are known to be affected structurally by water stress and in some cases the cellulose walls become striated and more fibrous (Sprent, 1972*a*; M. Engin, unpublished observations). The surface features of water stressed nodules are shown in Plate 30.2. Cells of both ridge and inter-ridge areas are more shrunken than in unstressed nodules, with the ridge cells showing signs of collapse. Whether these effects of desiccation alter permeability to gases is not known.

When nitrogen-fixing activity is reduced by water stress, there is a concomitant effect on respiration, as measured by oxygen uptake (Sprent, 1971*a*). This, coupled with effects on nodule fine-structure (Sprent, 1972*a*) and reports on effects of water stress on fine-structure and mitochondrial activity of maize roots (Nir, Klein & Poljakoff-Mayber, 1969, 1970; Nir, Poljakoff-Mayber & Klein, 1970) led us to examine the histochemistry of respiratory enzymes in stressed nodules (I. Marks, unpublished data). Some enzymes, such as succinic dehydrogenase are detectable in severely stressed nodules, long after nitrogen-fixing activity has been lost, whilst others, particularly cytochrome oxidase in bacteroids, appear more sensitive to stress. Detailed enzyme localisation may be made by electron microscopy after treatment with dilute fixative (Seligman *et al.*, 1968). Cytochrome oxidase is then detectable in control bacteroids but not in those from stressed nodules. This may reflect altered permeability in stressed bacteroids rather than a direct effect on enzyme activity. In either case it is the first demonstration that water stress directly affects bacteroids.

Waterlogging

Sprent (1969, 1971*b*), Schwinghamer, Evans & Dawson (1970) and Mague & Burris (1972) reported that waterlogging depresses nitrogen fixation, largely as a result of oxygen deficiency. Nodules of *P. vulgaris* and *G. max* formed under wet or waterlogged conditions show enhanced phellogen activity which produces greatly pronounced ridging and increased surface area (Plate 30.1*d*, *e* and *f*). Nodule number, size and water content are also affected by waterlogging. Table 30.2 shows

Table 30.2 *Effect of waterlogging on nodule number, size, water content and acetylene-reducing activity of* P. vulgaris *cv. Glamis. Values averaged over five strains of* R. phaseoli

Treatment	Age (days)	Nodule no./plant	Average nodule size (mg)	Water content (fresh wt: dry wt)	pmoles C_2H_4/mg fresh wt/min
Waterlogged	29	15.1	3.63	2.72	4.10
Waterlogged	37	44.2	3.65	7.46	6.09
Waterlogged	43	94.5	4.49	9.67	6.30
Control	29	53.5	3.76	2.54	11.73
Control	37	58.4	5.36	5.42	15.47
Control	43	77.5	6.38	6.74	15.44

typical data for *P. vulgaris*. Plants were grown under the same conditions as in the water stress experiment (Table 30.1) except that the water level in waterlogged pots was kept at the top of the sand. The smaller size of waterlogged nodules, coupled with enhanced ridge development give them a more favourable surface:volume ratio, but there is no evidence that this equips them to fix nitrogen any more efficiently under waterlogged conditions. In Table 30.3 acetylene reduction data for *P. vulgaris* grown with five different strains of *Rhizobium* are given. In all cases the nodules formed under water were less active than control nodules when incubated in air and were inferior to control nodules in their activity under water, except where the peat inoculant was used. The enhanced meristematic activity in waterlogged nodules may be caused by reduced oxygen tension in a manner akin

Table 30.3. *Rates of acetylene reduction (pmoles/mg fresh wt/min) in air and under water by* P. vulgaris *nodules formed in waterlogged or drained sand*

	Rhizobium strain				
	Peat inoculant	453b	458a	3601	3605
Waterlogged	0.77	1.61	1.58	1.33	1.40
Control	2.35	2.98	3.08	5.43	6.13
			Ratio $\frac{\text{rate in water}}{\text{rate in air}} \times 100$		
Waterlogged	15.2	23.5	16.5	33.6	22.9
Control	14.6	38.7	39.8	72.4	35.1

Gas phase for incubation under water: 90 % O_2+10 % C_2H_4. Forty-three day-old plants.

to that observed by Kessel & Carr (1972) for carrot callus tissue. The data in Table 30.3 show some variation amongst nodules formed with different strains of rhizobia in their ability to withstand waterlogging, but this needs confirmation.

Shading

This can occur in a number of ways. For example, in New Zealand *Lupinus arboreus* is commonly grown on sand dunes with *Pinus radiata.* The shading by the tree species is a major factor affecting the nodulation and nitrogen fixation of the lupin and its survival under the forest canopy (Sprent & Silvester, 1973). Fig. 30.2 shows a linear relationship between the logarithm of relative irradiance and acetylene-reducing activity per plant (Sprent, 1973). Shading principally reduced nodule number and size, and only the deepest shade affected nodule activity per unit weight of tissue.

Most plants shade themselves as they grow in size. R. J. Lawn & W. A. Brun (personal communication) have found that self-shading in soybean depresses nitrogen-fixing activity in older plants. Mutual shading by closely spaced plants will be discussed below under field trials. Self- and mutual-shading will vary with the morphology of the species or cultivar under test and all plants will not be expected to show similar responses to variations in light intensity. Data obtained with *V. faba* and *P. vulgaris* under identical controlled-environment facilities illustrate this point. Light intensity (4500 or 7000 lux) was

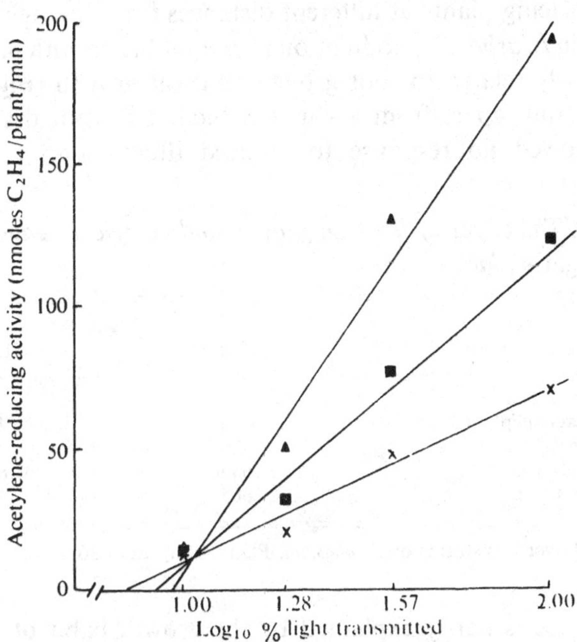

Fig. 30.2. Effect of shading on acetylene-reducing activity of plants of *Lupinus arboreus* at 4(×), 8(■) and 10(▲) weeks after beginning of treatment. Linear regression lines have been fitted through the data for each experiment.

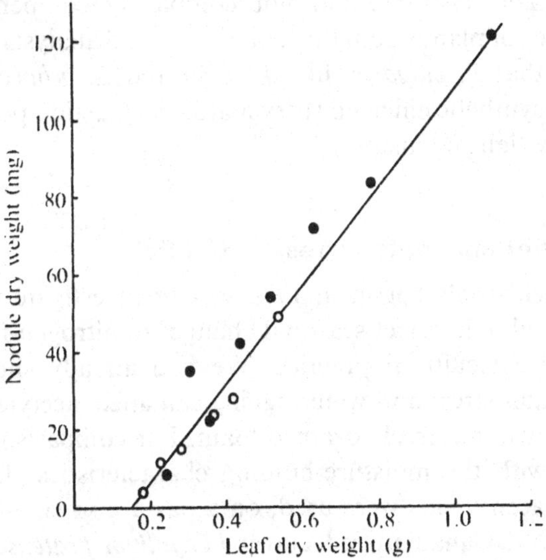

Fig. 30.3. Linear regression of nodule dry weight on leaf dry weight in *Vicia faba* cv. Herz Freya grown under 4500 (○) or 7000 (●) lux.

varied by placing plants at different distances from the light source. In *V. faba* (as in *L. arboreus*) nodulation is in equilibrium with plant growth, and is linearly related to root growth and leaf growth (Fig. 30.3). On the other hand, apart from a slight reduction in root dry weight, *P. vulgaris* showed no response to reduced illumination (Table 30.4).

Table 30.4 *The effect of light on growth and acetylene-reducing activity of* P. vulgaris *plants.*

	Light	
	High	Low
pmoles C_2H_4/mg/min	60.0	67.9
Nodule fresh weight/plant (mg)	486	415
Nodule number/plant	79	86
Shoot dry weight (mg)	659	625
Root dry weight (mg)	665	451

Data averaged over five strains of *R. phaseoli*. Plants sampled at 36 days.

This difference is partly explained by the growth habit of *P. vulgaris*. At least in the earlier stages of growth, low light intensity causes greater petiole elongation and raises the laminas so that all leaves become placed at about the same level nearer the light source and there is no self-shading. This response did not completely compensate for the greater distance of plants from the light source and analysis of dry weight figures show that *P. vulgaris*, like *V. faba* and *L. arboreus* shows increased photosynthetic efficiency (dry matter production per unit of light energy) at low light intensity.

Studies of light and water stress in the field

The aim of field trials begun in 1973 was to investigate the extent to which water and light affect seasonal changes in nitrogen-fixing activity under normal agricultural practice. We had already shown (Sprent, 1972*b*) that water stress and waterlogging can affect acetylene reduction by *V. faba* grown in a freely drained loam. For comparison, a clay soil was selected with the moisture-holding characteristics shown in Fig. 30.4. Two adjacent fields were used, on a pasture sown with a mixture containing 5.7 % Canadian red clover (*Trifolium pratense*) and 4.3 % New Zealand white clover (*T. repens*) and the other a field of peas (*Pisum sativum* cv. Victory Freezer). Acetylene-reducing activity, soil

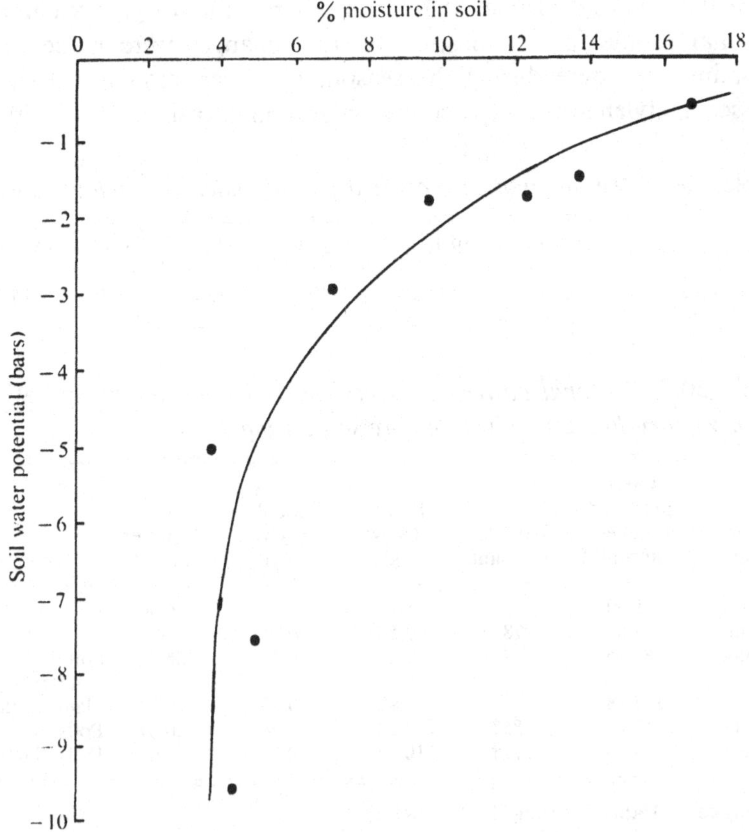

Fig. 30.4. Relationship between soil water potential and percentage moisture for the soil in which clover (Tables 30.5 and 30.6) and peas (Table 30.7) were grown.

moisture content and plant height (clover) or weight (pea) were measured every two weeks.

During March–April 1973 the weather was very dry and rates of acetylene reduction varied markedly across the pasture field with water content of the soil. Subsequently two plots were watered and one week later the rates of acetylene reduction in these plots had increased

Table 30.5. *Effect of irrigation on acetylene-reducing activity of nodules from grass/clover sward*

	Watered	Control
pmoles C_2H_4/mg/min	32.3	2.2
% water in soil	19.2	16.0
Soil moisture tension (bars)	0	−1.5

more than tenfold (Table 30.5). It is not possible to give a complete seasonal profile for the pasture because changes were made in the sampling procedure during the season. However, moisture stress depressed acetylene-reducing activity on several occasions. Table 30.6 of

Table 30.6. *Average moisture content of soil under grass/clover sward.*

Date	22 May	19 June	3 July	17 July	30 July	14 August
% moisture	12.16	22.85	9.98	16.20	13.76	15.60

Table 30.7. *Seasonal pattern of acetylene reduction in* Pisum sativum *cv. Victory Freezer grown on clay soil during 1973*

Sample Date	C_2H_4 produced (nmoles/ plant/min)	Nodule no./plant	Shoot dry wt (g)	Root+ nodule dry wt (g)	% water in soil	Comments
22 May	1.60	–	–	–	16.56	
5 June	54.04	78	0.50	0.11	–	
19 June	85.15	74	2.01	0.17	26.42	Soil partly waterlogged
3 July	147.48	77	4.92	0.23	16.72	Plants in flower
17 July	118.64	85*	8.22	0.24	20.67	Pods setting
30 July	69.96	111†	10.65	0.26	21.04	Pods filling

Sowing date 14 April, freezing (11–12 August).
* Includes some and † includes many green nodules.

the percentage water in pasture soil suggests that the soil was under moisture tension for a large part of the season and contained less water at each sampling than the pea field (Table 30.7). This was apparently due to poor moisture penetration in the pasture when plant height was above about 25 cm; rain frequently evaporated from the foliage before reaching the soil. The dense pasture would also have used more water than the more widely spaced pea crop. The pasture was cut twice during the season. Each cut was followed by reduced nodule activity, nodule senescence and regrowth four to five days later, a result resembling closely the data of Moustafa, Ball & Field (1969) for white clover in New Zealand.

Table 30.7 shows the seasonal pattern of activity for peas. Water shortage delayed seed germination, but no stress symptoms were observed during growth, and from the soil moisture data, none would have been expected. Thus we have a clear difference between the pea field and the neighbouring pasture. On 19 June the soil was partly waterlogged and nodule activity was lower than would have been expected, assuming nodule activity is related to plant growth. Acetylene-reducing activity per plant fell after flowering. Pate (1958*a*, *b*), using total nitrogen analysis found nitrogen fixation in *P. arvense* fell at flowering and in *Vicia sativa* at pod-fill. Sprent & Silvester (1973) found acetylene-reducing activity for *L. arboreus* fell at pod-fill, whereas for soybean and peanut (*Arachis hypogea*) activity was retained until senescence (Hardy *et al.*, 1971).

V. faba cv. Herz Freya (a type of field bean) was grown on soil similar to that of the earlier field trials (Sprent, 1972*b*). Two planting densities, 7 and 200 plants/m², were chosen to define upper and lower limits to the effects of shading, competition, nutrient and water utilisation for subsequent field trials. Seeds were sown on 23 March and plants were sampled every two weeks, commencing 1 May. Table 30.8 shows that

Table 30.8. *Effect of planting density on shoot and root dry weights (g) and nodule number of* Vicia faba *cv. Herz Freya*

	Plants (m^{-2})					
	200		7		200	7
Sample date	Shoot DW	Root DW	Shoot DW	Root DW	Nodule no.	
1 May	0.26	0.13	0.31	0.14	–	–
15 May	0.71	0.31	0.62	0.31	31	30
30 May	1.54	0.36	1.72	0.44	36	43
12 June	3.23	0.69	5.93	1.33	65	132
26 June	6.06	1.24	20.21	3.72	85	139
10 July	8.52	1.30	54.30	7.12	77	206
24 July	11.14	1.22	72.24	6.74	101	175
7 August	14.26	1.14	111.85	8.10	100	175
21 August	–	–	–	–	68	65

the effects of close planting on the vegetative growth of plants began to appear by 30 May. Subsequently the growth rate of the closely spaced plants fell increasingly behind that of the widely spaced plants. After 7 August storms caused the plants to become lodged and readings were discontinued after 21 August. Both groups of plants began to flower at about 12 June, pods were full by 7 August and then began to ripen. On 6 July the light intensity 25 cm above soil level in the close spacing was only 8 % of that in the wide spacing.

The acetylene-reducing activity per plant is given in Fig. 30.5, together with soil moisture data, given as % dry weight to include waterlogging as well as water deficits. For the widely spaced plants the seasonal pattern follows expectation with a maximum at about full flowering; the discontinuities in the curve at 30 May and 24 July were

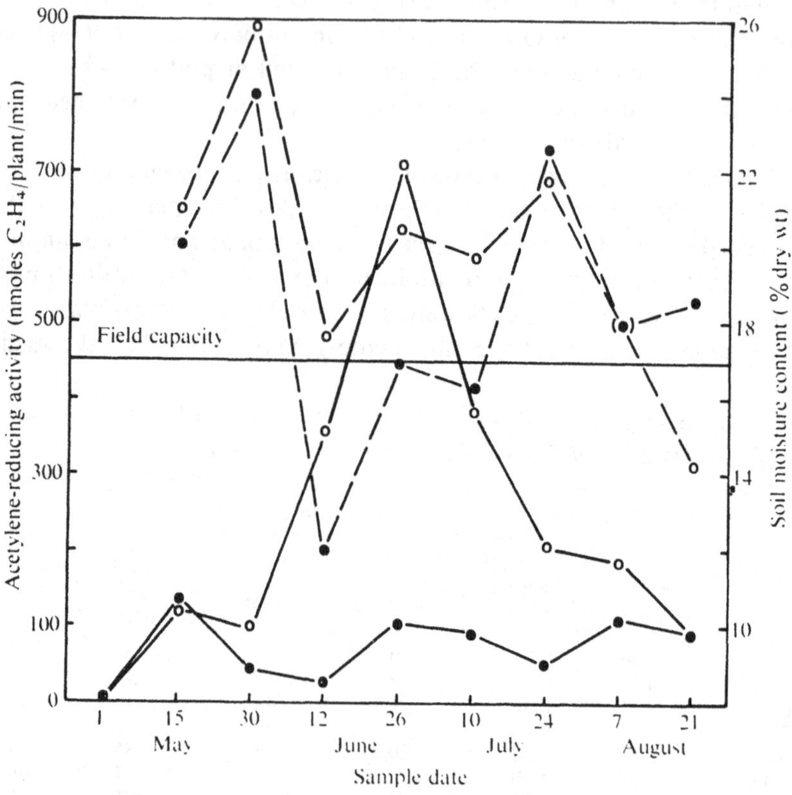

Fig. 30.5. Seasonal pattern of acetylene reducing activity of field beans (*V. faba* cv. Herz Freya) during 1973. Plants sown at two densities, 200 (●) and 7 (○) plants m^{-2}. Continuous lines for acetylene reduction, broken lines for soil water content.

due to waterlogging. Even at maturity, these plants had not formed a closed canopy and light penetration at all levels was good around mid-day. The seasonal pattern for the dense planting was quite different; although the plants flowered and set seed at the same time as those of sparse planting, the well-defined maximum was obscured and activity varied about a mean of 100 nmoles per plant per min throughout the season. The variation shows a close relationship with soil moisture content, which was generally lower in the dense than in the sparse planting. On 30 May and 24 July waterlogging and on 12 June water deficit were sufficient to depress nitrogen-fixing activity. If acetylene reduction per unit area of soil, rather than per plant is considered (Fig. 30.6) the dense planting gave much higher values than the sparse planting. Since the activity per plant does not decline after flowering in the dense planting, a high rate of activity per unit area is maintained for a longer period during the season than in the sparse planting. Thus,

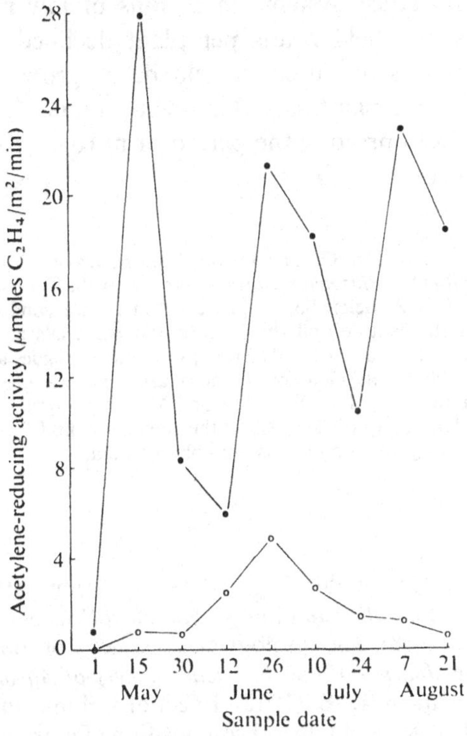

Fig. 30.6. Effect of planting density on acetylene reduction per unit area of soil. 200 (●) and 7 (○) plants m^{-2}.

although closely spaced plants clearly compete for light and water and probably other factors not measured here, for some crops the advantages of increased overall nitrogen fixation might be more important. We hope to extend these results to include yield analyses at different spacings, so that growth conditions which optimise utilisation of symbiotically fixed nitrogen for grain production can be found.

Conclusions

The above data show that water stress can seriously reduce nitrogen fixation by depressing the activity of existing nodules and by reducing nodulation. Waterlogging also reduces nodule formation and activity.

Shading reduces nodulation in relation to plant growth in *Lupinus arboreus* and in *Vicia faba*. Young plants of *Phaseolus vulgaris* were more tolerant to shading, probably because of their shoot morphology.

Nitrogen fixation in the field is closely related to soil moisture content in both clay and more freely drained soils. Moisture penetration was closely related to plant density, so that crowded plants are more likely to show moisture stress systems in regions of low rainfall. Nitrogen fixation in peas and field beans per plant declined when pods were developing. This was obscured in field beans grown at high density, when the activity per plant was also greatly reduced. Planting density could be adjusted to prolong the period of nitrogen fixation and maximise crop production.

The technical assistance of Mrs C. Greenwood, Miss E. Davidson and Mr S. Sinclair is acknowledged with thanks. *Rhizobium* cultures were from the Rothamsted collection and soybean seed from USDA Beltsville. Field bean trials were conducted at the Scottish Horticultural Research Institute and the pasture and pea trials at Mid Dod Farm, by Forfar, Angus. Thanks are due to Mr R. Thompson and colleagues and to Mr J. Kelman for their help in these field trials. Generous financial assistance was given by the Agricultural Research Council. I am grateful to Dr D. Idle and Miss P. Merrick of the Botany Department, Birmingham University for suggesting the replica method for SEM studies and to Dr I. Marks for allowing me to quote his unpublished data.

References

Bergersen, F. J. & Goodchild, D. J. (1973). Aeration pathways in soybean root nodules. *Australian Journal of Biological Sciences*, **26**, 729–40.

Bieberdorf, F. W. (1938). The cytology and histology of root nodules of some Leguminosae. *Journal of the American Society of Agronomy*, **30**, 375–89.

Bourget, S. J. & Carson, R. B. (1962). Effect of soil moisture stress on yield, water-use efficiency and mineral composition of oats and alfalfa grown at two fertility levels. *Canadian Journal of Soil Sciences*, **42**, 7–12.

Engin, M. & Sprent, J. I. (1973). Effects of water on growth and nitrogen-fixing activity of *Trifolium repens*. *New Phytologist*, **72**, 117–26.

Foulds, W. (1971). Effect of drought on three species of *Rhizobium*. *Plant and Soil*, **35**, 665–7.

Hardy, R. W. F., Burns, R. C., Herbert, R. R., Holsten, R. D. & Jackson, E. K. (1971). Biological nitrogen fixation: a key to world protein. *Plant and Soil*, Special Volume, 561–90.

Kessel, R. H. J. & Carr, A. H. (1972). The effect of dissolved oxygen concentration on growth and differentiation of carrot (*Daucus carota*) tissue. *Journal of Experimental Botany*, **23**, 996–1007.

Kilmer, V. J., Bennett, O. B. L., Stahly, V. F. & Timmons, D. R. (1960). Yield and mineral composition of eight forage species grown at four levels of soil moisture. *Agronomy Journal*, **52**, 282–5.

Mack, A. R. (1973). Soil temperature and moisture conditions in relation to the growth and quality of field peas. *Canadian Journal of Soil Science*, **53**, 59–72.

Mague, T. H. & Burris, R. H. (1972). Reduction of acetylene and nitrogen by field-grown soybeans. *New Phytologist*, **71**, 275–86.

Mériaux, S. (1972). Influence de la sécheresse sur la croissance, le rendement et la composition de la féverole. *Annals of Agronomy*, **23**, 533–46.

Moustafa, E., Ball, R. & Field, T. R. O. (1969). The use of acetylene reduction to study the effect of nitrogen fertilizer and defoliation on nitrogen fixation by field-grown white clover. *New Zealand Journal of Agricultural Research*, **12**, 691–6.

Nir, I., Klein, S. & Poljakoff-Mayber, A. (1969). Effect of moisture stress on submicroscopic structure of maize roots. *Australian Journal of Biological Sciences*, **22**, 17–33.

Nir, I., Klein, S. & Poljakoff-Mayber, A. (1970). Changes in fine structure of root cells from maize seedlings exposed to water stress. *Australian Journal of Biological Sciences*, **23**, 489–91.

Nir, I., Poljakoff-Mayber, A. & Klein, S. (1970). The effect of water stress on mitochondria of root cells: a biochemical and cytochemical study. *Plant Physiology*, Lancaster, **45**, 173–7.

Pate, J. S. (1958*a*). Nodulation studies in legumes. I. The synchronization of host and symbiotic development in the field pea *Pisum arvense* L. *Australian Journal of Biological Sciences*, **11**, 366–81.

Pate, J. S. (1958*b*). Nodulation studies in legumes. II. The influence of various environmental factors on symbiotic expression in the vetch (*Vicia sativa* L.) and other legumes. *Australian Journal of Biological Sciences*, **11**, 496–515.

Schwinghamer, E. A., Evans, H. J. & Dawson, M. D. (1970). Evaluation of effectiveness in mutant strains of *Rhizobium* by acetylene reduction relative to other criteria of N_2 fixation. *Plant and Soil*, **33**, 192–212.

Seligman, A. M., Karnorsky, M. J., Wasserkrug, H. L. & Hanker, J. S. (1968). Non-droplet ultrastructural demonstration of cytochrome oxidase activity with a polymerising osmiophilic reagent, diaminobenzidine (DAB). *Journal of Cell Biology*, **38**, 1–14.

Sprent, J. I. (1969). Prolonged reduction of acetylene by detached soybean nodules. *Planta, Berlin*, **88**, 372–5.

Sprent, J. I. (1971*a*). The effects of water stress on nitrogen fixing root nodules. I. Effects on the physiology of detached soybean nodules. *New Phytologist*, **70**, 9–17.

Sprent, J. I. (1971*b*). Effects of water stress on nitrogen fixation in root nodules. *Plant and Soil*, Special Volume, 225–8.

Sprent, J. I. (1972*a*). The effects of water stress on nitrogen fixing root nodules. II. Effects on the fine structure of detached soybean nodules. *New Phytologist*, **71**, 443–50.

Sprent, J. I. (1972*b*). The effects of water stress on nitrogen fixing root nodules. IV. Effects on whole plants of *Vicia faba* and *Glycine max*. *New Phytologist*, **71**, 603–11.

Sprent, J. I. (1973). Growth and nitrogen fixation in *Lupinus arboreus* as affected by shading and water supply. *New Phytologist*, **72,** 1005–22.

Sprent, J. I. & Silvester, W. B. (1973). Nitrogen fixation by *Lupinus arboreus* grown in the open and under different aged stands of *Pinus radiata*. *New Phytologist*, **72,** 991–1004.

Wilson, J. K. (1931). The shedding of nodules by beans. *Journal of the American Society of Agronomy*, **23,** 670–4.

Wilson, J. K. (1942). The loss of nodules from legume roots and its significance. *Journal of the American Society of Agronomy*, **34,** 460–71.

Note added in proof

Since this chapter went to press, we have shown more conclusively (Pankhurst & Sprent, 1975*a*) that gaseous diffusion into soybean nodules occurs at the lenticels and that these regions collapse under water stress. Evidence that oxygen diffusion into stressed nodules is restricted has also been obtained (Pankhurst & Sprent, 1975*b*) and nodules thus become partially anaerobic. This is reflected in an increased production of ethanol (Sprent & Gallacher, 1975).

References

Pankhurst, C. E. & Sprent, J. I. (1975*a*). Surface features of soybean root nodules. *Protoplasma*, in press.

Pankhurst, C. E. & Sprent, J. I. (1975*b*). Effects of water stress on the respiratory and nitrogen-fixing activity of soybean root nodules. *Journal of Experimental Botany*, **26,** 287–304.

Sprent, J. I. & Gallacher, A. (1975). Effects of water stress on respiratory pathways in nitrogen-fixing root nodules. *Abstracts of XII International Botanical Congress* (*Leningrad*), vol. II, p. 476.

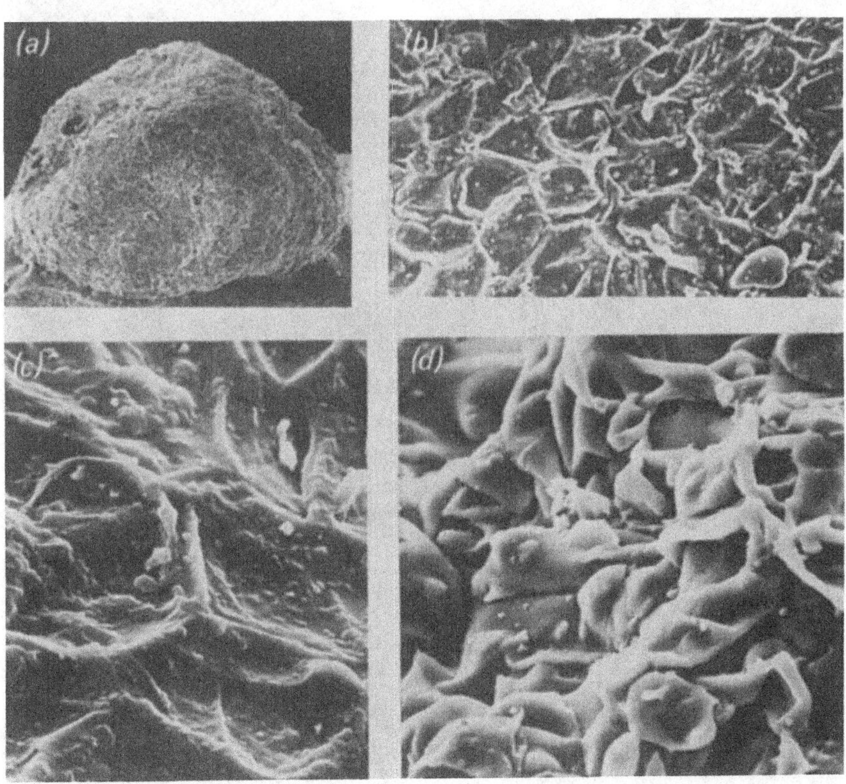

Plate 30.2. Scanning electron micrographs of surface of replicas of water-stressed soybean root nodules. (*a*) Whole nodule ×12. (*b*) Surface parenchyma cells ×115. (*c*) Surface parenchyma cells ×535. (*d*) Cells of nodule 'ridge', partially collapsed ×380.

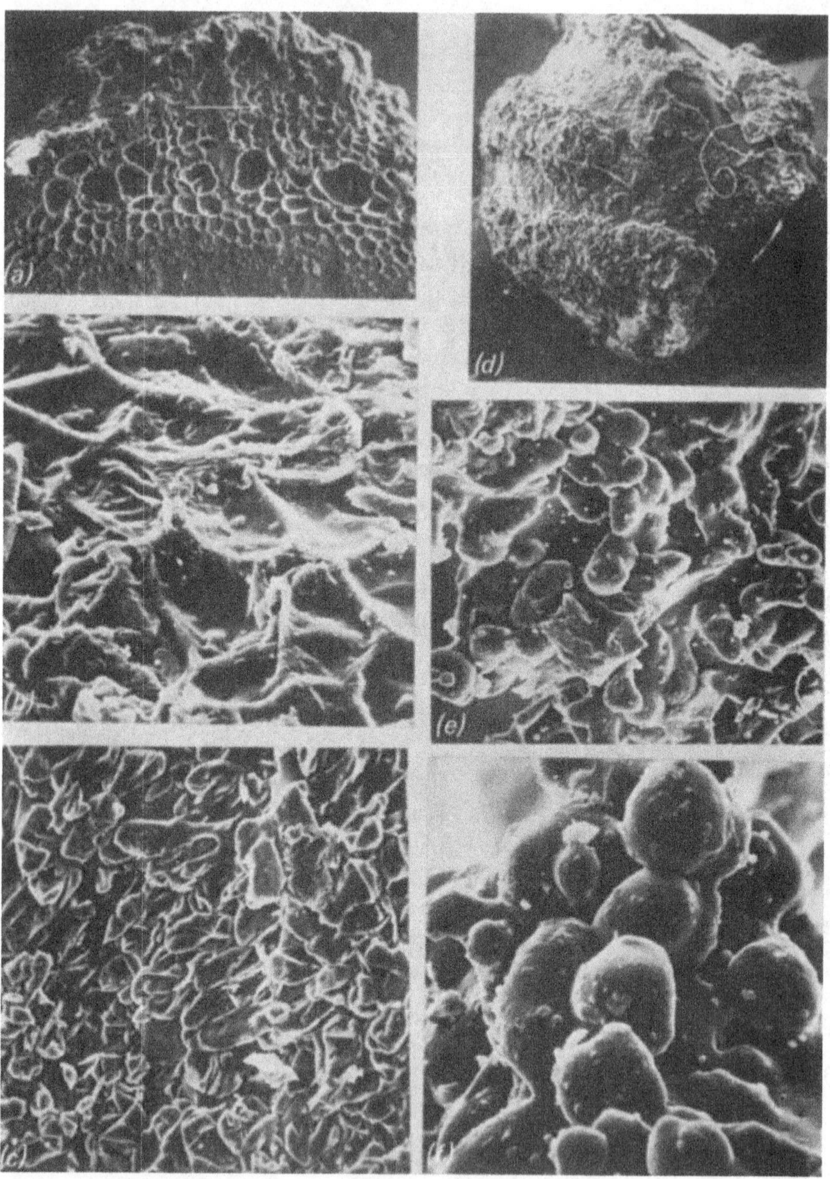

Plate 30.1. Scanning electron micrographs of replicas of soybean root nodules. (*a*), (*b*) and (*c*) controls; (*d*), (*e*) and (*f*) waterlogged nodules. (*a*) Cut surface ×140. Arrow shows highly localised activity of cork cambium. (*b*) Surface parenchyma ×275. (*c*) 'Ridge' ×110. (*d*) Whole waterlogged nodule. Note well-developed ridges ×10. (*e*) Ridge ×135. (*f*) Detail of ridge showing proliferation of cells ×325.

(*Facing p.* 420.)

31. Photosynthate as a major factor limiting nitrogen fixation by field-grown legumes with emphasis on soybeans

R. W. F. HARDY & U. D. HAVELKA

Leguminous plants contain higher concentrations of protein and consequently nitrogen than other crops. For example, each kilogram of soybean seed contains 60 to 70 g of nitrogen and the associated vegetative growth an additional 30 to 35 g, so that 540 to 630 kg nitrogen are required for a 6000 kg/ha yield. These nitrogen inputs, greater than for any other crop, may be provided by soil/fertilizer nitrogen and/or symbiotically fixed nitrogen.

Since 1967, we have developed and used the acetylene-reduction assay to make extensive measurements of nitrogen fixation by field-grown annual legumes, especially soybeans. An objective has been to establish a key to increased nitrogen input by the elucidation of factor(s) limiting for nitrogen fixation under field conditions. This report summarizes our results and those from other laboratories which lead us to propose photosynthate as a major limiting factor. Relevant information on non-legume symbionts is also included.

The limitation in photosynthate was overcome by carbon dioxide enrichment and thereby achieved the first major increase in nitrogen input into a field-grown annual legume; viz. a measured fivefold increase in nitrogen fixation and a calculated threefold decrease in use of soil nitrogen.

Results

Characteristics of nitrogen fixation

Typical profiles of nitrogen fixation by field-grown soybeans and peanuts shown in Fig. 31.1 (Hardy *et al.*, 1971) are provided as background. Nitrogen-fixing activity was measured biweekly throughout the growth cycle using the acetylene-reduction assay and expressed as N_2 fixed/plant/day and as total N_2 fixed/plant (Hardy, Burns & Holsten, 1973;

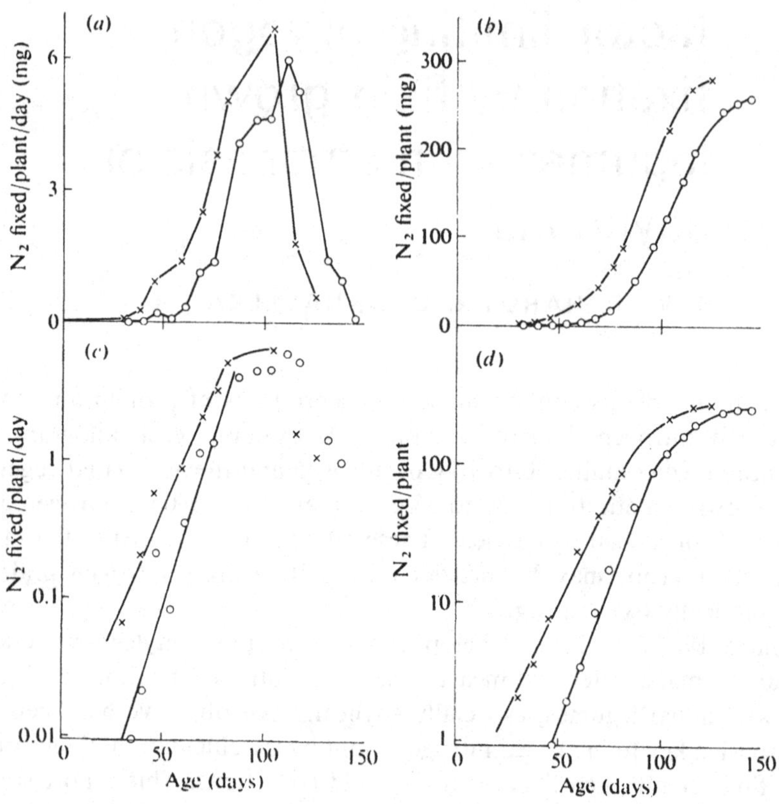

Fig. 31.1. Profiles of nitrogen fixation by field-grown soybeans (×) and peanuts (○). (*a*) and (*c*) N_2 fixed per plant per day, (*b*) and (*d*) N_2 fixed per plant, on linear and semi-logarithmic scales. After Hardy *et al* (1971).

Hardy & Holsten, 1975). The profiles are similar for both legumes, with more than 90 % of the nitrogen fixed during the last half of the growth cycle, the period of reproductive growth. Semi-logarithmic plots clearly demonstrate an exponential phase for the development of

Table 31.1. *Nitrogen fixation characteristics of field-grown legumes*

Legume	Exponential phase: Initiation age (days)	Exponential phase: Termination age (days)	Exponential phase: Increase (%/day)	N_2 fixed: mg N/ plant	N_2 fixed: kg N/ ha	% total nitrogen from N_2
Soybeans	26	90	8.3	260	84	25
Peanuts	42	97	8.5	221	–	25
Peas*	–	–	–	247	–	23

* T. LaRue (personal communication).

nitrogen-fixing activity. From the profiles are calculated: (1) age at initiation of exponential phase, (2) age at termination of exponential phase, (3) rate of increase during exponential phase and (4) total nitrogen fixed per season. Typical values are recorded in Table 31.1. Earlier initiation, delayed termination, and greater rate of increase could all increase total nitrogen fixed. For example, a delay of about nine days in the termination of the exponential phase or alternatively a similar advance in initiation could double the total nitrogen fixed. A part of the increase in nitrogen fixation by carbon dioxide enrichment is the product of both an extended exponential phase and an earlier apparent initiation produced by the increased nodule specific activity.

Proportion of nitrogen input from the atmosphere

We have determined the total nitrogen in the mature field-grown legume that is provided from N_2 for soybeans and peanuts while others have determined it for peas (T. LaRue, personal communication). Our results were obtained under normal field production conditions on non-nitrogen-deficient soils; for soybeans similar results were obtained for seven years in three different locations of the mid-Atlantic coast area of the United States. The average proportion from atmospheric nitrogen in all examples was about 25 % (Table 31.1), establishing soil nitrogen as the primary source. Several investigators have expressed surprise at the small size of the contribution by nitrogen fixation and completeness of nodule recovery has been questioned. In our experiments this was facilitated by irrigation of plots two to three days prior to collection and analysis. Our results are supported by other laboratories which found values for soybean from 10 % (Harper, 1973) to 35 % (Sloger *et al.*, 1975). On nitrogen-deficient soils the contribution from N_2 would be greater but this does not apply to most soils used for soybean production in the USA.

Possible factors limiting nitrogen fixation

Primary, secondary and tertiary factors controlling symbiotic nitrogen fixation are diagrammed in Fig. 31.2. The N_2 fixation reaction is catalyzed by the enzyme nitrogenase which couples ATP hydrolysis to ADP and P_i with electron transfer from a reduced electron donor, probably a ferredoxin or a flavodoxin, to reduce N_2 to $2NH_3$ (e.g. Hardy & Burns, 1973). Primary factors that could limit this reaction are concentration of nitrogenase, degree of saturation of nitrogenase by the substrates of the reaction–ATP, reduced electron donor and N_2 – and concentration of the product, ammonia. Evidence in the next section suggests that the concentration of nitrogenase is not of itself limiting and the low K_m of 0.02 to 0.05 atm for N_2 eliminates N_2 partial pressure as a practical limitation in the field. The large ATP requirement for in-vitro nitrogenase activity – most measurements indicate about 12 molecules of ATP per molecule N_2 fixed (e.g. Dixon, 1975) – suggests

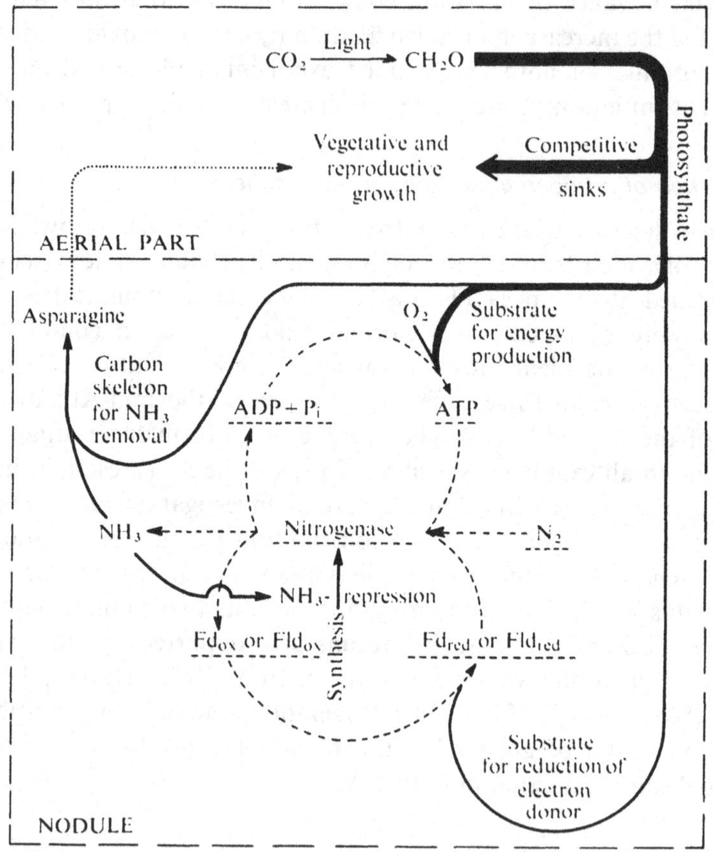

Fig. 31.2. Schematic diagram of symbiotic nitrogen fixation system to show possible limiting factors.

an energy source whose production is dependent in turn on reduced substrate and oxygen as a possible limitation. Reduction of N_2 requires six electrons provided by electron donors the reduction of which is dependent also on oxidation of reduced substrates. As well as sub-optimal concentrations of substrates, product accumulation may limit nitrogen fixation. Ammonia is a repressor of nitrogenase synthesis although not a feedback inhibitor of nitrogenase activity. Suitable carbon skeletons are necessary for ammonia incorporation prior to transfer from the nodule to the aerial part mainly in the form of asparagine (Pate, Walker & Wallace, 1965; Pate, 1968; Streeter, 1972*a*,*b*; Sloger, 1973). Photosynthate available to the nodule is a secondary factor common to the above three possible primary limitations since it provides the substrate for ATP generation, reduction of electron donors and removal of ammonia. Factors such as light, competitive sinks, P_{CO_2} etc. that limit photosynthate availability are considered as tertiary agents. Although the nitrogen-fixing activity of sliced nodules was low, an early report by Bach, Magee & Burris (1954) demonstrated that the addition of sugars to sliced nodules increased ^{15}N-enrichment from $^{15}N_2$ and is consistent with a limiting role for photosynthate. It is realized that factors additional to those listed above, such as ATP/ADP ratio, reduction of H_3O^+ to H_2 which may compete with N_2 for electrons etc. may also limit nitrogenase activity.

We believe the following results indicate that the amount of photosynthate available to the nodule may be a most significant factor limiting nitrogen fixation. It is not yet possible to specify which of its three listed roles is crucial; however, the substantial requirement for ATP for in-vitro nitrogenase activity as well as for NH_3-incorporation reactions might suggest this function, while the insensitivity of nitrogen fixation to treatments that increase soluble nitrogen in the nodule such as darkness and defruiting (Pate, Chapter 27) does not support ammonia removal as a limitation.

Nitrogenase – a non-limiting factor

Comparison of the effect of incubation temperature on nitrogen-fixing activity of isolated nitrogenase, nitrogen-fixing *Clostridium* cells and nodulated soybean roots indicate that nitrogenase is not the limiting factor for soybeans in the temperature range above 20 °C (Fig. 31.3). Isolated nitrogenase exhibits a biphasic Arrhenius plot with the break occurring around 20 °C; above that the rate of reaction increases less rapidly with increased temperature but is temperature-dependent. An identical Arrhenius plot is found for *Clostridium* cells (Hardy *et al.*, 1968; Burns, 1969). Nodulated soybean roots show a similar temperature response below 20 °C, but are essentially unaffected from 20 to 35 °C. Others have also found no consistent increase in legume N_2 fixation at temperatures above 20 °C (Gibson, 1971; Dart & Day, 1971).

It is concluded that nitrogenase concentration is not the factor limiting nitrogen fixation by soybean nodules above 20 °C although it may be limiting below that. Accordingly, elimination of the limiting factor should increase the specific activity of the nodule by increased saturation

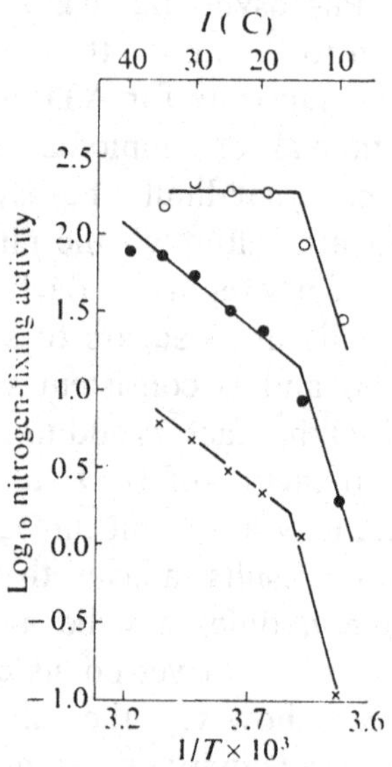

Fig. 31.3. Arrhenius plots of the effect of incubation temperature (T) on nitrogen fixation by isolated nitrogenase (●), nitrogen-fixing *Clostridium* (×) and soybean nodules (○). Note the dissimilar temperature responses between 20 °C and 35 °C. After Hardy *et al.* (1968).

of the excess nitrogenase and this is found with tertiary factors such as carbon dioxide enrichment and supplemental light.

Plant variety and bacterial strain

The nitrogen fixation profiles and nodular serology of several different varieties of soybeans with different flowering characteristics and maturation dates have been compared (Table 31.2; Hardy *et al.*, 1971, 1973). All values were obtained in one season from a single location. The characteristics of fixation vary greatly among the different varieties with total N_2 fixed ranging from 94 to 188 mg N/plant/season, while there is no correlation with the serological distribution of bacterial strains. For example, the serology of Dare and Adelphia are similar but the amount of N_2 fixed by Dare is almost twice that of Adelphia. These results suggest that the major limiting factor in the control of nitrogen fixation under field conditions may be found in the plant. This conclusion does not deny the importance of the microsymbiont but suggests that the underinvestigated role of the plant should receive more emphasis. The following section examines the limiting nature of one of these plant factors, photosynthate.

Table 31.2. *Nitrogen fixation characteristics for selected soybean varieties*

	Exponential phase			N_2 fixed	Serology % strains of *R. japonicum*				
Variety	Initiation age (days)	Termination age (days)	Increase (%/day)	(mg N/plant)	94	110	122	123	C_1
Adelphia	19	59	9.6	94	4	26	10	32	4
Verde	17	55	10.4	108	24	32	4	6	16
Delmar	27	69	10.4	147	20	22	6	8	18
Clark 63	28	73	10.4	149	14	28	12	12	10
Wayne	21	64	10.4	151	4	12	12	36	14
Dare	32	69	11.4	162	4	40	14	18	6
Kent	30	70	11.4	170	22	24	10	10	6
York	27	66	10.9	188	12	18	14	28	4

Photosynthate – a major limiting factor

Factors that affect photosynthate available to the nodule have been examined. These can be classified as light intensity, size of photosynthetic source, competitive sinks, carbon dioxide enrichment and photosynthate translocation. We will report the effect of variation of each of these parameters on nitrogen fixation by soybeans (and in some cases *Alnus*). In all examples nitrogen-fixing activity correlates directly with the amount of photosynthate available to the nodule.

Light intensity. Diurnal effects on specific nitrogen-fixing activity of soybean and alder nodules have been observed (Fig. 31.4) with maxima occurring near the period of maximum light intensity (Hardy *et al.*, 1968; Wheeler, 1969; Sloger *et al.*, 1975). Fifty percent shading imposed at the end of flowering in soybeans decreased fixation from 125 to 91 kg N/ha/season, while supplemental light increased fixation to 165 kg N/ha/season. Nodule specific nitrogen-fixing activity was similarly affected (Brun, 1972; Ham, Lawn & Brun, Chapter 20). These observations show a direct relationship between light intensity and nitrogen fixation and are interpreted as an expression of the amount of photosynthate.

Source size. The effect of four factors – defoliation, additional foliage, plant density, and lodging – related to source size have been measured. Nitrogen fixation by defoliated plants was reduced (Hardy *et al.*, 1968, Moustafa, 1969). In one study (Brun, 1972; Ham *et al.*, Chapter 20), partial defoliation by removal of two leaflets from each soybean leaf after flowering reduced fixation from 125 to 100 kg/N/ha, while in another (Streeter, 1973) increasing source size by grafting a

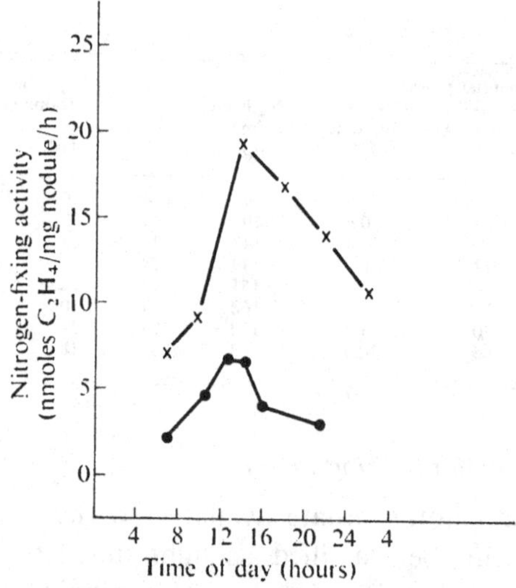

Fig. 31.4. Diurnal changes in nitrogen-fixing activity of soybean (×) and alder (●) nodules. After Hardy *et al.* (1968) and Wheeler (1969).

second top to a single root increased specific nitrogen-fixing activity of nodules by up to 100 % for a short time. Field situations affecting source size are planting density and lodging (Hardy, Havelka & Holsten, 1972). Nitrogen fixation per plant is increased in inverse

Table 31.3. *Planting density and nitrogen fixation*

	Row width (cm)		
Seeds/ha	76	51	38
		mg N_2 fixed/plant/season	
370 000	130	136	128
490 000	98	79	102
620 000	68	73	65
		kg N_2 fixed/ha/season	
370 000	48	50	47
490 000	48	39	50
620 000	42	45	40

proportion to plant density but N_2 fixed/ha/season is essentially independent of planting density (Table 31.3). Fixation by lodged soybean plants was less than one-third that of unlodged plants (Fig. 31.5; Table 31.4); this major decrease was produced by loss of exponential

Table 31.4. *Effect of lodging on nitrogen fixation characteristics*

Treatment	Exponential phase: Initiation age (days)	Termination age (days)	Increase (%/day)	N_2 fixed (mg N/plant)
Unlodged	32	77	11.4	275
Lodged*	32	60	11.4	82

* Lodging occurred at 56 days.

phase within only three to four days following lodging. These effects of source size are consistent with photosynthate as a limiting factor for nitrogen fixation.

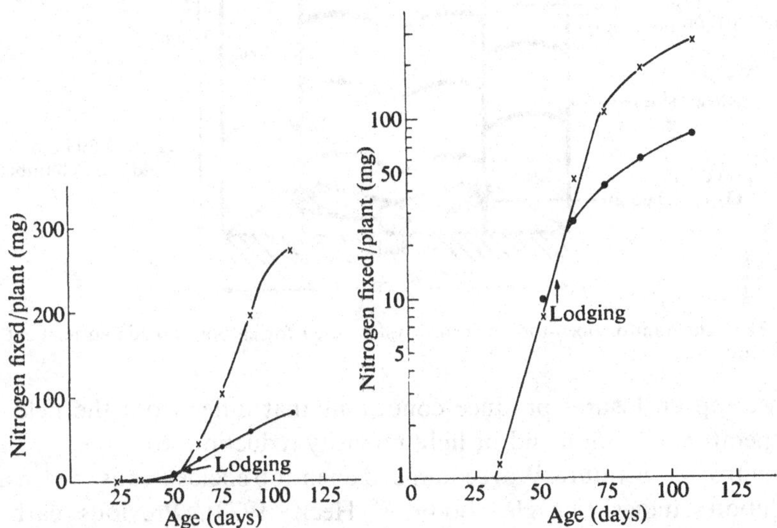

Fig. 31.5. Profiles of nitrogen fixation by lodged (●) and unlodged (×) soybeans. Loss of exponential phase occurs four days after lodging.

Competitive sinks. As a soybean plant matures, the reproductive sinks compete with the nodules for available photosynthate. Data on the partitioning of a pulse of fixed $^{14}CO_2$ by field-grown soybean plants at different stages of development show the successful competition of reproductive tissue with nodules for available photosynthate (Sweetser, personal communication). Pod removal from every other node of soybeans at the end of flowering increased nitrogen fixation since it presumably made more photosynthate available to the nodules (Brun, 1972, Ham *et al.*, Chapter 20).

Carbon dioxide enrichment. Recently, we have utilized carbon dioxide enrichment to increase photosynthate production and determine its effect on nitrogen fixation (Hardy & Havelka, 1973). Soybeans var. Kent were field-grown in open-top enclosures as shown in Fig. 31.6. These enabled us to duplicate field conditions of temperature and light while maintaining a carbon dioxide enriched environment, whereas

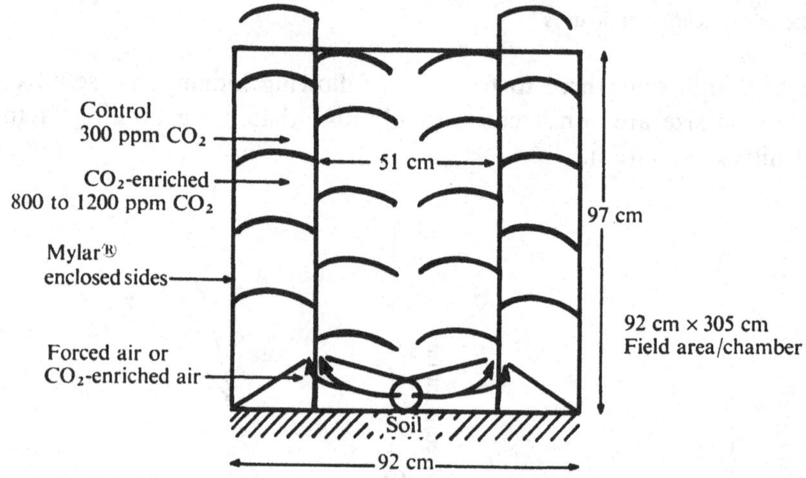

Fig. 31.6. Diagram of open-top enclosure system used for carbon dioxide enrichment of field-grown soybeans.

closed-top enclosures produce conditions that differ from the field in temperature elevation and/or light intensity reduction from the shading to control temperature. We recommend open-top enclosures as used in air pollution studies (Heagle, Body & Heck, 1973). Previous carbon dioxide enrichment experiments with soybeans in closed-top chambers demonstrated increased vegetative and reproductive growth but nitrogen fixation was not determined (Cooper & Brun, 1967). In our experiments the forced air supply to the experimental plants was injected with carbon dioxide during the day (0800 to 1700 h) from pre-flowering at 38 days of age to maturity, to provide 800 to 1200 ppm CO_2 within the canopy. This concentration required the addition of about 0.5 kg

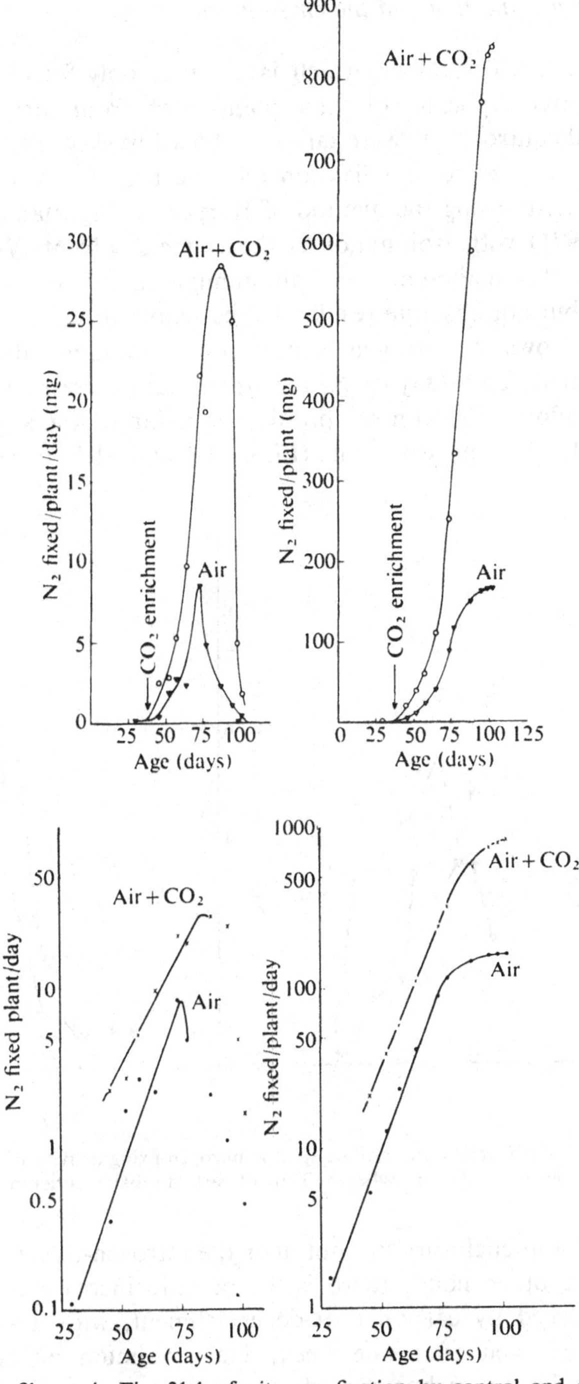

Fig. 31.7. Profiles as in Fig. 31.1 of nitrogen fixation by control and carbon dioxide enriched soybeans.

CO_2/hour for each chamber, which is practical only for experimental purposes. Three replicates of three plants each from carbon dioxide enriched and control plots were sampled at least weekly. The nodulated roots were analyzed for N_2 fixation and the tops for in-vivo nitrate reductase activity using the method of Harper & Hageman (1971) and Jaworski (1971) with isobutanol as the organic solvent. We are not satisfied with this method and are continuing to modify it. Accordingly, the relative but not absolute results will be reported.

Fig. 31.7 shows the nitrogen fixation profiles obtained expressed as N_2-fixing activity/plant/day or N_2 fixed/plant and corresponding semi-logarithmic plots. The control profile is similar to those previously determined for field-grown plants (Figs. 31.1 and 31.5) demonstrating

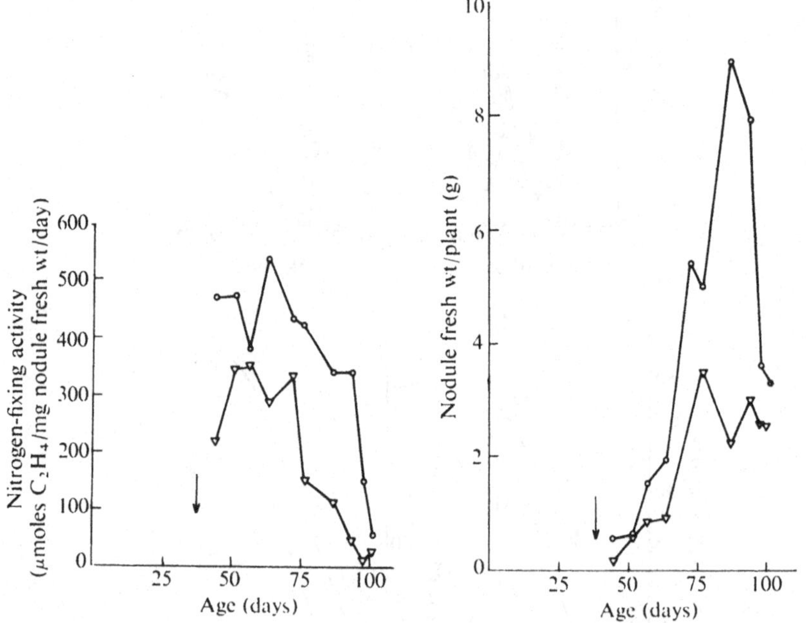

Fig. 31.8. Nodule fresh weight and nodule specific nitrogen-fixing activity of control (▽) and carbon dioxide enriched (○) soybeans. Time of carbon dioxide enrichment indicated by arrow.

that the open-top enclosure did not alter the nitrogen-fixing characteristics. On the other hand, there is a dramatic increase in nitrogen fixation produced by carbon dioxide enrichment, with a substantial effect observed even after one week. The maximum nitrogen-fixing activity/plant/day by carbon dioxide enrichment was over three times that of control and occurred at a somewhat later date. The exponential phase was extended by eight days to 84 days, thereby providing substantially more nitrogen from N_2 during the critical pod-filling stage. More N_2 was fixed in one week, viz. from 87 to 94 days of age, by enriched plants than by control plants during the complete growth cycle. Total nitrogen fixed per plant was 842 mg compared with 167 mg by the control. Although these experiments used isolated two-row plots of soybeans, other experiments in which the two-row experimental plots were surrounded by soybeans produced similar responses.

This large increase in total nitrogen fixation by carbon dioxide enrichment is caused by the delayed loss of exponential phase, the extension of maximum nitrogen-fixing activity towards maturation and increased specific activity. Nodule fresh weight was increased by an average of twofold to a maximum of 8.97 g fresh wt/plant at 87 days compared with 3.48 g at 77 days for the control. The nodule average specific nitrogen-fixing activity was doubled, with a maximum of 536 mg N_2 fixed/g fresh wt nodule/day at 64 days compared with 329 mg at 73 days for the control (Fig. 31.8). This increased specific activity is attributed to a more complete saturation of the suggested excess nitrogenase in the nodule.

While carbon dioxide enrichment increases nitrogen fixation, it may decrease nitrate reductase in the aerial part of the plant. Measurements of nitrate reductase by the in-vivo assay showed less total activity during the complete growth cycle for carbon dioxide enriched plots, which suggests that less nitrate may reach the aerial portion of the plant because more is reduced in the root by the additional photosynthate and/or less is taken up by the plant. Additional measurements are needed to clarify this situation.

The dry weight of the mature plants collected at 101 days was increased by carbon dioxide enrichment (Table 31.5). For example, reproductive tissue was 1.78 times the control. The nitrogen content was determined by Kjeldahl analysis. Percent nitrogen in reproductive tissue was unchanged while that of vegetative tissue was decreased slightly. Total nitrogen of both reproductive and vegetative tissue was increased to an average of 1.009 g N/plant for carbon dioxide enriched plants compared with 0.648 g N/plant for the control. Measured N_2 fixation accounts for only 26 % of the total nitrogen of control plants but 83 % of enriched plants; thus the contribution required from soil nitrogen to make up the balance decreased by 65 % with enrichment.

Finally, the density of plants at maturity was increased 11 %. Again similar increases have been observed for two-row experimental plots in the center of a soybean field as for the isolated two-row plots reported here. We attribute this increased density to the improved competitiveness of shaded plants growing in the carbon dioxide enriched environment. Total nitrogen, N_2 fixation and nitrogen from soil expressed on

Table 31.5. *Effects of carbon dioxide enrichment on soybean plants*

	Air	Carbon dioxide enriched air
Mass* (g dry wt/plant)		
Nodules	0.73	0.71
Roots	1.2	1.8
Stems	5.6	7.1
Leaves	4.3	5.4
Pods	8.8	15.7
Total	20.6	30.7
% nitrogen		
Nodules	3.1	4.1
Roots	1.1	1.1
Stems	1.5	1.3
Leaves	3.7	3.2
Pods	4.3	4.4
Nitrogen content (g N/plant)		
Nodules	0.017	0.028
Roots	0.013	0.020
Stems	0.083	0.089
Leaves	0.16	0.18
Pods	0.38	0.70
Total	0.648	1.009
N_2 fixed (g N/plant)	0.167	0.842
Number		
Plants/ha	455 000	505 000
Pods/plant	23	34
Nitrogen input (kg N/ha)		
Total	295	511
N_2 fixed	76	427
Nitrogen from soil†	219	511

* Determined at maturity at 101 days, for six replicates of three plants per sample for each treatment.
† Difference between N_2 fixed and total nitrogen.

an area basis are recorded in Table 31.5. Nitrogen from soil is the calculated difference between total nitrogen and N_2 fixation. Over 400 kg N_2/ha was fixed by carbon dioxide enriched plants to give a total nitrogen content of $>$500 kg N/ha compared with fixation of 76 kg N/ha and total nitrogen of $<$300 kg N/ha by controls.

Other effects of carbon dioxide enrichment were also noted. Control plants lodged about 15 days earlier than enriched plants. The increased resistance to lodging is probably a result of the increased stem size arising from incorporation of some of the extra photosynthate into stem fibers. This could be an attractive added benefit since lodging may be a major factor limiting soybean yields. Senescence was also delayed and the number of pods at maturity was increased.

In summary, these results show that carbon dioxide enrichment increases nitrogen fixation more than fivefold by: (1) doubling nodule specific activity, (2) doubling average nodule mass, and (3) extending exponential phase. The nitrogen input is 80+ % from N_2 for carbon dioxide enriched plants compared with 25 % for control. Total nitrogen was increased >70 % to over 500 kg N/ha. These observations on the effect of carbon dioxide enrichment on nitrogen input are interpreted as effects of photosynthate on N_2 fixation. Elevated P_{CO_2} increases net photosynthesis largely by decreasing photorespiration and thereby making more photosynthate available to the nodule. An alternative experimental procedure to decrease photorespiration is to grow plants in a subambient oxygen atmosphere. For example, root growth of soybeans has been increased up to sixfold in plants where the aerial portion was grown in an environment containing 5 % oxygen (Quebedeaux & Hardy, 1973). Similar effects on nitrogen fixation should be found with subambient P_{O_2} as for elevated P_{CO_2}.

Effects of carbon dioxide on nodule initiation and *Rhizobium* growth have been reported but appear to be unrelated to the results reported here. Mülder & Van Veen (1960) found an effect on nodule initiation in cultured red clover plants which was greater at pH 4.8 than 6.4. However, treatment of an acid soil with a carbon dioxide enriched gas phase did not produce a consistent effect on either nodulation or nitrogen fixation. Their carbon dioxide concentration was 5 % or fifty times as much as used in our experiments and they enriched the root gas phase while we enriched the leaf canopy gas phase. A requirement of the carbon dioxide in air for growth of various rhizobia has also been found, but beneficial effects of carbon dioxide enrichment of air were not reported (Lowe & Evans, 1962). A much earlier experiment found an increase in nodule number and nitrogen fixation with added carbon dioxide and attributed this to an effect on the C/N ratio (Wilson, 1941) but was not further pursued in recent times with the capability of improved measurements.

Photosynthate translocation. Stem-girdled non-legumes show a rapid decline in nitrogen-fixing activity (Wheeler, 1971) with over a 50 % decrease within two hours of girdling (Fig. 31.9). Similar results have been observed with soybeans (Havelka & Hardy, unpublished results). These results suggest the close-coupling of photosynthate movement to

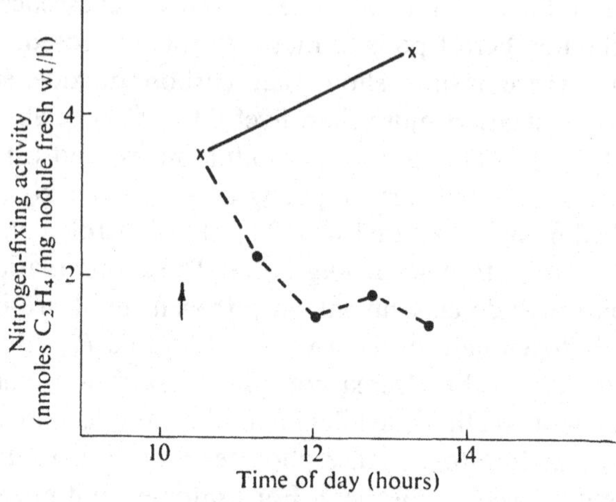

Fig. 31.9. Rapid decline of nitrogen-fixing activity of *Alnus* after stem-girdling (●) compared with a control (×). Girdling indicated by arrow. After Wheeler (1971).

the nodule and nitrogen-fixing activity and are consistent with the rapid response in specific nitrogen-fixing activity to carbon dioxide enrichment. We find that exposure of the canopy to 1500 ppm CO_2 for only six hours increased specific activity by more than 70 %.

Discussion

Table 31.6 summarizes the effect of various factors on symbiotic nitrogen fixation. Fertilizer nitrogen was not discussed in this article but it is well known that it decreases fixation and we have extensively examined this effect with the acetylene-reduction assay (Hardy *et al.*, 1971). Strain of rhizobia and variety of plant can both affect the amount of N_2 fixed. In this report we have focussed our attention on various factors that affect photosynthate available to the nodule, and all observations support the proposal of a direct relationship between amount of photosynthate and nitrogen-fixing activity. The most significant increase in fixation is produced by carbon dioxide enrichment. Seeking ways to increase total photosynthate so that more is available to the nodule without decreasing that available to other tissue is an attractive research goal. Decreasing photorespiration, a process that occurs in all legumes, is one possible approach. Alternatively, other respirations wasteful of photosynthate may be decreased. Root respiration may be a candidate since Minchin & Pate (1973) found that 35 % of the carbon gained by the pea shoot from the atmosphere during vegetative growth is lost by root respiration. Success in increasing

Table 31.6. *Summary of factors that affect nitrogen fixation*

Factors	Increased fixation	Decreased fixation
Fertilizer		NO_3^-, NH_4^+
Rhizobium		Strain
Plant		Variety
Factors that affect photosynthate availability		
Light	Day	Night
	Supplemental light	Shading
Source size	Additional foliage	Defoliation
	Low plant density	High plant density
		Lodging
Competitive sinks	Pod removal	Pod filling
CO_2	CO_2-enrichment	CO_2-depletion (?)
O_2	O_2-depletion (?)	O_2-enrichment (?)
Photosynthate translocation		Girdling

photosynthate should lead to a major increase in nitrogen input through a large increase in fixation. It is concluded that the route for increasing symbiotic nitrogen fixation is now more clearly defined than before.

References

Bach, M. K., Magee, W. E. & Burris, R. H. (1954). Nitrogen fixation by excised soybean root nodules. *Journal of Biological Chemistry*, **208**, 29–39.

Brun, W. A. (1972). Nodule activity of soybeans as influenced by photosynthetic source–sink manipulations. *Agronomy Abstracts*, p. 31.

Burns, R. C. (1969). The nitrogenase system from *Azotobacter*: activation energy and divalent cation requirement. *Biochimica et Biophysica Acta*, **171**, 253–9.

Cooper, R. L. & Brun, W. A. (1967). Response of soybeans to a carbon dioxide-enriched atmosphere. *Crop Science*, **1**, 455–7.

Dart, P. J. & Day, J. (1971). Effects of incubation temperature and oxygen tension on nitrogenase activity of legume root nodules. In: *Biological nitrogen fixation in natural and agricultural habitats* (ed. T. A. Lie & E. G. Mülder), *Plant and Soil*, Special Volume, 167–84.

Dixon, R. O. D. (1975). Relationship between nitrogenase systems and

ATP-yielding processes. In: *Nitrogen fixation in free-living micro-organisms* (ed. W. D. P. Stewart). Cambridge University Press, London. (In press)

Gibson, A. H. (1971). Factors in the physical and biological environment affecting nodulation and nitrogen fixation by legumes. In: *Biological nitrogen fixation in natural and agricultural habitats* (ed. T. A. Lie & E. G. Mülder), *Plant and Soil*, Special Volume, 139–52.

Hardy, R. W. F. & Burns, R. C. (1973). Comparative biochemistry of iron–sulfur proteins of nitrogen fixation. In: *Iron–sulfur proteins* (ed. W. Lovenberg), vol. 1, pp. 65–110. Academic Press, New York & London.

Hardy, R. W. F., Burns, R. C., Hebert, R. R., Holsten, R. D. & Jackson, E. K. (1971). Biological nitrogen fixation: a key to world protein. In: *Biological nitrogen fixation in natural and agricultural habitats* (ed. T. A. Lie & E. G. Mülder), *Plant and Soil*, Special Volume, 561–90.

Hardy, R. W. F., Burns, R. C. & Holsten, R. D. (1973). Applications of the acetylene–ethylene assay for measurement of nitrogen fixation. *Soil Biology and Biochemistry*, **5**, 47–81.

Hardy, R. W. F. & Havelka, U. D. (1973). Symbiotic N_2 fixation: multifold enhancement by CO_2-enrichment of field-grown soybeans. *Plant Physiology* (Supplement), **48**, 35.

Hardy, R. W. F., Havelka, U. D. & Holsten, R. D. (1972). Dinitrogen (N_2) fixation by field-grown soybeans: effect of variety, planting density and lodging. *Agronomy Abstracts*, p. 97.

Hardy, R. W. F. & Holsten, R. D. (1975). Methods for measurement of N_2 fixation. In: *Treatise on dinitrogen fixation* (ed. R. W. F. Hardy, A. Gibson & W. Silver). Wiley, New York. (In press.)

Hardy, R. W. F., Holsten, R. D., Jackson, E. K. & Burns, R. C. (1968). The acetylene–ethylene assay for N_2 fixation: laboratory and field evaluation. *Plant Physiology*, **43**, 1185–1205.

Harper, J. E. (1973). Symbiotic $N_2[C_2H_2]$ fixation and nitrate utilization capabilities of soybeans. *Agronomy Abstracts*, p. 34.

Harper, J. E. & Hageman, R. H. (1971). Canopy and seasonal profiles of nitrate reductase in soybeans (*Glycine Max* L. Merr). *Plant Physiology*, **49**, 146–54.

Heagle, A. S., Body, D. E. & Heck, W. W. (1973). An open-top field chamber to assess the impact of air pollution on plants. *Journal of Environmental Quality*, **2**, 365–8.

Jaworski, E. G. (1971). Nitrate reductase assay in intact plant tissue. *Biochemical and Biophysical Research Communications*, **43**, 1274–9.

Lowe, R. H. & Evans, H. J. (1962). Carbon dioxide requirement for growth of legume nodule bacteria. *Soil Science*, **94**, 351–6.

Minchin, F. R. & Pate, J. S. (1973). The carbon balance of a legume and the functional economy of its root nodules. *Journal of Experimental Biology*, **24**, 259–71.

Moustafa, E. (1969). Use of acetylene reduction to study the effect of nitrogen fertilizer and defoliation on nitrogen fixation by field-grown white clover. *New Zealand Journal of Agricultural Research*, **12**, 691–6.

Mülder, E. G. & Van Veen, W. L. (1960). The influence of carbon dioxide on symbiotic nitrogen fixation, *Plant and Soil*, **13**, 265–78.

Pate, J. S. (1968). Physiological aspects of inorganic and intermediate nitrogen metabolism (with special reference to the legume *Pisum arvense* L.). In: *Recent aspects of nitrogen metabolism in plants* (ed. E. J. Hewitt & C. V. Cutting), pp. 219–242. Academic Press, New York & London.

Pate, J. S., Walker, J. & Wallace, W. (1965). Nitrogen-containing compounds in the shoot of *Pisum arvense* L. II. The significance of amides and amino acids released from roots. *Annals of Botany*, **29**, 475–93.

Quebedeaux, B. & Hardy, R. W. F. (1973). Oxygen as a new factor controlling reproductive growth. *Nature, London*, **243**, 477–9.

Sloger, C. (1973). Assimilation of ammonia by glutamine synthetase and glutamate synthetase in N_2-fixing bacteroids from soybean nodules. *Plant Physiology* (Supplement), **48**, 34.

Sloger, C., Bezdicek, D., Milberg, R. & Boonkerd, N. (1975). Seasonal and diurnal variations in $N_2[C_2H_2]$-fixing activity in field soybeans. In: *Nitrogen fixation in free-living micro-organisms* (ed. W. D. P. Stewart), pp. 271–84. Cambridge University Press, London.

Streeter, J. G. (1972*a*). Nitrogen nutrition of field-grown soybean plants. I. Seasonal variations in soil nitrogen and nitrogen composition of stem exudate. *Agronomy Journal*, **64**, 311–14.

Streeter, J. G. (1972*b*). Nitrogen nutrition of field-grown soybean plants. II. Seasonal variations in nitrate reductase, glutamate dehydrogenase and nitrogen constituents of plant parts. *Agronomy Journal*, **64**, 315–19.

Streeter, J. G. (1973). Growth of two shoots on a single root as a technique for studying physiological factors limiting the rate of nitrogen fixation by nodulated legumes. *Plant Physiology* (Supplement), **48**, 34.

Wheeler, C. T. (1969). The diurnal fluctuation in nitrogen fixation in the nodules of *Alnus glutinosa* and *Myrica gale*. *New Phytologist*, **68**, 675–82.

Wheeler, C. T. (1971). The causation of the diurnal changes in nitrogen fixation in the nodules of *Alnus glutinosa*. *New Phytologist*, **70**, 487–95.

Wilson, P. W. (1941). *The biochemistry of symbiotic nitrogen fixation.* University of Wisconsin Press, Madison, Wisconsin.

PART V

Nitrogen-fixing symbioses in non-leguminous plants

Preface

The first chapter in this Part presents the results of a survey – in which some fifty botanists and workers in allied fields from some thirty countries participated – planned to secure further information on the occurrence of nitrogen-fixing root nodules on non-leguminous angiosperms under field conditions. IBP provided an almost unique opportunity for conducting such a survey. The more specialised studies described in the remaining chapters were in most cases sponsored by the appropriate National Committees for IBP. The chapters by Angulo *et al.*, Gardner, Wheeler, Akkermans and their respective collaborators are intimately related to the above survey, since they present detailed studies on the development, fine-structure, physiology and ecology of known examples of non-legume root nodules, especially those of *Alnus glutinosa*. In a wider-ranging study, Becking has examined various non-legume symbioses of the Indonesian flora for nitrogen fixation in the field, including the *Gunnera–Nostoc* symbiosis, which is also the special subject of Silvester's paper.

Two chapters are included which were not read at the Edinburgh meeting. That by van Hove is an appropriate inclusion since initially it was planned that a survey of the incidence of leaf nodules should figure in the Section PP–N programme. Actually this proposal was not pursued, partly because the initial appeal for collaborators met with a low response, but also because evidence mounted that leaf nodules were not the site of any significant fixation of nitrogen. Van Hove's chapter presents further evidence in that direction, and also includes a valuable survey of literature. The short chapter by Rodriguez-Barrueco and Bond highlights some of the handicaps faced by those who elect to work on non-legume root nodules.

G. Bond

32. The results of the IBP survey of root-nodule formation in non-leguminous angiosperms

G. BOND

1. Introduction

When this survey was begun in 1967 the only kind of nitrogen-fixing root nodules known to occur on non-leguminous angiosperms were of what are conveniently called the *Alnus* type, i.e. nodules showing the general structure and properties of those occurring on species of *Alnus* (alders), which are the longest-known and most-studied non-legume nodules. Since it is with such nodules that the survey has been mostly concerned, it will be appropriate first to recall their main features.

Alnus-type nodules commence as simple, swollen structures with a superficial resemblance to legume nodules, but as a result of repeated branching, coralloid nodule clusters are formed; these are perennial and over a period of years the clusters, often roughly spherical in shape, may attain a diameter of several centimetres. Each lobe of a nodule cluster shows an internal structure somewhat resembling that of a root, two of the differences being that there is a superficial periderm and an enlarged cortex in which some cells are filled with a dense growth of the nodule endophyte. The endophytes of *Alnus*-type nodules show a finely hyphal structure, the dimensions and some other features of the hyphae being similar to those of an actinomycete. The hyphae may produce structures called vesicles and granulae (see later chapters). *Alnus*-type nodules are nitrogen-fixing and, as in legumes, provide a source of combined nitrogen which enables the plant bearing them to grow vigorously in nitrogen-free rooting media. There is ample evidence of the ecological importance of this fixation of nitrogen by non-legume root nodules and of the possibility of its exploitation in forestry and to a lesser extent in agriculture.

By 1967 it was already known that some 119 species, distributed among 13 genera of woody, dicotyledonous angiosperms, were capable of bearing *Alnus*-type nodules. These species are listed in Table 32.1; it should be noted (1) that the sources of the records of nodulation were

Table 32.1. *Species known, prior to IBP, to bear* Alnus-*type nodules*

CORIARIA

C. angustissima, *C. arborea* Lindsay, *C. intermedia* Matsum., *C. japonica* A. Gray, *C. kingiana*, *C. lurida*, *C. myrtifolia* L., *C. plumosa*, *C. pottsiana*, *C. pteridoides*, *C. sarmentosa*, *C. thymifolia* Humb. & Bonpl.

ALNUS

A. cordata (Lois.) Desf., *A. crispa* (Ait.) Pursh., *A. firma* Sieb. et Zucc., *A. formosana* Mak., *A. fruticosa*, *A. glutinosa*, *A. hirsuta*, *A. incana*, *A. inokumai* Murai & Kusaka, *A. japonica* Sieb. et Zucc., *A. jorullensis* Kunth, *A. maritima* Nutt., *A. mollis*, *A. multinervis* Matsumura, *A. nepalensis*, *A. nitida*, *A. orientalis*, *A. rubra* Bong., *A. serrulata*, *A. sieboldiana* Matsum, *A. sinuata* Rydb., *A. tenuifolia* Sarg. var. *glabra* Call., *A. tinctoria* Sarg., *A. undulata*, *A. viridis* Regel, *A. rugosa*

MYRICA

M. adenophora Hance, *M. asplenifolia*, *M. carolinensis*, *M. cerifera* L., *M. gale* L., *M. javanica* Blume, *M. pensylvanica*, *M. pilulifera* Rendle, *M. pubescens* Willd., *M. rubra* Sieb. et Zucc., *M. sapida* var. *longifolia*, *M. serrata* Lam.

CASUARINA

C. cunninghamiana Miq., *C. equisetifolia*, *C. fraseriana*, *C. glauca*, *C. huegeliana*, *C. lepidophloia*, *C. littoralis* Salisb.*, *C. montana*, *C. muellerana* Miq.†, *C. muricata*, *C. nodiflora*, *C. pusilla* Macklin†, *C. quadrivalvis*, *C. stricta*, *C. sumatrana*, *C. tenuissima*, *C. torulosa* Ait.

HIPPOPHAË

H. rhamnoides L.

ELAEAGNUS

E. angustifolia L., *E. argentea* Pursh., *E. commutata* Bernh., *E. edulis*, *E. longipes* A. Gray, *E. macrophylla* Thunb., *E. multiflora*, *E. pungens*, *E. rhamnoides*, *E. umbellata* Thunb.

SHEPHERDIA

S. canadensis Nutt., *S. argentea*

CEANOTHUS

C. americanus L., *C. azureus*, *C. cordulatus* Kell., *C. crassifolius* Torr., *C. cuneatus* (Hook.) Nutt., *C. delilianus*, *C. divaricatus* Nutt., *C. diversifolius* Kell., *C. fendleri*, *C. foliosus* Parry, *C. fresnensis* Dudley, *C. glabra*, *C. gloriosus* Howell var. *exaltatus*, *C. greggii* Gray, *C. griseus* (Trel.) McMinn, *C. impressus* Trel., *C. incana* T. & G., *C. integerrimus* H. & A., *C. intermedius*, *C. jepsonii* Greene, *C. leucodermis* Greene, *C. microphyllus*, *C. oliganthus* Nutt., *C. ovatus*, *C. parvifolius* Trel., *C. prostratus* Benth., *C. rigidus* Nutt., *C. sanguineus* Pursh., *C. sorediatus* H. & A., *C. thyrsiflorus* Esch., *C. velutinus* Dougl.

DISCARIA

D. toumatou Raoul.

DRYAS

D. drummondii Richardson, *D. integrifolia* Vahl., *D. octopetala* L.

PURSHIA

P. glandulosa Curran, *P. tridentata* (Pursh.) DC.

CERCOCARPUS

C. betuloides

ARCTOSTAPHYLOS

A. uva-ursi (L.) Spreng.

* The presence of nodules in this species was recorded by Hannon (1956).

† Nodulation in these two species was recorded by Specht & Rayson (1957). The compiler is indebted to Dr P. D. Coyne for drawing his attention to the records for the three species to which these notes refer.

cited by Rodriguez-Barrueco (1968)* for all the species except three species of *Casuarina* for which sources are now cited in notes to the table, (2) that the lack of an authority for a specific name indicates that none was provided by the author of the original report, and (3) that the name *Casuarina triangularis* which appeared in some previous lists has been deleted since it cannot be traced in any taxonomic treatise.

It has to be recognised that in relatively few of the species listed in Table 32.1 is the statement that the nodules are of *Alnus* type based on direct evidence apart from their external form. However, such evidence does exist for the nodules of several species from each of the genera with numerous recorded species, and of at least one species from the remaining genera (except *Arctostaphylos*), making it reasonable to assume that the nodules of the other species have similar structure and properties.

The survey had three objectives, the first being to secure fuller information on the regularity of nodulation in the field of species already known – subject to the reservation made in the previous paragraph – to have the ability to form nodules of *Alnus* type, i.e. the species listed in Table 32.1. The previously existing information often referred to one region or locality, and even for the better-known species there had never been any concerted survey.

The second objective was to examine further species of the genera (Table 32.1) already known to include nodule-forming species. The need for this is indicated by the following summarised statement in which, for each genus, the number of species known prior to the survey to bear nodules is given, while in parentheses is the total number of species credited to the genus, according to Willis (1966).

Coriaria 12 (15), *Alnus* 26 (35), *Myrica* 12 (35), *Casuarina* 17 (45), *Hippophaë* 1 (3), *Elaeagnus* 10 (45), *Shepherdia* 2 (3), *Ceanothus* 31 (55), *Discaria* 1 (10), *Dryas* 3 (4), *Purshia* 2 (2), *Cercocarpus* 1 (20), *Arctostaphylos* 1 (70).

The third objective was to encourage a search for the presence of nodules in genera not previously known to include nodule-forming species; in particular it seemed that the Rhamnacae, a family of some sixty genera of which two (*Ceanothus* and *Discaria*) were already known to be nodule-forming, and the Rosaceae, comprising some one hundred

* This opportunity is taken of amending the citations made by Rodriguez-Barrueco (1968) in respect of the following species. The first published record of nodulation in *Myrica gale* should be credited to Brunchorst (1886), in *M. javanica* to von Faber (1925), in *Casuarina torulosa* to Hannon (1956) and in *Elaeagnus angustifolia* and *E. argentea* to Brunchorst (1868–8).

genera of which *Dryas*, *Purshia* and *Cercocarpus* had been shown to be nodule-forming, might include further examples.

An explanation of the purposes and importance of the survey, a description of typical non-legume nodules, and a list (similar to that given in Table 32.1) of species already known to be nodule formers, were sent to botanists and workers in related fields who, it was hoped, would collaborate in the survey. The chairmen of IBP National Committees were asked to help in locating further potential collaborators.

Collaborators were asked to take special care over the identification of plant species, and to indicate where herbarium specimens had been deposited; also to state how many plants were inspected and how many (if any) were nodulated, and to provide a brief description of nodules found. Details of the habitat and soil type favoured by plants found to be nodulated were requested, and also a note of associated plants. Collaborators were warned to ascertain that any nodular structures found were not due to an infestation of the roots by nematodes or other animals.

Out of a considerably larger number who were approached, some fifty people collaborated in the survey, their names and locations being listed below. A few of them have published their findings independently; these have been included in the results of the survey, with appropriate reference to the publication. It was a disappointment that no collaborators were found in some large areas of the world, a situation to which several factors contributed.

2. Collaborators in the survey

Dr C. D. Adams, Department of Botany, University of the West Indies, Mona, Kingston 7, Jamaica

Dr A. D. L. Akkermans, Laboratorium voor Microbiologie der Landbouwhogeschool, Wageningen, The Netherlands

Professor Dr Bernard Boullard, Laboratoire de Botanique et de Biologie végétale, Faculté des Sciences de Rouen, 76 Mont-Saint-Aignan, France

Dr T. V. Callaghan, Department of Botany, The University, Manchester, UK

Mr H. D. L. Corby, Department of Botany, University College of Rhodesia, Salisbury, Rhodesia

Dr P. D. Coyne, Department of Forestry, The Australian National University, Canberra, Australia

Mr A. C. Crundwell, Department of Botany, University of Glasgow, Glasgow, UK

Dr D. R. Dietz, Rocky Mountain Forest and Range Expt. Station, Rapid City, South Dakota, USA.

Dr H. J. Evans, Department of Botany and Plant Pathology, Oregon State University, Corvallis, Oregon, USA

Dr D. E. Eveleigh, Department of Biochemistry and Microbiology, Rutgers University, New Brunswick, NJ, USA
Dr G. Fåhraeus, Institute of Microbiology, Royal Agricultural College, Uppsala, Sweden
Mr D. B. Fanshawe, Ministry of Rural Development, Forest Department, Research Division, Kitwe, Zambia
Dr R. J. Fessenden, Faculty of Forestry, University of Toronto, Canada
Mr W. Foulds, Dudley College of Education, Dudley, Worcestershire, UK
Professor C. H. Gimingham, Department of Botany, The University, Aberdeen, UK
Dr U. Granhall, Institute of Microbiology, Royal Agricultural College, Uppsala, Sweden
Professor N. Grobbelaar, Department of Plant Physiology and Biochemistry, University of Pretoria, Pretoria, Republic of South Africa
Dr Nola J. Hannon, School of Botany, University of New South Wales, Kensington, Australia
Mr B. Hansen, Botanical Museum, The University, Copenhagen, Denmark
Mrs E. B. Hidajat, Department of Biology, Institute of Technology, Bandung, Indonesia
Dr D. J. Hill, Department of Plant Biology, University of Newcastle-upon-Tyne, UK
Dr R. A. Howard, Arnold Arboretum, Harvard University, Jamaica Plain, Mass., USA
Professor Anne Johnson, Biology Department, Nanyang University, Singapore
Professor H. Kayacik, I. Ü. Orman Fakültesi, Botanik Kürsüsü, Büyükdere/ Istanbul, Turkey
Dr A. G. Khan, Institute of Biological Sciences, The University, Islamabad, Pakistan
Mr M. Lalonde, Microbiology Department, Macdonald College, Quebec, Canada
Dr Gloria Lim, Botany Department, University of Singapore, Singapore
Dr H. McAllister, The Botanic Garden of Liverpool University, Neston, Cheshire, UK
Dr H. S. McKee, Centre National de la Recherche Scientifique, Service Forestier, Nouméa, Nouvelle Caledonie
Dr Anne H. Mackintosh, Botany Department, Glasgow University, Glasgow, UK
Professor P. Mikola, Department of Silviculture, Helsinki University, Helsinki, Finland
M. Claude Moureaux, ORSTOM, Centre de Dakar-Hann, Dakar, Sénégal
Mr J. W. Parham, The Herbarium, Department of Agriculture, Suva, Fiji
Professor J. S. Pate, Botany Department, The Queen's University, Belfast, UK
M. Guy Pizelle, Faculté des Sciences, Université de Nancy, Nancy, France
Dr V. J. Radulović, Laboratory of Microbiology, Faculty of Agriculture, Sarayevo, Yugoslavia
Mr B. W. Ribbons, Botany Department, Glasgow University, Glasgow, UK
Professor J. C. Ritchie, Life Science Division, Scarborough College, University of Toronto, West Hill, Ontario, Canada
Dr C. Rodriguez-Barrueco, Centro de Edafologia y Biologia Aplicada, Salamanca, Spain
Dr Dorothy E. Shaw, Department of Agriculture, Stock and Fisheries, Konedobu, Papua, Territory of Papua and New Guinea
Dr W. B. Silvester, Botany Department, The University of Auckland, Auckland, New Zealand
Professor A. Skogen, Botanical Museum, University of Bergen, Norway

Mrs Sri S. Soemartono, Botany Department, Bogor Agricultural University, Bogor, Indonesia (Received grant towards expenses from central IBP funds.)
Professor W. D. P. Stewart, Department of Biological Sciences, The University, Dundee, UK
Mr U. Suriawiria, Biology Department, Institute of Technology, Bandung, Indonesia
Miss A. C. Tallantire, Botany Department, Makerere University College, Kampala, Uganda
Mr K. Thompson, Botany Department, Makerere University College, Kampala, Uganda
Dr M. J. Trinick, CSIRO, Western Australian Laboratories, Wembley, W. Australia
Dr S. Uemura, Government Forest Experiment Station, Meguro-ku, Tokyo, Japan
Mr G. J. White, Department of Botany, University of New England, Armidale, NSW, Australia
Professor A. G. Wollum, Department of Soil Science, North Carolina State University at Raleigh, N. Carolina, USA
Professor C. T. Youngberg, Department of Soil Science, Oregon State University, Corvallis, Oregon, USA

3. Regularity of nodulation in species already known to be capable of bearing *Alnus*-type nodules

Here the concern is with those species named in section 1 that are already recorded as having the ability to form nodules of *Alnus*-type. The genera in respect of whose species information was received during the survey will be considered in turn.

Genus Coriaria

Pate recorded that a plant of *C. japonica* growing in the Dublin Botanic Garden (Ireland) bore no nodules. Rodriguez-Barrueco found that a bush of *C. myrtifolia* in the Botanic Garden, University of Coimbra, Portugal, was without nodules although growing close to a well-nodulated specimen of *C. nepalensis* – a species to be considered in section 4.

Genus Alnus

Fahraeus reported that the examination of several plants of *A. glutinosa* (L.) Gaertn. at each of seven sites in the southern half of Sweden revealed no instance of a lack of nodules; the same was true for *A. incana* (L.) Moench. examined at four sites. Mikola reported a similar experience in his examination of hundreds of trees of *A. glutinosa* and of *A. incana* in southern Finland, as also did Skogen in respect of these species in coastal regions of central Norway, and Hansen with reference to *A. glutinosa* in Denmark. Akkermans (1971) found nodules to be present on all trees of the latter species at eight sites of varied nature in the Netherlands, though none was found on submerged roots or on roots growing in mud. Bond and his past and present research associates have always found nodules to be present on trees of *A. glutinosa* in western Scotland. Pate reported that in many examinations

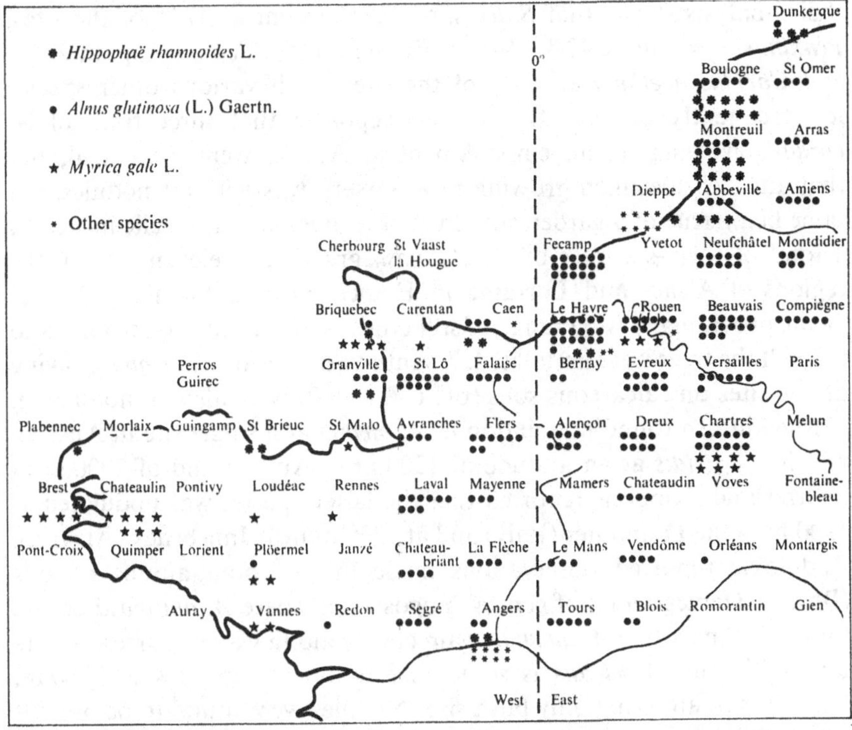

Fig. 32.1. Map supplied by Professor Dr Bernard Boullard showing the sites in northwest France at which nodulation was inspected by his group.

of *A. glutinosa* in various parts of Ireland nodules had always been found; *A. incana* and *A. rubra* growing at Tollymore Park, Co. Down were well-nodulated. Foulds examined 101 trees of *A. glutinosa* at different sites in Staffordshire, England and found all to be nodulated.

In a most comprehensive survey Boullard examined a total of some 650 trees of *A. glutinosa* at 244 sites in northwest France (Fig. 32.1). A very wide range of soil types was included, with soil pH ranging from 4.0 to 8.0. At most sites nodulation was reported as good, even at pH 4.0, with nodule clusters ranging up to 7 cm in diameter, but at fifteen sites nodulation was found to be poor and at five sites (involving fourteen trees, or about 2 % of the total examined) no nodules could be found. There is no obvious reason for the last finding; the lack of nodules was not associated with any particular soil type or pH. At most sites nodule lobes were cut open and the internal colour noted, this being recorded as a definite pink in many instances. The additional presence of mycorrhizal roots was noted at a number of sites. Lists of species associated with the alders were included in the report, and their analysis shows that *Salix* spp. were present at 47 % of the sites, *Fraxinus excelsior* at 42 %, *Urtica dioica* at 37 %, *Populus* spp. at 29 %, and *Corylus avellana* at 26 % of the sites, with various other species less frequently present. It was also reported that three trees of *A. cordata* growing on an embankment at Angers were nodulated, but that another specimen growing in a nursery was without nodules, the same being true of a garden specimen of *A. multinervis*. Pizelle inspected a total of forty-six trees of *A. glutinosa* growing at eleven sites in the regions of Alsace and Lorraine in France, also in the Ile d'Oléron, again providing a wide range of soil types, with pH varying from 4.8 to 8.2. All the trees were nodulated. Twenty-two trees of *A. incana* growing at five sites on calcareous soils (pH 6.7. to 8.0) were also all nodulated.

Mackintosh found nodulation in *A. incana* at a single site in Austria, and in *A. viridis* at an altitude of 1200 m in Austria and of 2000 m in Switzerland. Ribbons reported that the latter species was nodulated at 1900 m in the Dolomites (Italy) and at 1300 m near Innsbruck (Austria). Radulović reported observations made in the mountainous regions (Bosnia, Herzegovina) of central Yugoslavia. There *A. glutinosa* occurs up to 1000 m, while *A. incana*, again chiefly along water courses, occurs up to 1700 m. *A. viridis* is sometimes found at altitudes of 2000 m, though then attaining only bush size. Nodules were found to be present on all three species in all localities. Rodriguez-Barrueco recorded that *A. glutinosa* is of common occurrence in north and south Portugal, and in his experience is always nodulated, as is also true of the same species in Spain. Kayacik reported that *A. glutinosa* is regularly nodulated in Turkey, while Crundwell found that trees of *A. orientalis* Dcne. growing along the margins of a stream in the Mŭgla province of Turkey were likewise nodulated.

Khan (1971) found that both plants that he examined of *A. nitida* Endl. growing in the mountains of north Pakistan were nodulated. Soemartono recorded that her examination of six trees of *A. japonica* growing near Bogor, Indonesia, showed that all were nodulated; in the same area the two trees of *A. maritima* that were examined were both nodulated. Uemura (1971) confirmed the regular presence of nodules in Japan on *A. firma*, *A. hirsuta* Turcz., *A. japonica*, *A. pendula* Matsum. (= *A. multinervis*) and *A. sieboldiana*.

In Canada Ritchie found sparse nodulation on *A. crispa* growing on recent alluvium in the Mackenzie delta, while Lalonde reported species to be well-nodulated over a large area at Tadoussac, Quebec, also on dry sites all around Lake St John, Quebec, at Perce on the Gaspe peninsula and in the Magdalen Islands, Quebec, and also that *A. rugosa* was always nodulated at many wet sites in the Montreal and Quebec regions.

Genus Myrica

Van Ryssen and Grobbelaar (1970) recorded that *M. pilulifera* and *M. serrata*, for which the previous records referred only to Rhodesian specimens, are also nodulated in South Africa.

Hidajat reported that of six inspected trees of *M. javanica* growing on a mountain slope in the Bandung region of Indonesia all were nodulated; associated plants included species of *Eupatorium*, *Melastoma*, *Imperata*, *Gaulteria* and *Lantana*. In respect of the same region Soemartono also reported that the three trees of *M. javanica* examined were all nodule-bearing. Uemura (1971) confirmed the regular nodulation of *M. rubra* in Japan, where the species eventually makes a very large tree.

Ritchie inspected plants of *M. pensylvanica* Loisel. growing on dunes in Nova Scotia and found them all to be well-nodulated, as were also plants of *M. asplenifolia* L. (= *Comptonia peregrina* (L.) Coult.) in the same province; plants of *M. gale* were examined at three sites in Canada, near Halifax, Nova Scotia, Inuvik, NWT, and Riding Mt National Park, Manitoba, nodulation being found to be sparse or absent except at the first site. Fessenden reported that in his examination of *M. asplenifolia* at numerous sites in the provinces of Ontario and Quebec, Canada, and also in the Adirondack Mountains, USA, he had not encountered any instance of a lack of nodules. Lalonde recorded *M. pensylvanica* to be nodulated in the Magdalen Islands, Quebec, and *M. asplenifolia* to be nodulated over large areas near Lake St John, Quebec, while plants of *M. gale* growing near lakes and rivers to the north and northeast of Quebec City and in the Magdalen Islands were similarly nodulated. Eveleigh recorded abundant nodulation in *M.*

pensylvanica growing on sand dunes at various sites in New Jersey and at Nantucket and Cape Cod, USA. In his report from Jamaica, Adams indicated that the taxonomy of *Myrica* in the Antilles as a whole requires further study, and that although in the past three other species have been proposed in addition to *M. cerifera*, their validity is doubtful; two saplings which he named as *M. cerifera* L. *sens. lat.* and which were growing in Portland Parish, Jamaica, both proved to be nodulated.

Rodriguez-Barrueco reported that all eight plants of *M. gale* examined in the Beira Litoral region of Portugal were nodulated; he also noted that bushes of *M. cerifera* and *M. pensylvanica* in the botanic garden of Coimbra University bore nodules. Pizelle examined plants of *M. gale* at five sites in France, all being found to be nodulated except that at a site in Ardennes only five of twenty plants inspected bore nodules. Boullard inspected one to six plants of *M. gale* at each of thirty-three sites in northwest France (Fig. 32.1); the soils ranged from wet peat to drier peat and sandy-clay types, with a pH range of 4.0 to 6.0. For many of the sites nodulation was described as very good, but at five sites not all the plants were nodulated, these instances not being associated with any particular soil type. Mycorrhizal roots were also noted at some sites. *Molinia caerulea* was associated with *M. gale* at 64 % of the sites examined by Boullard, *Sphagnum* spp. at 36 %, *Salix* spp. at 33 %, *Betula pubescens* at 33 %, and *Erica* spp. at 27 %.

Gimingham sent data obtained by his students for *M. gale* growing at Bettyhill, Scotland, which suggested that the water content of the soil affects nodulation, since at rather dry sites only about half of the plants bore nodules, whereas at wetter places all plants were nodulated, and the nodules were more numerous and larger. Pate, in numerous inspections of *M. gale* at various sites in Ireland, found no instance of a lack of nodules. Fåhraeus reported that the examination of *M. gale* at four sites in southern Sweden showed nodules to be present in all cases; Mikola reported similarly in respect of ten plants of that species growing near Helsinki, Finland, as did Hansen for five plants examined in a fen in Jylland, Denmark, and Skogen in respect of plants in the coastal region of central Norway.

Genus Casuarina

The consideration of this genus is affected by the review of its taxonomy that is now under way. Mr P. D. Coyne has kindly advised the compiler of some of the attendant changes in nomenclature. Of the species listed in Table 32.1, it is now proposed that *C. lepidophloia* F. Muell. should be re-named *C. cristata* Miq., *C. montana* should become *C. junghuhniana* Miq., *C. muricata* should be placed in *C. equisetifolia* L., *C. quadrivalvis* should become *C. stricta* Ait., and *C. tenuissima* should be placed in *C. torulosa.*

White reported that five trees of *C. cristata* growing near Condobolin, NSW, Australia, were all nodulated, but that he has failed to find nodules on many trees of this species in other parts of NSW; nodulation appeared to occur only where the trees are subject to occasional flooding from some near-by river or stream. The same collaborator reported that the many trees of *C. cunninghamiana* that he examined at numerous natural sites in NSW were all nodulated. The 3–4 trees of *C. glauca* that he examined at each of three sites (Yamba and Urunga, NSW and Caloundra, Queensland) were all nodulated, this still being true of trees growing right down to the edge of the Pacific Ocean at the last site. A single tree of *C. torulosa* in a forest on Dorrigo Mountain, NSW, proved on examination to be nodulated. The inspection of 1–2 trees of *C. littoralis* at each of three sites in NSW revealed nodules on about half of them, but the nature of the soil made thorough excavation difficult.

Coyne reported that trees of *C. cunninghamiana*, *C. glauca* and *C. cristata* were examined at 100 sites in Australia. For the first two species nodules were found at practically every site, but at only nine for *C. cristata*: he pointed out that the upper soil at the sites for the last species was extremely dry, and that nodules might have been present at deeper, moister levels, which can scarcely be reached by hand excavation. Trees of *C. stricta* examined at two sites and of *C. torulosa* at one site were nodulated.

Hannon reported that further examination of trees of *C. littoralis* growing near Sydney, NSW had confirmed her earlier report (Table 32.1) of the presence of nodules on this species, though she stressed that sites where the nature of the soil made excavation relatively easy had been selected. Three trees of *C. glauca* growing near Sydney were found to be nodulated; the associated plants were species of *Juncus* and *Cladium*. Suriawiria provided further information about the above two species; all ten trees of *C. littoralis* that were inspected near Liverpool, NSW bore nodules, though they could be found on about one-third of the larger number of trees of *C. glauca* that were examined in the same region.

Shaw confirmed the presence of nodules on saplings of *C. equisetifolia* growing near Konedobu, Papua. Parham reported that two of the four trees of *C. nodiflora* Forst. f. that he examined near Viti Levu, Fiji bore nodules; associated plants include species of *Agathis*, *Trichospermum*, *Endospermum*, *Hernandia*, *Planchonella* and *Psychotria*.

Soemartono reported that of eighty-three trees of *C. equisetifolia* growing on latosol soil in the Bogor district of Indonesia, nodules could be found on three only; in the same district all the seventy-two trees of *C. sumatrana* Jungh. that she examined were nodulated (Fig. 32.2). Six trees of *C. montana* Ait. (see above) again in the same district

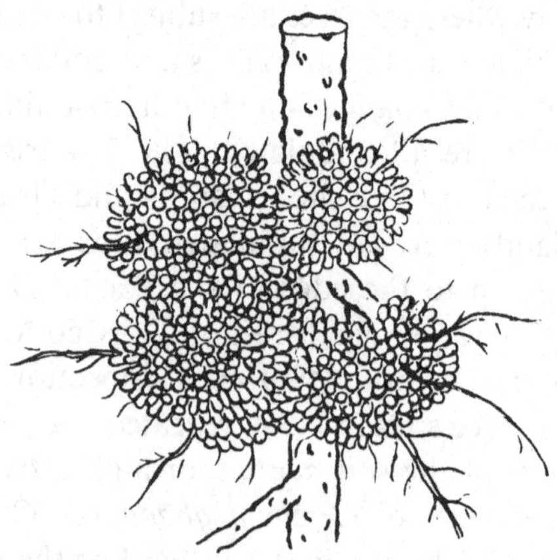

Fig. 32.2. Sketch sent by Mrs Sri S. Soemartono showing nodule clusters from an Indonesian tree of *Casuarina sumatrana*. (× 0.875)

were inspected and all found to be nodulated. Hidajat examined two trees of *C. cunninghamiana*, seven of *C. equisetifolia* and one of *C. junghuhniana* growing in the Bandung region of Indonesia and found all to be nodulated. Johnson and Lim state that *C. equisetifolia* is regularly nodulated in the Singapore region.

Khan (1971) states that three trees of *C. cunninghamiana* growing as introductions in Pakistan proved to be nodulated. Moureaux reported that in Sénégal, where *C. equisetifolia* has been extensively planted, the trees are regularly nodulated. Stewart reported to the same effect in respect of *C. cunninghamiana* planted in Morocco. On the other hand Rodriguez-Barrueco found no nodules on trees of *C. equisetifolia* and *C. glauca* growing in the botanic garden of Coimbra University, Portugal.

Genus Hippophaë

Mikola, Fåhraeus and Hansen reported that *H. rhamnoides* was found to be regularly nodulated at various coastal sites in Finland, Sweden and Denmark respectively, and Skogen that the same is true in Norway, though no nodules were present at a particular site where the plants were growing vigorously in a nitrogen-rich swamp of pH 4.5–5.0. He drew attention to the common association of nitrophilous species with nodulated *Hippophaë* plants, including species of *Urtica*, *Filipendula*, *Cochlearia*, *Humulus* and *Stellaria*. Akkermans (1971) found nodules to be present at two coastal sites in the Netherlands, though large, old plants bore very few. Mackintosh found nodules present on *H. rhamnoides* beside Loch Fyne, Scotland, as did Pate on sand dunes in Donegal and Co. Down, Ireland. Boullard examined 1–5 plants of *H. rhamnoides* at each of thirty-two natural sites on the north and west coasts of France (Fig. 32.1); nodules were present at thirty sites but could not be found at the remaining two sites. The soil pH ranged from 6.0 to 8.8 at the various sites, nodules being present over the whole of that range. The largest nodule clusters found were 6 cm in diameter. Over seventy other species of flowering plants were encountered as associates of the *Hippophaë*, those appearing most frequently being *Ligustrum vulgare* (at 44 % of the sites), *Ammophilia arenaria* (47 %), *Solanum dulcamara* (22 %) and *Echium vulgare* (19 %). Planted specimens of *H. rhamnoides* in gardens were examined at five sites, nodules being found at two only. Also in France, Pizelle examined 4–15 plants of the above species at each of four sites in Ile d'Oléron, Côtes du Nord and Bas Rhin (2) respectively; all the plants were found to be nodule-bearing. Here again the associated plants were very varied and included species of *Pinus*, *Phragmites*, *Pteridium*, *Silene*, *Betula*, *Clematis* and grasses.

Rodriguez-Barrueco reported nodulation in *H. rhamnoides* at an inland site in the Huesca district of Spain; a plant in the botanic garden of Coimbra University, Portugal, was also well-nodulated. Mackintosh noted that specimens planted along an autobahn near Vienna were nodulated, as did Kayacik in respect of plants in a nursery at Bahçeköy in Turkey, and Khan (1971) in respect of three plants growing in the Himalayan region of Pakistan.

Genus Elaeagnus

Uemura (1971) confirmed the regular nodulation of *E. macrophylla*, *E. multiflora* Thunb., *E. pungens* Thunb. and *E. umbellata* in Japan. Khan (1971) found that plants of *E. umbellata* in the Himalayan region of Pakistan and of *E. angustifolia* in the Quetta region of the same country were nodulated.

Ritchie reported that six plants of *E. commutata* growing in aspen parkland near Neepewa, Manitoba, Canada were devoid of nodules; of four plants examined in Riding Mt National Park in the same province, three were moderately nodulated. Dietz found that five bushes of *E. commutata* growing in the central Black Hills of South Dakota, USA were all nodulated, though the species is not native to the area.

Rodriguez-Barrueco reported that nodules were found on specimens of *E. angustifolia* growing near Madrid; the species is to some extent naturalised in Spain. In northwest France Boullard examined a few plants of the non-native *E. angustifolia*, *E. edulis* Siebold ex E. May, *E. pungens* and *E. umbellata*; good nodulation was present in some cases, but in others it was absent. Pate reported that specimens of *E. pungens* and *E. argentea* growing in his own garden bore nodules.

Genus Shepherdia

Lalonde found nodules to be present on *S. canadensis* at several sites on the River St Lawrence below Quebec, also near Lake St John, Quebec, on the Gaspe peninsula and the Magdalen Islands. Ritchie examined ten plants of *S. canadensis* growing on an unstable river bank in the Mackenzie delta, Canada, and found fair nodulation on eight plants, the other two being without nodules. Of three plants of *S. argentea* Nutt. growing in aspen parkland near Glenboro, Manitoba, no nodules could be found on two, the remaining plant being sparsely nodulated. Dietz reported that six of the twelve plants of *S. canadensis* that he examined in the Black Hills Forest, S. Dakota, were nodulated; all six plants that were examined of *S. argentea*, growing under very harsh conditions on a white clay soil in a Badlands formation in S. Dakota, bore nodules.

Genus Ceanothus

The only observation relating to plants growing naturally came from Dietz who reported that no nodules could be found on *C. ovatus* Desf. at a site in S. Dakota. As regards plants growing as introductions in countries where the genus is unrepresented, Pate reported that *C. rigidus* is sometimes nodulated in Ireland, Boullard that a plant of *C. azureus* Desf. in a French garden was nodulated, but that one of *C. impressus* was not; Uemura (1971) recorded that plants of *C. americanus* raised from seed in a Japanese nursery formed nodules.

Genus Dryas

Plants of *D. octopetala* were examined by Gimingham and by McAllister at several sites in Scotland, by Pate at sites in Ireland, by Rodriguez-Barrueco in the Pyrenees near Huesca, Spain, by Mackintosh in the Rax Alps, Austria, and by Granhall at Abisko, Sweden, while Uemura (1971) made a thorough inspection of plants of *D. octopetala* L. var. *asiatica* Nakai on mountains in Japan; none of these collaborators found fleshy, coralloid nodules as described by Lawrence *et al.* (1967) for Alaskan material, though mycorrhizal roots were noted. Ritchie found no nodules on the four plants of *D. integrifolia* that he inspected on the Caribou Hills, Mackenzie River, Canada, neither did Callaghan after much time spent in excavating plants of that species growing on heaths and fell-fields at Disko, Greenland. However, Lalonde found nodules to be present on *D. drummondii* on the banks of the Grand River, Gaspe peninsula, Quebec, though few were found until a depth of about 50 cm had been reached.

Genus Arctostaphylos

Plants of *A. uva-ursi* were inspected by Gimingham, Bond and McAllister at various Scottish sites, and by the last-named at a site in northwest Spain; also, by Pate at three sites in Ireland, by Pizelle on Ile d'Oléron, France, by Mackintosh at a site in Austria, by Ritchie at sites in Nova Scotia and NWT, Canada, by Youngberg at many sites in Oregon, USA, and by Dietz in S. Dakota, USA. Fleshy nodules as described by Allen, Allen & Klebesadel (1964) for Alaskan material were not found by any of these collaborators. Granhall reported that at Abisko, Sweden, he had only once seen hypertrophies on the roots of this species, in the form of small black structures present mostly near the root tips.

4. Records of nodulation in further species of genera already known to include species bearing *Alnus*-type nodules

The genera in which species additional to those already recorded to bear nodules have been inspected during the survey will be considered in turn.

Genus Coriaria

Rodriguez-Barrueco reported that a specimen of *Coriaria nepalensis* Wall. growing in the botanic garden of Coimbra University, Portugal, was nodulated. This species is of course not native in Portugal, but the report indicates that the species has the capacity to nodulate.

Genus Alnus

Uemura (1971) recorded that the following additional species or varieties of *Alnus* were nodulated in Japan: *A. fauriei* Lev. et Vnt., *A. matsumurae* Callier, *A. maximowiczii* Callier, *A. mayrii* Callier, *A. serrulatoides* Callier, *A. trabeculosa* Handel-Mazzett, *A. japonica* (Thunb.) Steudel var. *arguta* (Regel) Callier. According to information provided by the collaborator, none of these plant names is a synonym for a species previously recorded to bear nodules, except that by some taxonomists the plant here named as *A. maximowiczii* has been regarded as a variety of *A. crispa*. Boullard also recorded that a specimen of *A. maximowiczii* growing in a French garden was nodulated.

Genus Myrica

Van Ryssen and Grobbelaar (1970) recorded that nodules were present in South Africa on the following additional species of *Myrica*: *M. brevifolia* E. May ex C. DC., *M. cordifolia* L., *M. diversifolia* Adamson, *M. humilis* Cham. & Schlechtdl., *M. integra* (A. Chev.) Killick, *M. kraussiana* Buching. ex Meisn. and *M. quercifolia* L. Thompson reported that he found *M. kandtiana* Engl. and *M. salicifolia* Hochst. ex A. Rich. to be abundantly nodulated in Uganda. Corby recorded *M. microbracteata* Weim. to be nodulated in Rhodesia, and also referred to another plant growing there which is considered to be a distinct species of *Myrica* though so far only temporarily designated as 'Corby 2170'; after a long trek a necessarily hurried examination of this 'species' failed to reveal nodules, but rooted cuttings formed nodules in the greenhouse after suitable inoculation. Fanshawe reported the presence of nodules in Zambia on plants which he identified as *M. conifera* Burm. f.; however in the opinion of Killick (1969) the preferred name for this plant is *M. serrata* Lam. Plants under the latter name have been previously recorded as nodulated in Rhodesia and South Africa (see p. 451).

Soemartono reported that the examination of six trees growing in the district of Bogor, Indonesia, which she identified as *M. esculenta* Ham. showed all to be nodulated. According to the compilers of the *Index Kewensis* 'Buch.-Ham ex D. Don' is a more correct statement of the authority for that name, while according to the same source *M. nagi*

Thunb. is the more correct name for this species, but to the compiler's knowledge there is no previous record of nodulation under either specific name. The trees examined were growing on latosol soil at an altitude of 250 m, associated plants being species of *Caladium*, *Paspalum*, *Nephrolepis*, *Costus* and *Axonopus*. Lim reported that nodules were also present on *M. esculenta* on Singapore island in young plants growing in secondary forest. Uemura (1971) recorded the presence of nodules in Japan on plants of *M. gale* L. var. *tomentosa* C. DC.

Rodriguez-Barrueco examined four bushes (height 5 m) of a plant which he identified as *M. faya* Ait. growing with conifers on a sandy soil near Mira in the province of Beira, Portugal, and found all to be nodulated. A specimen of the same species growing in the botanic garden of Coimbra University was also found to be nodulated. This appears to be the first record of nodulation in this species, which is native to the Azores, and is at least naturalised and possibly native in parts of Portugal, the Canary Islands and Madeira.

Evans reported the presence of nodules on *M. californica* Cham. & Schlecht. growing on the west coast of the USA.

Genus Casuarina

Reference has already been made to the current revision of the taxonomy of this genus, and to the element of temporary confusion thereby arising, but it does appear that certain first records of nodulation have been secured during the survey.

White reported from Australia the presence of nodules on all the three plants (shrubs) that he identified as *C. rigida* Miq. They were growing in association with *Eucalyptus* species on a granitic soil east of Guyra, NSW. The single shrubby plants that he identified as *C. nana* Sieb. *ex* Spreng and *C. distyla* Vent. respectively, both growing near Katoomba, NSW, were also nodulated. Coyne looked unsuccessfully for nodules on the further Australian species *C. inophloia* F. Muell. & F. M. Bailey and *C. luehmannii* R. T. Baker, but he again made the point that the specimens were growing in a sandy soil the upper layer of which was almost permanently dry, and that nodules might have been present at moister but inaccessible lower levels. Dr Coyne also reported that the presence of nodules in *C. decussata* Benth. is recorded on a herbarium specimen of that species collected by Willis in Western Australia and now in the National Herbarium of Victoria, Melbourne.

McKee reported the presence of nodules on *C. deplanchei* Miq.

growing near Nouméa, New Caledonia, while Soemartono sent a similar report for *C. rumphiana* Miq. of which two trees growing in latosol soil in the Bogor district of Indonesia were examined; associated plants included species of *Nephrolepis*, *Erigeron*, *Eupatorium* and *Costus*. *C. papuana* S. Moore growing on ultra-basic soil (very poor in plant nutrients) on San Cristobal, Solomon Islands, was found by Hill to be well-nodulated.

It seems probable that the above reports include records of seven species not hitherto recorded as nodule-bearing.

Genus Elaeagnus

Hidajat reported that two out of three trees identified as *E. conferta* Roxb. proved to bear nodules; they were growing on andosolic soil on a mountain slope in the Bandung district of Indonesia, associated plants being species of *Altingia*, *Ageratum*, *Pandanus*, *Rubus* and *Psychotria*. Soemartono reported that the nine plants that were examined of *E. latifolia* L. (a climbing species) growing on a regosol soil at 1700 m in the Bogor district of Indonesia were all nodulated; associated plants included species of *Ficus*, *Castanea*, *Leea* and *Curculigo*.

Uemura (1971) recorded the presence of nodules in Japan on the following species: *E. glabra* Thunb., *E. matsunoana* Makino, *E. murakamiana* Thunb., *E. yoshinoi* Makino, and also on two varieties of *E. multiflora* Thunb., namely var. *augustifolia* Makino et Nemoto and var. *hortensis* Maxim.

Boullard reported that a plant growing in a French garden and identified as *E. reflexa* E. Morr. & Dcne. was nodule-bearing; however according to Index Kewensis this name is a synonym of *E. umbellata*, a species already recorded to be nodule-bearing.

The above reports add six to the list of recorded nodulating species in this genus.

Genus Discaria

Hannon reported that no nodules could be found on *D. pubescens* Druce growing as a native in the Canberra and Armidale districts of Australia, though the search was made difficult by the habit of the species of developing its roots in the cracks of the rocks on which it grows.

Bond found that a potted plant of a *Discaria* species in the Glasgow Botanic Gardens was well-nodulated. Unfortunately it has so far been impossible to identify the species, but the plant had been recently acquired by the curator (Mr E. W. Curtis) from the Royal Botanic Garden, Edinburgh, where it had been raised from seed collected in the wild in Chile. This appears to be the first evidence that at least one of the South American species of this genus has the capacity to form nodules. *D. toumatou,* which as noted in section 1 is the only species previously recorded to bear nodules, is confined to New Zealand.

Genus Cercocarpus

Youngberg reported that though he failed to find nodules on *C. ledifolius* Nutt. in the field in central Oregon, further study (Youngberg & Hu, 1972) showed that young plants of this species nodulated when planted in a soil from another source. It may be noted that Hoeppel & Wollum (1971) recorded nodulation in *C. montanus* Raf. and *C. paucidentatus* Britt.

5. Results of the search for nodulation in other species, mostly in genera closely related to those forming *Alnus*-type nodules

Following the suggestion made at the beginning of the survey (p. 445) a number of collaborators have inspected the root systems of further members of the Rosaceae and of the Rhamnaceae. The results are presented below.

Family Rosaceae

Wollum could find no nodules on a species of *Cowania* which is native in the southwest of the USA. This genus has been regarded as closely related to *Purshia.* Dietz was similarly unsuccessful in a search for nodules on two species native to S. Dakota, USA, namely *Rosa woodsii* Lindl. and *Rubus strigosus* Michx.

Soemartono examined several species of *Rubus* growing in the Bogor district of Indonesia and found that all four plants that were examined of one of the species, namely *Rubus ellipticus* Sm., bore what appeared to be root nodules. These particular plants were growing at an altitude of about 1500 m, on a regosol soil, and were associated with species of *Eupatorium*, *Impatiens* and *Elastostemma.* Mrs Soemartono sent a sketch (see Fig. 32.3) showing the nodules to consist of repeatedly forking, slender lobes, with the added presence of nodule roots analogous to those of species of *Myrica* or *Casuarina*; her examination of sections of the nodules showed that their formation had not been induced by any nematode or similar agent*, and that certain of the nodule cells appeared to contain a tangled mass of very fine hyphae. Mrs Soemartono sent preserved nodules and also seed of this *Rubus* to the compiler, and the further information gained by the use of these will be provided in section 7.

Fig. 32.3. Sketch of nodule clusters on *Rubus ellipticus* in Indonesia, drawn by Mrs Sri S. Soemartono. (× 0.5)

On the receipt of the above report Bond and Mackintosh examined Scottish material of the moorland species *Rubus chamaemorus* L. (some of it procured with the help of Mr B. W. Ribbons), also of *R. saxatilis* L. and of one of the species in the *R. fruticosus* complex; no nodules were found.

Family Rhamnaceae

Pizelle examined a total of sixteen trees of *Rhamnus frangula* L. at seven sites in France, without finding nodules; Khan (1971) had the same experience in his inspection of ten plants of *Rhamnus pentapomica* Parker growing in Pakistan, and has since reported that no nodules have been found on numerous plants of *Zizyphus jujuba* Lam. and of *Z. rotundifolia* Lam. examined in that country. Tallantire examined plants growing in Uganda of *Rhamnus prinoides* L'Hérit., *Gouania longispicata* Engl. and *Maesopsis eminii* Engl., but observed no nodulation. Hannon reported an absence of nodules on plants of *Cryptandra*

* It may be noted that Professor Anne Johnson reported a widespread occurrence of nematode nodules in the Singapore region.

amara Sm. and *Pomaderris lanigera* (Andr.) Sims growing in their natural habitats in the Sydney area; she examined very carefully excavated root systems of several plants each of *Pomaderris elliptica* Labill., *P. apetala* Labill. and *Spyridium ulicinum* Benth. sent from Tasmania, again with negative results. Dr Hannon also found that plants of *Alphitonia excelsa* (Fenzl.) Benth., native in the Sydney area, developed no nodules when grown from seed in the glasshouse in habitat soil for two to three years. Silvester reported that plants of *Pomaderris kumeraho* A. Cunn. and *P. phylicifolia* Lodd. ex Link, which often appear as pioneers in New Zealand, were non-nodulated.

On the other hand, in August 1972 Bond examined the root system of a potted plant labelled *Colletia cruciata* in the Glasgow Botanic Gardens to which his attention had been drawn by Mr E. W. Curtis, the Curator, and found plentiful nodulation. The nodules were branched and pinkish-white in colour, dotted with darker areas presumed to be lenticels (see Plate 32.1*a*). The plant had been grown from seed received from the Mainz Botanic Garden. The genus *Colletia*, which is South American and includes some seventeen species, has many points of resemblance to the genus *Discaria* (known to include at least two nodulating species, as noted earlier), but shoots of the above plant sent by Mr Curtis to the Herbarium at the Royal Botanic Gardens, Kew, were confirmed as belonging to a *Colletia*, the species being probably *C. paradoxa* (Spreng.) Escalante (syn. *C. cruciata* Gillies ex Hook.), though this could not be finally verified until flowers became available. Nodules were also found on plants growing in a border in a glasshouse at the Glasgow Botanic Gardens which were again certified by Kew to be a *Colletia* species, probably *C. armata* Miers. No nodules could, however, be found on an old plant of *Colletia paradoxa* (4 m in height) growing in another glasshouse. Further information about these *Colletia* nodules will be provided in section 7 and in the Addendum.

Other families

McKee reported that nodules had not been found on the rather recently named plant *Canacomyrica monticola* Guillaumin, native to New Caledonia and belonging to the Myricaceae; he also kindly sent seed of the species to the compiler so that further tests for a nodulating capacity might be made, but unfortunately they did not germinate.

Silvester examined a number of New Zealand plants which, though unrelated to known root nodule-bearing species, show pioneer properties in colonising sand dunes and dry river beds, and thus might be suspected of having some special source of nitrogen; these comprised members of the Myrtaceae (species of *Leptospermum* and *Metrosideros*) and of the Rubiaceae (species of *Coprosma* and *Nertera*). No root nodules were found.

Howard examined the root systems of a considerable range of species growing in Puerto Rico, some of the families represented being Piperaceae, Urticaceae, Myrtaceae, Melastomaceae and Ericaceae, together with many others. No examples of nodulation were found.

6. The occurrence on non-leguminous angiosperms of nitrogen-fixing nodules not of *Alnus*-type

Trinick has discovered that a species of the genus *Trema*, a member of the Ulmaceae, bears nodules of legume-type. Several months ago he sent details of the work to the compiler and said that, if desired, his findings could be included with the other results of the survey. Since an account of the work has now been published (Trinick, 1973) only the main features will be noted here.

Trema is a genus of some forty species occurring in tropical and subtropical regions of Asia, Africa and America, usually as an understorey in forests. Trinick found that root nodules were present on *T. cannabina* growing as weeds in tea plantations in New Guinea.* Sections showed that the structure of the nodules was similar to that of the legume-type, with a central area of cells packed with bacteria, a peripheral vascular system, and a parenchymatous cortex.

An organism closely resembling a slow-growing strain of *Rhizobium* in cultural characteristics was isolated from the nodules without difficulty, and proved able to induce nodulation in fresh *Trema* seedlings. These nodulated plants grew well in a nitrogen-free medium, and the nodules were active in acetylene reduction.

The same isolate proved able to nodulate four species of legumes – two species of *Vigna* and two of *Macroptilium* – out of nineteen species tested, selected from various cross-inoculation groups. Agglutination tests confirmed that the bacteria in all these nodules were identical with the strain originally isolated from *Trema*. Comments on these unexpected findings will be offered in section 8.

Notice should also be taken of the work of Klevenskaya (1971) on nodular hypertrophies occurring on the roots of the grass *Calamagrostis*

* In Trinick (1973) the plant studied was named as *Trema aspera* Decaisne, but Dr Trinick has since informed the compiler that the correct name of the plant is *Trema cannabina* Laur. var. *scabra* (Bl.) de Wit.

arundinacea (L.) Roth. in certain regions of Siberia, carried out as part of the Russian contribution to IBP. The nodules are some 4 mm in diameter, white or pink in colour when young, differing in their structure from both legume- and *Alnus*-type nodules. A bacterial endophyte is said to be present in their cortical cells, but no photographic evidence of this is provided; organisms isolated from the nodules were found to resemble a rhizobium in cultural characteristics but failed to induce nodulation in test plants. Percent nitrogen content in the nodules was considerably higher than in other parts of the plant. The exposure of whole plants, or of detached nodules (in some cases after a preliminary surface-sterilisation) to excess ^{15}N gave some evidence of the occurrence of fixation in the nodules, but not of benefit to the plant. In view of the potential significance of the occurrence of nitrogen-fixing nodules on a grass it is hoped that this work will be continued; in particular, tests of the ability of nodulated plants to grow in a nitrogen-free rooting medium would be of much interest.

7. Observations made in the compiler's laboratory on some species involved in the survey

With the help of seed and endophyte sources sent to the compiler by collaborators, it has been possible for him to grow nodulated plants of a few of the species mentioned above, to test the nodules for fixation, and to acquire other information concerning them.

Observations on Myrica faya

The discovery of nodules on this species is reported on p. 459. Dr C. Rodriguez-Barrueco sent seed and habitat soil to the compiler, and young plants raised from the seed formed nodules about four weeks after the application of a suspension of habitat soil to the roots, and remained nodule-free without that inoculation. As shown by Plate 32.1(*b*) and (*c*), nodulated plants grew vigorously in nitrogen-free water culture or Peralite, whereas nodule-free plants made little growth. A plant harvested 16 months after nodulation showed a leading shoot 110 cm tall, a dry weight of 102 g for top and roots and of 4.7 g for the nodules, while the total nitrogen content of the plant was 1.12 g. Obviously a vigorous fixation of nitrogen proceeds within the nodules. Upward-growing nodule roots were produced in abundance in these cultured plants of *M. faya*, as in all other species of *Myrica* that have been studied. It was also found that *M. faya* responded favourably to inoculation from *M. cerifera*, but that while an *M. gale* inoculum induced nodulation in *M. faya*, the nodules conveyed no benefit to the plant, in keeping with previous cross-inoculation trials (Mackintosh & Bond, 1970).

Observations on Myrica javanica

Becking (1970) reported briefly that nodulated plants of this species in pot culture showed evidence of fixation. Mrs E. B. Hidajat, who confirmed the presence of nodules on this species in Indonesia (section 3), sent seed to the compiler, and in his laboratory Mrs S. Mian found that young plants formed nodules in response to inoculation from several other species of *Myrica*, the subsequent growth of the plants indicating that good fixation was occurring. No source of the normal endophyte for *M. javanica* was available, but it can be concluded that nodules induced by it would also be nitrogen-fixing. Nodules induced by an *M. gale* inoculum again proved to be ineffective in fixation.

Observations on Myrica asplenifolia

Although this North American species has long been known to bear nodules, until recently there has been little experimental study of them. Fessenden, Knowles & Brouzes (1973) have now made an extensive study of the nodules under field conditions, and have confirmed that they are nitrogen-fixing, but the writer knows of no previous growth trials with control over the access of combined nitrogen in order to assess the value of the nodules to the plant. Dr R. J. Fessenden sent underground stem cuttings of the species and nodules to the compiler. Plants raised from the cuttings formed nodules two weeks after the application of a crushed-nodule inoculum to the roots and the nodulated plants proved able to grow strongly in water culture without combined nitrogen (Plate 32.1*d*). A plant harvested three months after inoculation had a leading shoot of height 46 cm, a shoot and root dry matter of 16.6 g and nodule dry matter of 0.58 g; percent nitrogen in the combined shoot and root dry matter was 1.79 and in the nodules 3.08, while the total nitrogen content of the plant was 316 mg, compared with the 20 mg of combined nitrogen supplied in the early establishment of the plant. The compiler's colleague Dr C. T. Wheeler found that the nodules, but not the roots, were active in acetylene reduction; the nodules produced 8.1 μmole ethylene/h/g fresh weight, which was similar to the value for *M. gale* nodules assayed at the same time.

Sections of the nodules showed that the endophyte was confined to the middle cortex where the cells were radially elongated. Upward growing roots were produced in abundance by the nodule clusters.

Observations on Rubus ellipticus

The external appearance of the preserved nodules of this species sent by Mrs Soemartono to the compiler was exactly as she had described (section 5). Transverse sections of nodule lobes were prepared, the structure found being shown in Plate 32.2(*a*); it conforms to the typical pattern of an *Alnus*-type nodule in that there is a superficial periderm, a broad cortex in which some cells are enlarged and contain a micro-organism, and a central stele. The enlarged infected cells are seen to form a layer 1–2 cells deep in the middle cortex, recalling a distribution found in most species of *Myrica.*

Unfortunately the method of fixation of the nodules did not permit of the preparation of sections suited to examination under the electron microscope, but after the scrutiny (with the help of Mrs S. Mian) of sections and squash preparations of the nodules under the highest power of the light microscope, the compiler is satisfied that the infected cells contain a dense growth of a finely hyphal organism, as suggested by Mrs Soemartono, the diameter of the hyphae being of the order of 0.5 μm. No vesicles comparable with those of *Alnus* nodules have been detected.

The seed also sent by Mrs Soemartono was sown in 1971, and since no habitat soil or fresh nodules were available, the seedlings were set up in greenhouse potting soil of low nitrogen content and grown on for 18 months. By that time the shoots were some 50 cm high and showed symptoms of nitrogen deficiency. Periodic examination of the roots showed that no nodules had appeared, permitting the conclusion that infection from the soil is necessary to nodulation, and that the required organism was absent from the soil employed.

The compiler tried to arrange for living material of the species to be sent from Indonesia to some centre where the nodules could be tested for nitrogen fixation but this proved to be impossible.

Observations on Colletia paradoxa

Nodules taken from the potted plant in the Glasgow Botanic Gardens on which they were first observed (section 5) were fixed in glutaralde-hyde–osmic acid, and sections for examination under the light and electron microscope were prepared by the compiler's colleague, Dr B. G. Bowes. Examination under the light microscope showed that the general structure was that of an *Alnus*-type nodule, with greatly enlarged cells with dense contents present in the middle cortex of the nodule (Plate 32.2*b*). Under somewhat higher magnification these cells were seen to contain numerous structures resembling the vesicles of *Alnus*, though somewhat smaller. Under the electron microscope this basic similarity was confirmed, since as seen in Plate 32.2(*c*) the structures in question (1.5–2 μm in diameter) proved to be compartmented, as in *Alnus*: fine, septate hyphae, their diameter of the order of 0.5 μm, are also clearly visible in the plate.

As a first step in the examination of the nodules for nitrogen fixation, nodules from the plant at the Botanic Gardens were assayed by Dr C. T. Wheeler by the acetylene method, at the rather unfavourable season (in the United Kingdom) of late October. A mean production of 1.6 μmole ethylene/h/g fresh weight of nodules was shown, while roots showed no reduction of acetylene.

As a second step, long-term growth trials were set up in spring, 1973, with the help of a number of rooted cuttings from the original plant kindly provided by Mr Curtis. These were set up in water culture in Crone's solution made up to a nitrogen-free formula, though a small addition of combined nitrogen was made to the cultures at this stage. After the cuttings were well-established they were transferred to nitrogen-free solution and inoculated by applying an aqueous extract of the soil from the original pot to their roots. Nodules appeared in 3–4 weeks and were similar to those shown in Plate 32.1(*a*). Nodulated plants proved to be able to grow well in the nitrogen-free rooting medium (Plate 32.2*d*). A similar plant was harvested after growth in nitrogen-free water culture for 7 months counted from the time of nodule initiation, when the height of the shoot was 5 cm. During the preceding 3 months a total of 10 mg of ammonium-nitrogen had been supplied, but even assuming its complete uptake the content of the plant at nodule initiation could not have exceeded 15 mg. At harvest the height of the leading shoot was 42 cm and the total dry weight of the plant 2.80 g; of this the nodules, which had formed at 37 sites, accounted for 0.18 g. Percent nitrogen in top and root combined was 2.80, in nodules 4.07. The total nitrogen content of the plant was 81 mg, indicating that at least 66 mg had been fixed.

8. Review of the results of the survey

The information presented in section 3 obviously puts our knowledge of the regularity with which species known prior to the survey to be capable of nodulation are actually nodulated under field conditions, on to a much firmer footing. Thus we now know that in Europe *Alnus glutinosa* is almost unfailingly nodulated, while *Myrica gale* is not quite so regularly nodulated. The new information in section 3 concerning the regularity of nodulation in *Myrica* species in eastern North America and in Indonesia is very welcome.

In the same section the reports concerning the large and important genus *Casuarina* reveal a degree of uncertainty concerning the regularity of nodulation in species recorded to have a nodule-forming capacity. Thus in Australia, while nodules can nearly always be found on some species, in others they frequently appear to be absent. As noted above, the view has been advanced that nodules are probably present in these latter instances but are situated too deeply to be revealed by conventional excavation. There is also the report by a collaborator in Indonesia that nodules could be found on only 4 % of the considerable number of trees of *C. equisetifolia* that were examined in a particular locality. In view of this situation, caution must be exercised at present in assessing the contribution of fixed nitrogen to the biosphere by some species of *Casuarina*.

The searches made by several collaborators for the presence of nodules on species of *Dryas* (see section 3) seem to confirm that although nodules can be found in Alaska and in some areas of Canada, they cannot be found in Europe, nor in Japan. However, the report by Mr Lalonde that at a Canadian site the nodules were rarely encountered until a depth of about 50 cm had been reached may be significant. In Lawrence *et al.* (1967) it was emphasised that in Alaska *Dryas* nodules cannot be found near to the soil surface and are mostly at a depth of 15 cm or more, but it now seems that where the depth of soil permits, the nodules may occur at much lower levels. The compiler has arranged for the excavation of *D. octopetala* plants at a Scottish site to a greater depth than hitherto (see Addendum).

The data reported in section 3 for *Arctostaphylos uva-ursi* seem to indicate a situation broadly parallel to that existing for *Dryas* species, since the presence of nodules has been detected in Alaska but not elsewhere. Here, however, there is at present no hint of the reason for these discrepant findings.

The observations presented in section 4 bring some improvement to the situation existing at the beginning of the survey, namely that only 119 out of a total of 342 species credited to the 13 genera known to include some nodulating species (section 1) had actually been examined for the presence of nodules. A dramatic rectification of this situation would be difficult to effect, since a common reason for the lack of information concerning so many species is the remoteness and inaccessibility of their places of growth.

As is indicated in Table 32.2, as a result of the survey records of nodulation have been obtained in a further 36 species. That the number of recorded species of *Myrica* has been more than doubled is a most welcome result for which the credit goes chiefly to collaborators in Africa, though the tracking down of a rather elusive European species

Table 32.2. *Number of additional species, in genera previously known to include species forming* Alnus-*type nodules, recorded during survey to be nodule-bearing*

Genus	Species complement*	No. of species recorded to bear nodules	
		Before IBP survey	At end of survey
Coriaria	15	12	13
Alnus	35	26	32
Myrica	35	12	26
Casuarina	45	17	24
Hippophaë	3	1	1
Elaeagnus	45	10	16
Shepherdia	3	2	2
Ceanothus	55	31	31
Discaria	10	1	2
Dryas	4	3	3
Purshia	2	2	2
Cercocarpus	20	1	4†
Arctostaphylos	70	1	1
Totals	342	119	157

* According to Willis (1966).
† Including the two species reported by Hoeppel & Wollum (1971).

(*M. faya*) is also a pleasing achievement. The failure to find nodules on some of the further species of *Casuarina* that were examined reinforces the reserved attitude taken towards this genus on p. 469. The situation in the genus *Elaeagnus* has been only moderately improved as a result of the survey, principally because there were no collaborators in the appropriate regions.

The discovery of nodulation in a South American species (so far unnamed) of *Discaria* prompts the hope that some botanist in that continent will investigate the several species of this genus that occur there. However, the nodulating habit may not be a generic characteristic since, as noted in section 4, Dr Hannon was unable to find nodules on an Australian species. The evidence now available suggests that the ability to form nodules is a generic attribute in *Coriaria*, *Alnus*, *Myrica*, *Shepherdia*, *Dryas* and *Purshia*.

Turning to section 5, the fact that the very limited examination of further genera of the Rosaceae during the survey yielded a record of nodulation in a further genus (*Rubus*) suggests that there must be further examples awaiting discovery. It is notable that whereas the previously known nodulating genera in this family (p. 466) are rather closely related and were all placed in the tribe Dryadeae by Engler, *Rubus* is placed elsewhere in the family. The finding that nodules were absent in other *Rubus* species in Indonesia and in Scotland shows that nodulation is not a generic character, though fresh evidence to this effect was scarcely necessary since several *Rubus* species are of course in cultivation, and their root systems thus open to inspection at transplanting.

Also in section 5 it is recorded that during the survey species of nine further genera of the Rhamnaceae were examined, with the result that nodulation was found in one of them (*Colletia*). Here again it seems likely that further examples of nodulation will eventually be found in the 50 so far unexamined genera. The presence of nodules in *Colletia* is not really surprising in view of its close relation to *Discaria*; an examination in their native habitats of the various South American species of *Colletia* would obviously be valuable.

The apparent absence of nodules on *Canacomyrica* (section 5) is of interest since it suggests that rather small morphological differences in the plant are accompanied by an absence of the nodule-forming habit that characterises the other genus in the Myricaceae (*Myrica*).

A general point concerning *Alnus*-type nodules is the occasional development of nodules in species growing as introductions in countries where the genus of the species is unrepresented. Space considerations forbid more than passing reference to this problem. The nodulation of *Casuarina* species in Africa, of *Ceanothus* species in Ireland, France and Japan, of *Discaria* and *Colletia* species in Glasgow Botanic Gardens, all raise the problem of the source of the endophyte. Doubtless these plants were all raised from seed in the country of introduction, and although the presence of the endophyte as a surface contaminant on the seed may be mentioned as a possible source, laboratory experience is that plants raised from seed which have not been surface-sterilised very rarely develop nodules without deliberate inoculation. Still, that remains the most likely explanation.

All the foregoing part of this section refers to *Alnus*-type nodules. The discovery of legume-type nodules on a non-legume, noted in section 6, obviously has far-reaching implications. The long-standing belief that *Rhizobium* symbioses only with legumes has to be discarded. Moreover it is inconceivable that Dr Trinick should have stumbled across the only example of the symbiosis of *Rhizobium* with a non-legume.

The compiler is grateful to Mr A. A. Bullock, formerly of the Herbarium, Royal Botanic Gardens, Kew, and to Dr C. Rodriguez-Barrueco for their assistance in drafting the documents sent to collaborators at the beginning of the survey. The British National Committee for IBP provided a grant which enabled the compiler to have an assistant for one year, and he is very appreciative of the many ways in which Dr Anne H. Mackintosh, who acted in that capacity, facilitated the organisation of the survey and assisted in the laboratory work arising from it. The compiler is grateful to Dr F. J. Bergersen and Dr R. F. Tarrant for special help in securing collaborators, to Mr T. N. Tait for photography, and to the University of Glasgow for defraying the considerable postal expenses incurred during the survey.

References

Akkermans, A. D. L. (1971). Nitrogen fixation and nodulation of *Alnus* and *Hippophaë* under natural conditions. PhD Thesis. University of Leiden.

Allen, E. K., Allen, O. N. & Klebesadel, L. J. (1964). An insight into symbiotic nitrogen-fixing plant associations in Alaska. In: *Science in Alaska* (ed. G. Dahlgren), pp. 54–63. Anchorage, Alaska.

Becking, J. H. (1970). Plant–endophyte symbiosis in non-leguminous plants. *Plant and Soil*, **32**, 611–54.

Brunchorst, J. (1886). Die Structur der Inhaltskörper in den Zellen eineger Wurzelanschwellungen. *Bergens Museums Aarsberetning*, pp. 233–48.

Brunchorst, J. (1886–8). Über einige Wurzelanschwellungen, besonders die jenigen von *Alnus* und den *Elaeagnaceen. Untersuchungen aus dem Botanische Institut Tübingen*, **2**, 150–77.

Engler, A. (1964). *Syllabus der Pflanzenfamilien*, 12th edition (ed. H. Melchior), vol. 2. Gebrüder Borntraeger, Berlin.

Fessenden, R. J., Knowles, R. & Brouzes, R. (1973). Acetylene–ethylene assay studies on excised root nodules of *Myrica asplenifolia* L. *Proceedings of the Soil Science Society of America*, **37**, 893–98.

Hannon, N. J. (1956). The status of nitrogen in the Hawkesbury sandstone soils and their plant communities in the Sydney district. I. The significance and level of nitrogen. *Proceedings of the Linnaean Society of New South Wales*, **81**, 119–43.

Hoeppel, R. E. & Wollum, A. G. II. (1971). Histological studies of ectomycorrhizae and root nodules from *Cercocarpus montanus* and *Cercocarpus paucidentatus*. *Canadian Journal of Botany*, **49**, 1315–18.

Index Kewensis (1893–1970). Parts 1–4 and Supplements 1–14. Clarendon Press, Oxford.

Khan, A. G. (1971). Root nodules in non-leguminous plants in West Pakistan. *Pakistan Journal of Botany*, **3,** 71–8.

Killick, D. J. B. (1969). The South African species of *Myrica. Bothalia*, **10,** 5–17.

Klevenskaya, I. L. (1971). The root nodules of *Calamagrostis arundinacea* (L.) Roth. In: *The recent achievements in the study of the biological nitrogen fixation* (ed. E. N. Mishustin), pp. 100–5. Publishing House Nauka, Moscow.

Lawrence, D. B., Schoenike, R. E., Quispel, A. & Bond, G. (1967). The role of *Dryas drummondii* in vegetation development following ice recession at Glacier Bay, Alaska, with special reference to its nitrogen fixation by root nodules. *Journal of Ecology*, **55,** 793–813.

Mackintosh, A. H. & Bond, G. (1970). Diversity in the nodular endophytes of *Alnus* and *Myrica. Phyton*, **27,** 79–90.

Rodriguez-Barrueco, C. (1968). The occurrence of nitrogen-fixing root nodules on non-leguminous plants. *Botanical Journal of the Linnaean Society*, **62,** 77–84.

Specht, R. L. & Rayson, P. (1957). Dark Island heath (Ninety-Mile Plain, South Australia). III. The root systems. *Australian Journal of Botany*, **5,** 103–14.

Trinick, M. J. (1973). Symbiosis between *Rhizobium* and the non-legume, *Trema aspera. Nature, London*, **244,** 459–60.

Uemura, S. (1971). Non-leguminous root nodules in Japan. *Plant and Soil*, Special Volume, 349–60.

Van Ryssen, F. W. J. & Grobbelaar, N. (1970). The nodulating and nitrogen fixing ability of South African *Myrica* species. *South African Journal of Science*, **66,** 22–5.

von Faber, C. (1925). Untersuchungen über die javanischen Solfataren-Pflanzen. *Flora*, N.F., **18/19,** 89–110.

Willis, J. C. (1966). *Dictionary of the flowering plants and ferns*, 7th edition revised by H. K. Airy Shaw. Cambridge University Press, London.

Youngberg, C. T. & Hu, L. (1972). Root nodules on mountain mahogany. *Forest Science*, **18,** 211–12.

Addendum

Dr P. S. Nutman reported that *Myrica kraussiana* (see p. 458) growing at an altitude of 1053 m on Table Mountain, South Africa was well-nodulated.

In a further examination of *Dryas octopetala* (see pp. 457 and 469) Gimingham excavated roots at Bettyhill, Scotland to a depth of 1.5 m without finding nodules.

Uemura & Sato (in: *Nitrogen fixation and nitrogen cycle* (ed. H. Takahashi), JIBP Synthesis Volume 12, University of Tokyo Press, 1975) have reported nodulation in *Elaeagnus formosana* Nakai, *E. montana* Makino and *E.*

numajriensis Makino. These appear to be first records, additional to those credited to Uemura on p. 460 above.

The identification of the nodulated plant found in the Glasgow Botanic Gardens as *Colletia paradoxa* (pp. 463 and 467) has been confirmed by the Herbaiium at Kew Gardens.

Plate 32.1. (*a*) Root nodules of *Colletia paradoxa* from plant in the Glasgow Botanic Gardens. (*b*) Plants of *Myrica faya*, one year old, grown from seed in Peralite free of combined nitrogen, without (left) and with (right) nodules. (*c*) A nodulated plant of *Myrica faya* after 16 months growth from seed in water culture free of combined nitrogen. (*d*) A nodulated plant of *Myrica asplenifolia* grown for three months after inoculation in water culture free of combined nitrogen.

(*Facing p.* 474.)

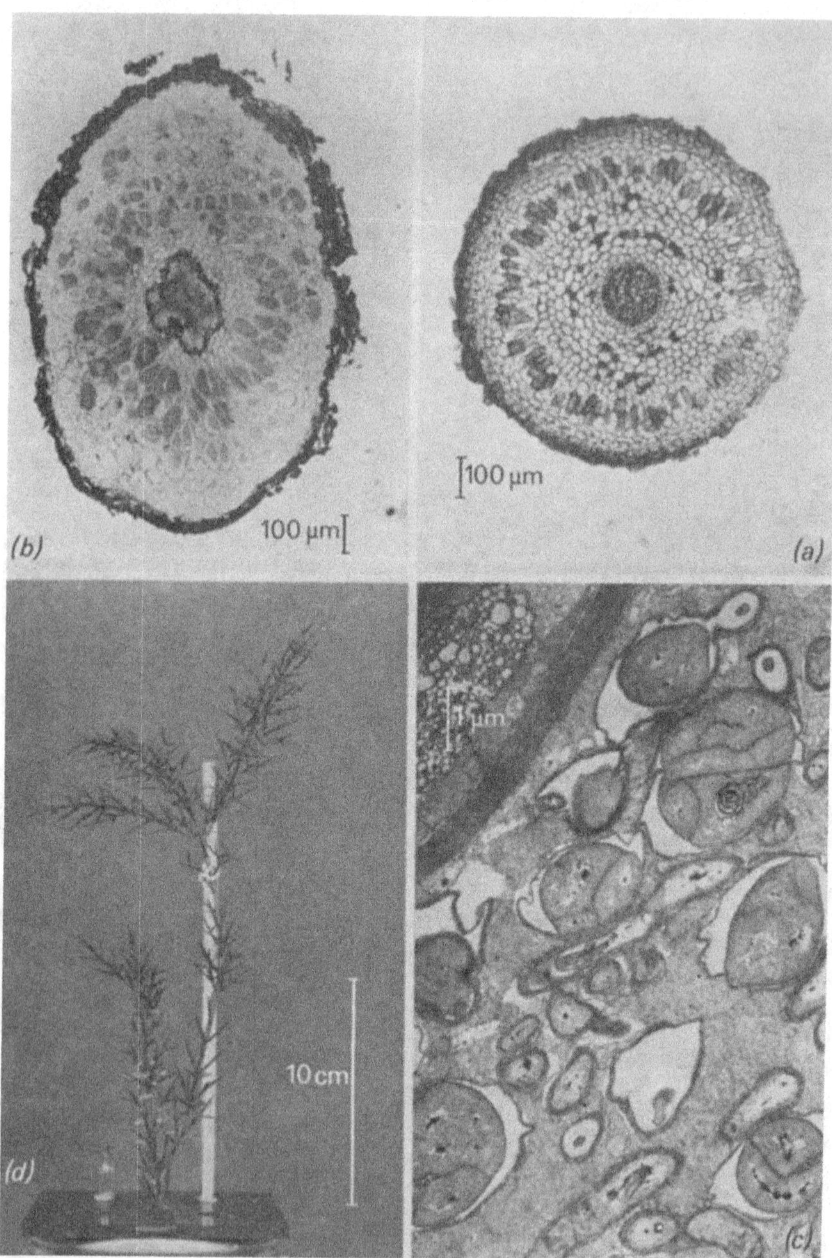

Plate 32.2. (*a*) Transverse hand-section of a lobe of a nodule cluster of *Rubus ellipticus*. The endophyte occurs in the layer of radially enlarged cells in the cortex. For details see text. (*b*) Transverse section of a nodule lobe of *Colletia paradoxa*. The endophyte is present in the enlarged cells with dark-stained contents in the cortex. There is a superficial periderm. (*c*) Part of an infected cell from a section of a nodule lobe of *Colletia paradoxa* seen under the electron microscope. Endophytic hyphae cut in various planes are present, also the larger, roughly spherical vesicles showing internal septa. (*d*) A nodulated plant of *Colletia paradoxa* grown in nitrogen-free water culture for 7 months after inoculation. At inoculation there was a single shoot 5 cm tall.

33. Symbiotic interactions in non-leguminous root nodules

A. F. ANGULO, C. van DIJK & A. QUISPEL

The comparison of symbiotic nitrogen fixation in leguminous and non-leguminous plants to identify points of difference and similarity can provide important information about those symbiotic interactions that are general and essential prerequisites for the occurrence of symbiotic nitrogen fixation.

It is for this reason that in the research group on biological nitrogen fixation in Leiden, where attention is given mainly to leguminous plants, some research is also being done on non-leguminous root nodules, especially those of *Alnus glutinosa* Gaertn. This work was promoted by a grant from the Dutch IBP Committee, covering the period 1966–70, to A. D. L. Akkermans, whose results were published in 1971. The present article is mainly concerned with more recent research performed as a continuation of this original IBP work. Of the studies now described, those on infection and nodule formation were performed by the first author (A.F.A.), and most of the work on the life cycle of the endophyte by the second (C.v.D.). For a fuller description of these studies reference should be made to Angulo (1974).

The infection process

In non-leguminous plants, just as in leguminous species, infection is preceded by deformation of root hairs (Hiltner, 1903; Pizelle, 1972). The characteristic zones with deformed and reduced root hairs made it easier to detect the very first phases of infection. Moreover, infections proved to be limited to those parts of the root that were elongating or had just elongated at the onset of the infection. This was demonstrated by time-lapse photographs of inoculated root systems growing between two glass slides.

In such very early phases of the infection process in *Alnus* we see that the endophyte, in the form of hyphae, passes through a deformed root hair and from there penetrates the cells of the cortex. Our observations confirm the earlier reports of Pommer (1956) and Taubert (1956). Under the influence of the penetrating hyphae, the cortical cells close to the infected cells start to divide (Plate 33.1). In fact, cell divisions are only observed in the cells ahead of the penetrating hyphae, penetration itself being restricted to the now elongating cells formed by these cell divisions. This process is essentially comparable with the situation seen during the early phases of nodulation and growth of leguminous root nodules, and therefore appears to be a general aspect of infection in both types of root nodule symbiosis.

There is, however, one marked difference between leguminous and non-leguminous nodule formation. Whereas in leguminous nodules cell divisions spread to other cells and finally produce the growing root nodule, in *Alnus* the meristematic activity is initially restricted to the immediate vicinity of the infection and causes only a slight swelling of the root. We shall call such swellings the primary nodules. The final root nodule is only formed if in the neighbourhood of these endophyte-filled cortical cells, cell divisions in the pericycle lead to the formation of what in the beginning looks like a normal lateral root primordium. The endophyte penetrates the cells of these primordia in the way described by Taubert (1956), after which a true nodule primordium develops, one which can be distinguished from the uninfected normal root primordium by its more rounded appearance.

This raises the question of whether the successful formation of a nodule in *Alnus* is dependent on the accidental presence of a normal root primordium somewhere in the vicinity of the primary nodule or whether the formation of these primordia is initiated by the infected cells of the primary nodule. An attempt was made to solve this problem by a precise analysis of the number and distribution of lateral roots, root primordia, and nodule primordia in inoculated plants as compared with roots and root primordia in uninoculated control plants. To avoid the influence of differences in nitrogen nutrition, all plants were cultivated in nutrient solutions containing 0.05 g KNO_3/l. This concentration was chosen because small fluctuations in nitrogen content in this region did not influence either nodulation or lateral root formation.

Simple counts of roots and root and nodule primordia sufficed to show that the formation of root nodules did not coincide with any reduction of root formation, as was to be expected if the nodule primordia were the result of infection of normal root primordia. The distribution pattern of root and nodule primordia along the root system was determined by measuring the distance between roots, between nodules, and between nodules and neighbouring roots, allowing for the radial angle between the different outgrowths. This was done by grouping the observations according to five radial angles, i.e. 0, 45, 90,

135 and 180 degrees. No difference was observed, except for the very closest spacings, between the root pattern of inoculated and uninoculated plants or between the patterns of roots and root nodules. This means that nodules may be formed much closer together and much closer to neighbouring roots than is usual for normal root development. This is in accordance with microscopic observations in the young nodules, which show that several nodule primordia are often formed in a very short part of the root, a phenomenon which is rarely seen in normal root formation.

We conclude from these observations that the occurrence of the cell divisions in the pericycle near the primary nodule are at least partly attributable to the influence of the primary nodule, since the number of sites of division exceeds that observed in the absence of infection.

This conclusion raises another question, i.e. whether such primordia are always initiated underneath or near the primary nodule. Investigation of the endophyte cells in the cortex of root regions where many nodule primordia developed indeed showed that in most cases these primordia were formed in the centre of endophyte concentrations in the cortex. However, endophyte development was sometimes observed in parts of the cortex where no primordium had been formed, which indicated that the mere presence of infected cells is not the sole inductive factor in the formation of nodule primordia.

Some of the microscopical observations point in the same direction. If the primordia were initiated only by the infected cells of the primary nodule, we could expect them to be formed underneath the cells filled with endophytes. Although this happens in some cases, most of the primordia are formed in a lateral position or even at the opposite side of the root. Such effects could result from the action of growth substances. Libbenga & Bogers (1974) explained the initiation of cell divisions in peas by the interaction of gradients of auxins, and probably cytokinins, produced by the infected cells in conjunction with another as yet unidentified factor or factors arising from the central cylinder. A similar explanation might hold for either the initiation of cell divisions in the cortex forming the primary nodule or the initiation of cell divisions in the pericycle forming the nodule primordium. It is generally observed that cell divisions leading to normal lateral root formation are stimulated by auxins, and *Alnus* does not appear to be an exception to this rule. Earlier observations by Dullaart (1970) demonstrated hyperauxiny in fully developed *Alnus* nodules. It remains for further studies to show whether the same holds for the primary nodule as well, and how the plant itself co-determines the occurrence of cell divisions.

The life cycle of the endophyte

All penetration of host cells is accomplished by actinomycete-like prokaryotic hyphae. As noted, these hyphae penetrate the root hairs and proceed further into the cortical cells of the primary nodule. Up to about two or three weeks after the infection of the root hairs, the endophyte in the primary nodule is still present mainly in the form of these hyphae, the cortical cells being completely filled with clusters of hyphae growing around the swollen nucleus of the host cells. Thereafter, the characteristic vesicles are formed by the endophyte in all cells, even by hyphae still present in the infected root hairs. In the nodule proper, cells close behind the apical meristem become infected, and initially contain only hyphae, but in cells just below this zone vesicle formation occurs. Initially the vesicles are formed only at the cell surface; later they may be found closer to the central parts as well.

As first described by Becking, De Boer & Houwink (1964) for *Alnus*, the vesicles are characterised by remarkable, usually incomplete septa partially dividing the nuclear material, by many plasmalemmosomes, and the so-called striated bodies (see Gardner, Chapter 34).

With regard to the symbiotic interactions, the contact between the surface of the endophyte cells and the host cytoplasm deserves special attention. Under the electron microscope this surface shows the thin cell envelope of the endophyte surrounded by an amorphous layer of varying thickness. This layer is in turn surrounded by a membrane of obvious host origin (van Dijk & Merkus, unpublished observations). Where hyphae penetrate the host cells, the amorphous layer is contiguous with the penetrated cell wall, which indicates that the amorphous layer is deposited by the host. Since this layer is also found on the vesicles, we must assume that even the vesicles are still surrounded by a deposit of cell wall material from the host. If this is indeed the case, we are concerned here with a distinct developmental difference between leguminous and non-leguminous root nodules, because the infection thread of the leguminous root nodules is dissolved during liberation of the bacteria, which subsequently are surrounded only by a host plasma membrane.

The endophyte of *Alnus* occurs in still another form, depending on strain characteristics of the infecting endophyte (see Akkermans & van Dijk, Chapter 36). This form has been described under various names;

bacteroids, bacteria-like cells, and granules. Until more is known about their function, we prefer to use the last of these terms, which is purely descriptive.

The granules consist of small, thick-walled cells with a variably irregular outline in cross-section. It certainly is not justifiable to consider them as characteristic only of the older parts of the nodules. Although they are generally formed somewhat later than the vesicles, they can be observed in the young cells, in which the first vesicles are formed. The observations of Käppel & Wartenberg (1958) on granule formation were confirmed and extended. Formation occurs in the intercellular spaces or within cells, in the latter case starting with the thickening of certain hyphae. In the next phase not only longitudinal but also transverse cell divisions occur. This leads to the so-called spindles, which grow out by further cell divisions, the small endophyte cells being separated from each other by a thickening mucilaginous mass (Plate 33.2). The resulting mass of cells is always surrounded by a layer of cell wall material deposited by the host, which separates the mass of granules from the living cytoplasm with its hyphae and vesicles. Only when the host cell dies off does this surrounding cell wall dissolve, after which the granules fill the now empty dead host cell. After deterioration of the endophyte cells in the older parts of the nodule, the granules remain as the only vital parts of the endophyte. This gives the erroneous impression that they are preferentially formed in the older parts of the nodules.

There is another very important problem with regard to our understanding of symbiotic interactions. There can be no doubt that the hyphal form is the original form of the endophyte, since it is only in this form that the endophyte penetrates the host cells and growth in these cells initially only takes place in this form. Both vesicles and granules are formed at a later stage in the development of the symbiosis. The formation of different cell types after a certain growth period, e.g. fructifications, is a quite normal phenomenon in growing microorganisms. On the other hand, the formation of these structures may be the consequence of the influence of the host cells on the endophyte. This problem is directly related to the question of the possible functions of vesicles and granules. Although it is likely that a definite answer to these questions will only be given after successful cultivation of the endophyte, we can nevertheless try to obtain some indirect evidence.

It is interesting to compare the formation of vesicles with the formation of bacteroids in leguminous root nodules. Both are formed at a moment when the initial growth of the micro-organisms in the host cells comes to an end. This end of the multiplication period certainly is influenced to some degree by the host, and in this respect we may regard the formation of vesicles as the result of an effect of the host on the endophyte. However, we cannot say with certainty whether the formation of the vesicles is merely the result of the cessation of multiplication or is caused by a more specific influence of the host on the endophyte. It is not unusual in micro-organisms for certain types of cells, such as fructifications, to be formed only after vegetative growth has stopped. On the other hand, we know of no other micro-organisms of the prokaryotic type that forms cells of this remarkable kind. Their structure, with all the evidence of high metabolic activity, excludes their being considered merely as involution forms.

The concept that the formation of vesicles is indeed in some way analogous to the formation of bacteroids in leguminous root nodules leads to the further possibility that they might be the active sites of nitrogen fixation. Akkermans (1971) tried to determine in ^{15}N fixation experiments whether the ^{15}N could be recovered in the vesicles after separation by selective filtration through nylon gauze. Although some ^{15}N was recovered from the vesicle suspensions, the amount was small compared with the amount in other fractions. If nitrogen was fixed in the vesicles, it must have been rapidly transported to other fractions. Reduction of acetylene in suspensions of clusters of vesicles could not be demonstrated by Akkermans.

Using cytochemical methods with tetrazolium salts, Akkermans (1971) demonstrated that the clusters of vesicles were the sites of the highest reducing activity in the nodules. The formation of the formazan crystals could even be observed inside the vesicles. Of course, the presence of highly reducing conditions does not prove that nitrogen is reduced as well. On the other hand, we know that nitrogen fixation demands strongly reducing conditions.

We must realise that this method is essentially a comparative one. All living cells reduce tetrazolium salts under appropriate conditions. Under exclusion of oxygen, e.g. in a hydrogen atmosphere, hyphae and granules reduce tetrazolium salts as well, a phenomenon rarely observed under aerobic conditions. Angulo (1974) applied this method to study the cells of the primary nodule in which vesicles had been just formed. In oxygen-free preparations the hyphae and vesicles formed tiny formazan crystals after the addition of neo-tetrazolium, but in some of the probably ripest vesicles a thick black precipitate was formed, thus showing that vesicle development coincides with an increase in the reducing potentialities. We can only speculate about the origin of these reducing conditions. The greater dimensions of the vesicles may promote the formation of special reducing sites. The striated bodies, which appear to be highly specific for the vesicles, also deserve special attention. It may be that they have some function in nitrogen reduction, though they could also be products of nitrogen enrichment.

There is no reason to believe that the formation of spindles and granules is associated with the end of cell multiplication by thc endophyte. On the contrary, their formation by definite cell multiplications gives the impression that they represent a kind of escape from the limitations set by the symbiosis. They are formed at sites where they are separated from the remaining parts of the cell or even in intercellular spaces, and remain in a healthy condition after all other phases of the endophyte have died off. Their absence in many nodules and the finding that there are no differences in average nitrogen fixing capacity between nodules with or without granules, means that they cannot be assigned an important role in nitrogen fixation. Clusters of granules are not characterised by pronounced reducing conditions. In our laboratory van Hiele (unpublished observations) obtained infection and nodule formation with isolated groups of granules in which no hyphae or vesicles could be found, though it is very difficult to exclude their presence beyond all doubt. If the infectivity of granules could be demonstrated with certainty, there would be every reason to consider them true spores. This might be helpful in elucidating the taxonomic position of the endophytes. In actinomycetes the formation of spores by cell divisions in longitudinal and transverse directions is described only for the family of the Dermatophylaceae. Extracts of *Alnus* endophyte suspensions showed the presence of mesodiaminopimelic acid (m-DAP), but no LL-DAP was found. If we take these characteristics as the key to the families of the Euactinomycetes, as proposed by Lechevalier & Lechevalier (1960), we are immediately led to the family of the Dermatophylaceae.

Cultivation of the endophyte

Advance in the elucidation of the symbiotic interactions and understanding of the prerequisites for nitrogen fixation in this type of symbiosis can only be expected after axenic cultures of the endophytes become available. The problems involved in obtaining such cultures are still unsolved. Axenic inoculation material can be obtained either by selective incubation, as used by Quispel (1954*a*,*b*, 1955, 1960), or by the dilution technique used by Lalonde & Fortin (1972), but neither of these methods absolutely guarantees the absence of contaminating organisms. In our experiments we sometimes observed growth of actinomycetes in certain nutrient solutions after inoculation with nodule material obtained by selective incubation, but such solutions were not infective. Other authors have made similar observations. It is conceivable that the endophyte loses its infective capacity during growth in nutrient solutions. It would be extremely interesting if it could be shown that such cultures are identical with the endophyte, but for further experimental work only really infective cultures will be useful. Experiments demonstrating an increase in infective capacity during incubation (Quispel, 1960) are still difficult to reproduce and are open to the criticism that cell fragmentation or sporulation without further growth can also lead to an increase in the number of infective particles.

A combination of different methods based on microscopical observations and measurement of infective capacity may ultimately yield convincing results. In such efforts due account can be taken of our improved understanding of symbiotic relations and the vitality and growth *in vivo* of the different phases in the life cycle of the endophyte.

References

Angulo, A. F. (1974). La formation des nodules fixateurs d'azote chez *Alnus glutinosa* (L.) Vill. PhD Thesis, University of Leiden.

Akkermans, A. D. L. (1971). Nitrogen fixation and nodulation of *Alnus* and *Hippophaë* under natural conditions. PhD Thesis, University of Leiden.

Becking, J. H., De Boer, W. E. & Houwink, A. L. (1964). Electron microscopy of the endophyte of *Alnus glutinosa. Antonie van Leeuwenhoek, Journal of Microbiology and Serology*, **30**, 343–76.

Dullaart, H. (1970). The auxin content of root nodules and roots of *Alnus glutinosa* (L.) Vill. *Journal of Experimental Botany*, **21**, 975–84.

Hiltner, L. (1903). Über die biologischen und physiologischen Bedeutung der endotrophen Mykorhiza. *Naturwissenschaftliche Zeitschrift für Land- und Forstwirtschaft*, **17**, 9–25.

Käppel, M. & Wartenberg, H. (1958). Der Formenwechsel des *Actinomyces alni* Peklo in den Wurzeln von *Alnus glutinosa* Gaertn. *Archiv für Mikrobiologie*, **30**, 46–63.

Lalonde, M. & Fortin, J. A. (1972). Formation des nodules racinaires axéniques chez *Alnus crispa* var. mollis. *Canadian Journal of Botany*, **50**, 2597–600.

Lechevalier, H. A. & Lechevalier, M. P. (1970). A critical evaluation of the genera of aerobic Actinomycetes. In: *The actinomycetes* (ed. H. Prawser), pp. 393–403, Fischer Verlag, Jena.

Libbenga, K. R. & Bogers, R. J. (1974). Root nodule morphogenesis. In: *Biology of nitrogen fixation* (ed. A. Quispel), pp. 430–72. North Holland Publishing Co., Amsterdam.

Pizelle, G. (1972). Observations sur les racines des plantules d'Aune glutineux (*Alnus glutinosa* Gaertn.) en voie de nodulation. *Bulletin de la Société botanique de France*, **119**, 571–80.

Pommer, E. H. (1956). Beiträge zur Anatomie und Biologie der Wurzelknöllchen von *Alnus glutinosa* Gaertn. *Flora, Jena*, **143**, 603–34.

Quispel, A. (1954*a*). Symbiotic nitrogen fixation in non-leguminous plants. I. Preliminary experiments on the root-nodule symbiosis of *Alnus glutinosa*. *Acta Botanica Neerlandica*, **3**, 495–511.

Quispel, A. (1954*b*). Symbiotic nitrogen fixation in non-leguminous plants. II. The influence of the inoculation density and external factors on the nodulation of *Alnus glutinosa* and its importance to our understanding of the mechanism of the infection. *Acta Botanica Neerlandica*, **3**, 512–32.

Quispel, A. (1955). Symbiotic nitrogen fixation in non-leguminous plants. III. Experiments on the growth *in vitro* of the endophyte of *Alnus glutinosa*. *Acta Botanica Neerlandica*, **4**, 671–89.

Quispel, A. (1960). Symbiotic nitrogen fixation in non-leguminous plants. V. The growth requirements of the endophyte of *Alnus glutinosa*. *Acta Botanica Neerlandica*, **9**, 380–96.

Taubert, H. (1956). Über den Infektionsvorgang und die Entwicklung der Knöllchen bei *Alnus glutinosa* Gaertn. *Planta, Berlin*, **48**, 135–56.

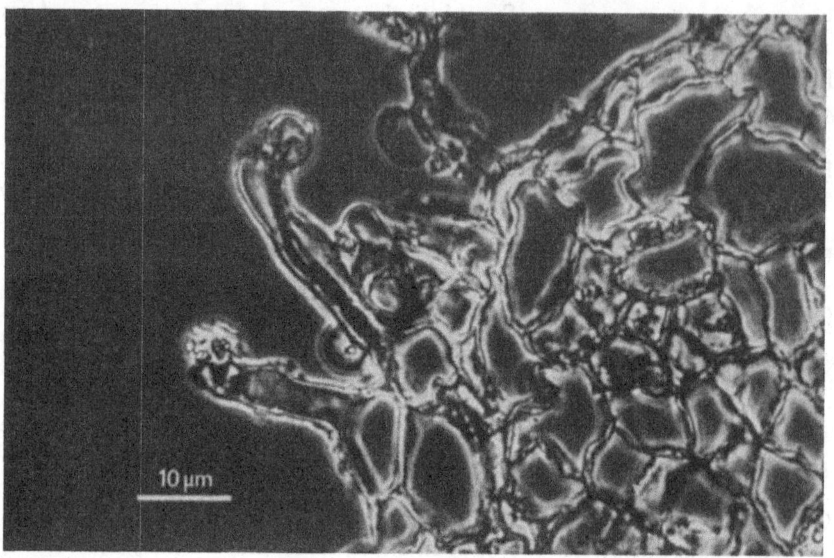

Plate 33.1. Phase-contrast photograph of the early infection in root of *Alnus glutinosa*. Note the hyphae in the central curled root hair and the group of dividing cells in the cortex.

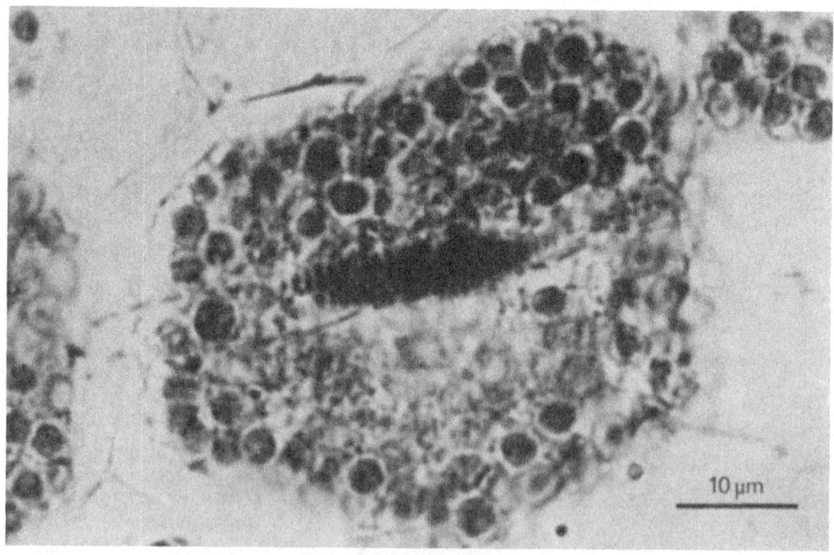

Plate 33.2. Formation of a spindle in an infected cell of the cortex of *Alnus glutinosa*. Note the septate hypha which connects the cells and thickens to a spindle in the centre of a cluster of hyphae and vesicles. Stained with toluidine-blue, section 10 μm thick, cut in Epon.

34. Ultrastructural studies of non-leguminous root nodules

ISOBEL C. GARDNER

Although electron microscopy is proving of the greatest assistance in the investigation of non-leguminous root nodules, studies by that technique have been concentrated so far almost exclusively on the nodules of *Alnus glutinosa*, apart from a description by Gatner & Gardner (1970) and Gardner & Gatner (1973) of the nodules of *Hippophaë rhamnoides*. This chapter reports results of further studies on the nodules of the latter species, together with observations on those of *Myrica gale* L. and *M. cerifera* L., *Ceanothus velutinus* Dougl. and *Casuarina cunninghamiana* Miq.

The preparation of ultra-thin sections of these nodules presents many problems. It is difficult to devise fixation procedures which are satisfactory both for the endophyte and the host plant tissue, or even for the different stages in the development of the endophyte itself. In addition the resin or tannin deposits regularly present in the uninfected cells of these nodules create sectioning problems. The deposits become very hard after normal fixation and appear to resist the embedding process. Where the substances exist as discrete droplets in the vacuoles (Plate 34.1*a*), a fall-out of these during sectioning can result in contamination of other cells. When larger deposits of these hardened substances are present, shattering of sections is a common experience.

Endophyte hyphae

The hyphae of the various endophytes here studied resemble those of the *Alnus glutinosa* endophyte and thus present a similar ultrastructure irrespective of their host. In diameter they range from 0.5 to 1.0 μm (Table 34.1). The hyphae of the *Hippophaë rhamnoides* endophyte serve as an example (Plate 34.1*b*, *c* and *d*). They are septate and exhibit true branching. They are bounded by a cell wall inside which lies a plasma membrane and they contain granular cytoplasm with typical bacterial nucleoid regions which possess no nuclear membranes. Plasmalemmasomes (mesosomes) are conspicuous features of the hyphae (Plate 34.1*c*). Such structures, which are characteristic of Gram-positive bacteria, have been most extensively studied in the bacilli (Rogers,

1970), but have also been reported in filamentous bacteria (Glauert & Hopwood, 1960; Gordon & Edwards, 1963; Chen, 1964; Bradley & Ritzi, 1968). Details of a typical plasmalemmasome are shown in Plate 34.1(*d*). As in the bacilli it consists of two parts; a membrane sac which is continuous with the endophyte plasma membrane delimits the

Table 34.1. *Endophyte characteristics in various non-leguminous nodules*

Host plant	Hyphae	Vesicles	Bacteroids
Alnus glutinosa	0.7–1.0 μm	Spherical 4–5 μm	Oval 1.0 μm diam. 1.3 μm long
Hippophaë rhamnoides	0.6 μm	Spherical 3–4 μm	Rod-shaped 0.9 μm diam. 2.0 μm long
Myrica gale	0.7 μm	Club-shaped 1.5 μm diam 4.5 μm long	Oval 1.0 μm diam. 1.5 μm long
Myrica cerifera	1.0 μm	Club-shaped 2.0 μm diam. 10.0 μm long	Rod-shaped 0.5 μm diam. 1.0 μm long
Ceanothus velutinus	0.6 μm	Pear-shaped 2.3 μm diam. 2 5 μm long	Oval 0.9 μm diam. 1.2 μm long
Casuarina cunninghamiana	0.6 μm	Club-shaped 1.0 μm diam. 2.0 μm long	Oval 1.0 μm diam. 1.4 μm long

structure and within this lies a membrane system which in part appears lamellar and in part vesicular. Becking, De Boer & Houwink (1964) recorded plasmalemmasomes in the *Alnus glutinosa* endophyte and concluded that different types existed. However, it is now generally agreed that the appearance of plasmalemmasomes can vary considerably, depending on the preparative techniques used prior to sectioning (Burdett & Rogers, 1970). The function of these membranous organelles is still not clear. That they are oxidative organelles equivalent to the mitochondria of the eucaryotes has recently been questioned by Burdett & Rogers (1972) but they do seem to be involved in septum formation and in the export of some exocellular enzymes.

These hyphal characters are strongly indicative of the actinomycetal nature of the endophytes of these non-leguminous nodules. Hyphae of the *Coriaria myrtifolia* endophyte were shown by Becking (1970a) to possess a similar ultrastructure and a preliminary report by Chandler &

Dart (1971) on the *Casuarina cunninghamiana* endophyte would also agree with the above although no micrographs were presented. Such a micrograph (Plate 34.1*e*) is of special interest because the only previous detailed cytological studies of nodules of this genus were those of Miehe (1918) and McLuckie (1923) who considered the endophyte to be a fungus and a rhizobium-like bacterium respectively.

Endophyte vesicles

The spherical vesicles formed by the endophyte of *Hippophaë rhamnoides* are somewhat smaller than those of *Alnus glutinosa* (Table 34.1) but structurally they are very similar. When mature they exhibit random septation (Plate 34.2*a*). The septa divide the vesicles into distinct subunits each with its own plasma membrane and nucleoid regions. Frequent plasmalemmasomes occur either within the subunits or associated with the septa. An electron transparent region or 'halo' surrounds mature vesicles. At an even later stage in vesicle development striated bodies appear within the subunit cytoplasm (Plate 34.2*b*). Only one such body has been seen to occur per subunit, and its appearance seems to coincide with the condensation of the cytoplasm and the shrinkage of the plasma membrane from the vesicle walls. Similar striated bodies were recorded by Becking *et al.* (1964) in *Alnus* vesicles. The vesicles originate as swellings of the hyphal tips (Plate 34.2*c*) and in the young stage are non-septate and are not surrounded by a 'halo'. They grow to full size before septation begins and Plate 34.2*d* shows clearly that the septa arise as invaginations of the vesicle wall. Gardner & Gatner (1973) have shown that the degree of septation increases with vesicle maturity and not with vesicle size (Becking *et al.*, 1964).

Studies of the nodules of *Myrica* species by previous authors with the light microscope and by Silver (1964) of *M. cerifera* nodules with the electron microscope had shown that the hyphal tips of the endophytes of these species were also enlarged, but only to a degree sufficing to produce a club-shaped structure. Electron microscopic studies by the present author reveal that these clubs in *M. gale* and *M. cerifera* are highly septate and that each subunit so formed has its own plasma membrane and nucleoid region. Longitudinal sections through mature clubs are shown in Plate 34.3 (*a*), (*b*) and (*c*), while Plate 34.3 (*d*) illustrates the correlation between the formation of the septa and the division of the nuclear material. Well-defined striated bodies develop within the subunits of older clubs (Plate 34.3*e*). This evidence shows clearly that in their essential features the clubs of these *Myrica* endophytes closely resemble the spherical vesicles formed by the endophytes of other genera and the author considers that the term vesicle should in fact be extended to these *Myrica* structures in future.

A light microscopic study of the nodules of *Ceanothus velutinus* and *C. sanguineus* by Furman (1959) showed that many of the endophyte hyphae terminated in spherical vesicles about 2.0 μm in diameter. Electron microscopy of the nodules of *C. velutinus* (Plate 34.3*f* and *g*) confirms Furman's observations and shows that the vesicles differ from those of *Hippophaë* and *Alnus* in being separated from the parent hyphae by a transverse septum. A more striking difference from the vesicles of other genera, however, is that in none of the nodule material examined by the author – drawn both from water-cultured and soil-grown plants – was any septation of the vesicles themselves observed, although some of the vesicles (Plate 34.3*g*) are judged to be mature by virtue of the electron transparent 'halo' which surrounds them.

The endophyte in *Casuarina* nodules has swollen club-shaped tips to the hyphae (Schaede, 1962; Becking, 1970*b*). These clubs in *Casuarina cunninghamiana* are very slender (Table 34.1). Ultrastructural evidence (Plate 34.4*a*) indicates the vesicular nature of these clubs, numerous septa dividing them into individual subunits as in the endophytes of *Myrica* described previously. In this greenhouse material of *Casuarina* the frequency of vesicles within the host cells was much lower than in the other genera examined. The reason for this is not immediately obvious since such nodules fixed nitrogen efficiently.

In all nodules here investigated, the vesicles, like the hyphae, are entirely surrounded by an electron dense 'capsule'. In the mature vesicle, however, the 'capsule' is separated from the vesicle wall by a prominent electron transparent 'halo' – a feature also found by Becking *et al.* (1964) and Gardner (1965) in *Alnus glutinosa*. Similar transparent zones surrounding the bacteria in some preparations of legume nodules have been attributed to shrinkage of the bacteria by permanganate fixation (Dixon, 1969). The zone in the non-legumes is unlikely to be an artefact because it is not seen around the young vesicles but becomes progressively more conspicuous as the degree of septation increases and is seen after fixation in a range of different fixatives. Draper & Rees (1973) have shown that the zone surrounding *Mycobacterium lepraemurium* in mouse liver cells contains a peptidoglycolipid, and the featureless appearance of this zone in electron micrographs is attributed to the dissolution of the lipid complex during dehydration and embedding. That a similar situation could exist in the non-legumes is indicated by Plate 34.4 (*b*) where the 'halo' around one vesicle still contains traces of electron dense material.

Endophyte bacteroids

The third endophytic form, the bacteroid, was found to be present in the cells or the intercellular spaces of the nodules of all species examined by the author. Gatner & Gardner (1970) and Akkermans (1971) failed to find bacteroids in *Hippophaë* nodules, but Gardner & Pho (unpublished observations) have now located them and find them to occur in greatest numbers in the nodule periderm layer.

The bacteroids from all the species possess a similar ultrastructure although they vary somewhat in size and shape (Table 34.1). Young bacteroids are thin-walled and Plate 34.4 (*c*) shows longitudinal and transverse sections of such bacteroids in *Hippophaë*. Within the thin cell wall the plasma membrane encloses a dense cytoplasm in which nucleoid regions can just be discerned while electron translucent granules are conspicuous. As the bacteroids age the cell wall thickens (Plate 34.4*d*). Ultrastructurally the mature bacteroids closely resemble *Azotobacter* cysts. The latter are resting cells which are considerably more resistant than vegetative cells to deleterious physical and chemical agents (Socolofsky & Wyss, 1962) and to desiccation (Stevenson & Socolofsky, 1972). These cysts also possess electron translucent granules which have been identified as poly β-hydroxybutyric acid and it has been shown that their viability depends upon the levels of this reserve substance in them (Sobek, Charba & Foust, 1966). The chemical nature of the granules of nodule bacteroids has not yet been established, but by analogy with the *Azotobacter* cysts it seems likely that here also they are food reserves. Thus it would appear that the bacteroids are well adapted to the role suggested for them, namely that of being the particular endophytic structure which, after nodule decay, persists in the soil and provides for the infection of further generations of host plants (Rodriguez-Barrueco, 1968; Akkermans & van Dijk, Chapter 36).

The relationship of the bacteroid to the other endophytic forms is not clearly understood. Hawker & Fraymouth (1951) believed that bacteroids emanated from vesicles, while Käppel & Wartenberg (1958) considered them to be of hyphal origin, a view shared by Becking *et al.* (1964) and by Angulo, van Dijk & Quispel (see Chapter 33). Becking *et al.* argued that since vesicles had only incomplete internal walls they could not produce bacteroids. The micrographs presented here, however, offer substantial evidence in favour of complete septation in the mature vesicles. This is particularly clear in old vesicles (Plates 34.2*b* and 34.3*e*) where the subunit cytoplasm has just begun to condense and the plasma membrane is drawn away from the walls. Indeed this is regarded (Gardner & Pho, unpublished observations) as being the first step in the transformation of vesicles into bacteroids. It is also at this stage of vesicle development that striated bodies become prominent within the subunit cytoplasm. The function of such bodies is as yet unknown. Plate 34.4 (*e*) shows a vesicle at a later stage in transformation. The subunit cytoplasm has further condensed, obscuring the striated bodies, and electron translucent granules are present. The subunits may now develop into mature bacteroids *in situ* or they may be shed as thin-walled bacteroids which mature individually in the host cells or intercellular spaces. This process of vesicle transformation has been seen not only in *Hippophaë* but also in *Myrica gale* and *M. cerifera*. In *M. gale* the occurrence of increasingly large club-shaped masses of young bacteroid (Plate 34.4*f*) suggests that such bacteroids are capable of division. By maturity they completely fill the host cell (Plate 34.5*a*) which at this stage is dead. In *Casuarina* nodules it appears that the endophyte hyphae can transform directly into bacteroids (Plate 34.5*b*). The dense cytoplasm and the presence of electron translucent granules in these hyphae suggest that this process is similar to that for vesicle transformation.

Spratt (1912) reported that 'spherical bodies' (vesicles) could be induced to transform into 'bacilli' when nodule slices were incubated in a simple synthetic medium with aseptic precautions. Investigation, at the ultrastructural level, of tissues incubated in Spratt's medium revealed that after two days the first 'induced' bacteroids had formed, while after seven days abundant bacteroids were present in the nodule tissues. The sequence of events leading to the formation of these bacteroids from vesicles in *Hippophaë* (Plate 34.5*c*, *d* and *e*) was as described previously for normal tissues. In similar incubation tests with *Myrica cerifera* nodule slices, electron translucent granules and striated bodies were also found together in the hyphae (Plate 34.5*f*). Such a phenomenon was not found under normal conditions in the nodules of this species. It would therefore appear that in the non-legume nodules, bacteroids are normally produced by the vesicles but can, under certain circumstances, also be produced by the hyphae.

Host cell ultrastructure

Entry of the endophyte into a host cell evokes a rapid response. Starch grains are conspicuous in the amyloplasts in recently infected cells (Plate 34.6*a*) but they are rapidly utilised (Plate 34.6*b*). As endophyte growth proceeds there is a marked increase in the numbers of mitochondria, ribosomes and Golgi bodies in the host cell and a great proliferation of endoplasmic reticulum (Plate 34.6*c*, *d* and *e*). An uninfected cell, still unvacuolated, is estimated to contain about 500 mitochondria, while the number in a young infected cell which has undergone a twofold increase in size ranges up to 2000. These mitochondria appear to be randomly distributed through the host cytoplasm. The disappearance of starch and the rise in mitochondrial numbers is indicative of the increased demand for energy in the infected cells. The mitochondria are often very elongate and appear to divide by constriction; they are still quite numerous in cells containing mature vesicles, though little endoplasmic reticulum is then present. Another feature of the infected cell is its lobed nucleus. The attendant proliferation of the nuclear membrane seems to be directly connected with that of the endoplasmic reticulum (Plate 34.6*d*).

Except perhaps in very early stages of infection and in cells containing mature bacteroids, a membrane regularly separates the endophyte from the host cell cytoplasm (Plates 34.1*c* and *d*, 34.2*d* and 34.6*f*). This membrane is best seen in osmium or glutaraldehyde/osmium fixed tissues, whereas permanganate tends to destroy it. Such a membrane was observed by Becking *et al.* (1964) in *Alnus glutinosa*, by Gatner & Gardner (1970) in *Hippophaë rhamnoides* and has also been seen in the other species now studied. A similar membrane is present in many other symbiotic and parasitic associations. Its biological significance in legume nodules was discussed by Nutman (1963). That its production is a host cell defence mechanism against invasion by a foreign organism is generally agreed but the origin of the membrane is uncertain. Some authors propose that in the legumes it arises *de novo* or from the endoplasmic reticulum while others contend that it arises by endocytosis of the host cell plasma membrane as the rhizobia enter the cell (see Dixon, 1969). Plate 34.6*f* shows that in *Myrica gale* nodules a continuity exists between the host cell plasma membrane and the membrane surrounding the endophyte, indicating that the latter is 'plasma membrane-like'. The proliferation of the endophyte within the host cell must necessitate the production of large quantities of this type of membrane. The greatly increased amount of endoplasmic reticulum in a recently infected cell, noted above, is of interest here. In animal cells complex whorls or 'finger print' configurations of endoplasmic reticulum (Plate 34.6*e*) are thought to reflect rearrangements in the components of the membrane in response to changes in the intracellular environment (Fawcett, 1961; Arstila & Trump, 1968), while Clowes & Juniper (1968) have recorded similar whorls in injured plant cells. It is also well established that endoplasmic reticular membrane can transform to the plasma membrane type within the Golgi apparatus (Grove, Bracker & Morré, 1968). Perhaps, then, in the non-legume nodule the membrane surrounding the endophyte has its origin in the endoplasmic reticulum.

Enzyme studies

In *Alnus glutinosa* citrulline is the predominant free amino acid of all organs (Miettinen & Virtanen, 1952). Tracer studies (Leaf, Gardner & Bond, 1958; Gardner & Leaf, 1960) showed that the synthesis of citrulline in alder involved the addition of newly fixed ammonia and carbon dioxide to ornithine, probably by means of ornithine carbamayl-transferase. Plate 34.7 (*a*) to (*c*) demonstrates the location of this enzyme in alder nodules at an ultrastructural level using the method of Merker & Spors (1969). After a short incubation enzyme activity is located in the host cell mitochondria and the endoplasmic reticulum. No indication of enzyme activity was found in the endophyte at this stage. This is in agreement with the data of Merker & Spors (1969) for rat liver cells. After longer incubation periods enzyme activity was found in the cytoplasm of the endophyte hyphae and in the subunit cytoplasm of the vesicles. Since citrulline is synthesised from newly fixed ammonium nitrogen such results would suggest that nitrogen fixation can occur in both hyphae and vesicles. Akkermans (1971) and Angulo, van Dijk & Quispel (Chapter 33) suggest that the occurrence of reducing conditions in the vesicles indicates that they are sites of nitrogen fixation though fixation in the hyphae as well is not excluded.

That ATP is required for the process of biological nitrogen fixation is now well established. Where this ATP is generated within the cells is not clear. The localisation of ATPase activity in tissues of *Alnus glutinosa* nodules was investigated using the modified Wachstein-Meisel medium of Farquhar & Palade (1966). The results are presented in Plate 34.7 (*d*) and (*e*). In the endophyte, enzyme activity appears to be confined to the walls of the hyphae and the walls and septa of the vesicles. It would appear that either the ATP cannot penetrate the endophyte or it is being completely utilised in the walls. The great activity associated with uptake of metabolites by the endophyte and removal of fixation products from the endophyte would perhaps point to the latter explanation being more probable.

Scanning electron microscope studies

Scanning electron microscope pictures of nodular tissues are presented in Plate 34.8. The resolution and the depth of field of this instrument permits a three-dimensional examination of the endophyte within the host cells. In Plate 34.8 (*a*) to (*d*) the hyphae and vesicles with their hyphal attachments are seen in cells of *Alnus glutinosa* nodules, while Plate 34.8 (*e*) shows part of a cell containing mainly bacteroids but with a few vesicles as well. A portion of a cell from a *Myrica gale* nodule is seen in Plate 34.8 (*f*) and the flattened nature of the club-shaped vesicles of this endophyte is evident. The vesicles in both species examined appear to have a smooth external surface except in the instances when they are covered with a layer of host cytoplasm. Certainly no indication of the internal septation of the vesicles can be observed.

Professor G. Bond kindly provided nodular material of *Myrica cerifera*, *Ceanothus* and *Casuarina* used in these studies. I wish to thank Drs E. M. S. Gatner and L. K. Pho for permission to use material from their PhD theses. The photographs for Plate 34.8 were taken on the Stereoscan in the Department of Biological Sciences, University of Dundee, and I am indebted to Professor W. D. P. Stewart for this facility and to Miss E. Davidson for technical assistance.

References

Akkermans, A. D. L. (1971). Nitrogen fixation and nodulation of *Alnus* and *Hippophaë* under natural conditions. PhD Thesis, University of Leiden.

Arstila, A. U. & Trump, B. F. (1968). Studies on cellular autophagocytosis. *American Journal of Pathology*, **53**, 687–94.

Becking, J. H. (1970*a*). Plant–endophyte symbiosis in non-leguminous plants. *Plant and Soil*, **32**, 611–54.

Becking, J. H. (1970*b*). *Frankiaceae* Fam. Nov. (Actinomycetales) with one new combination and six new species of the genus *Frankia* Brunchorst 1886, 174. *International Journal of Systematic Bacteriology*, **20**, 201–20.

Becking, J. H., DeBoer, W. E. & Houwink, A. L. (1964). Electron microscopy of the endophyte of *Alnus glutinosa*. *Antonie van Leeuwenhoek, Journal of Microbiology and Serology*, **30**, 343–76.

Bradley, S. G. & Ritzi, D. (1968). Composition and ultrastructure of *Streptomyces venezuelae*. *Journal of Bacteriology*, **95**, 2358–64.

Burdett, I. D. J. & Rogers, H. J. (1970). Modification of the appearance of mesosomes in sections of *Bacillus licheniformis* according to fixation procedures. *Journal of Ultrastructural Research*, **30**, 354–67.

Burdett, I. D. J. & Rogers, H. J. (1972). Structure and development of mesosomes studied in *Bacillus licheniformis* strain 6364. *Journal of Ultrastructural Research*, **38**, 113–33.

Chandler, M. R. & Dart, P. J. (1971). *Casuarina* nodules. Report of the Rothamsted Experimental Station for 1971, part 1, p. 99.

Chen, P. L. (1964). The membrane system of *Streptomyces cinnamonensis. American Journal of Botany*, **51**, 125–32.

Clowes, F. A. L. & Juniper, B. E. (1968). *Plant cells.* Blackwell Scientific Publications, Oxford.

Dixon, R. O. D. (1969). Rhizobia. *Annual Review of Microbiology*, **23**, 137–58.

Draper, P. & Rees, R. J. W. (1973). The nature of the electron transparent zone that surrounds *Mycobacterium lepraemurium* inside host cells. *Journal of General Microbiology*, **73**, 79–83.

Farquhar, M. G. & Palade, G. E. (1966). ATPase localisation in amphibian epidermis. *Journal of Cell Biology*, **30**, 359–79.

Fawcett, D. W. (1961). The membranes of the cytoplasm. *Laboratory Investigations*, **10**, 1162–72.

Furman, T. E. (1959). The structure of the root nodules of *Ceanothus sanguineus* and *Ceanothus velutinus*, with special reference to the endophyte. *American Journal of Botany*, **46**, 698–703.

Gardner, I. C. (1965). Observations on the fine structure of the endophyte of the root nodules of *Alnus glutinosa* (L.) Gaertn. *Archiv für Mikrobiologie*, **51**, 365–83.

Gardner, I. C. & Gatner, E. M. S. (1973). The formation of vesicles in the development cycle of the nodular endophyte of *Hippophaë rhamnoides* L. *Archiv für Mikrobiologie*, **89**, 233–40.

Gardner, I. C. & Leaf, G. (1960). Translocation of citrulline in *Alnus glutinosa. Plant Physiology*, **35**, 948–50.

Gatner, E. M. S. & Gardner, I. C. (1970). Observations on the fine structure of the root nodule endophyte of *Hippophaë rhamnoides* L. *Archiv für Mikrobiologie*, **70**, 183–96.

Glauert, A. M. & Hopwood, D. A. (1960). The fine structure of *Streptomyces coelicolor.* 1. The cytoplasmic membrane system. *Journal of Biophysical and Biochemical Cytology*, **7**, 479–88.

Gordon, M. A. & Edwards, M. R. (1963). Micromorphology of *Dermatophilus congolensis. Journal of Bacteriology*, **86**, 1101–15.

Grove, S. N., Bracker, C. E. & Morré, J. D. (1968). Cytomembrane differentiation in the endoplasmic reticulum–Golgi apparatus–vesicle complex. *Science*, **161**, 171–83.

Hawker, L. E. & Fraymouth, J. (1951). A re-investigation of the root nodules of species of *Elaeagnus*, *Hippophaë*, *Alnus* and *Myrica* with special reference to the morphology and life histories of the causative organism. *Journal of General Microbiology*, **5**, 369–86.

Käppel, M. & Wartenberg, H. (1958). Der Formenwechsel des *Actinomyces alni* Peklo in den Wurzeln von *Alnus glutinosa* Gaertner. *Archiv für Mikrobiologie*, **30**, 46–63.

Leaf, G., Gardner, I. C. & Bond, G. (1958). Observation on the composition and metabolism of the nitrogen-fixing nodules of *Alnus*. *Journal of Experimental Botany*, **9**, 320–34.

McLuckie, J. (1923). Studies in Symbiosis. IV. The root nodules of *Casuarina cunninghamiana* and their physiological significance. *Proceedings of the Linnean Society of New South Wales*, **48**, 194–205.

Merker, H. J. & Spors, S. (1969). Electron microscopic demonstration of ornithine carbamoyltransferase in rat liver. *Histochemie*, **17**, 83–8.

Miehe, H. (1918). Anatomische Untersuchung der Pilzsymbiose bei *Casuarina equisetifolia* nebst einigen Bemerkungen über das Mykorhizenproblem. *Flora*, **11/12**, 431–49.

Miettinen, J. K. & Virtanen, A. I. (1952). The free amino acids in the leaves, roots and root nodules of the alder. *Physiologia Plantarum*, **5**, 540–9.

Nutman, P. S. (1963). Factors influencing the balance of mutual advantage in legume symbiosis. In: *Symbiotic associations*, 13th Symposium of the Society for General Microbiology (ed. P. S. Nutman & B. Mosse), pp. 51–71. Cambridge University Press, London.

Rodriguez-Barrueco, C. (1968). The occurrence of the root-nodule endophytes of *Alnus glutinosa* and *Myrica gale* in soils. *Journal of General Microbiology*, **52**, 189–94.

Rogers, H. J. (1970). Bacterial growth and the cell envelope. *Bacteriological Reviews*, **34**, 194–214.

Schaede, R. (1962). *Die pflanzlichen Symbiosen*. Gustav Fischer Verlag, Stuttgart.

Silver, W. S. (1964). Root nodule symbiosis. I. Endophyte of *Myrica cerifera* L. *Journal of Bacteriology*, **87**, 416–21.

Sobek, J. M., Charba, J. F. & Foust, W. N. (1966). Endogenous metabolism of *Azotobacter agilis*. *Journal of Bacteriology*, **92**, 687–95.

Socolofsky, M. D. & Wyss, O. (1962). Resistance of the *Azotobacter* cyst. *Journal of Bacteriology*, **84**, 119–24.

Spratt, E. R. (1912). The morphology of the root tubercles of *Alnus* and *Elaeagnus* and the polymorphism of the organisms causing their formation. *Annals of Botany*, **26**, 119–28.

Stevenson, L. H. & Socolofsky, M. D. (1972). Encystment of *Azotobacter vinelandii* in liquid culture. *Antonie van Leeuwenhoek, Journal of Microbiology and Serology*, **38**, 605–16.

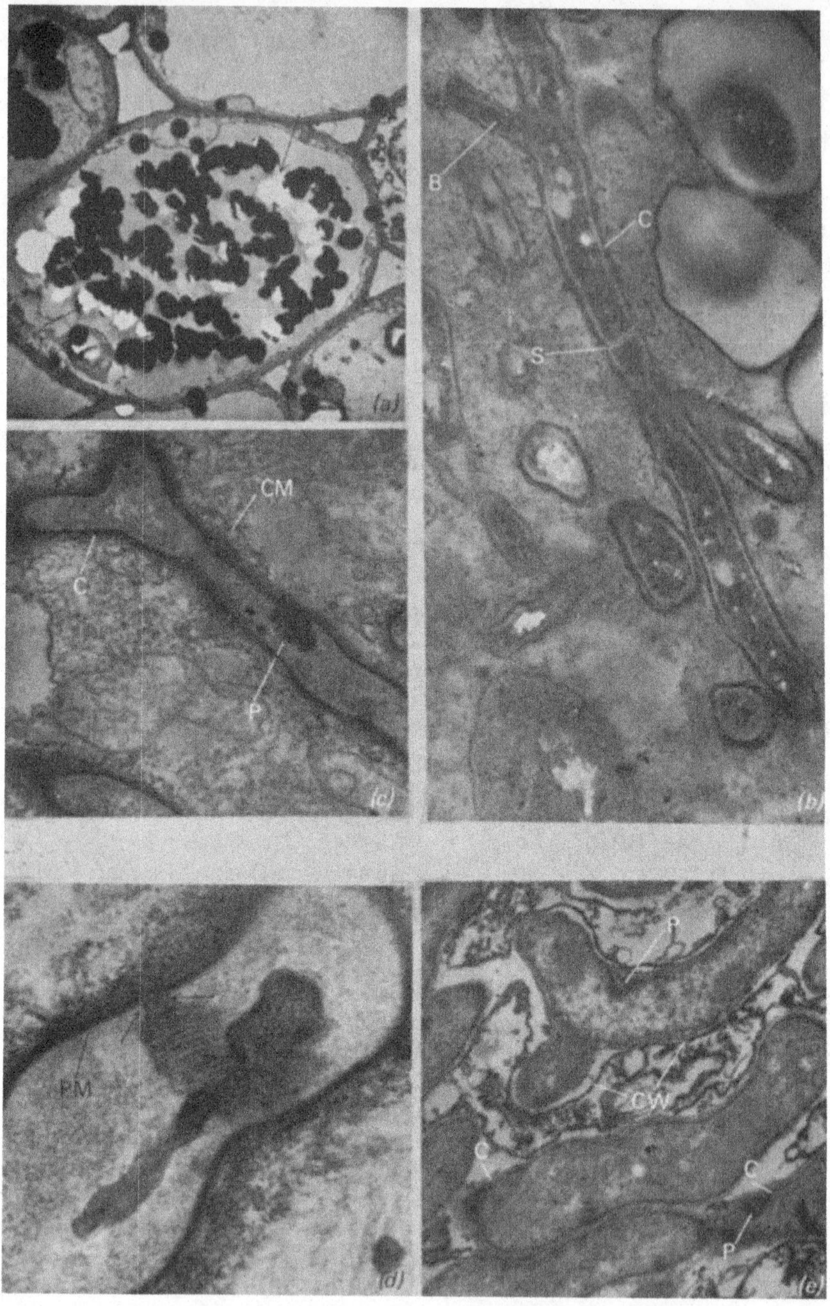

Plate 34.1. (*a*) Uninfected cells from a *Casuarina cunninghamiana* nodule showing the presence of discrete droplets of resin or tannin in the large central vacuole. Such inclusions complicate fine-sectioning and arrows indicate regions where these hard granules have fallen out of the section. Osmium fixation. ×1100. (*b*) Part of an infected cell from a young *Hippophaë rhamnoides* nodule (18 days after inoculation). Endophyte hyphae permeate the host cell cytoplasm. They are septate (S), show true branching (B) and are surrounded by a 'capsular' material (C). Glutaraldehyde/osmium fixation. ×4230. (*c*) Hyphal elements in a newly infected cell of *Hippophaë rhamnoides* nodule. C, 'capsular' material; CM, cytoplasmic membrane surrounding the endophyte; P, endophyte plasmalemmasome. Osmium fixation. ×3380. (*d*) Portion of an endophyte hypha as in (*c*) but at high magnification to show the details of the plasmalemmasome. Arrows indicate the continuity of this organelle with the endophyte plasma membrane (PM). Osmium fixation. ×23200. Reproduced with permission from Gatner & Gardner (1970). (*e*) Details of the hyphae in *Casuarina cunninghamiana* nodule cell. CW, host cell cytoplasmic membrane; C, capsular material; P, plasmalemmasome. ×4230.

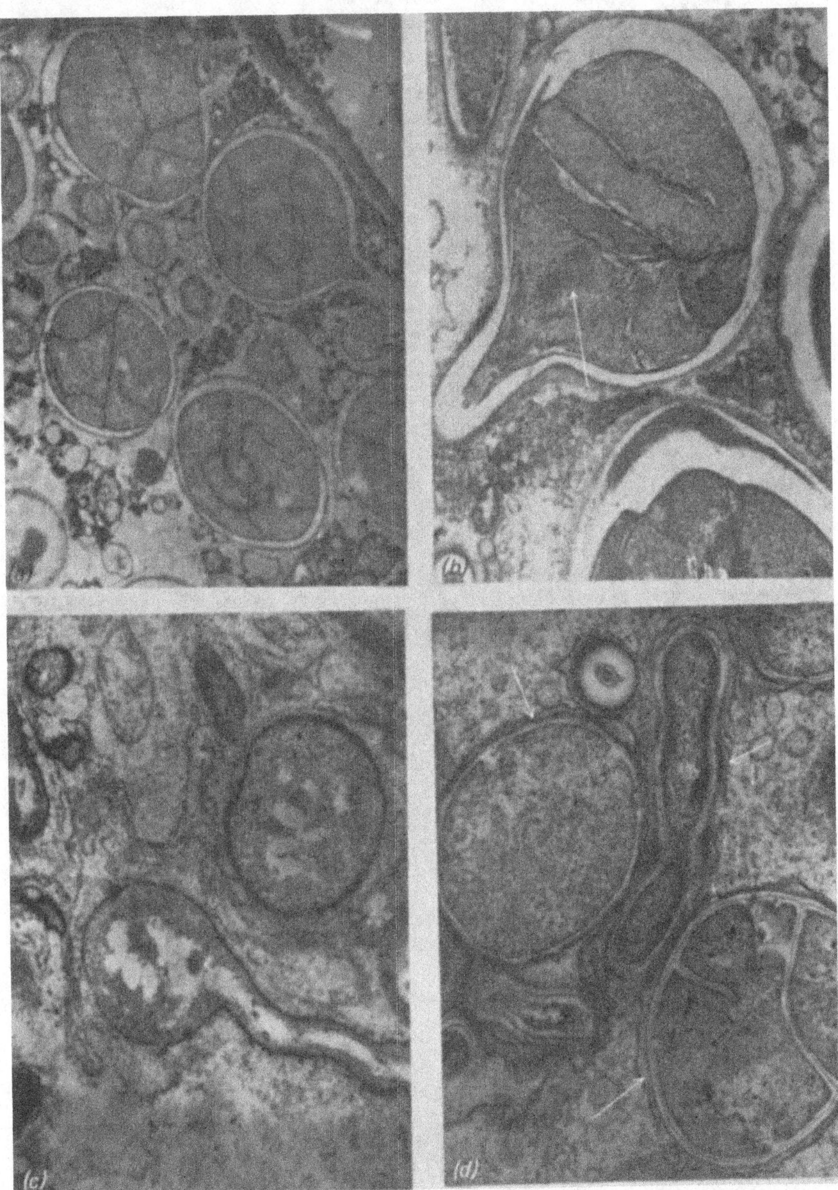

Plate 34.2. (*a*) Part of a cell from a *Hippophaë rhamnoides* nodule containing mature endophyte vesicles. These vesicles exhibit random septation and are 3–4 μm in diameter. Osmium fixation. × 2240. (*b*) Section of a *Hippophaë rhamnoides* vesicle which has attained an even later stage of development than those shown in (*a*). The subunit cytoplasm is condensed and the plasma membrane is separating from the vesicle wall. An arrow indicates the striated body in one of the subunits. Glutaraldehyde fixation. × 6100. (*c*) Young vesicles develop as swellings of the hyphal tips in the nodule cells of *Hippophaë rhamnoides*. Permanganate fixation. × 2400. (*d*) fully grown vesicles showing the first stage of septum formation. The septum arises as an invagination of the vesicle wall. Arrows indicate the membrane which surrounds the endophyte and separates it from the host cell cytoplasm. *Hippophaë rhamnoides* tissue. Osmium fixation. × 3540.

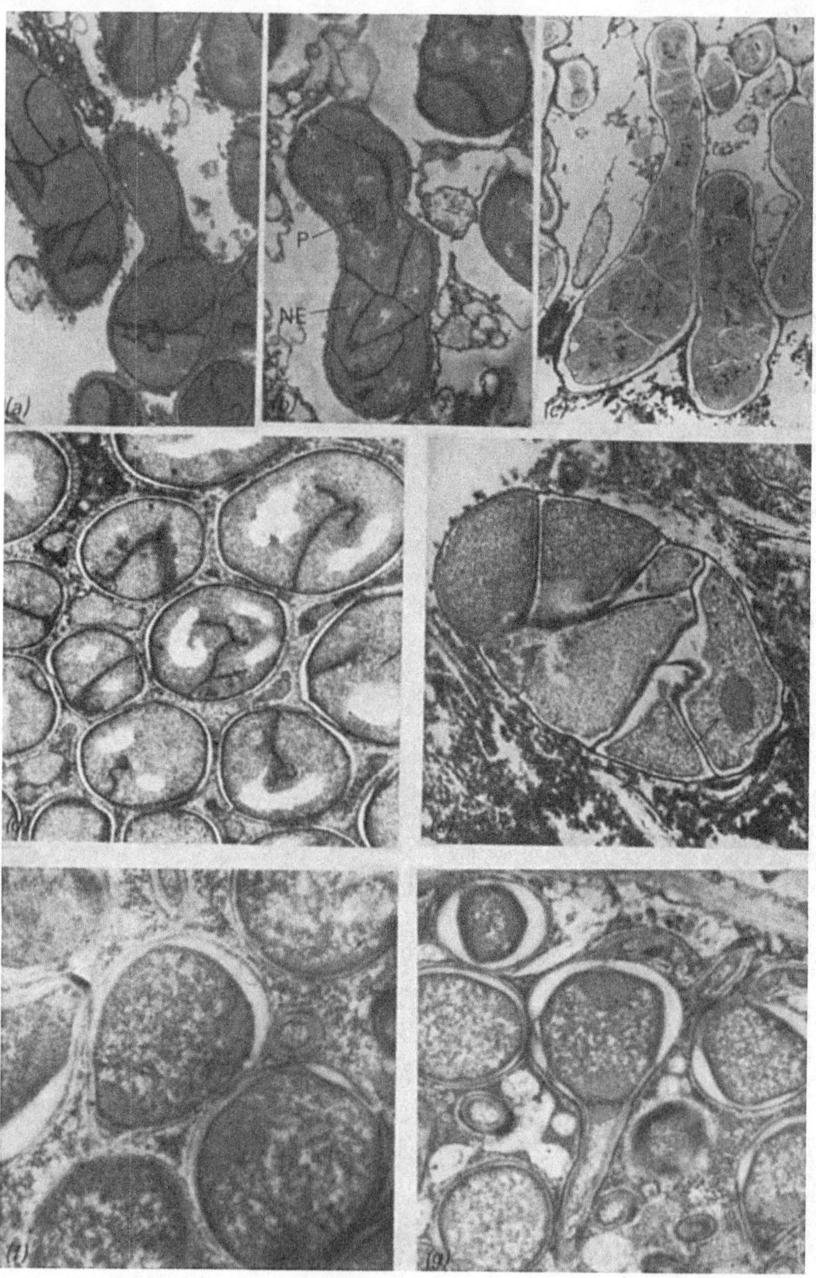

Plate 34.3. (*a*) Longitudinal section of developing club-shaped vesicles of the *Myrica gale* endophyte. Osmium fixation. ×4200. (*b*) Longitudinal tangential section through a *Myrica gale* endophytic vesicle. Random septation divides the vesicle into subunits. NE, nucleoid regions; P, plasmalemmasomes. Osmium fixation. ×4200. (*c*) Longitudinal section of vesicles of the endophyte of *Myrica cerifera* nodules. Glutaraldehyde/osmium fixation. ×2750. (*d*) Transverse section through the tips of developing vesicles of *Myrica cerifera* showing the correlation between division of the nuclear regions and septation. ×4750. (*e*) A late stage in the life of a vesicle in *Myrica cerifera* nodule cell. The subunit cytoplasm is condensing and drawing away from the vesicle walls. Arrow indicates a striated body within the subunit cytoplasm. ×5500. (*f*) Pear-shaped vesicles of the *Ceanothus* endophyte. These are surrounded by an electron transparent 'halo' but lack internal septation. Glutaraldehyde/osmium fixation. ×3800. (*g*) *Ceanothus* endophyte vesicles develop as swellings of the hyphal tips. A septum separates the vesicle from the parent hypha. Glutaraldehyde/osmium fixation. ×3300.

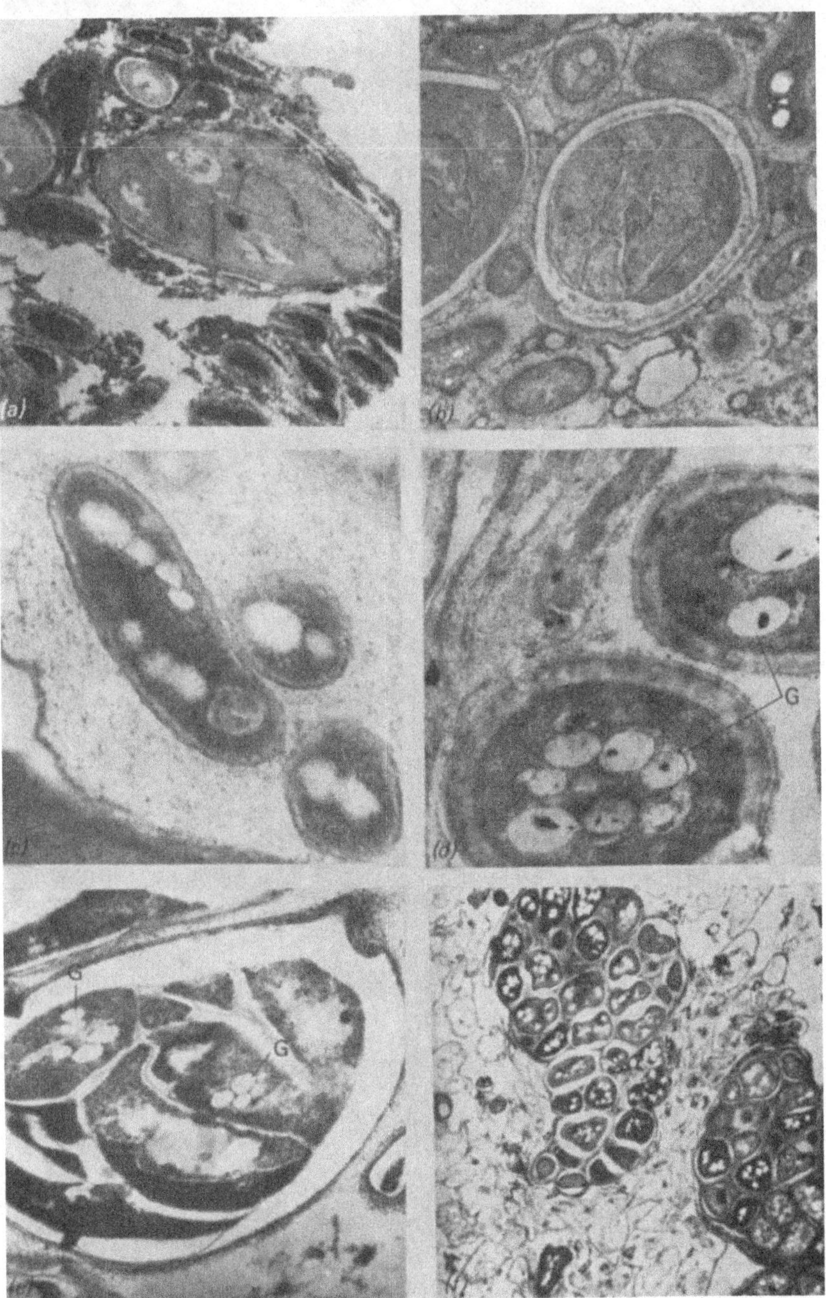

Plate 34.4. (*a*) Longitudinal section through a club-shaped vesicle of the *Casuarina cunninghamiana* endophyte. Permanganate fixation. × 5000. (*b*) Section of a mature vesicle from a *Hippophaë rhamnoides* nodule. Traces of an electron dense material are evident in the vesicle 'halo' region. Osmium fixation. × 2500. (*c*) Young bacteroids in longitudinal and transverse section within a cell from the periderm region of a *Hippophaë rhamnoides* nodule. Glutaraldehyde/osmium fixation. × 17600. (*d*) Mature thick-walled bacteroids from *Hippophaë rhamnoides* nodules. Prominent electron translucent granules (G) occur within the dense cytoplasm. Osmium fixation. × 11400. (*e*) Early stages in the transformation of vesiculra subunits into bacteroids in *Hippophaë rhamnoides* root nodule. The subunit cytoplasm is condensed and darkly staining and contains electron translucent granules (G). Glutaraldehyde fixation. × 8500. (*f*) Club-shaped masses of developing bacteroids in *Myrica gale* nodule. Empty digested endophyte elements surround these masses in the dead host cells. Osmium fixation. × 2100.

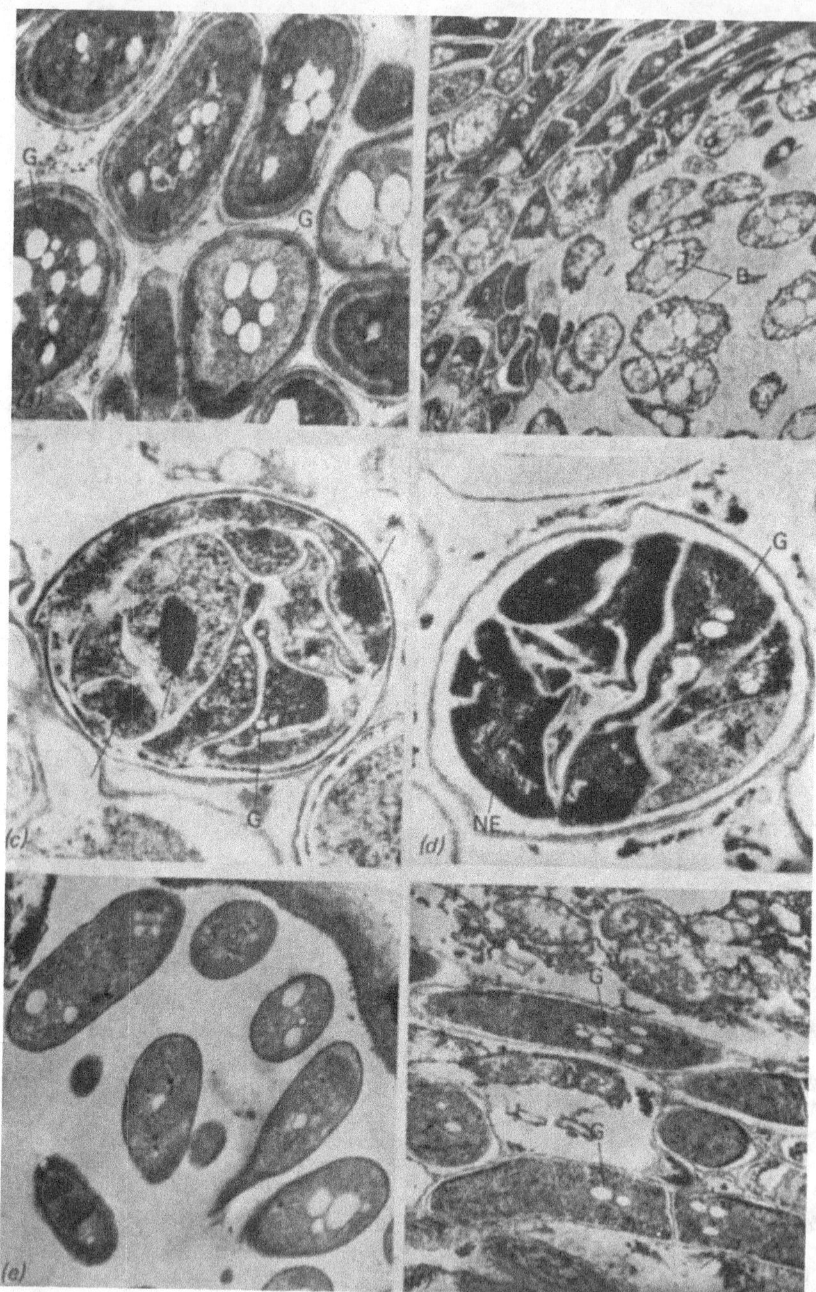

Plate 34.5. (*a*) Mature thick-walled bacteroids in *Myrica gale* nodule cells. These bacteroids contain a dense cytoplasm with nucleoid regions and prominent electron transparent granules (G). Osmium fixation. ×7200. (*b*) Endophyte filaments transforming into bacteriods (B) within a *Casuarina cunninghamiana* nodule cell. Permanganate fixation. ×3800. (*c*) Vesicle from *Hippophaë rhamnoides* nodule tissue incubated for 7 days to induce bacteroid formation. Small electron translucent granules (G) can be seen in the subunits and arrows indicate the striated bodies. Glutaraldehyde fixation. ×9300. (*d*) Vesicle from *Hippophaë rhamnoides* nodule tissue incubated for 7 days to induce bacteroid formation. The cytoplasm of the subunits is condensed and darkly staining. G, electron translucent granules; NE, nucleoid material. Glutaraldehyde fixation. ×5500. (*e*) Young thin-walled bacteroids induced in *Hippophaë rhamnoides* nodule tissues after a 12-day incubation period. Glutaraldehyde fixation. ×4200. (*f*) Endophyte filaments showing the first stages in the transformation into bacteroids in *Myrica cerifera* nodule tissues incubated for 7 days. G, electron translucent granules; arrow indicates a striated body. Glutaraldehyde fixation. ×3400.

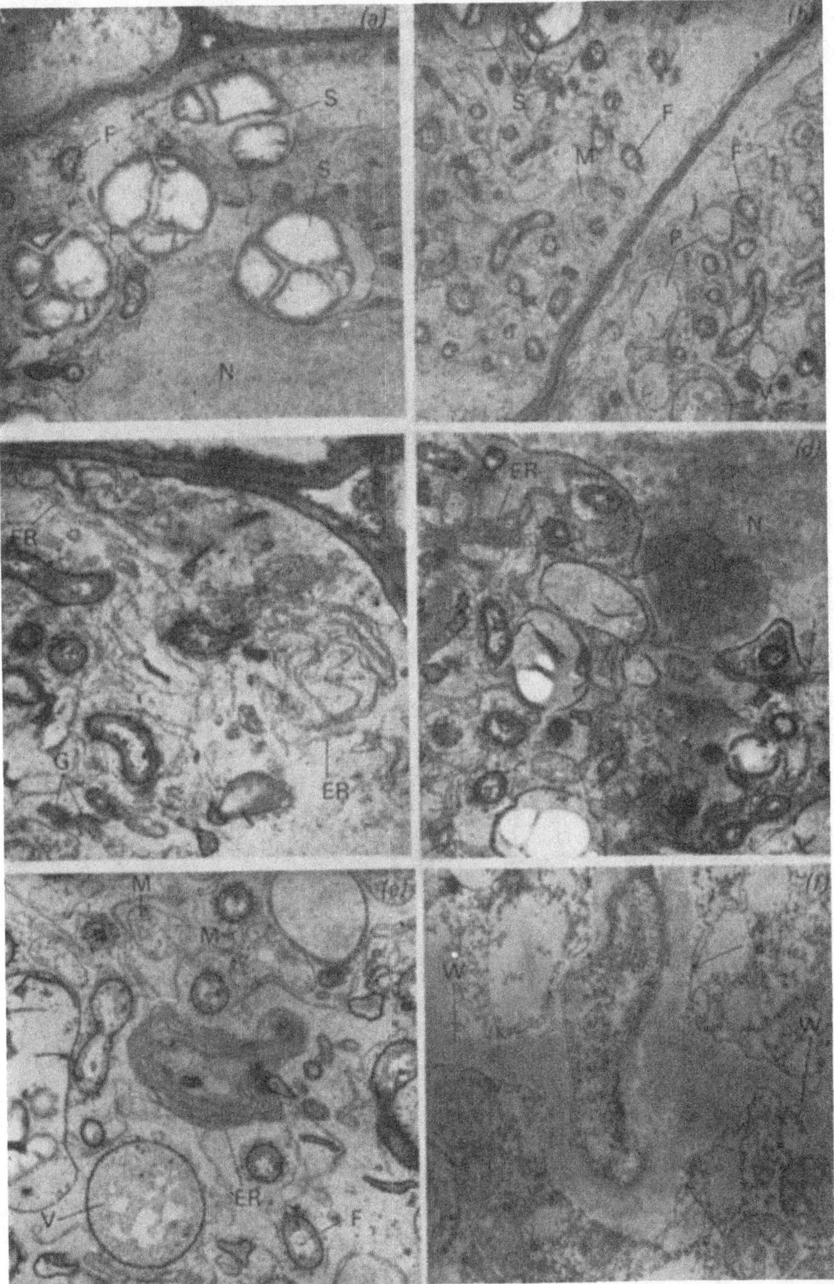

Plate 34.6. (*a*) Part of a nodule cell from *Hippophaë rhamnoides* showing a very early stage of infection. A few endophytic filaments (F) occur in the host cytoplasm which contains numerous starch plastids (S). N, host cell nucleus. Permanganate fixation. ×1050. (*b*) cells from young nodule of *Hippophaë rhamnoides*. The cell on the right shows a more advanced stage of infection than that on the left. F, endophyte filaments; V, developing vesicle; P, empty plastids; S, starch plastids; M, mitochondria. Permanganate fixation. ×1250. (*c*) Part of a young infected cell from a *Hippophaë rhamnoides* nodule showing abundant endoplasmic reticulum (ER) and Golgi bodies (G) in the host cytoplasm. Permanganate fixation. ×2100. (*d*) Part of an infected cell from a young *Hippophaë* nodule showing the host nucleus (N). The arrows indicate points of continuity between the nuclear membrane and the endoplasmic reticulum (ER). Permanganate fixation. ×1700. (*e*) Proliferation of the endoplasmic reticulum (ER) in a young infected cell from a *Hippophaë rhamnoides* nodule. F, endophyte filaments; V, developing vesicle; M, mitochondria. Permanganate fixation. ×2100. (*f*) Endophyte filament penetrating the host cell wall (W) in *Myrica gale* nodular tissue. Note the continuity between the host cell plasma membrane and the membrane surrounding the endophyte (arrowed). Osmium fixation. ×2100.

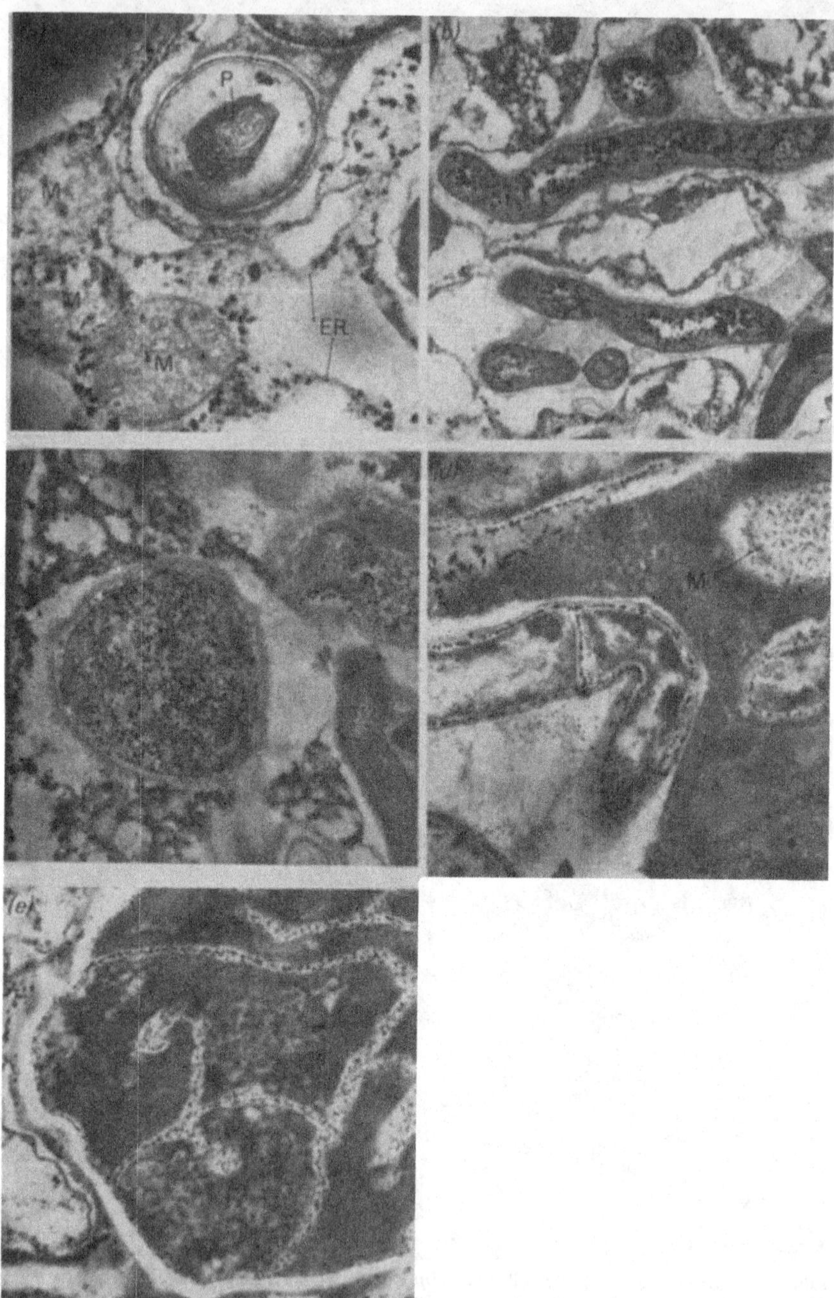

Plate 34.7. (*a*) *Alnus glutinosa* nodule tissue incubated for ornithine carbamoyltransferase. The reaction product occurs in the host cell mitochondria (M) and the endoplasmic reticulum (ER) after a 10 min incubation. Only slight activity can be seen in the endophyte hypha. Note the complex plasmalemmasome (P). ×8500. (*b*) *Alnus glutinosa* tissues incubated for ornithine carbamoyltransferase. After 30 min incubation the reaction product is found in the cytoplasm of the endophytic hyphae. ×2100. (*c*) Ornithine carbamoyltransferase activity located in the cytoplasm of the vesicular subunits after incubation of *Alnus glutinosa* nodule tissue as in (*b*). ×3380. (*d*) *Alnus glutinosa* nodule tissue incubated for ATPase. The activity is localised in the host cell mitochondria (M) and the endophyte cell wall. ×4230. (*e*) *Alnus glutinosa* nodule tissue incubated for ATPase. Within the endophyte vesicles the enzyme activity is associated with the walls and the septa. ×10500.

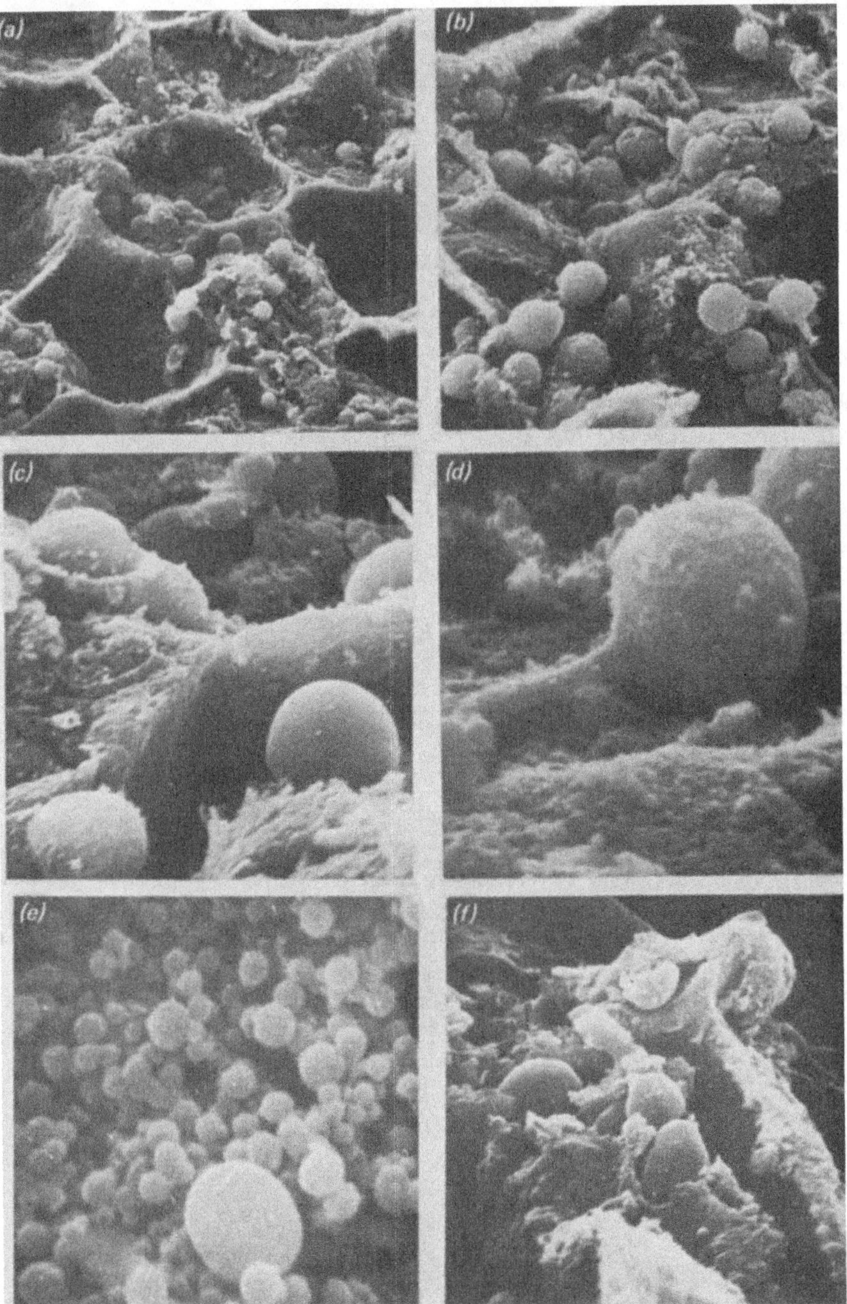

Plate 34.8. (*a*) Scanning electron micrograph of a thick section of a root nodule of *Alnus glutinosa*. Endophyte hyphae and vesicles can be seen within the nodule cells. ×1270. (*b*) Scanning micrograph of the hyphae and spherical vesicles of the endophyte in *Alnus glutinosa* nodule cells. The attachments of the hyphae to the vesicles can be clearly seen. ×2540. (*c*) Higher magnification of the vesicular stage of the *Alnus* endophyte. ×7200. (*d*) Higher magnification of a vesicle with attached hypha within a cell of an *Alnus glutinosa* nodule. This vesicle would appear to be covered with a layer of host cytoplasm. ×10600. (*e*) Scanning micrograph of part of an *Alnus glutinosa* nodule cell containing the bacteroidal stage of the endophyte. Within such cells occasional large vesicles can be seen. ×6300. (*f*) Scanning electron micrograph of a thick section of nodular cells of *Myrica gale*. The flattened club-shaped vesicles can be seen at the tips of the endophyte hyphae. ×3400.

35. Nitrogen fixation in root nodules of alder and pea in relation to the supply of photosynthetic assimilates

C. T. WHEELER & ANN C. LAWRIE

The importance of photosynthetic products for nitrogen fixation has been established for a number of leguminous plants, although non-leguminous species have been less well studied. Wheeler & Bowes (1974) showed that the rate of acetylene reduction by young alder plants was dependent on the light intensity in which they were grown, while complete darkening virtually eliminated nitrogenase activity within 24 hours (Wheeler, 1971).

The rapidity with which a change in light intensity affects fixation must depend primarily upon the rate of translocation of photosynthates from the shoot to the nodules and the extent to which reserve substances can be utilised to support fixation when supplies of newly synthesised assimilates are reduced. The detection of radioactivity in the nodules of some species within an hour of feeding the shoots $^{14}CO_2$ (e.g. Pate, 1962, for pea; Small & Leonard, 1969, for pea and clover; Wheeler, 1971 for alder) shows that translocation can be a rapid process, although the continued rise in the radioactivity of the nodules for a substantial period subsequent to the exposure of the shoot system to $^{14}CO_2$ shows that only a small proportion of the assimilates produced at any one time may be translocated immediately to the nodules. The role of reserve substances in the nodules in nitrogen fixation has not been clearly established, although in darkened alder plants Wheeler (1971) found that the content of 'reserve' (perchloric acid-soluble) carbohydrates did not decrease significantly during the first 24 hours of darkening despite a rapid decline in soluble sugar and nitrogenase activity at this time.

These features of the metabolism of the plant are likely to be of importance in the diurnal fluctuations in nitrogen fixation which have been reported for some species and in which both fluctuations in light and temperature have been implicated as causal factors. In legumes, most work has been on the soybean. Thus Hardy *et al.* (1968) found greater activity in nodulated roots during the day than at night, and Bergersen (1970) also described variations in the activity of nodulated roots which closely followed changes in light intensity. Though Bergersen found little variation in the activity of the nodules when detached from the roots at different times of day prior to testing for acetylene reduction, Mague & Burris (1972) observed diurnal fluctuations with detached nodules in which maximum activity was correlated with both temperature and light intensity.

Diurnal fluctuations in fixation have also been detected in some non-leguminous species. Stewart, Fitzgerald & Burris (1967) showed that the rate of acetylene reduction by detached nodules of *Comptonia peregrina* (= *Myrica asplenifolia*), sampled from the field at 13.30 hours and 18.5 °C was twice that of nodules sampled at 07.30 hours and 17.5 °C, although the difference between comparable samples in ^{15}N enrichment was much less. However, again in detached nodules of *M. asplenifolia* under field conditions, Fessenden, Knowles & Brouzes (1973) were unable to detect diurnal variations in acetylene reduction. These authors suggested that this might be due to the substantial reserves of carbohydrates in the rhizome system of this species and stressed the variability of the results obtained with field grown material. Bond (1971) described diurnal fluctuations in glasshouse-grown *Casuarina* plants in which the rate of fixation at midday was more than twice that in the early morning and late evening, a large part of the difference being attributed to the effect of temperature. In contrast, Wheeler (1969) observed diurnal fluctuations in glasshouse-grown plants of *Alnus glutinosa* and *Myrica gale*, with a maximum in nitrogenase activity at about midday, a large part of which was thought to be due to changes in light intensity since the air temperature was kept roughly constant during the experiment. It was suggested that diurnal variations in fixation might also be found under natural conditions of growth where the effect of temperature would reinforce the light effect, but Akkermans (1971) was unable to detect significant differences between the activity of alder nodules harvested from fully grown trees at night-time and during the day.

In this chapter diurnal fluctuations in the rate of acetylene reduction which have been found both in alder and pea are compared and discussed in relation to the effects of light and temperature. Studies on the translocation of photosynthates to alder nodules are described and the distribution and accumulation of radioactive photosynthates within the nodules, as determined by soluble compound micro-autoradiography, are compared for the two species in relation to the fixation of nitrogen (see also Pate, Chapter 27 and Hardy & Havelka, Chapter 31).

Materials and methods

The culture of glasshouse plants of *Alnus glutinosa* (L.) Gaertn. and of *Pisum sativum* L. cv. 'Alaska', inoculated with Rothamstead strain 1202 of *Rhizobium leguminosarum*, the estimates of light intensity and the assay of acetylene reduction by nodulated roots using a Pye series 104 gas chromatograph were carried out essentially as described previously (Wheeler, 1971; Lawrie & Wheeler, 1973). The small stature (20 cm) and short life span (6 weeks) of the pea plants used in these studies was thought to be due mainly to the high temperatures often experienced in the glasshouse.

The exposure of plants to $^{14}CO_2$, the assessment of the radioactivity of aqueous ethanolic extracts of the nodules and determination of the ethanol-soluble carbohydrate content of the nodules are also described in the above two publications. Micro-autoradiographs were prepared using a technique suitable for soluble compounds (Bowen *et al.*, 1972) as described by Lawrie & Wheeler (1973).

Results

Diurnal fluctuations in acetylene reduction

Results which have been obtained with alder are summarised in Table 35.1. Significant fluctuations in nitrogenase activity were detected in two experiments with the rate at midday in one instance more than double, and in the other 40 % higher than at night. In the third experiment, which was carried out on a dull, rainy day, there was no significant difference between the rate at midday and midnight. Experiments with second-year alder failed to show significant differences between the rate of acetylene reduction at midday and midnight, even when the day was sunny.

In the pea, in contrast to the results obtained with alder, highest rates of acetylene reduction were usually found at night, although a peak of activity was often detected about midday also. Fig. 35.1 is typical of experiments showing an evening peak alone and shows that the rate of reduction fell during the morning and early afternoon to a minimum at 16.00 h. The rate then increased to reach a maximum at 20.00 h which was seven times the minimum rate. In this experiment, light intensity was at a maximum at 12.00 h and maximum temperature at

14.00 h. The temperature at the time of highest nitrogenase activity was 18.2 °C, which was 15 °C lower than the maximum recorded at 14.00 h. Since the optimum temperature for acetylene reduction by the detached nodulated roots of the pea plants used in these studies was about 25–30 °C (Lawrie, 1973), nitrogenase activity could well have been

Table 35.1. *Daily fluctuations in acetylene reduction by the nodules of first-year and second-year alder plants*

Date	Time of day (h)	C_2H_4 (μmole/g fresh weight nodules/h)	Least significant difference between means, $P = 0.05$	Temperature variation in glasshouse during experiment (°C)	Total sunshine (h)
First-year alders					
21 August 1969*	12.00	6.5	0.60	19–23	4.6
	20.15	3.1			
1 September 1969†	12.00	8.8	1.88	20–3	8.2
	24.00	6.2			
12 July 1973	13.00	3.41	1.82	19–21	Nil
	24.00	4.33			
Second-year alders					
29 August 1969	13.00	2.22	0.41	20–4	10.1
	24.00	1.92			
12 July 1973	13.00	1.92	0.25	19–21	Nil
	24.00	1.74			

* Data from Wheeler (1969) and † Wheeler (1971). In the other experiments, acetylene reduction by the nodules of 18 first-year or 15 second-year alder plants was measured on each occasion over a 1 h period in an incubator at 24.5 °C. Sunshine data were provided by the Glasgow Weather Centre.

inhibited by the high temperatures recorded about 14.00 h. In a second experiment, conducted on a cooler day, two peaks of nitrogenase activity were detected (Fig. 35.2). The first peak at 12.00 h occurred just before maxima in light intensity and temperature. The second, more prominent peak again was found just before midnight. In both experiments, the evening peaks in nitrogenase activity followed 2 h after the ethanol-soluble carbohydrate content of the nodules reached a maximum. Changes in the carbohydrate content and nitrogenase activity of the nodules could not be absolutely correlated in any experiment.

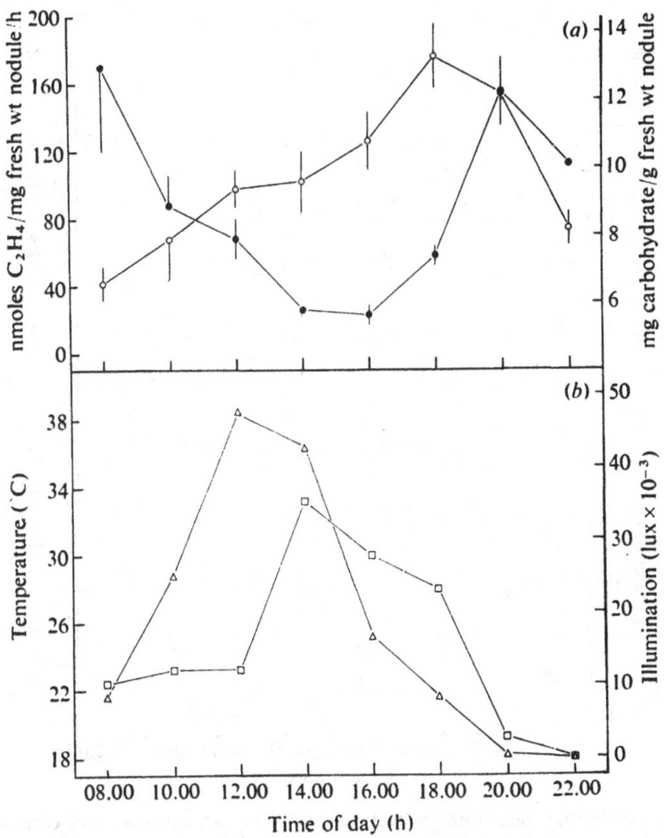

Fig. 35.1. (*a*) Diurnal variations in acetylene-reducing activity (●) and ethanol-soluble carbohydrate content (○) of nodules from 3-week-old pea plants, in an experiment carried out on 2 May 1973. Each point on the acetylene-reduction graph is based on the assay of the nodulated roots of sixteen plants in groups of four, the assays being made at the prevailing temperature of the glasshouse. Subsequently the nodules from the assayed root systems were detached and analysed for ethanol-soluble carbohydrates. The vertical bars indicate the standard errors of the means. (*b*) Records of variations in light intensity (△) and air temperature (□) in the glasshouse during the course of the above experiment.

Translocation of ^{14}C-labelled photosynthates

Fig. 35.3 shows that when alder shoots were exposed to $^{14}CO_2$ for 40 min at different times of day the radioactivity of ethanolic extracts of the nodules harvested immediately after exposure reached a maximum at

13.35 h which was about six times that at 10.00 h and 16.45 h. There was little difference in the radioactivity of extracts from nodules harvested at different times of day 130 min after exposure to $^{14}CO_2$, but after 240 min maximum radioactivity was found at 18.10 h. Similar

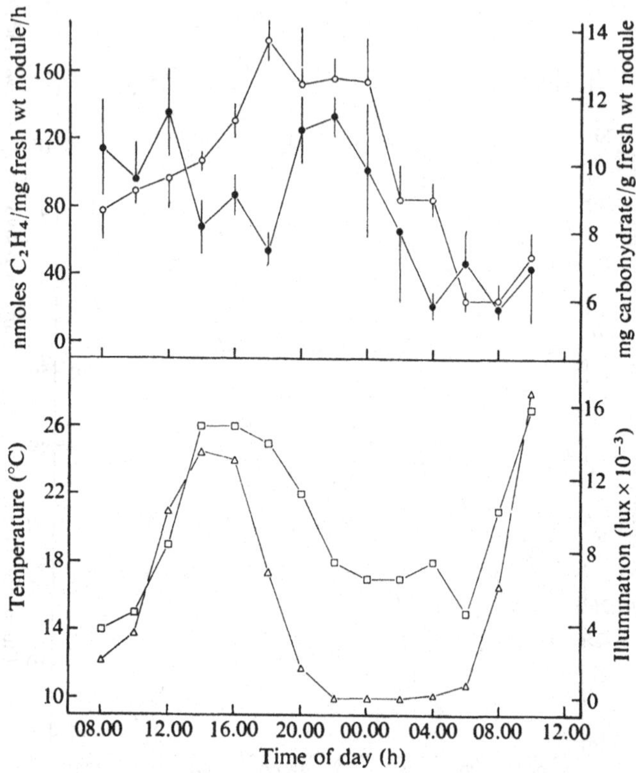

Fig. 35.2. (*a*) Diurnal variations in acetylene-reducing activity (●) and ethanol-soluble carbohydrate content (○) of nodules from 3-week-old pea plants, in experiment carried out on 18–19 May 1973. The procedure was the same as in the experiment of 2 May (see Fig. 35.1). The vertical bars again indicate the standard errors of the means. (*b*) Records of variations in light intensity (△) and air temperature (□) in the glasshouse during the course of the above experiment.

changes were found in extracts of roots harvested at the same time as the nodules (Fig. 35.3) where, immediately after exposure to $^{14}CO_2$, the radioactivity at 13.35 h was nearly twice that at 10.00 h and five times that at 16.45 h. After 130 min, the radioactivity at different times of day was roughly constant but after 240 min maximum radioactivity was found at 18.10 h. The actual amount of ethanol-soluble carbohydrates in the roots (Table 35.2) was at a maximum during the early part of the night, as reported earlier for the nodules (Wheeler, 1971).

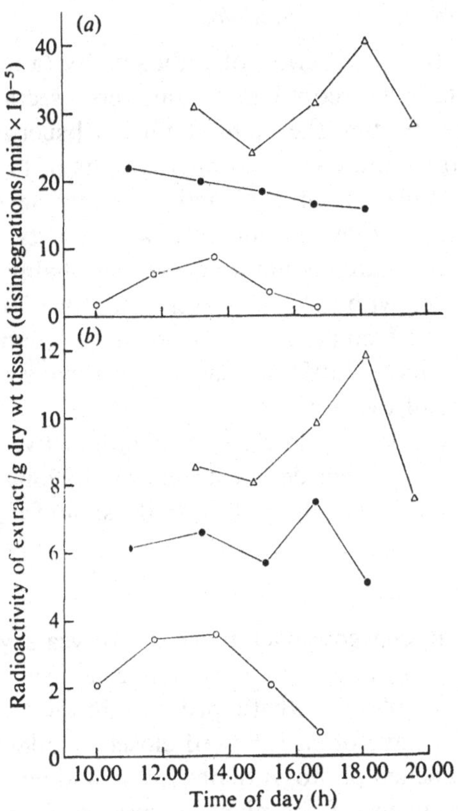

Fig. 35.3. Radioactivity of ethanolic extracts of alder nodules (*a*) and roots (*b*) harvested immediately (○), 130 min (●) or 240 min (△) after 40 min photosynthesis by the plants in an atmosphere containing $^{14}CO_2$ (200 μCi/10 l container) at different times of day. Experiment conducted in the glasshouse on 28 August, 1969 under conditions described by Wheeler (1971).

Table 35.2. *Daily fluctuations in the ethanol-soluble carbohydrate content of roots of first-year alder plants*

Time of harvest (h)	10.45	12.30	14.30	20.30	00.15	03.30
Mean carbohydrate content (mg/g dry weight roots)	6.14	5.90	6.43	8.78	9.03	11.70

Experiment of 21 August 1969. The day was sunny with some light cloud and the temperature in the glasshouse was maintained between 19–23 °C. Roots from 3 batches of 4 plants were harvested for analysis at each time of day. Least significant difference between means ($P = 0.05$) = 1.75.

Micro-autoradiography of alder nodules

Plate 35.1 shows the distribution of radioactivity (appearing as white dots in dark field microscopy) in a transverse section of a nodule harvested 5 h after feeding the plant shoot [^{14}C] sucrose. Comparison of the phase contrast and dark field micrographs of this section shows considerable accumulation of radioactivity over the stele, especially the phloem. Although large amounts of radioactivity were associated with the uninfected, starch-containing cells, probably relatively little of this was associated with the starch itself since over 70 % of the radioactivity of the nodules could be extracted in 80 % ethanol 24 h after feeding ^{14}C to the shoots. Little radioactivity was shown by the infected cells at the corners of the section.

The accumulation of only low levels of radioactive photosynthates in the infected cells is further demonstrated by the longitudinal section of a nodule taken 4 h after feeding $^{14}CO_2$ to the shoot for 1 h (Plate 35.2).

Discussion

Results obtained at Glasgow over the last five years with both alder and pea have confirmed that nodules, active in fixation, are notable metabolic 'sinks' for photosynthetic products in the plant. Changes in nitrogenase activity have often followed closely similar changes in the accumulation of labelled photosynthates in the nodules, e.g. during the growth and development of the plant (Lawrie & Wheeler, 1973), but the degree to which the two processes are correlated when there are relatively rapid changes in the availability of photosynthates, e.g. as may occur during diurnal fluctuations in light intensity, is less clear. The correlation between the midday maximum in acetylene reduction (Table 35.1) and the arrival of new photosynthates in the nodules of young alder plants in the glasshouse on sunny days (Fig. 35.3) suggests that changes in light intensity can affect the rate of nitrogen fixation quite rapidly, but the absence of significant changes in older plants (Table 35.1) and in field material (Akkermans, 1971) suggests that the response to variations in light intensity may also involve a number of other factors, e.g. the conditions under which the plants are grown, distances over which photosynthetic products must be translocated, competition within the plant from other metabolic 'sinks' and possibly the levels of reserve substances in the plant.

The detailed study of the time course of translocation of ^{14}C-labelled photosynthates from the shoots to the nodules and roots of young alder plants (Fig. 35.3) suggests that although translocation of assimilates may continue for a considerable time after their synthesis, the amount of photosynthate translocated is not uniform over this period. The data suggest that there may be two main 'waves' of translocation to the nodules. The first wave of new photosynthates is maximal at midday and, as mentioned above, may support enhanced rates of fixation at this time under suitable conditions. The absence of a peak in the accumulation of labelled photosynthates in the nodules 130 min after feeding $^{14}CO_2$ suggests that the translocation of photosynthates synthesised at different times of day may be relatively constant after this time interval. The second peak in radioactivity of the nodules 240 min after feeding $^{14}CO_2$ over the midday period presumably reflects the arrival in the nodules of a more slowly translocated wave of photosynthates. It seems probable that this second wave contributes to the overnight accumulation of carbohydrates but is apparently unable to stimulate the rate of fixation in alder nodules. The similarity of the above data with changes in the ethanol-soluble carbohydrate content (Table 35.2) and accumulation of ^{14}C-labelled photosynthates in root samples (Fig. 35.3) harvested at different times of day shows how translocation to the nodules reflects that to the roots, although the higher specific activity of the nodules at all times of harvest emphasises the greater potential of the nodules as a metabolic sink.

In pea plants grown in nitrogen-free culture in the glasshouse where temperature and light were allowed to fluctuate naturally, the diurnal fluctuations in nitrogen fixation were often quite different from those previously reported for soybean and alder, although it is possible that the somewhat reduced growth of the pea plants used in these studies, coupled with the relatively high respiratory rate of their nodules (3.5 μl O_2 consumed/mg fresh weight detached nodules/h) may have produced a greater dependence on supplies of new photosynthates than in more normal pea plants.

Two basic types of fluctuation were found in pea (Figs. 35.1 and 35.2) although there were small variations on the basic patterns in replicate experiments, possibly as a result not only of differences in the weather conditions on the day of the experiment but also because of differences in day length (time of year) and fluctuations in sunlight and temperature on days prior to the experiment. A maximum in fixation was always found during the early part of the night. This usually corresponded to maximum levels of ethanol-soluble carbohydrates in the nodules, with a lag of 2 h, suggesting that it may depend on the arrival of a more slowly translocated wave of carbohydrates, as found in alder. This possibility is supported by measurements of the radioactivity of the nodules made over a 24 h period following exposure of the shoots to $^{14}CO_2$, which showed that although the radioactivity of pea nodules harvested immediately after exposure of the shoots to $^{14}CO_2$ for 1 h at different times of day reached a maximum at 13.00 h, the radioactivity of the nodules harvested at any one time of day continued to increase for 4 to 8 h after feeding (Lawrie, 1973). These findings, together with similarities between the two species in the accumulation of ethanol-soluble carbohydrates in the nodules (Wheeler, 1971, Figs. 35.1 and 35.2) suggest that the basic pattern of translocation of photosynthates to the nodules may be rather similar even though the effect on acetylene reduction in the two species is different.

The second, usually smaller, peak in acetylene reduction in pea nodules (Fig. 35.2) was only detected on cooler days, when changes in temperature were favourable to increased nitrogenase activity. Its absence on other occasions could be attributed to high temperatures experienced over the midday period which would suppress rather than stimulate nitrogenase activity. Observations made on a number of different occasions suggest that changes in temperature favourable to increased nitrogenase activity could be the major cause of the midday maximum, with changes in light intensity reinforcing the temperature effect.

The rise in the carbohydrate content of the nodules during the morning, even when the rate of acetylene reduction was falling rapidly as often happened when a midday maximum in acetylene reduction was not detected, shows that carbohydrate levels in the pea nodule, as in alder nodules (Wheeler, 1971), cannot be absolutely correlated with changes in nitrogenase activity. This suggests that a significant proportion of the accumulating carbohydrates is not used in support of nitrogen fixation, but in other cellular activities, e.g. growth. This possibility is supported by micro-autoradiography of both pea and alder nodules taken after feeding the shoot $^{14}CO_2$ or [^{14}C] sucrose.

Micro-autoradiographs of alder nodules always showed a large accumulation of ^{14}C-labelled photosynthates in the stele, presumably reflecting the import and export of labelled metabolites. Although high concentrations of radioactivity were often detected in uninfected cells containing storage deposits such as starch and tannin, evidence (p. 501) indicated that a substantial proportion of the accumulated photosynthates must still have been in a soluble, readily metabolised form. The absence of any large accumulation of radioactivity in the infected cells (Plate 35.2) suggests that the photosynthates translocated to these cells must be used quickly to support the metabolic activities of the cell and endophyte.

The accumulation of photosynthates in the relatively large proportion of uninfected tissue situated between the cells containing the vesicular form of the endophyte in the alder nodules may partially isolate the accumulating sugars from the nitrogen-fixing sites which, on the basis of their reduction of tetrazolium salts, are thought to be the vesicles (Akkermans, 1971). This compartmentation of the accumulating sugars could be responsible for the absence of any increase in nitrogenase activity in response to the higher carbohydrate levels at night in this species. The occurrence of the midday maximum in nitrogen fixation in the first year alder plants – which coincides with the arrival of increased amounts of new photosynthates at this time, but a lower ethanol-soluble carbohydrate content of the nodules – can only be explained satisfactorily on this basis if the distribution and utilisation in the nodules of immediately translocated photosynthates is different from those translocated after a delay. It has not been possible to test this by these techniques because the radioactivity of the nodules immediately after exposure of the shoots to $^{14}CO_2$ for short periods is insufficient to be detected by micro-autoradiography. After exposure to $^{14}CO_2$ for 90 min, however, the distribution of radioactivity in nodule sections showed no difference from results presented here except for a reduction in the number of silver grains seen in the micro-autoradiographs.

In pea nodules, Lawrie & Wheeler (1973) showed that the proportion of uninfected tissue accumulating photosynthates is much less than in alder nodules. No accumulation was detected in the outer, uninfected cells of the cortex and there were few uninfected cells in the central region of the nodules examined. Most of the accumulation of radioactive assimilates occurred in the vascular traces, the meristematic region of the nodules and in the central infected mass of tissues. In this region, cells in which the bacteroids were still increasing in number accumulated more photosynthates than the cells packed with bacteroids which contributed most to nitrogen fixation, suggesting that in the former cells a significant proportion of the photosynthates may be used to support growth and multiplication of the bacteroids while in the latter cells photosynthates are rapidly used in support of the metabolism of the host cell and bacteroids, as noted above for alder. However, the close association of infected cells in the central region of the pea nodule suggests that accumulating carbohydrates could be much more readily transferred between cells, and thus be available to stimulate fixation, than in the alder nodule.

In conclusion, it is apparent from these studies that different plants may well respond differently to relatively short-term changes in supplies of photosynthetic assimilates such as occur naturally during a 24 h period. The age and size of the plant and the conditions of growth are also likely to be important factors in determining the nature and extent of the response to a change in the supply of photosynthates to the nodules.

References

Akkermans, A. D. L. (1971). Nitrogen fixation and nodulation of *Alnus* and *Hippophäe* under natural conditions. PhD Thesis, University of Leiden.

Bergersen, F. J. (1970). The quantitative relationship between nitrogen fixation and the acetylene reduction assay. *Australian Journal of Biological Sciences*, **23**, 1015–25.

Bond, G. (1971). Root-nodule formation in non-leguminous angiosperms. *Plant and Soil*, Special Volume, 317–24.

Bowen, M. R., Wilkins, M. B., Cane, A. R., & McCorquodale, I. (1972). Auxin transport in roots. VIII. The distribution of radioactivity in the tissues of *Zea* root segments. *Planta*, **105**, 273–92.

Fessenden, R. J., Knowles, R., & Brouzes, R. (1973). Acetylene–ethylene assay studies on excised root nodules of *Myrica asplenifolia* L. *Proceedings of the Soil Science Society of America*, **37**, 893–8.

Hardy, R. W. F., Holsten, R. F. P., Jackson, E. K. & Burns, R. C. (1968). The acetylene–ethylene assay for nitrogen fixation; laboratory and field evaluation. *Plant Physiology, Lancaster*, **43**, 1185–1207.

Lawrie, A. C. (1973). Nitrogen fixation in nodulated legumes in relation to the assimilation of carbon. PhD Thesis, University of Glasgow.

Lawrie, A. C. & Wheeler, C. T. (1973). The supply of photosynthetic assimilates to nodules of *Pisum sativum* L. in relation to the fixation of nitrogen. *New Phytologist*, **72**, 1341–8.

Mague, T. H. & Burris, R. H. (1972). Reduction of acetylene and nitrogen by field-grown soybeans. *New Phytologist*, **71**, 275–86.

Pate, J. S. (1962). Root-exudation studies on the exchange of C^{14}-labelled organic substances between the roots and shoot of the nodulated legume. *Plant and Soil*, **17**, 333–56.

Small, J. G. & Leonard, D. A. (1969). Translocation of ^{14}C-labelled photosynthate in nodulated legumes as influenced by nitrate nitrogen. *American Journal of Botany*, **56**, 187–94.

Stewart, W. D. P., Fitzgerald, G. P. & Burris, R. H. (1967). *In situ* studies on N_2 fixation using the acetylene reduction technique. *Proceedings of the National Academy of Sciences, USA*, **58**, 2071–8.

Wheeler, C. T. (1969). The diurnal fluctuation in nitrogen fixation in the nodules of *Alnus glutinosa* and *Myrica gale*. *New Phytologist*, **68**, 675–82.
Wheeler, C. T. (1971). The causation of the diurnal changes in nitrogen fixation in the nodules of *Alnus glutinosa*. *New Phytologist*, **70**, 487–95.
Wheeler, C. T. & Bowes, B. G. (1974). Effects of light and darkness on nitrogen fixation by root nodules of *Alnus glutinosa* in relation to their cytology. *Zeitschrift für Pflanzenphysiologie*, **71**, 71–5.

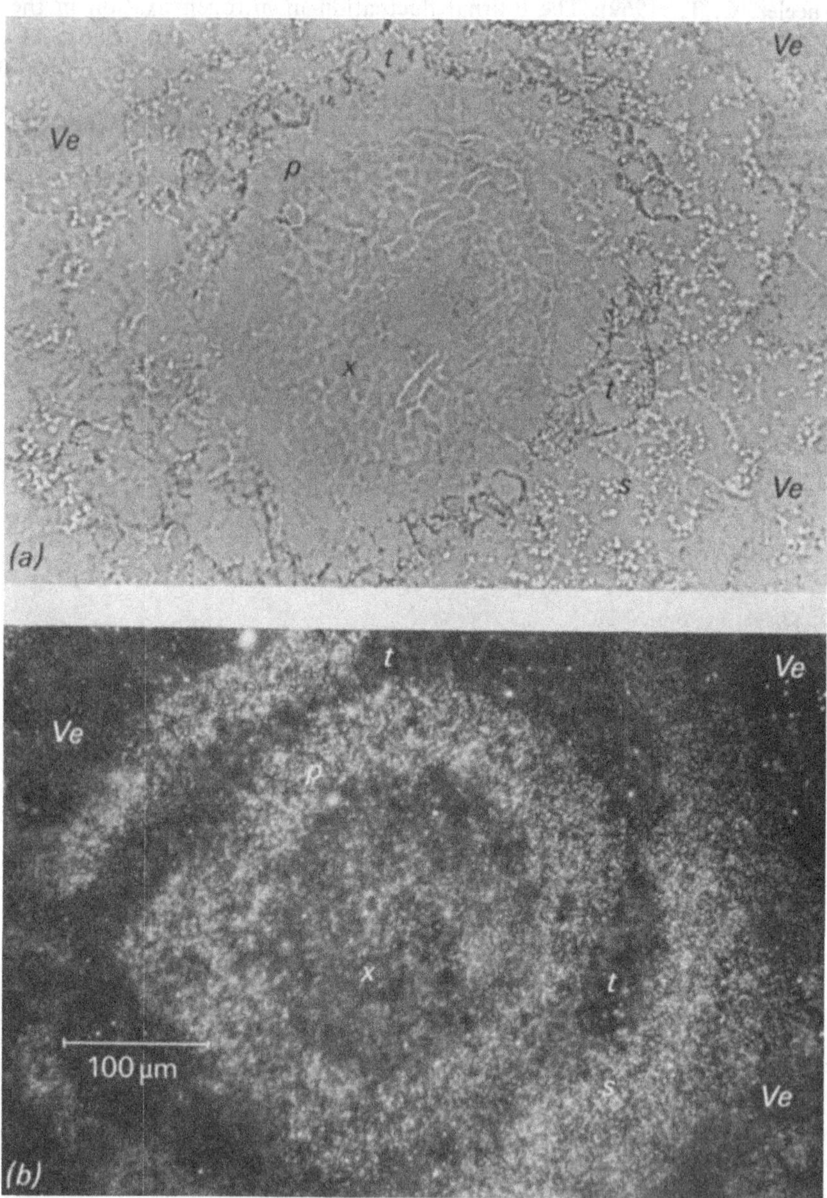

Plate 35.1. Phase contrast (*a*) and dark field (*b*) micrographs of a transverse section through an alder nodule, harvested 5 h after feeding the shoot [^{14}C]sucrose. Accumulation of radioactivity is particularly evident over the xylem (*x*) and phloem (*p*) of the stele, also the starch-containing, uninfected cells (*s*) just outside the stele. There is little accumulation in the cortical cells containing the endophyte in the vesicular form (*Ve*) or over the rings of tannin-containing cells in the endodermal region (*t*).

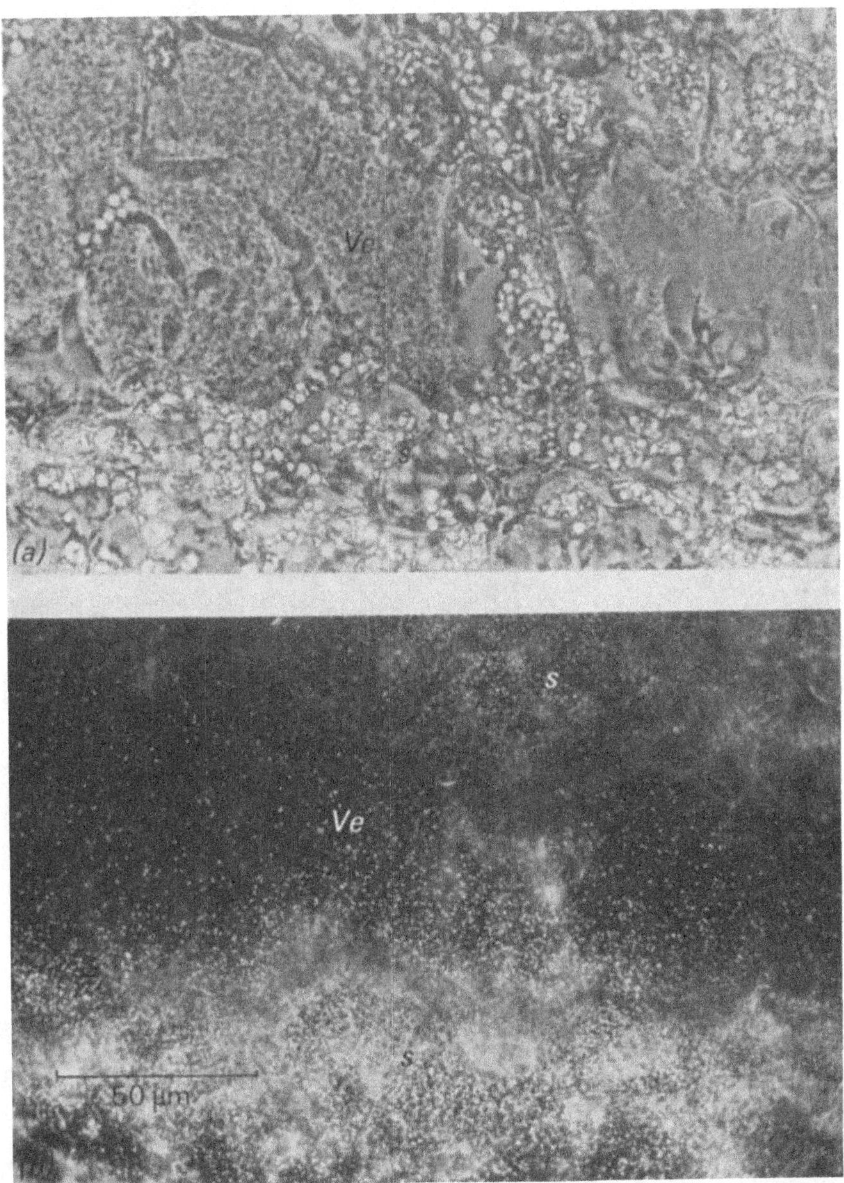

Plate 35.2. Phase contrast (*a*) and dark field (*b*) micrographs of part of the cortex in a longitudinal section through an alder nodule, harvested 4 h after feeding the shoot $^{14}CO_2$ for 1 h. These show more clearly how there is very little accumulation of radioactivity in the infected, vesicular cells (*Ve*), and again that there is considerable accumulation in the starch-containing, uninfected cells (*s*).

36. The formation and nitrogen-fixing activity of the root nodules of *Alnus glutinosa* under field conditions

A. D. L. AKKERMANS & C. van DIJK

In The Netherlands, as in northern Europe generally, *Alnus glutinosa* (L.) Gaertn. is a native and abundant species, and has also been planted in many parts of that country. The trees grow well on wet clay and acid peat soil down to a pH of about 4, and also on drier soils of pH up to 8. On peat soil the species is an important pioneer in the succession from the original reed-land vegetation to the wet forest type – the alder grove. This vegetation is considered to be the final stage in the hygrosere in mesotrophic to eutrophic, stagnant waters.

Although it is very probable that as a result of the nitrogen-fixing activity of their root nodules, black alder trees contribute substantially to the nitrogen status of the soil, previous estimates of the magnitude of the contribution have been based on data obtained for laboratory-grown plants (Ferguson & Bond, 1953; Virtanen, 1957). In the present article the results of studies on the growth of nodules on trees in the field, and on their fixing activity, are presented, together with observations on some factors affecting the distribution of the nodule endophyte in the soil. Fuller information on some of the aspects to be dealt with, particularly as regards the techniques employed, is to be found in Akkermans (1971).

Nodulation in the alder

The pattern of distribution of the nodules on the root system and their abundance are affected by several environmental factors. Depending on the humidity of the soil at least four different types of distribution have been distinguished, as follows:

(*a*) In alder groves with a very high water level the trees are surrounded by water and the roots are concentrated around the trunk of the tree. Nodules are present on the superficial roots in the clump and are covered by a moss layer which protects them against drying.

Roots growing in water or in permanently water-logged soils are not nodulated.

(*b*) In alder groves on quaking bogs with a water level 10–20 cm below the soil surface, the trees form a floating layer of intermingled roots in the upper 50 cm of soil. The nodules are present just above the mean water level within a radius of 5 m from the trunk.

(*c*) On drier soils the root system develops more vertically, and most nodules are present in the upper 30 cm of soil.

(*d*) On trees along ditches and brooks most of the nodules are situated superficially at the side of the water-course. This nodulation pattern has been observed even when the trees were growing several metres from the water channel.

These observations show clearly that the nodules mostly occur in the well-aerated, humid, upper soil layer, while on permanently submerged roots nodules have never been found.

The total quantity of nodules on a tree is a function of its age. Information on this aspect was obtained by excavating trees of different ages. The results (Table 36.1) show a rapid increase in the number and dry

Table 36.1. *Growth and nodulation data for alder trees of different ages growing on peat soil in 'De Wieden', The Netherlands*

Age of trees (yr)†:	3	4–5	7–8	15–20
Number of trees harvested	7	6	3	3
Mean height of stem (cm)	30	54	310	600
Mean dry wt stem and roots per tree (g)	1.7	23.1	1723	⩾10 000
Mean dry wt leaves per tree (g)	1.0	7.6	287	1 570‡
Mean dry wt living nodules per tree (g)	0.06	0.4	36	245
Mean dry wt living nodules as % of mean total dry wt of tree	3	1	2	⩽2
Mean number of nodules per tree:				
Healthy	16	70	160	575
Partly necrotic	0	0	32	51
Wholly necrotic	0	15	34	182
Total	16	85	226	808
Necrotic as % of total	0	18	15	23

* Table 36.1–36.5 are modified from Akkermans (1971).
† Number of annual rings in the stem 50 cm above soil surface.
‡ Based on one tree only.

weight of nodules with increasing tree age. However the total weight of living nodules relative to the weight of the tree remains in the range of 1–3 %. Similar values were obtained by Bond (1958) for young alder plants grown in a glasshouse, suggesting that the weight of nodules of the alder is more or less constant in relation to its size throughout its life. The presence on the root systems of both living and dead nodules of different ages and sizes suggests a continual formation and death of nodules.

More detailed information on the growth of the nodules was obtained by estimating the age of the nodules from the number of annual rings in the xylem of the roots just behind the nodules. Table 36.2 shows the number and weight of nodules of different ages on a twenty-year-old tree. With increasing nodule age the total number per age class decreases

Table 36.2. *Age analysis of the nodule population of a twenty-year-old alder tree**

Age class of the nodules (yr)	Number of nodules		Nodule dry wt (g)		Estimated annual production of nodule tissue (g dry wt)	
	Counted in sample†	Calculated for tree	Mean per nodule	Total for tree	Per nodule	Whole tree‡
1–2	69	324	0.018	5.8	0.018	6
2–3	95	447	0.072	32.2	0.054	24
3–4	62	291	0.188	54.7	0.116	34
4–5	37	174	0.465	80.9	0.277	48
5–6	12	56	1.170	65.5	0.705	39
6–7	2	9	0 440	4.0	–	–
7–8	1	5	0.100	0.5	–	–
Totals	278	1306		243.6		151

* Only healthy, necrosis-free nodules are considered in this table.
† All the healthy nodules on the tree were detached and counted, but only a representative sample comprising 21 % of the total number was analysed for age.
‡ Calculated by multiplying the estimated annual production of nodule tissue for a particular age class by the total number of nodules in the same class.

almost continuously, while the total dry weight per class increases until decay favours nodule growth. Nodules aged 4–5 years make the most important contribution to the weight of nodules per tree. Observations on many other trees showed that nodules only occasionally reach an age of 10 years. The annual production of nodule tissue can be calculated from the average growth rate of the nodules and the numbers of nodules of different ages (Table 36.2). In this tree an apparent production rate of $151 \times \frac{100}{244} = 62\ \%$ is indicated. Similar calculations for nine other trees aged 8–25 years showed an annual production rate between 50 and

60 %. Due to necrosis the actual annual increase in living nodule tissue will be less than the above estimates.

For the purpose of later calculations it was necessary to know the total amount of nodules in the whole ecosystem. Data presented in Table 36.3 indicate that the dry weight of nodules in a particular alder grove amounted to 454 kg/ha. A similar study was made for a mixed

Table 36.3. *Total amount of healthy nodules in an alder grove in 'De Wieden'*

Age of trees (yr):	5–8	8–15	15–20	Totals
Number of trees/100 m^2	37	32	15	84
Number of trees in which the total amount of nodules was determined	9	3	3	15
Mean total nodule dry wt per tree (g)	3	36	219	
Calculated nodule dry wt (kg/ha)	11	115	328	454

stand of alder and willow (*Salix cinerea*) on wet clay soil. The alders grew at a density of 625 trees/ha, and had a mean nodule mass of 134 g/tree, corresponding to 84 kg/ha. In Oregon, Zavitkovski & Newton (1968) estimated nodule dry weights in seven- and thirty-year-old stands of *Alnus rubra* to be 117 and 244 kg/ha respectively.

Nitrogen-fixing activity of the nodules

In considering how determinations of the fixing activity of field nodules could best be made, initial observations were made on nodules still attached to the tree. Roots bearing nodules were sealed into 'Saran' plastic bags or glass bottles, and then incubated in a mixture of 10% acetylene in air. The acetylene-reducing activity, measured by analysing small gas samples every 10 minutes, showed a linear rate for at least 3 hours. The severance, at a distance of 2–5 cm from the nodule, of the connecting root did not affect the rate of ethylene formation during the subsequent 3 hours. Considering also that the oxygen level in the container (19 %) was similar to that experienced by nodules in soil, it was concluded that fixation rates shown by nodules on severed roots approximate to those shown by the same nodules *in situ* in the soil. Complete excision of nodules from the roots often resulted in a marked reduction of the fixation rate, sometimes by as much as 50 %, especially when the nodules were small. Therefore in routine field assays the nodules were left attached to a root piece 2–5 cm long. Within 5 minutes after cutting such specimens from the tree, they were enclosed in 20 ml glass vessels and placed at 21 ± 1 °C, which was known to be near optimal for fixation in alder nodules. After 10 minutes adaptation time acetylene was injected to a concentration of 10 % in air. After 30 or 60 minutes the reaction was stopped by adding dilute sulphuric acid. Gas analyses were made within two days after incubation. The value obtained in these assays will be termed the 'potential activity'. By reference to a record of soil temperatures at the field site kept throughout the year, the actual activity of nodules whose potential activity was known could be calculated.

It soon became obvious that a serious difficulty arises in attempting to determine the fixation of nitrogen associated with alder trees by means of assays on nodule samples, namely that the activity of different samples, even when the nodules are of similar age and taken from the same tree, varies very greatly. A fiftyfold variation is not unusual. Subsequently Waughman (1972), again with *Alnus glutinosa*, has reported a similar situation, while Fessenden, Knowles & Brouzes (1973) have had the same experience with field nodules of *Myrica asplenifolia*. While various reasons for this suggest themselves, in the present study it was established that one contributory factor is the morphology of the nodule. As shown in Plate 36.1, for some reason certain nodules are composed of relatively slender lobes, whereas in others the lobes are much more laterally distended. When ethylene assays were made on numerous samples of nodules of selected type, the mean ethylene production was 1.4 μmoles/g dry wt/h where the lobes were 1–1.5 mm in diameter, rising as diameter increased to 7.8 μmoles for the nodules with the thickest lobes (diameter 3.5–5.0 mm). Another factor affecting variation in nodule activity is possibly the molybdenum content of the nodules, which when determined in ten field nodules from the same tree by the method of Sseekaalo & Johnson (1969), was found to vary from 0.02 to 0.16 μg/g fresh weight. The acetylene-reducing power of the same nodules showed a certain correspondence with the molybdenum content, though the correlation coefficient (0.54) just failed to attain significance at $P = 0.05$.

In practice the only way of coping with the above variability is to ensure that mean values for nodule activity at a given site and at a particular time of year are based on the assay of large numbers of nodule samples. Table 36.4 presents data for acetylene reduction by nodule samples collected from an alder grove at regular intervals throughout 1969. On each occasion about fifty nodule samples were assayed. Even at the favourable temperature prevailing during assay, no activity could be found until the end of April, i.e. about the same time as leaf emergence. Potential activity rose rapidly in early May to attain a value which, as indicated in the table, showed little further change until about the time of leaf-fall in early November. Before the middle of that month activity had returned to zero. It should be noted that the values

Table 36.4. *Estimation of amount of nitrogen fixed per unit nodule dry weight in 1969 in an alder grove in 'De Wieden'*

				Estimated actual acetylene reducing activity		
Month	No. of days per interval	Mean soil temp. over interval (°C)	Potential activity (μmol C_2H_4/g nodule dry wt/ day)*	% of potential activity‡	μmol C_2H_4/g nodule dry wt/ interval	μmol C_2H_4/g nodule dry wt/ month
April	1	10	0.2	31	0	0
May	15	11		38	561	
	16	13		60	945	1506
June	15	14		70	1033	
	15	15		80	1181	2214
July	31	16		86	2623	2623
August	31	17	98.4†	92	2806	2806
September	15	17		92	1358	
	15	16		86	1269	2627
October	8	15		80	630	
	15	13		60	886	
	8	11		38	299	1815
November	3	10	3.8	31	4	4
	12	10	0.0		0	0
	15	7	0.0		0	0
					Total for year	13595
					Equivalent amount of nitrogen (see text)	0.127 g

* Activity at 21 ± 1 °C.
† Over the period May–October inclusive nodule activity was assayed on each of five occasions. The means for the different occasions did not differ significantly. For purposes of calculation for this period the overall mean for the five occasions was used, namely 4.1 μmoles C_2H_4/g nodule dry wt/h, or 98.4 μmoles/day.
‡ Calculated from the known effect of temperature on nodule activity.

for potential activity per day were calculated on the belief that the results of measurements of activity made during daylight hours were valid also for night-time, since prior tests had failed to reveal any clear diurnal variation in activity of these field nodules. As indicated in the table, after allowing for the effect of prevailing soil temperatures and then calculating on to a monthly basis, the production during the year of 0.0136 mole ethylene/g nodule dry matter is indicated, or, assuming a correspondence of $\frac{1}{3}$ mole N_2 fixed to 1 mole C_2H_4 produced, a fixation of 0.127 g nitrogen. Since previous observations (Table 36.3) had shown that in this same alder grove the dry weight of nodules/ha amounted to 454 kg, suitable calculation indicates a fixation of 58 kg nitrogen/ha during the year in question.

It is to be expected that alder trees growing in eutrophic woodland soils will, in addition to the nitrogen gained from their nodules, take up combined nitrogen from the soil. Data indicating this were obtained. Thus by measuring the fixing activity of a large number of nodule samples from a particular twenty-year-old tree at different times of year, and by finding the total amount of nodules on the tree, a fixation of

Table 36.5. *Contribution of nitrogen fixation to the total uptake of nitrogen by alder trees in the field*

Age of the trees studied (yr):	3	4–5	7–8	15–20
Dry wt of nodules per tree (g)	0.06	0.4	36	245
Total nitrogen per tree (g):				
Stem and roots	0.107	0.23	17.2	⩾100
Nodules	0.002	0.01	1.1	7
Leaves	0.030	0.22	8.6	47
Estimated annual increase in nitrogen per tree (g):				
Stem, roots + nodules	0.01	0.20	6.2	⩾12
Leaves	0.03	0.22	8.6	47
Total	0.04	0.42	14.8	⩾59
Nitrogen fixed annually per tree (g)*	0.01	0.05	4.7	32

* Based on a mean of 0.13 g nitrogen fixed/g dry wt nodules/yr (Table 36.4).

22.4 g nitrogen/yr for the tree was estimated. However by autumn the total amount of nitrogen in the leaves alone was 51 g, and though some of the excess may have come from nitrogen stored in the tree over winter, it seems very likely that most of it was current uptake of combined nitrogen. Some further calculations are presented in Table 36.5, which concerns the same trees as in Table 36.1. From the data provided it appears that in these particular trees nodular fixation was contributing 12–54 % of the total annual nitrogen demand of the trees. Zavitkowski & Newton (1968) calculated that at high soil nitrogen levels less than 20 % of the nitrogen in young *Alnus rubra* plants could be credited to fixation.

The occurrence of the *Alnus glutinosa* endophyte in soils

In The Netherlands, trees of *A. glutinosa* are generally well-nodulated, and although the frequency of nodules may vary to some extent, nodule-free trees have never been observed, though they have occasionally been reported elsewhere (see Bond, Chapter 32, p. 449). Since there is no evidence that the endophyte is seed-borne, the infection of young plants must be dependent on its presence in the soil. Lack of evidence of any ability of the endophyte to actually grow in the soil makes it likely that it is present in a resting stage, such as a non-growing hyphal form or a spore-like body. There is some evidence, though chiefly circumstantial, that these spore-like bodies or ' granula ' which the endophyte produces in some cells of the alder nodule are the forms which, on nodule decay, become incorporated into the soil and ensure the nodulation of future plants (see Angulo *et al.*, Chapter 33 and Gardner, Chapter 34). In that event the input of infective particles into a soil is determined by the death rate of nodules and the amount of infective particles that have been formed within them. Factors tending to reduce the stock of infective particles in the soil include their death or their disappearance due to soil predators or in drainage waters.

The authors have made estimates of the minimum numbers of infective particles in both root nodules and soil samples by inoculating endophyte-free alder seedlings with suspensions of the crushed nodules or of the soil, using a modification of the method of Quispel (1954) which is based on the consideration that each nodule that forms on the test plants is indicative of the presence of one infective particle in the inoculum applied.

Schaede (1933) first noticed that in *A. glutinosa* two types of nodule could be distinguished on the basis of the presence or absence of the granula mentioned above. It seems that this may be attributable to the existence of two strains of the endophyte. By using the method referred to above, the authors found the minimum number of infective particles in healthy nodules without granula (Gr^-) to be 10^4/g nodule fresh weight, the corresponding result for healthy nodules containing granula (Gr^+) being 10^6–10^7. This shows that although Gr^- nodules still have infective power, that of Gr^+ nodules is 100–1000 times greater, supporting the idea that the granula are infective particles.

Similar tests were carried out with necrotic nodules of both types, care being taken to use only nodules which appeared to be externally intact, so that minimum loss of endophyte material could have taken place before collection. Results so far obtained show that the infecting capacity of necrotic Gr^- nodules is 10–100 times greater than that of the corresponding healthy nodules, i.e. infectiveness seems to increase during nodule decay. Necrotic Gr^+ nodules showed a somewhat smaller infective capacity than healthy ones of that type. These results seem to suggest that the eventual contribution of infectiveness to the soil by the two types of nodule is less disparate than the experiments with healthy nodules, together with considerations of the morphology of the endophyte, would indicate. More detailed work on this aspect is in progress.

An attempt to gain information on the extent of the release of infective bodies into the soil by nodule decay has been made on the basis of the nodulation data already provided in respect of a particular tree in Table 36.2. If it is assumed that after the tree had attained a reasonable size the number of new nodules formed each year was of the same order, then if there were no decay of nodules, the number in each class (third column from left in the table) would also be fairly constant. That the numbers were found to diminish is attributed to nodule decay. Suitable calculations on that basis from the data provided in the table, with the added information that all the nodules mentioned in the table were collected from the upper 10 cm of soil over an area of 25 m^2, suggests that about 160 g of nodule dry matter was lost annually by the tree to 2.5 m^3 of soil. The nodules on this particular tree were all of Gr^- type, and on the basis of the infective capacity of the nodules it was calculated that nodule necrosis on the scale just indicated would release some 2000 infective particles per gram dry soil annually. Direct estimation of the infectiveness of four samples (each composed of ten cores) of soil taken from the upper 10 cm of soil in the above area, after removal of the surface litter, indicated the presence of a minimum of 350–1000 infective particles per gram of dry soil.

Experiments have been carried out which were designed to give information on the capacity of the infective particles of the alder endophyte to survive in the soil. Nodule homogenates have been added to samples of forest top-soil which were then incubated under controlled conditions for long periods. Results to date have shown a decrease in infectiveness on the part of the soil of at least 90 % per year, pointing to a low survival capacity of the infective particles in these particular soils, although the data of Rodriguez-Barrueco (1968) indicated that at least in some soils the infective agents of the endophyte can survive for a good many years.

Work on the above aspects is continuing and will be described in more detail elsewhere.

References

Akkermans, A. D. L. (1971). Nitrogen fixation and nodulation of *Alnus* and *Hippophaë* under natural conditions. PhD Thesis, University of Leiden.

Bond, G. (1958). Symbiotic nitrogen fixation by non-legumes. In: *Nutrition of the legumes* (ed. E. G. Hallsworth), pp. 216–31. Butterworth & Co., London.

Ferguson, T. P. & Bond, G. (1953). Observations on the formation and function of the root nodules of *Alnus glutinosa* (L.) Gaertn. *Annals of Botany*, New Series, **17,** 175–88.

Fessenden, R. J., Knowles, R. & Brouzes, R. (1973). Acetylene–ethylene assay studies on excised root nodules of *Myrica asplenifolia. Proceedings of the Soil Science Society of America*, **37,** 893–8.

Quispel, A. (1954). Symbiotic nitrogen fixation in non-leguminous plants. II. *Acta Botanica Neerlandica*, **3,** 512–32.

Rodriguez-Barrueco, C. (1968). The occurrence of the root-nodule endophytes of *Alnus glutinosa* and *Myrica gale* in soils. *Journal of General Microbiology*, **52,** 189–94.

Schaede, R. (1933). Über die Symbionten in den Knöllchen der Erle und des Sanddorns und die cytologischen Verhältnisse in ihnen. *Planta*, **19,** 389–416.

Sseekaalo, H. & Johnson, R. M. (1969). Determination of molybdenum in biological materials with 2-amino-4-chlorobenzenethiol. *Journal of the Science of Food and Agriculture*, **20,** 581–3.

Virtanen, A. I. (1957). Investigations on nitrogen fixation by the alder. II. *Physiologia Plantarum*, **10,** 164–9.

Waughman, G. J. (1972). The effect of varying oxygen tension, temperature and sample size on acetylene reduction by nodules of *Alnus* and *Hippophaë. Plant and Soil*, **37,** 521–8.

Zavitkovski, J. & Newton, M. (1968). Effect of organic matter and combined nitrogen on nodulation and nitrogen fixation in red alder. In: *Biology of alder* (ed. J. M. Trappe *et al.*), pp. 209–23. Northwest Scientific Association Symposium Proceedings, 1967. US Dept of Agriculture, Portland, Oregon.

Plate 36.1. The variation in the thickness of the lobes between nodules from one alder tree.

(*Facing p.* 520)

37. Endophyte adaptation in *Gunnera–Nostoc* symbiosis

W. B. SILVESTER

Blue-green algae form a considerable variety of symbioses with multicellular plants from bryophytes to angiosperms, but only in *Gunnera* does the alga penetrate the host cell. The symbiosis was first described by von Reinke (1873) and the endophyte later identified as *Nostoc punctiforme* (Harder, 1917; Winter, 1935). N_2 fixation has been confirmed for the alga in pure culture (Drewes, 1928) and in symbiosis (Silvester & Smith, 1969). All ten New Zealand species of *Gunnera* normally contain the alga (Batham, 1943; Philipson, 1964; Silvester & Smith, 1969) and it is probable that all forty species in the genus house *Nostoc*. Both the host plant (von Reinke, 1873; Jonnson, 1894) and the endophyte (Harder, 1917; Bowyer & Skerman, 1968) are capable of independent growth.

Nostoc invades subapical excretory glands of the host forming dense nodules within the stem at the bases of leaves (Silvester & McNamara unpublished observations). The algae penetrate host cells at an early stage (Schaede, 1951; Neumann, Ackermann & Jacob, 1970) and it is reported that the ultrastructure of endophytic *Nostoc* does not differ from free-living cells in *G. chilensis* (Neumann *et al.*, 1970).

Symbiotic nitrogen fixation is considerable, with *Nostoc* being capable of supplying the total requirements of the host plant; in *G. dentata* fixation was estimated to be 72 kg N/ha/yr (Silvester & Smith, 1969).

The nodules are served by a well-developed vascular system (Batham, 1943) and fixed nitrogen is rapidly translocated from endophyte to host leaf and stem (Silvester & Smith, 1969). In pure culture *N. punctiforme* can grow in the dark on a variety of organic substrates (Harder, 1917) and it has been assumed (Schaede, 1962) that in *Gunnera*, where the nodules are embedded in host tissue and often buried in leaf bases, the algae are entirely dependent on host carbohydrate.

Materials and methods

Plant material

Infected plants collected in the field were maintained in sand culture watered weekly with a nitrogen-free nutrient solution. The plants spread rapidly by stolons and remained infected. Five local species, namely *G. arenaria*, Cheesem., *G. albocarpa* (Kirk) Ckn., *G. strigosa* Col., *G. dentata* Kirk, and *G. prorepens* Hook. f., have been grown successfully, with *G. albocarpa* the most vigorous.

Isolation of alga

Nostoc was isolated from *G. arenaria* on to agar slopes containing nitrogen-free nutrient solution (Allen & Arnon, 1955) using the methods of Bowyer & Skerman (1968). This isolate is designated *Nostoc* ex *G.* and for comparison *Nostoc punctiforme* (Kutzing) Hariot, Culture No. 1453/3 from the Cambridge University culture collection, was used.

Analytical

Nitrogen was determined by standard semi-micro Kjeldahl techniques using a copper–selenium catalyst, protein being estimated as total $N \times 6.25$.

Acetylene reduction assays were conducted in serum bottles by addition of 10 % C_2H_2 to air and subsequent estimation of ethylene by gas chromatography. Assays were made of a variety of material; whole plants, stems containing nodules, excised nodules or stem slices through nodules. Assays in the light were conducted under cool white fluorescent lamps; light intensities (400–700 nm) were measured with an Eppeley thermopile and are expressed in W/m^2 (1000 lux = 4.55 W/m^2).

Algae were extracted from nodules by crushing the latter inside a nylon organdie bag, washing the extracted material into a centrifuge tube, spinning at 4000 r.p.m. and resuspending the upper algal layer.

Pigment was extracted from the algal pellet by grinding with washed alumina in a pestle and mortar. Phycocyanin was extracted in 0.1 M Tris and chlorophyll in methanol (Ogawa & Shibata, 1965). Total nodule chlorophyll was estimated by grinding nodules in methanol in a Duall all-glass Potter type homogeniser (Kontes Glass Ltd) and absorbance at 666 nm for chlorophyll *a* and 625 nm for phycocyanin measured with a Spectronic 700 spectrophotometer.

Carbon fixation was measured by suspending algae or nodules in 1 mM [^{14}C]sodium bicarbonate (1.0 μCi per vial) and incubating in light or dark. Tissue was extracted by grinding with alumina in methanol/chloroform/formic acid/water; 12/5/1/2 (v/v) and the residue re-extracted in methanol/formic acid/water; 40/1/59. Chloroform and water in ratio of 3:1 was added to the combined supernatants to give a two-phase separation (Bieleski & Turner, 1966). Use of these acid-extracting fluids ensured liberation of all uncombined sodium bicarbonate. The aqueous phase was then added to a toluene/PPO/POPOP plus Triton scintillation fluid (Turner, 1971) and counted in a scintillation spectrometer (Packard tri-carb).

Oxygen balance was measured with a Clarke type oxygen electrode (Beckman Instruments Ltd). Tissue was suspended in nitrogen-free nutrient solution (Allen & Arnon, 1955) in a 5 ml or 10 ml Perspex cell which was water-jacketed and maintained at 20 °C.

Tests with ^{15}N were conducted in a gas mixture of $Ar/O_2/N_2$; 60/20/20 (v/v) with 36 atoms per cent excess ^{15}N, the tissue being subsequently Kjeldahled and nitrogen gas liberated from lithium hypobromite and analysed in an AEI MS10 mass spectrometer.

Electron microscopy

Tissue was fixed for 1 h in 2.5 % glutaraldehyde (phosphate-buffered pH 7.5), washed in buffer and post-fixed in 1.3 % phosphate-buffered osmic acid. Material was then dehydrated through an alcohol series, embedded in Epon and polymerised for 36 h at 60 °C. Sections were cut at 60–90 nm on an LKB ultra-microtome and post-stained by immersion in a 1:1 mixture of 4 % aqueous uranyl acetate and absolute ethanol, washed, dried and further stained for 30 seconds in Venable & Coggeshalls lead citrate stain. Sections were viewed with a Philips EM-200 electron microscope.

'Thick-thin' sections (1 μm) were also cut from similar Epon-embedded blocks, stained with 1 % methylene blue in 0.1 N potassium hydroxide and viewed with a phase-contrast light microscope to determine heterocyst frequency.

Results

Symbiotic nitrogenase activity

For comparative purposes it is sufficient to express nitrogenase activity in terms of nodule weight or total nitrogen content. However, the estimation of absolute activity became essential when it was noted that very high reducing activities were being occasionally recorded. On a per unit total nodule protein basis, *Gunnera* nodules were found to possess significantly higher specific activity than *Nostoc* extracted from *Gunnera* (*Nostoc* ex *G.*) (Table 37.1). Examination showed that only a proportion

Table 37.1. *Nitrogenase activity of free-living* Nostoc *isolated from* Gunnera (*assayed on agar slopes*) *and of symbiotic* Nostoc *in* G. albocarpa

Material	nmoles C_2H_4/mg protein/min (mean ± standard error)
Nostoc ex *Gunnera* on agar	1.50±0.53
G. albocarpa whole stems	3.19±0.27*
G. albocarpa excised nodules	3.20±0.28*
G. albocarpa sliced nodules	2.50±0.18*

All assays performed under fluorescent light at 14 W/m² (c. 3000 lux).
* Expressed as rate per mg total nodule protein.

of the *Gunnera* nodule tissue volume is occupied by *Nostoc* and thus the absolute activity in terms of algal protein must be significantly greater than the figures given in Table 37.1.

Table 37.2. *Estimation of specific activity of endophytic* Nostoc *based on protein nitrogen of extracted algae or on the correlation between algal chlorophyll and protein content*

Species and method	nmoles C_2H_4/mg protein/min (range)
Extracted algae	
G. prorepens	(a) 11.7–56.6
	(b) 21.0–29.5
	(c) 16.4
G. dentata	18.3
G. albocarpa	20.6
Chlorophyll method	
G. albocarpa excised nodules	(a) 5.0–11.6
	(b) 8.0–16.3
G. albocarpa nodule slices	(a) 11.0–21.1
	(b) 7.6–12.6
Published results	
Nostoc sp.*	1.62
Algal bloom*	1.32–2.08
Nostoc entophytum†	0.9
Nostoc muscorum†	3.7
Various algae‡	0.4–7.4

All incubated in light. Published results for free-living blue-green algae are shown for comparison.
* After Stewart *et al.* (1967).
† After Stewart *et al.* (1968).
‡ After Hardy *et al.* (1973).

To gain a better estimate of specific activity two approaches were made. Firstly, after acetylene reduction assay, the algae were extracted from nodules by the procedure already described, and analysed for nitrogen. However, complete recovery of algae was never achieved, since the debris pellet always enclosed a varying proportion of *Nostoc*. In consequence this technique tended to over-estimate specific activity. Secondly, a pure algal suspension was extracted from nodules as above and separate aliquots used for chlorophyll and nitrogen analyses. Two separate extracts made by this technique gave absorbances at 666 nm (A_{666}) of 0.201 and 0.210 for 100 μg of algal total nitrogen. Thus after acetylene reduction assays, chlorophyll analysis of nodules could reliably predict algal protein.

Table 37.2 shows the specific activity of *Gunnera* nodules using the above techniques and comparisons made with published data for specific activity of a variety of algal species. As expected, the extracted alga technique gave very high activities and the more reliable chlorophyll extraction method confirmed that the specific activity of endophytic *Nostoc* was some five to ten times greater than free-living *Nostoc* and significantly greater than other species of blue-green algae. To confirm this result, parallel $^{15}N_2$ and C_2H_2 reduction experiments were

Table 37.3. *Determination of* $C_2H_2:N_2$ *conversion factor*

	nmoles/mg protein/min		
Conditions	N_2	C_2H_2	$C_2H_2:N_2$
Light	1.32	5.60	4.2
Dark	0.17	0.71	4.2

Excised nodules of *G. albocarpa* incubated for 1 hour in either $^{15}N_2$ or C_2H_2 under identical conditions. Gas phase $Ar/O_2/N_2$ or C_2H_2; 60/20/20 in light 14 W/m^2 or dark.

run to determine the ratio of $C_2H_2:N_2$ in this material. Previous determinations (Hardy, Burns & Holsten, 1973) yielded ratios varying from 3.2 for algae to 4.3 for soil samples, in reasonable agreement with the expected value of 3.0. for *G. albocarpa* nodules, the value of 4.2 (Table 37.3) confirms the high specific activity of *Nostoc* nitrogenase.

Ultrastructure of developing nodules

The stolon development of *Gunnera* allows one to trace nodule development from apical, recently formed nodules through to older senescent nodules. In this developmental series there is a distinct change in nodule colour from very deep green, almost black, to much lighter grey-green where chlorophyll is breaking down.

Tracing the activity of glands and their nitrogen content revealed a distinct and repeatable pattern in which by far the greatest activity was found in younger subapical nodules (Fig. 37.1). Most of the present work has been done on such nodules.

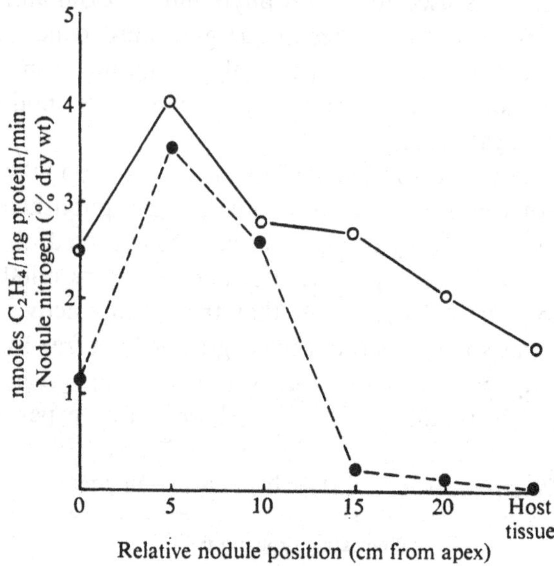

Fig. 37.1. Nitrogenase activity (●---●) and nitrogen content (○——○) of *G. albocarpa* nodules taken from various positions along advancing stolon. Young apical nodules are at position 0 and older nodules are designated by distance from apex.

The ultrastructure of nodules along the same developmental series was studied to see if there was a structural basis for the high nitrogenase activity and subsequent drop. The most striking event noted is the marked increase in heterocyst frequency along the series. At an early stage of development (Plate 37.1) apical nodules already possess a high heterocyst frequency. Plates 37.2 and 37.3 show further stages in development of symbiosis and represent nodules of greatest activity. Heterocysts are well-developed and characterised by much wall deposition, reduced cellular inclusions, reduced membrane organisation and presence of polar bodies. A very high frequency of paired heterocysts is evident. At later stages (Plate 37.4) cells start to disintegrate – the few vegetative ones remaining are heavily invested with inclusions while heterocysts are the first cells to break down into characteristic sickle-shaped cells. At this stage nitrogenase activity is negligible (Fig. 37.1), host cytoplasm is showing signs of degeneration, and algal cells appear to be becoming occluded within additional walls often continuous with host cell wall. Even in young nodules heterocyst frequency is high (Table 37.4) and rises rapidly to c. 60 % at which point nodule activity has fallen markedly (Fig. 37.1) and heterocyst degeneration has begun (Table 37.4, Plate 37.4).

Table 37.4. *Heterocyst frequency (percentage of total cells) determined by phase contrast microscopy of methylene blue-stained 'thick-thin' sections on a series from young apical to senescent basal nodules. (Means of six fields from* G. albocarpa *nodules.)*

Nodule development (cm from apex)	Heterocyst frequency (%); mean and range	Degenerate cell frequency (%)
Free-living *Nostoc*	5.2 (4.2–7.1)	0
Apical nodules	23 (18–35)	0
5 cm	32 (25–40)	0
10 cm	41 (33–51)	0
15 cm	63 (59–70)	5
20 cm	65 (60–81)	15

Metabolic activities of nodules

Effect of light on nitrogenase. In the intact plant, nodules are buried beneath a considerable number of cell layers, and additionally are often covered by leaf bases and soil, so that much of their metabolic activity may be considered as heterotrophic. To test this, whole plants of *G. albocarpa* were grown under constant environment conditions of 12 hours light and 12 hours dark for 10 days. Nitrogenase activity was assayed in the light and at intervals in an extended dark phase for 24 hours thereafter. Upon darkening there was an initial rapid drop in activity to about 80 % of light values (Fig. 37.2) which was originally thought to be a direct effect of light on algal glands. However, this is now thought to be a temperature effect, since in the light, direct radiant energy increases the temperature within the 500 ml incubating jars by 4–5 deg C although these experiments were conducted in cabinets whose air temperature did not fluctuate more than 1 deg C. Despite this, the results show that there is little change in nitrogenase activity over a 24-hour dark period. Other, less sophisticated experiments (Silvester & McNamara, unpublished data) in which the nodules on

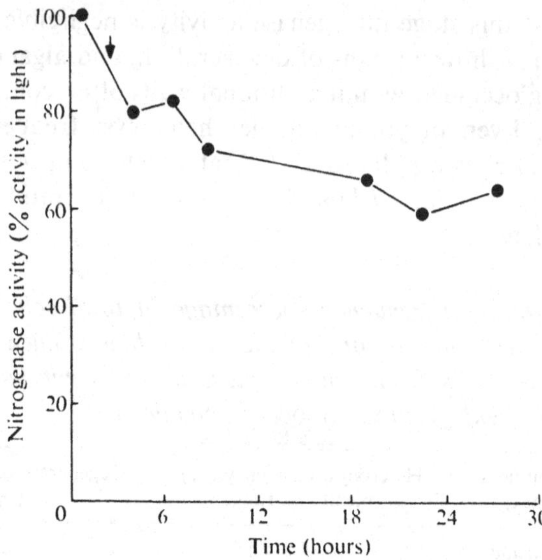

Fig. 37.2. Nitrogenase activity of intact *G. albocarpa* plants. Plants grown in constant environment (12 h day, light intensity 12 W/m², 22 °C) and then darkened at arrow. Nitrogenase activity expressed as percentage of activity in full light.

G. dentata plants were protected from light for up to six weeks, confirmed that the alga exists heterotrophically and that nitrogenase activity and chlorophyll content are maintained under extended dark conditions.

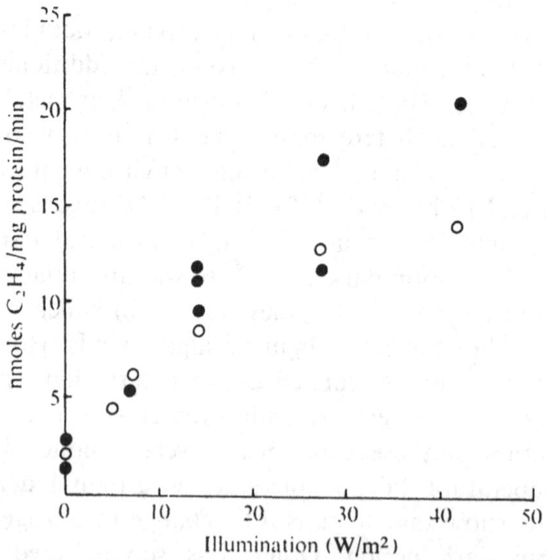

Fig. 37.3. Nitrogenase activity of excised glands (○) and nodule slices (●) at varying light intensities. Results of four experiments expressed per unit algal protein.

Despite the normal heterotrophic existence of symbiotic *Nostoc*, excised nodules respond very rapidly to light with some tenfold increase in nitrogenase activity between dark nodules and those at 40 W/m² (Fig. 37.3), the response being almost linear over the range 0–40 W/m²

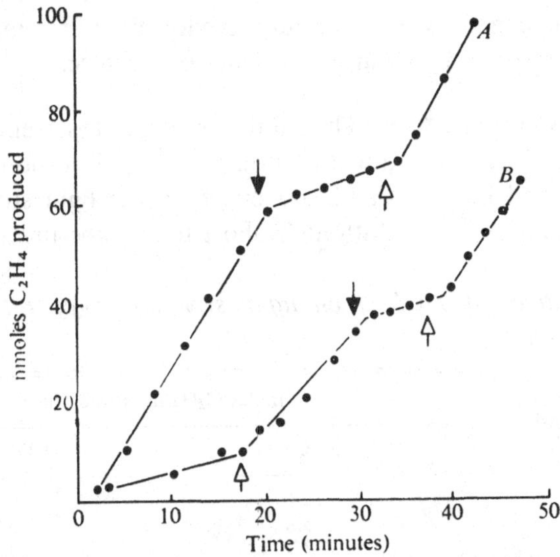

Fig. 37.4. Nitrogenase response of isolated *G. albocarpa* nodules to light and dark conditions. Experiment *A*, nodules kept in light (27 W/m²) for 40 minutes then gassed with acetylene. Experiment *B*, nodules kept in dark for 60 minutes prior to addition of acetylene. Appropriate arrows indicate change in conditions.

with some evidence of saturation at the highest light intensity. Further, the response to light is rapid, the change-over being accompanied with little lag whether going from light to dark or dark to light (Fig. 37.4).

Table 37.5. *Heterotrophic nitrogenase activity in* Nostoc

Species	Fructose	Light/dark	nmoles C_2H_4/ mg/protein/min
N. punctiforme	+	L	5.5
N. punctiforme	−	L	1.4
N. punctiforme	+	D	4.3
N. punctiforme	−	D	0
Nostoc ex *Gunnera*	+	L	2.7
Nostoc ex *Gunnera*	−	L	1.1
Nostoc ex *Gunnera*	+	D	1.2
Nostoc ex *Gunnera*	−	D	0

Nostoc punctiforme and *Nostoc* ex *Gunnera* grown on agar minus and plus 0.3 % fructose. Dark tubes assayed 90 h after darkening, light tubes incubated at 4 W/m².

Heterotrophic nitrogenase in Nostoc. To test the heterotrophic growth of *Nostoc* both *Nostoc* ex G. and *N. punctiforme* were grown on agar slopes with or without 0.3 % fructose and assayed for nitrogenase in light and dark. For both species there was a large enhancement of activity in the light with addition of fructose (Table 37.5) and, of greatest significance, a fairly high activity of nitrogenase after a prolonged dark period only in the presence of fructose.

Effect of DCMU on nodules. The inhibitor of photosystem 2, DCMU (3,3,4-dichlorophenyl-1, 1-dimethyl urea), was used on nodule slices of *G. prorepens* and had no significant effect on the light stimulation of nitrogenase (Table 37.6). Although short-term experiments often do

Table 37.6. *Effect of DCMU on light stimulation of nitrogenase in* G. prorepens

Light intensity (W/m^2)	nmoles C_2H_4/mg protein/min	
	−DCMU	+DCMU
30	10.1	10.2
9	5.8	5.7
2	3.7	3.5

10^{-4} DCMU added to nodule slices suspended in nutrient solution. Acetylene reduction followed for 2 h at three light intensities.

not show an effect of DCMU on the nitrogenase of free-living algae (Cox & Fay, 1969), in longer experiments of one hour or more and in the presence of oxygen DCMU normally has an inhibitory action (Lex, Silvester & Stewart, 1972).

Pigment composition and oxygen production. As noted already, pigment extraction was conducted by methanol extraction of alumina-ground cells followed by extraction in 0.1 M Tris buffer or alternatively grinding in buffer followed by addition of chloroform to afford a two-phase separation of the water-soluble phycocyanin and lipid-soluble chlorophyll. In all cases it was impossible to detect phycocyanin by spectrophotometer analysis. Parallel extractions of *Anabaena cylindrica* and *Nostoc punctiforme* showed comparable absorbence of chlorophyll at 666 nm and phycocyanin at 625 nm.

To test further for the absence of phycocyanin, excised nodules and nodule slices were suspended in nitrogen-free nutrient solution, and oxygen balance followed with an oxygen electrode. The results (Fig. 37.5) show that light had no influence on oxygen exchange, and that photosynthetic oxygen production is entirely inhibited in symbiotic *Nostoc*. Respiration rate of excised nodules varied from 0.05–0.25 mg

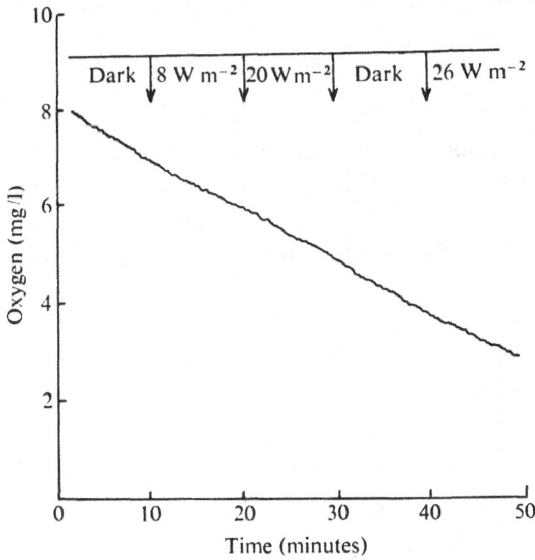

Fig. 37.5. Oxygen exchange of excised nodules of *G. albocarpa*. Nodules (80 mg fresh weight) were suspended in 5 ml of medium in a water-jacketed Perspex cell fitted with an oxygen electrode, and illuminated and darkened at intervals.

O_2/g fresh wt/h while excised nodules under the same conditions showed an eightfold increase in nitrogenase activity between dark and 30 W/m^2.

Carbon fixation. Excised nodules of *G. albocarpa* and free-living *Nostoc punctiforme* were exposed to ^{14}C as sodium bicarbonate. The results (Table 37.7) show a virtual absence of ^{14}C incorporation by *Gunnera*

Table 37.7. *Uptake of* ^{14}C *by excised* G. albocarpa *nodules and by free-living* Nostoc punctiforme

Material	Light/dark	c.p.m./min exposure $\times 10^{-3}$
G. albocarpa nodules	L	0.58
G. albocarpa nodules	D	0.32
N. punctiforme	L	250
N. punctiforme	D	1.15

Tissue given 1.0 μCi ^{14}C as $NaHCO_3$ at 1 mM and incubated in light (27 W/m^2) or dark at 25 °C.

nodules in the light. Whereas in free-living algae there is a 200-fold increase in ^{14}C uptake in the light, in nodules there is less than a twofold increase. This very small amount of carbon fixation may be associated with the *Nostoc* cells or may occur in the small amount of host tissue surrounding each nodule, which as superficial herbaceous stem tissue does contain chloroplasts.

Effect of oxygen tension. When the intact stem system of *Gunnera* is assayed for nitrogenase in the dark, optimum activity is achieved only at a P_{O_2} level of 0.4 or above (Fig. 37.6) while, when excised, nodules have an optimum P_{O_2} at near 0.3. Light-incubated excised glands, however, showed an entirely different response, since highest activity was achieved

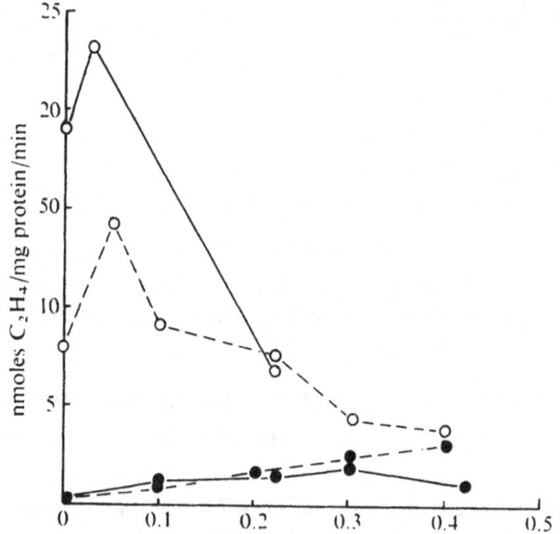

Fig. 37.6. Effect of oxygen tension on nitrogenase activity of *G. albocarpa*. Nodules exposed to various levels of oxygen in argon in the dark (closed circles) or in light of intensity 27 W/m² (open circles). Ten per cent acetylene was included in each gas mixture. Dark-incubated material was excised glands (●——●) or intact plants (●---●). Light-incubated material was excised glands in both experiments, in one case from plants grown in high light intensity (○—○) and in the other from those grown at a lower light (○---○). See also text.

at very low P_{O_2}, while atmospheric oxygen levels caused a large inhibition of nitrogenase activity. Fig. 37.6 also indicates a large effect of pre-treatment of plants, since, at favourable oxygen levels, the activity in nodules from plants growing in a glasshouse receiving 70 % of winter light considerably exceeded that of material from plants growing in a shaded house receiving only 15–20 % of daylight. It appears that the level of previously fixed carbohydrate controls the amount of light-stimulated nitrogenase activity.

Hydrogen inhibition. It has been reported (Silvester & Smith, 1969) that hydrogen inhibited acetylene reduction in the *Gunnera–Nostoc* symbiosis. At that time parallel ^{15}N tests showed identical inhibition of nitrogen fixation. However, we have been unable to repeat the hydrogen inhibition of acetylene reduction using a different source of hydrogen and must conclude that the initial cylinder of hydrogen was contaminated.

Discussion

Three major and hitherto undescribed modifications are undergone by *Nostoc* in this symbiosis. As compared with free-living *Nostoc*, there is firstly an eightfold increase in heterocyst frequency, secondly the alga loses detectable phycocyanin and hence its capacity to fix carbon and liberate oxygen, and thirdly there is at least a tenfold increase in nitrogenase activity based on algal protein.

The disposition of algae within *Gunnera* cells and the mode of infection were studied by Schaede (1951), while the conclusion of the only report on ultrastructure (Neumann *et al.*, 1970) was that there is no difference in structure between free-living and endophytic *Nostoc*. The latter authors described cell penetration whereby the alga becomes encapsulated in a portion of host cell wall and plasmalemma, with the subsequent dissolution of the encircling cell wall leaving the alga in plasmalemma envelopes. We have been unable to observe host cell wall penetration but there is no doubt that the algae are contained within a membrane of host origin. We have observed that towards the end of their active life algal cells then become enclosed within a wall which appears continuous with host cell wall, and it is possible that the events described by Neumann *et al.* (1970) were also concerned with degeneration rather than infection.

The increase in heterocyst frequency represents a dramatic modification of algal structure which was not recorded by Neumann *et al.* (1970) nor shown in their electron micrographs for *G. chilensis*. In free-living algae heterocyst frequency increases during growth of a culture and may be artificially increased by low ambient nitrogen conditions or low light (Fogg, 1949), and by low oxygen (Kulasooriya, Lang & Fay, 1972). It is probable that all of these conditions occur in *Gunnera* symbiosis. In lichens, the only other symbiosis in which this problem has been studied, low oxygen and nitrogen conditions probably again prevail (Millbank, 1972) but the alga must receive light and there is no change in heterocyst frequency (Griffiths, Greenwood & Millbank, 1972). It is therefore tempting to postulate that lowered light or darkness is the major factor in increasing heterocyst frequency in *Gunnera*.

In older nodules heterocyst frequency may reach over 90 % in individual cells although the overall frequency seldom exceeds 60 %. At the stage when activity has declined appreciably (Fig. 37.1) algal cells are already breaking down to give characteristic sickle-shaped cells, while at later stages when the nodules are grey-green in colour and of very low activity a large proportion of nodule cells contain only degenerate algal cells.

Coupled with the increase in heterocyst frequency is the loss in phycocyanin and of photosynthetic activity. Heterocysts are known to lack phycocyanin (Fay, 1969) and be incapable of carbon fixation (Stewart, Haystead & Pearson, 1969). However in the case of *Gunnera* there is no detectable photosynthetic activity even when 70 % of the cells are vegetative. Thus under these symbiotic conditions it seems likely that all cells are behaving as heterocysts at least in respect of photosynthetic ability.

The specific activities of nitrogenase recorded here for symbiotic *Nostoc* are up to ten times greater than those recorded for free-living *Nostoc*. It is difficult to make truly comparative tests, but under optimum light conditions in air, specific activities up to 20 nmoles C_2H_4/mg algal protein/min (= 0.5 nmoles C_2H_4/mg nodule fresh wt/min) have been achieved (Table 37.2), while under lowered oxygen tensions values three times greater than air values have been obtained (Fig. 37.6). Relatively high specific activities have been recorded both for symbiotic algae (Millbank, 1972) and for free-living algae (Stewart & Pearson, 1970) where in both cases rates of about 9 nmoles/mg algal protein/min were obtained and explained in terms of oxygen tension. In *Gunnera*, however, the explanation may lie in one or more of the following:

1. Heterocyst frequency

In free-living algae the evidence suggests that the heterocyst is the site of nitrogenase (Fay *et al.*, 1968; Stewart *et al.*, 1969). The correlation between heterocyst frequency and nitrogenase is very good in free-living algae over a range of 2.5–12.3 % heterocyst frequency, while Jewell and Kulasooriya (1970) have estimated the activity of nitrogenase as 1.5–20.6 nmoles C_2H_4/mg heterocyst protein/min, which is a similar range to that found for endophyte *Nostoc* on a total algal protein basis.

2. Heterotrophic activity

Nostoc spp. and especially *N. punctiforme* are able to utilise a variety of carbon compounds including polysaccharides (Harder, 1917) in the dark. *Nostoc* in *Gunnera* is undoubtedly reliant upon the host for a carbohydrate source. Fructose stimulates nitrogenase activity in free-living *Nostoc* in the light and entirely supports it in the dark and we may expect a similar result in symbiosis.

3. Oxygen tension

Lowered oxygen tension is important for optimum activity of algal nitrogenase (Stewart & Pearson, 1970). However in the dark (Fig. 37.6) the optimum oxygen tension is above 40 % in intact plants and at about 30 % for excised nodules indicating a considerable oxygen barrier in the tissue or a high respiratory oxygen demand. In the light the optimum oxygen tension for excised nodules is near zero (Fig. 37.6), a finding which resembles that for free-living algae (Stewart & Pearson, 1970). The symbiotic system does provide a barrier to oxygen penetration but in the intact plant this is probably a disadvantage as the nitrogenase activity is largely heterotrophic and unsaturated with oxygen. However, as an explanation for the high activity in the light the oxygen effect may be contributory.

Nitrogenase in symbiotic *Nostoc* represents a simplified system which is ideal for basic study. Photosystem 2 is absent, there is negligible carbon fixation and it is uncomplicated by the presence of photorespiration. Cox & Fay (1969) showed that *Anabaena cylindrica* nitrogenase was largely unaffected by the absence of photosystem 2 but was still stimulated by light. They concluded that the source of reductant was respiration and that ATP was derived from cyclic photophosphorylation. This primary involvement of photosystem 1 has been confirmed by nitrogenase action spectra (Fay, 1970) and by lack of Emerson enhancement (Lyne & Stewart, 1973). In *Gunnera* the complete absence of photosystem 2 provides further confirmation of the role of photosystem 1. Lex & Stewart (1973) have provided an alternative suggestion that photosystem 1 may also provide a source of reductant from previously fixed carbon as well as a source of ATP. Cox & Fay (1969) have shown that pyruvate is able to support nitrogenase activity in the light in the absence of oxygen and it is possible that the reaction involves pyruvate:ferredoxin oxidoreductase (Leach & Carr, 1971). In *Gunnera* the negligible activity in the absence of light and oxygen implicates reduced ATP synthesis and the immediate response to light a response to shortage of ATP.

As noted, in symbiotic *Nostoc* there is a dramatic effect of oxygen on nitrogenase similar to that found in free-living algae. This effect is not on nitrogenase itself because that oxygen effect is reversible and dark nitrogenase activity is not saturated until 30 % oxygen. Neither can this oxygen effect be explained in terms of photorespiration (Lex *et al.*, 1972) as in *Gunnera* there is no light stimulation of oxygen uptake (Fig. 37.5). It must be concluded that the oxygen effect is directly on the photosystem 1 activity and that in the absence of carbon fixation this will not be on the build up of carbon reserves or photorespiration of a photosynthetic product.

The simplified system offered by *Gunnera–Nostoc* symbiosis opens up a number of possibilities for the study of several basic aspects of algal nitrogenase. In particular, aspects of heterocyst development and the ATP/reductant supply to nitrogenase may well be studied more conveniently in the symbiotic condition.

I am grateful to Dr S. Bullivant for use of electron microscope facilities and to Mr G. Grayston for his assistance with this aspect. Also to Miss M. Patten for technical assistance This work was financed by grants from The University Grants Committee of New Zealand

References

Allen, M. B. & Arnon, D. I. (1955). Studies on nitrogen-fixing blue green algae. I. Growth and nitrogen fixation by *Anabaena cylindrica* Lemm. *Plant Physiology*, **30**, 366–72.

Batham, E. J. (1943). Vascular anatomy of New Zealand species of *Gunnera*. *Transactions of the Royal Society of New Zealand*, **73**, 209.

Bieleski, R. L. & Turner, N. A. (1966). Separation and estimation of amino acids in crude extracts by thin layer electrophoresis and chromatography. *Analytical Biochemistry*, **17**, 278–93.

Bowyer, J. W. & Skerman, V. B. D. (1968). Production of axenic cultures of soil-borne and endophytic blue green algae. *Journal of General Microbiology*, **54**, 299–306.

Cox, R. M. & Fay, P. (1969). Special aspects of nitrogen fixation by blue-green algae. *Proceedings of the Royal Society, London*, Ser. B, **172**, 357–66.

Drewes, K. (1928). Über die Assimilation des Luftstickstoffs durch Blaualgen. *Zentralblatt für Bakteriologie und Parasitenkunde*, Abt. II, **76,** 88–101.

Fay, P. (1969). Cell differentiation and pigment composition in *Anabaena cylindrica. Archiv für Mikrobiologie*, **67,** 62–70.

Fay, P. (1970). Photostimulation of nitrogen fixation in *Anabaena cylindrica. Biochimica et Biophysica Acta*, **216,** 353–6.

Fay, P., Stewart, W. D. P., Walsby, A. E. & Fogg, G. E. (1968). Is the heterocyst the site of nitrogen fixation in blue-green algae? *Nature, London*, **220,** 810–12.

Fogg, G. E. (1949). Growth and heterocyst production in *Anabaena cylindrica.* Lemm. II. In relation to carbon and nitrogen metabolism. *Annals of Botany*, **13,** 241–59.

Griffiths, H. B., Greenwood, A. D. & Millbank, J. W. (1972). The frequency of heterocysts in the *Nostoc* phycobiont of the lichen *Peltigera canina* Willd. *New Phytologist*, **71,** 11–13.

Harder, R. (1917). Ernahrungsphysiologische Untersuchungen an Cyanophyceen, hauptsächlich dem endophytischen *Nostoc punctiforme. Zeitschrift für Botanik*, **9,** 145–242.

Hardy, R. W. F., Burns, R. C. & Holsten, R. D. (1973). Applications of the acetylene–ethylene assay for measurement of nitrogen fixation. *Soil Biology and Biochemistry*, **5,** 47–81.

Jewell, W. J. & Kulasooriya, S. A. (1970). The relation of acetylene reduction to heterocyst frequency in blue-green algae. *Journal of Experimental Botany*, **21,** 874–80.

Jonnson, B. (1894). Studier ofver algaparasitism hos *Gunnera* L. *Botaniska Notiser Lund*, **1933,** 1–20.

Kulasooriya, S. A., Lang, N. J. & Fay, P. (1972). The heterocysts of blue-green algae. III. Differentiation and nitrogenase activity. *Proceedings of the Royal Society, London*, Ser. **B, 181,** 199–209.

Leach, C. K. & Carr, N. G. (1971). Pyruvate:ferredoxin oxidoreductase and its activation by ATP in the blue-green alga *Anabaena variabilis. Biochimica et Biophysica Acta*, **245,** 165–74.

Lex, M., Silvester, W. B. & Stewart, W. D. P. (1972). Photorespiration and nitrogenase activity in the blue-green alga *Anabaena cylindrica. Proceedings of the Royal Society, London*, Ser. **B, 180,** 87–102.

Lex, M., & Stewart, W. D. P. (1973). Algal nitrogenase, reductant pools and photosystem I activity. *Biochimica et Biophysica Acta*, **292,** 436–43.

Lyne, R. A. & Stewart, W. D. P. (1973). Emerson enhancement of carbon fixation but not of acetylene reduction (nitrogenase activity) in *Anabaena cylindrica. Planta*, **109,** 27–38.

Millbank, J. W. (1972). Nitrogen metabolism of lichens. IV. The nitrogenase activity of the *Nostoc* phycobiont in *Peltigera canina. New Phytologist*, **71,** 1–10.

Neumann, von D., Ackerman, M. & Jacob, F. (1970). Zur feinstructur der endophytischen cyanophyceen von *Gunnera chilensis* Lam. *Biochemie und Physiologie der Pflanzen*, **161,** 483–98.

Ogawa, T. & Shibata, K. (1965). A sensitive method for determining chlorophyll in plant extracts. *Photochemistry and Photobiology*, **4,** 193–200.
Philipson, W. R. P. (1964). Morphogenesis as affected by associated organisms. *Phytomorphology*, **14,** 163–98.
Schaede, R. (1951). Über die Blaualgensymbiose von *Gunnera. Planta*, **39,** 154–70.
Schaede, R. (1962). *Die Pflanzlichen Symbiosen.* Gustav Fischer Verlag, Stuttgart.
Silvester, W. B. & Smith, D. R. (1969). Nitrogen fixation by *Gunnera–Nostoc* symbiosis. *Nature, London*, **224,** 1321.
Stewart, W. D. P., Haystead, A. & Pearson, H. W. (1969). Nitrogenase activity in heterocysts of blue-green algae. *Nature, London*, **224,** 226–8.
Stewart, W. D. P. & Pearson, H. W. (1970). Effects of aerobic and anaerobic conditions on growth and metabolism of blue-green algae. *Proceedings of the Royal Society, London*, Ser. B, **175,** 293–311.
Turner, J. C. (1971). *Sample preparation for liquid scintillation counting.* The Radiochemical Centre Ld., Review No. 6. Whitefriars Press, London.
von Reinke, J. (1873). *Morphologische Abhandlungen.* Leipzig.
Winter, G. (1935). Über die Assimilation des Luftstickstoffs durch endophytische Blaualgen. *Beitrage zur Biologie der Pflanzen*, **23,** 295–335.

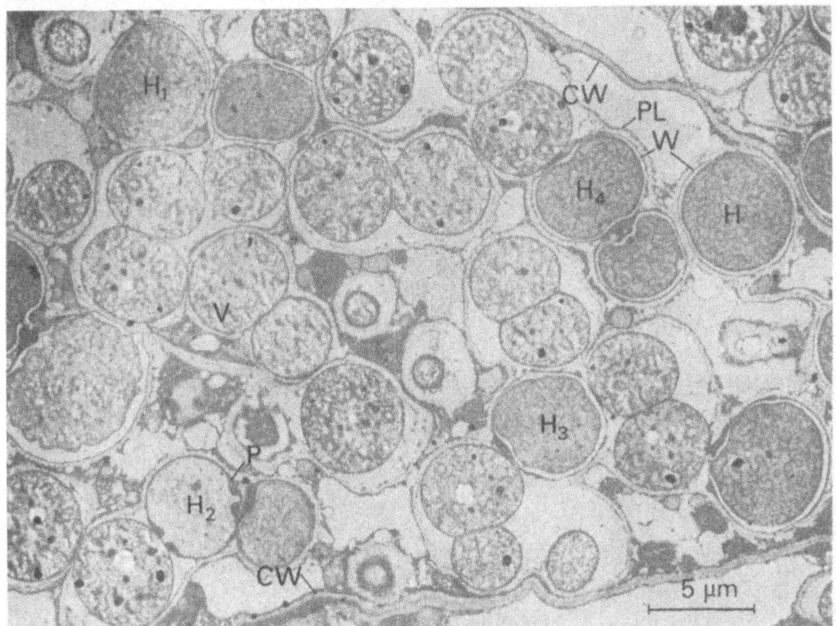

Plate 37.1. Electron micrograph of early subapical nodule of *G. albocarpa*. Heterocysts (H) can be distinguished by their thick wall (W), reticulate internal membrane structure and thickened polar wall (P). Vegetative cells possess a more highly organised thylakoid system concentrated at the periphery of the cell (V) and a number of cell inclusions. Stages in heterocyst development can be seen in which wall formation is complete but thylakoid organisation is retained (H_1) through to mature heterocyst (H_4). Host cell wall (CW) and plasmalemma (PL) enclose the *Nostoc* cells. Note several paired heterocysts; heterocyst frequency 25 %.

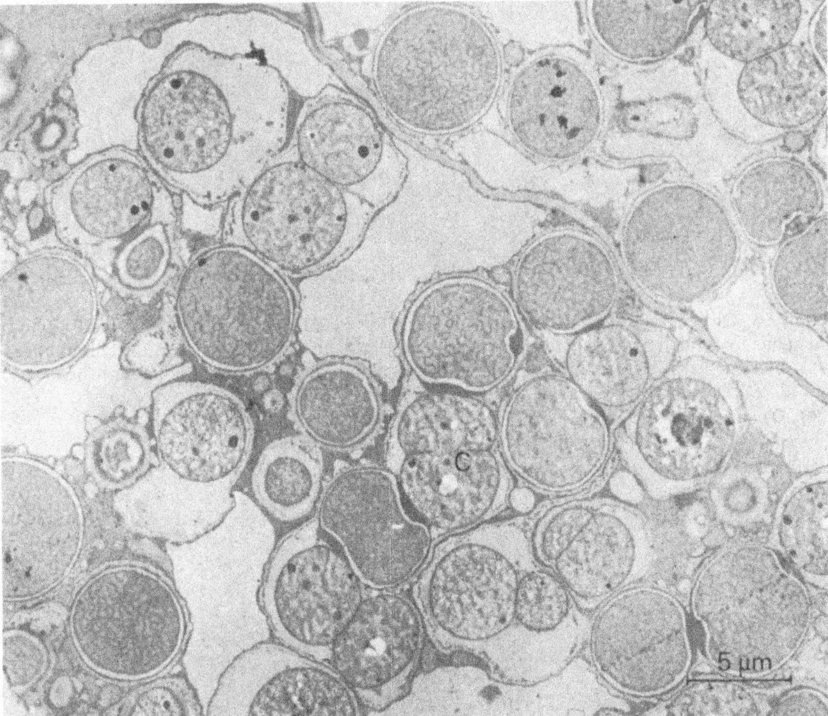

Plate 37.2. Older subapical nodule with heterocyst frequency increased to c. 40 %. Even at this stage cell division in vegetative cells is occurring (C).

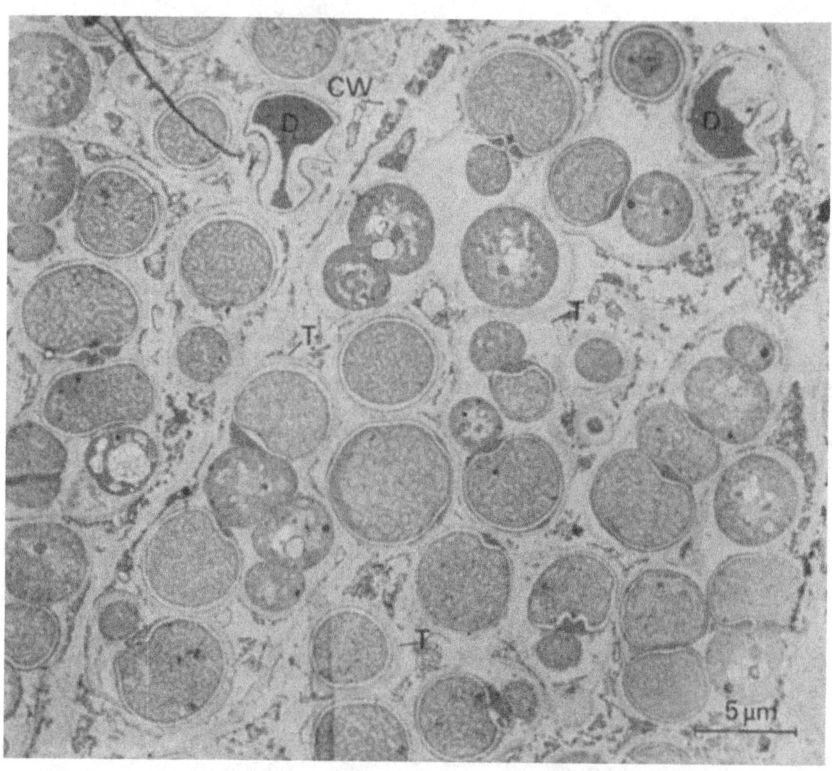

Plate 37.3. Old nodule, 15 cm from apex, with heterocyst frequency of c. 60 %. Host cytoplasm has lost much of its substructure at this stage and algal cells are becoming surrounded by a thickened wall (T) of host origin which in places appears continuous with host cell wall (CW). Degenerate cells now appearing (D).

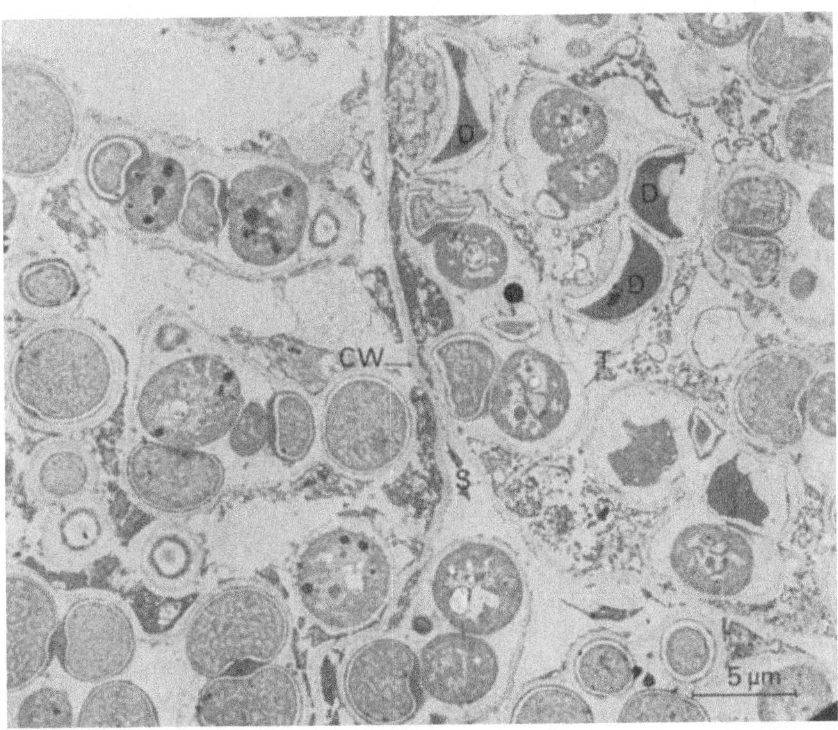

Plate 37.4. Many degenerate cells (D) occur in this senescent nodule which was 25 cm from apex. Crescent-shaped cells are typical of this degeneration. Groups of algal cells are now cut off into small internal 'host cells' whose walls (T) may be continuous with the original host cell wall (CW), e.g. at (S). At this stage heterocyst frequency is 60 % or more but nitrogenase activity negligible.

38. Nitrogen fixation in some natural ecosystems in Indonesia

J. H. BECKING

Although there have been many reports stressing the importance of biological nitrogen fixation in tropical regions, most of the data presented have been of a qualitative nature or based on laboratory experiments. The main objective of the present study was to measure nitrogen-fixing capacity in the field under ambient conditions, using the acetylene reduction method. The nitrogenase activity of the tropical ecosystems provided by *Azolla pinnata* and *Gunnera macrophylla* was measured; for both of these there is previous evidence of an ability to fix nitrogen as a result of the presence of a symbiotic blue-green alga. In addition, the supposed nitrogen fixation in *Podocarpus* root nodules, leaf nodules and also in the phyllosphere has been evaluated.

Methods

Since acetylene was not available commercially in Indonesia, and could not be transported by air for security reasons, it was generated from commercial calcium carbide. The acetylene was stored in 300 ml gas bottles provided with stopcocks at both ends and a side entrance closed by a rubber 'Suba-Seal' plug. The gas storage bottle was connected to a second bottle of the same size partly filled with water, to allow the pressure of the acetylene to be restored to 1 atm after usage.

Acetylene reduction assays were conducted with complete *Azolla* plants, tissue fragments of *Gunnera* stem, or excised *Podocarpus* roots enclosed in 70 ml glass vessels (Fig. 38.1*a*) with taps at the top and bottom and also side entrances, sealed with serum caps, which were used for acetylene injection and gas sampling. The complete assembly as used in the field with vessels of this type is shown in Plate 38.2(*d*). In the leaf nodule and phyllosphere assays the plant material was enclosed in 500 ml Perspex vessels equipped – when assays involving shoots still attached to the plant were to be carried out – with a split rubber stopper at the base, into which the shoot was sealed (Fig. 38.1*b*).

For acetylene introduction and gas sampling gas-tight syringes (Clark Hamilton, Switzerland) were used. The gas samples were stored in round-bottomed, soft-glass tubes, adjusted to contain precisely 2 ml by pushing a rubber septum down to the appropriate point from the open end. The part of the tube above the septum was then drawn out into capillary form. Prior to the introduction of the gas sample the pressure in the storage tube was reduced to 0.5 atm by inserting the needle of a syringe into the tube and withdrawing the plunger to the

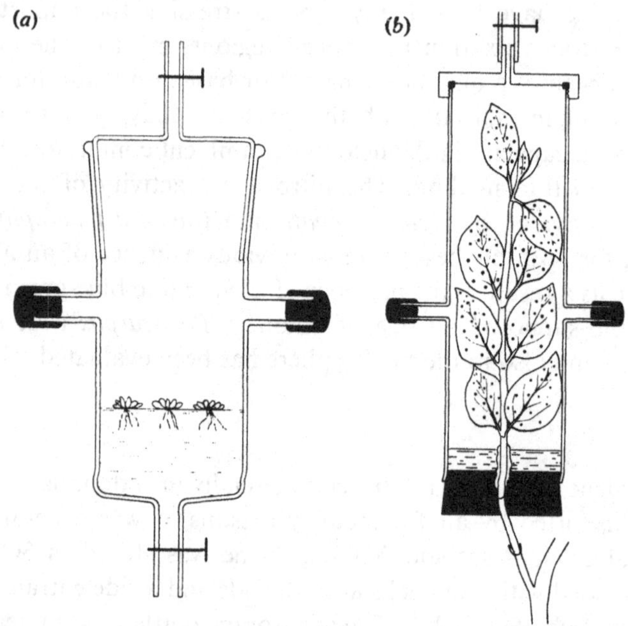

Fig. 38.1. (*a*) Glass vessel (70 ml) used in the acetylene assays for the enclosure of plant samples of small bulk. (×0.8). (*b*). Perspex vessel (500 ml) used for the enclosure of bulkier samples, including shoots still attached to the plant. (×0.07).

2 ml mark on the syringe. The latter was then replaced by another syringe containing a 1 ml gas sample which was injected into the storage tube. After withdrawal of this syringe the capillary end of the storage tube was sealed in a 'Primus' stove.

Acetylene and ethylene were determined by gas chromatography after the author's return to The Netherlands, using a Varian-Aerograph Model 1740 gas chromatograph fitted with a hydrogen-flame ionisation detector and a column of Porapak T (80/100 mesh, Waters Associates Inc., Mass., USA), and operating at 90 °C. All data are means of duplicate samples. Protein was estimated from N×6.25.

Data obtained

Results from Azolla pinnata *R. Br.*

This floating fern, belonging to the Salviniaceae, is an important constituent of the vegetation invading wet rice fields in Indonesia (Plate 38.1*a*). *Azolla* leaflets contain a large cavity communicating with the exterior by a pore on the underside (Fig. 38.2), and inhabited by a

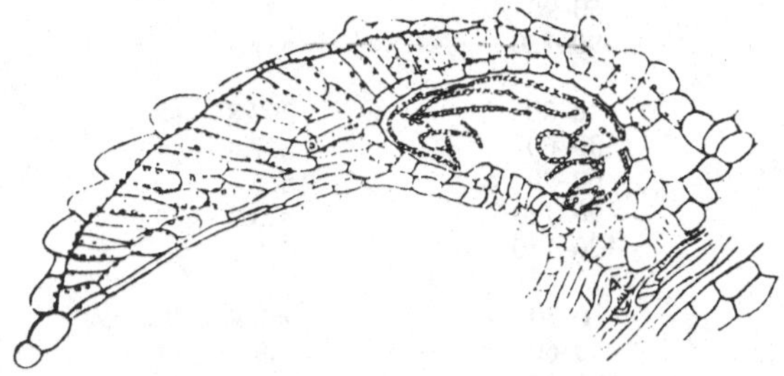

Fig. 38.2. Section through the dorsal lobe of an *Azolla* leaf showing the leaf chamber containing *Anabaena* filaments, also the site of the opening at the ventral side. After Strasburger (1873). (× 145).

blue-green alga, *Anabaena azollae* Strasburger, which lives symbiotically with the fern. Microscopic examination of crushed fronds revealed chains of typical *Anabaena* cells including many heterocysts (Plate 38.1*b*).

The exposures to acetylene were carried out at the Muara Experimental Fields, Bogor. *Azolla* plants were floated in rice-field water at the prevailing temperature of 28–30 °C, in the vessels already described (Fig. 38.1*a*), and the latter exposed to sunlight during the incubation period. The results of three typical experiments are shown in Table 38.1. It will be seen that in each experiment ethylene production over successive 30 minute periods stayed fairly constant except in the final period at the lowest level of acetylene supplied. The rather small differences in the acetylene levels provided in these experiments had no obvious effect on ethylene production.

In addition the *Anabaena* was isolated from *Azolla* fronds and cultured separately on an inorganic medium. Table 38.2 shows data for nitrogenase activity in the cultured alga. Here there is a clear time-lag in ethylene release, perhaps because it was not possible to shake the serum bottles mechanically during the exposure period. Also ethylene production was substantially increased by raising the acetylene level from 7.4 to 14.0 %. Considering only the data over the second hour at the 14.0 % level, and comparing them with those in Table 38.1, it might seem that

Table 38.1. *Acetylene reduction by* Azolla pinnata *plants in symbiosis with* Anabaena*

Per cent of C_2H_2 in air	Incubation time (min)	C_2H_4 produced (nmoles/mg sample N/30 min)	C_2H_4 (nmoles/mg sample protein/ min)
8.5	0–30	1285.9	6.9
	30–60	1074.0	5.7
	60–90	531.5	6.2
	90–120	531.5	3.2
10.1	0–30	992.8	5.3
	30–60	837.4	4.5
	60–90	815.1	4.3
	90–120	823.3	4.4
	120–150	1018.1	5.4
12.6	0–30	967.8	5.2
	30–60	949.0	5.1
	60–90	704.3	3.8
	90–120	1129.8	6.0
	120–150	1247.7	6.7

* The experiments were carried out over the midday period. Every 30 min, 1 ml CO_2 was injected into the 70 ml vessel containing the *Azolla*.

nitrogenase activity in the isolated alga was of the same order as that for the alga living symbiotically, until it is noted that for the latter ethylene production was related to whole plant protein (i.e. fern plus algal) rather than to algal protein only. Calculations based chiefly on the

Table 38.2. *Acetylene reduction by free-living* Anabaena *isolated from* Azolla *fronds**

Per cent of C_2H_2 in air	Incubation time (min)	C_2H_4 produced (nmoles/mg sample N/30 min)	C_2H_4 (nmoles/mg sample protein/min)
7.4	0–30	103.21	0.6
	30–60	338.80	1.8
	60–90	602.04	3.2
	90–120	596.81	3.2
14.0	0–30	172.77	0.9
	30–60	437.50	2.3
	60–90	735.19	3.9
	90–120	969.25	5.2

* The alga was grown in a nitrogen-free medium (Allen & Arnon, 1955), washed and re-suspended in similar, fresh medium. Aliquots of 5 ml algal suspension were assayed in the laboratory in 30 ml serum bottles at 29 °C and 19 000 lux.

estimate that the algal cavity accounts for one-third to one-quarter of the leaf volume suggest that on the basis of algal protein alone the symbiotic activity would be up to twice that of the isolated alga.

An estimate of the contribution of *Azolla* to the nitrogen economy of rice soils has been attempted. The density of the fern in the rice fields was measured by the use of a square wire frame of 5 cm edge, equipped with a handle, which was placed on the water surface at several sites where the fern completely covered the water. The *Azolla* plants within the frame were harvested and analysed. They showed a mean dry weight of 58 ± 3.6 mg and a protein content of 11.6 ± 0.3 mg/25 cm^2. The total of nine estimates of nitrogenase activity in the *Azolla* (of which three have been detailed in Table 38.1) showed that the least active samples produced about 3 nmoles ethylene and the most active ones about 6 nmoles, both per mg protein per minute, and calculations were made for each of these levels. It was assumed that 1.5 mole ethylene produced corresponded to 1 mole nitrogen fixed, and it was taken into account that daylength in Indonesia is constantly about 12 hours, that due to the presence of the rice plants and other factors the *Azolla* usually does not attain more than a 50 % cover of the water, and that it develops practically from zero after the three-month fallow period. It was concluded that under Indonesian conditions fixation by the *Azolla* amounts to 62–125 kg N/ha/annum, depending on which level of nitrogenase activity is accepted (see above).

Results from Gunnera macrophylla *Blume*

The nitrogenase activity of this species, which is in the family Haloragaceae, was assayed at a natural habitat at an altitude of 1650 m near the Cibeureum waterfalls, Cibodas Nature Reserve, Mt Gedeh, W. Java. The species forms a stout herb with long-stalked (60–80 cm), rhubarb-like leaves ($20–25 \times 36–40$ cm) reaching a height of some 1.2 m. It forms a dense storey under *Myrica javanica*, *Schima wallichii* spp. *noronhae*, *Vaccinium varingiæfolium*, *Leptospermum flavescens* and some other montane tree species, especially along brooks and places of seepage (Plate 38.2*a* and *b*).

The examination of the *Gunnera* plants at this locality showed that their stems invariably contained a blue-green alga. As described by previous authors the alga (a species of *Nostoc*) was located in special glands, of which three were always associated with one petiole base, one gland lying centrally below the petiole base and two laterally to it (Fig.

38.3). From the glandular region at the periphery of the stem the alga formed fan-like outgrowths into the colourless host tissue (Plate 38.2*c*). Microscopic examination of the *Nostoc* showed the presence of many heterocysts, and the frequency of these was still higher in older parts of the stem, where the *Nostoc* glands retain their vivid green colour, suggesting that in spite of leaf decay the glands in these parts remain active in nitrogen fixation.

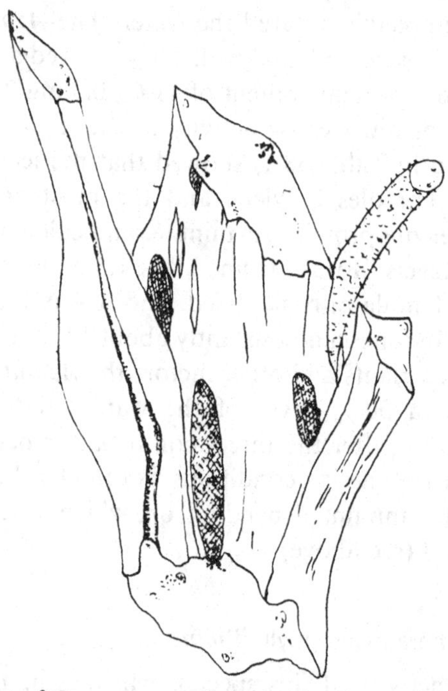

Fig. 38.3. Part of stem of *Gunnera macrophylla* showing the position of the *Nostoc* glands (cross-hatched) and, at the cut end of the stem, the fan-like ingrowths of *Nostoc* from the periphery of the gland. Field drawing, Cibeureum waterfalls, W. Java. (×0.7).

Nostoc glands with some surrounding tissue were excised from the stems and were incubated in air with added acetylene under the ambient conditions of light and temperature (18 °C) within 15 minutes after excision. Twenty glands were placed in each vessel of 70 ml. The results of two experiments are shown in Table 38.3. Ethylene was produced at a fairly steady rate, though the rate tended to rise with time. Nitrogenase activity appears to have been somewhat greater at the higher level of acetylene.

Here again an estimate of the nitrogen contribution by the symbiosis to the ecosystem has been attempted. The original data on which Table 38.3 was based showed that with 7.6 % acetylene provided, the sample of twenty *Nostoc* glands produced ethylene at an average rate of 6202 nmoles/h over the period of the experiment, while with 9.4 % of acetylene supplied the corresponding figure rose to 10 836. Examination of the *Gunnera* plants showed that during its development a plant produces fifteen leaves, and although by the time the plant is mature the oldest of the leaves have senesced, evidence that the *Nostoc* glands

Table 38.3. *Acetylene reduction by* Gunnera macrophylla *tissue fragments containing* Nostoc *glands**

Per cent of C_2H_2 in air	Incubation time (min)	C_2H_4 produced (n/moles/mg sample N/30 min)	C_2H_4 (nmoles/mg sample protein/min)
7.6	0–30	99.7	0.53
	30–60	115.0	0.60
	60–90	125.9	0.67
	90–120	142.1	0.76
	120–150	154.8	0.83
9.4	0–30	188.7	1.0
	30–60	169.5	0.90
	60–90	227.7	1.21
	90–120	184.8	0.99
	120–150	266.1	1.42

* Experiments were carried out between 08.30 and 11.00 h. In each experiment the total fresh weight of the excised tissue was about 9 g, the dry weight about 1 g.

associated with the bases of these leaves remain active has already been noted. A *Gunnera* plant with fifteen leaf insertions possesses forty-five *Nostoc* glands. It was also observed that the usual cover of *Gunnera* was 2 to 3 plants/m^2. Assuming again that 1.5 mole ethylene corresponds to 1 mole nitrogen, and also that fixation at night is negligible owing to the lack of light and the low ambient temperature (8–10 °C), and taking into account that owing to extensive mist formation at the site the effective daylength is about 10 hours, then suitable calculation indicates that with 2.5 *Gunnera* plants/m^2 fixation will approximate to 12–21 kg N/ha/annum.

Results from Podocarpus *species*

The root nodules of two species indigenous to Indonesia, namely *Podocarpus rumphii* Blume and *P. koordersii* Pilger, were subjected to the acetylene reduction test. Trees of these species in the Botanical Gardens, Bogor, were found to be abundantly nodulated. Adhering soil was shaken off from pieces of nodulated roots, and then, without washing, samples of 2.0–2.5 g fresh weight were exposed within 10 minutes after excision to acetylene levels of 6.3 and 8.8 % in air. The tests were conducted soon after midday, and ethylene production was followed for 2 hours. With *P. rumphii* an extremely small amount of ethylene (2 nmole/mg sample nitrogen) was produced in the first hour, and none in the second. For *P. koordersii* in the first hour ethylene production was slightly higher (5 nmole/mg sample nitrogen), and continued at that level over the second hour. These amounts are quite negligible in comparison with those recorded above for *Azolla* and *Gunnera*.

Results from leaf nodules

The acetylene reducing capacity of two Indonesian members of the Rubiaceae, namely *Psychotria bacteriophila* Val. and *Pavetta sylvatica* Blume, was investigated. The material used was taken from plants in the Botanical Gardens, Bogor, which had profusely nodulated leaves. Shoots still attached to the plants were, as already noted, sealed into Perspex containers of 500 ml capacity provided with a split rubber stopper at the base (Fig. 38.1*b*). An acetylene level of 5.8 % in air was provided in the container, and exposure was for 2 hours under the ambient conditions of sunlight and temperature (32 °C). In order to provide carbon dioxide for photosynthesis a CO_2-buffer (0.1 M $NaHCO_3$ and 0.1 M Na_2CO_3 mixed in equal parts) was added to the container to maintain a concentration of 0.03 % in the atmosphere. The experiments were conducted during a period of good photosynthetic activity (13.15–15.15 h). Each enclosed shoot bore 14–15 leaves with a total area of 290–380 cm^2 as determined planimetrically from shadow prints of the leaves on photographic paper. Since counts showed that the *Psychotria* leaves bore an average of 11.9 ± 1.7 nodules/cm^2, the corresponding figure for *Pavetta* being 1.7 ± 0.1, it is obvious that a considerable number of nodules was included in the sample vessels.

In none of the experiments could any ethylene production be detected.

Results from the phyllosphere

Here the species examined were *Theobroma cacao* L. and the fern, *Drymoglossum piloselloides* Presl. Material of the former was obtained from a tree growing in the shade of larger trees in the Botanical Gardens, Bogor, while that of the epiphytic fern was collected from tree-trunks in the same Gardens. The 500 ml Perspex containers were again employed, with acetylene levels of 6.1 and 7.9 % provided; also the CO_2-buffer. Experiments with *Theobroma* were made both with shoots still attached to the plant and with excised shoots, in each case the number of leaves included being 5–6, with a total leaf area of 450–550 cm^2. In the *Drymoglossum* test complete plants were placed in the container, a total of 83 leaves with a combined area of 123 cm^2 being included. The exposures were made during the period 14.00–16.00 hours under ambient conditions of sunlight and temperature (27 °C). It should also be noted that the *Theobroma* leaves, though fully-grown, were without any encrustation of epiphyllae such as lichens, blue-green algae and fungi as is sometimes observed in older cacao leaves, and that the *Drymoglossum* leaves were also free of such growths as observed by the naked eye. Enrichment cultures made from the *Theobroma* leaves in nitrogen-free nutrient solution (Becking, 1961) showed bacterial and fungal growth. From some of these enrichment cultures *Beijerinckia* could be isolated, though counts of this organism on the leaf surface were not made.

With both species a small production of ethylene was detected during the first hour of exposure (3 nmoles with *Theobroma*, 15 nmoles with *Drymoglossum*; both per 10 cm^2 of leaf surface), but during the second hour production fell to insignificant levels.

Discussion

The nitrogenase activity found in *Anabaena* isolated from *Azolla* fronds (Table 38.2) was relatively high. Stewart, Fitzgerald & Burris (1968) have reported for various blue-green algae, including *Anabaena*, activities of 1.0–2.0 nmole C_2H_4/mg protein/min, seldom rising as high as 4.0–5.0. There is no doubt that the activity of the alga when in symbiosis with *Azolla* was still higher, although, as explained, an accurate comparison was not possible owing to analytical difficulties. One reason for this greater activity might be that the *Azolla* augments the supply of carbohydrates to the alga, while another could be a rapid removal of fixed nitrogen from the vicinity of the alga by the fern, since W. D. P. Stewart (personal communication) has found a greater efficiency in blue-green algae when grown in continuous culture.

Saubert (1949), on the basis of laboratory experiments (involving Kjeldahl analyses) carried out in Indonesia, calculated that the fixation of nitrogen associated with *Azolla pinnata* in the field might amount to

312 kg/ha/annum. Though this is considerably higher than the present author's estimate, a reason for preferring the latter is that in its calculation full account was taken of the limitations to activity arising under field conditions; on the other hand it was based on an indirect method of measuring fixation. Saubert (1949) also referred to reports that *Azolla* had long been used in Tonkin and Thailand as green manure for rice fields, while more recently Bui Huy Dap (1967) reported similarly in respect of Vietnam. Olsen (1970) introduced *Azolla caroliniana* from the USA into Denmark in an attempt to provide a weed-excluding cover to lakes and ponds, and measured a nitrogen fixation of up to 95 kg/ha in one summer.

The symbiotic relation between *Gunnera macrophylla* and *Nostoc* had previously been studied chiefly from morphological and cytological angles (Treub, 1882; Merker, 1889; Miehe, 1924; Baas Becking, 1947). The data now provided indicate that the *Nostoc*-filled glands of the plant are centres of active nitrogen fixation, as shown by Silvester & Smith (1969) for New Zealand species of *Gunnera.* On the basis of laboratory experiments the latter authors estimated that under field conditions in New Zealand the fixation associated with *Gunnera dentata* might be of the order of 72 kg N/ha/annum. This considerably exceeds the present author's estimate for *G. macrophylla*, one possible explanation of this being that the dense creeping habit of the New Zealand species may result in a greater amount of alga per unit of soil area than in the case of the massive *G. macrophylla* plants. (See also Silvester, Chapter 37.)

The data presented for *Podocarpus* provide no evidence of any significant nitrogenase activity in the root nodules. Though the effect of surface sterilisation was not actually tested, the small reduction of acetylene observed in some experiments could have been due to bacteria associated superficially with the nodulated roots. Silvester & Bennett (1973) reported that the small ethylene production observed in nodulated podocarp roots ceased after surface sterilisation.

The literature on the functional significance of the leaf nodules present in some members of the Rubiaceae contains discrepant results. Some old claims for the occurrence of nitrogen fixation within them (see Schaede, 1962, also van Hove in this volume p. 555) seemed to have been supported by Silver, Centifanto & Nicholas (1963), who reported that *Psychotria* plants grew well in a nitrogen-free rooting medium, and by Grobbelaar, Strauss & Groenewald (1971) who by means of growth experiments and acetylene tests found evidence of fixation in *Pavetta* spp., though at a very low rate. On the other hand, Becking found no evidence of fixation in ^{15}N and acetylene tests on leaves of *Psychotria* or in growth experiments. The present study shows that also under the tropical conditions to which these plants are habituated no evidence of nitrogen fixation could be obtained.* Becking (1971) provided some evidence that the value of the nodules to the plant lay rather in the production by the nodular bacteria of hormones, especially a cytokinin-like substance.

Data claimed to show the occurrence of nitrogen fixation in phyllosphere associations have been presented by Edmisten & Harrelson (1967), Edmisten (1970) and Jones (1970). The present author's experience is that under the normal tropical conditions the leaves of cacao and of *Drymoglossum* showed initially a very low level of acetylene reduction, but that even this was not sustained. These particular plant species have been described by Ruinen (1953, 1956, 1961) to be particularly rich in phyllosphere micro-organisms which are at least capable of nitrogen fixation, and at present it is not clear why expectations based on those reports were not realised when actual tests for fixation were made.

This study has been made possible by a grant from the Netherlands Foundation for the Advancement of Tropical Research (WOTRO), The Hague, The Netherlands. I gratefully acknowledge the eminent technical assistance of Miss J. J. van der Kaa and Mr W. F. Pieters in acetylene and ethylene determinations by gas chromatograpny. Also I wish to thank Mr Adi Nurhadi for technical assistance in the field and Dr R. J. Fessenden for suggestions on which the method used for storage of gas samples was based. I am greatly indebted to Professor G. Bond for valuable criticism of the manuscript.

References

Allen, M. B. & Arnon, D. I. (1955). Studies on nitrogen-fixing blue-green algae. I. Growth and nitrogen fixation by *Anabaena cylindrica* Lemm. *Plant Physiology*, **30**, 366–72.

Baas Becking, L. G. M. (1947). Note on the endophyte of *Gunnera macrophylla* Bl. *Biologisch Jaarboek Dodonaea*, **14**, 93–6.

Becking, J. H. (1961). Studies on nitrogen-fixing bacteria of the genus *Beijerinckia*. I. Geographical and ecological distribution in soils. *Plant and Soil*, **14**, 49–81.

Becking, J. H. (1971). The physiological significance of the leaf nodules of *Psychotria*. *Plant and Soil*, Special Volume, 361–74.

Bui Huy Dap (1967). Some characteristic features of rice growing in Vietnam. In: *Agricultural problems*, No. 2, *Rice*. Vietnamese Studies, No. 13, pp. 38–66. China Books, San Francisco.

* At the Edinburgh meeting Professor Silver withdrew his claim for the occurrence of nitrogen fixation in *Psychotria* leaf nodules.

Edmisten, J. A. (1970). Preliminary studies of the nitrogen budget of a tropical rain forest. In: *A tropical rain forest* (ed. H. T. Odum & R. F. Pigeon), pp. H211–H215. Division of Technical Information, US Atomic Energy Commission, Oakridge, Tennessee.

Edmisten, J. A. & Harrelson, M. A. (1967). Nitrogen fixation by epiphyllae at El Verde. In: *The rain forest project.* Annual Report of the Puerto Rico Nuclear Center, University of Puerto Rico.

Grobbelaar, N., Strauss, J. M. & Groenewald, E. G. (1971). Non-leguminous seed plants in southern Africa which fix nitrogen symbiotically. *Plant and Soil,* Special Volume, 325–34.

Jones, K. (1970). Nitrogen fixation in the phyllosphere of the Douglas Fir, *Pseudotsuga douglasii. Annals of Botany,* NS **34,** 239–44.

Merker, P. (1889). *Gunnera macrophylla* Bl. *Flora, Marburg,* NR **47,** 211–32.

Miehe, H. (1924). Entwicklungsgeschichtliche Untersuchung der Algensymbiose bei *Gunnera macrophylla* Bl. *Flora, Jena,* **17,** 1–15.

Olsen, G. (1970). On biological nitrogen fixation in nature, particularly in blue-green algae. *Comptes rendus des travaux du Laboratoire Carlsberg,* **37,** 269–83.

Ruinen, J. (1953). Epiphytosis. *Annales Bogoriensis,* **1,** 101–57.

Ruinen, J. (1956). Occurrence of *Beijerinckia* species in the 'phyllosphere'. *Nature, London,* **177,** 220–1.

Ruinen, J. (1961). The phyllosphere. I. An ecologically neglected milieu. *Plant and Soil,* **15,** 81–109.

Saubert, G. G. P. (1949). Provisional communication on the fixation of elementary nitrogen by a floating fern. *Annals of the Royal Botanic Gardens, Buitenzorg,* **51,** 177–97.

Schaede, R. (1962). *Die pflanzlichen Symbiosen,* 3rd edition, revised by F. H. Meyer. Gustav Fischer Verlag, Stuttgart.

Silver, W. S., Centifanto, Y. M. & Nicholas, D. J. D. (1963). Nitrogen fixation by the leaf-nodule endophyte of *Psychotria bacteriophila. Nature, London,* **199,** 396–7.

Silvester, W. B. & Bennett, K. J. (1973). Acetylene reduction by roots and associated soil of New Zealand conifers. *Soil Biology & Biochemistry,* **5,** 171–9.

Silvester, W. B. & Smith, D. R. (1969). Nitrogen fixation by *Gunnera–Nostoc* symbiosis. *Nature, London,* **224,** 1231.

Stewart, W. D. P., Fitzgerald, G. P. & Burris, R. H. (1968). Acetylene reduction by nitrogen-fixing blue-green algae. *Archiv für Mikrobiologie,* **62,** 336–48.

Strasburger, E. (1873). *Ueber Azolla.* Hermann Dabis, Jena.

Treub, M. (1882). Korte botanische aantekeningen. *Nederlands Kruidkundig Archief,* Serie 3, **2,** 404–8.

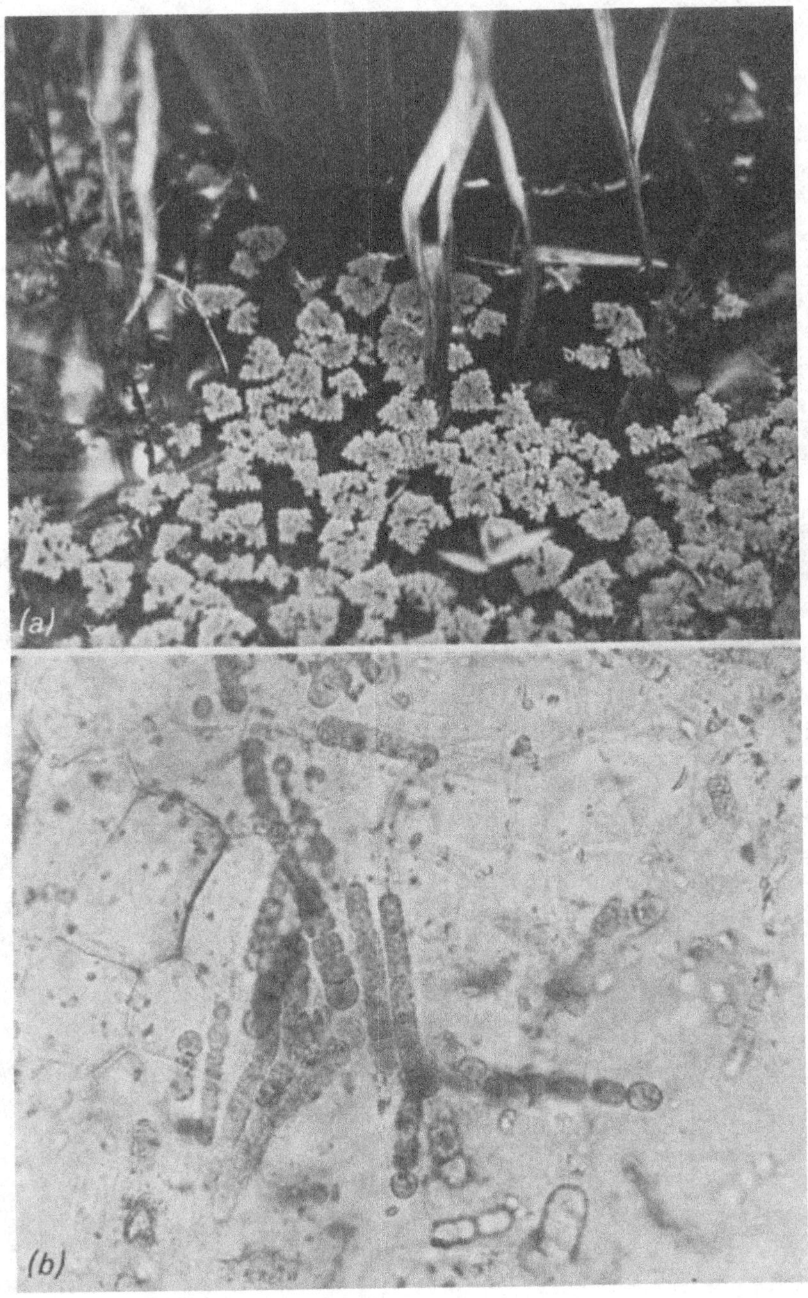

Plate 38.1. (*a*) *Azolla pinnata* plants floating on the water surface of an irrigated rice field. (×0.7). (*b*) *Anabaena azollae* from the leaf cavity as observed in a squashed frond of *Azolla pinnata*. (×600).

(*Facing p.* 550)

Plate 38.2. (*a*) and (*b*) Natural stands of *Gunnera macrophylla* at the Cibeureum waterfalls, Mt Gedeh, W. Java. ($a \times 0.06$, $b \times 0.07$). (*c*) Cut surface (transverse) of stem of *Gunnera macrophylla* showing two *Nostoc* glands. The algal growths appear black in the photograph. ($\times 0.9$). (*d*) Experimental set-up used in the field experiments. Visible are the acetylene storage vessel connected to a second vessel for adjustment of pressure, and the 70 ml all-glass vessels in which the plant material was enclosed during assay. ($\times 0.11$).

39. Bacterial leaf symbiosis and nitrogen fixation

C. van HOVE

In recent years fresh evidence has become available on the question (first asked by von Faber in 1911) of whether fixation of nitrogen occurs in the bacterial leaf nodules present on certain tropical plants. Since subsection PP-N of IBP has had some interest in this problem (see the preface to this Part of the volume), it may serve a useful purpose if the author, before presenting his own data, attempts a critical review of previous work on this aspect. The best-known examples of leaf nodules are those provided by species of *Pavetta* and *Psychotria* (both members of the Rubiaceae), *Ardisia* (Myrsinaceae) and to a lesser degree *Dioscorea*, a monocotyledonous genus, where the bacteria occur in cavities in the elongated tip of the leaf.

Observations on bacteria isolated from the nodules or from seeds

The bacterium isolated from *Pavetta* nodules by von Faber (1912) and called by him *Mycobacterium rubiacearum*, and those by Rao (1923) from the same plant genus, were reported to grow in a nitrogen-free medium; those isolated by Grobbelaar, Strauss & Groenewald (1971) were shown to fix nitrogen, while Bettelheim, Gordon & Taylor (1968) on the basis of immunofluorescence tests held the *Pavetta* symbiont to be similar to that of *Psychotria* which was reported to fix nitrogen (see below).

Von Faber (1914) reported direct evidence of fixation by his isolate from *Psychotria*, which he held to be identical with that previously obtained from *Pavetta* (von Faber, 1912). A bacterium isolated from *Psychotria* by Silver, Centifanto & Nicholas (1963) and Centifanto & Silver (1964), identified by them as *Klebsiella rubiacearum*, was able to grow in nitrogen-free media and shown to fix nitrogen by the use of ^{15}N, as Bettelheim *et al.* (1968) similarly reported for their isolate from *Psychotria*, identified as *Chromobacterium lividum*.

In the case of *Ardisia*, Miehe (1912, 1914) and de Jongh (1938) were unable to grow their isolated bacteria, which they named as *Bacillus follicola* and *Alcaligenes* respectively, in nitrogen-free media, although on such a medium Hanada (1954) successfully cultured a *Xanthomonas* claimed to be the *Ardisia* symbiont. The organism isolated by Bettelheim *et al.* (1968) from *Ardisia* was considered from varied evidence to be identical with that from *Psychotria*, and was again said to be nitrogen-fixing.

Orr (1923) and Lieske (1926) reported that bacteria isolated from the leaf acumen of *Dioscorea* were able to grow in nitrogen-free media.

It should be pointed out that some of the above authors provided no evidence that the organisms studied were really the nodule endophytes. This was the case with the two isolates from *Dioscorea* just mentioned, which, moreover, from the descriptions given by the authors, were clearly different organisms. In the studies of von Faber, de Jongh, Hanada and Silver successful re-inoculation tests were reported, though these are not completely convincing, considering the very limited numbers of plants used and the known natural reversion processes; those reported by Silver could not be repeated by Becking (1971). Furthermore, the fact that the six identified bacteria belong to five or six families in three different orders, strongly suggests that the majority of these isolates do not correspond with the endophyte, since it is unlikely that the similarities found between the *Pavetta*, *Psychotria* and *Ardisia* symbioses could exist if the endophytes were so varied. As noted, Bettelheim *et al.* (1968) provided good evidence that their isolates from those three plant genera were identical, and though their isolations were actually made from germinating seeds, immunofluorescence tests indicated that they were also the occupants of the leaf nodules. From the present standpoint, however, it is unfortunate that those authors give no details in their paper of the tests which showed the isolates to be nitrogen-fixing.

Growth experiments with nodulated plants in nitrogen-free media

Three *Pavetta* plants grown by von Faber (1912) did not show symptoms of nitrogen deficiency after three months in a nitrogen-free rooting medium, and were reported to have increased their nitrogen content by 99, 70 and 97 mg respectively during this period. These plants were, however, smaller than control plants provided with combined nitrogen (mean height: 145 mm versus 170 mm) and contained less nitrogen (mean nitrogen content: 107 mg versus 163 mg). There was one very disturbing feature in these results: from von Faber's description of these plants it may be assumed that the fresh weight was approximately 3 to 6 g per plant; 107 and 163 mg nitrogen thus represent between 1.8 and 5.4 % of the fresh weight! *Pavetta* seedlings cultured by Grobbelaar *et al.* (1971) for fourteen weeks without combined nitrogen made poor growth (far inferior to that of plants given combined nitrogen) and showed severe symptoms of nitrogen starvation; a small increase in nitrogen content was reported, which, however, cannot be linked unequivocally with the nodule endophyte in view of the lack of information on precautions taken to exclude contaminants.

Humm (1944), Becking (1971) and Silvester & Astridge (1971) obtained only poor growth and symptoms of nitrogen deficiency in *Psychotria* plants cultured in nitrogen-deficient soils, whereas Silver *et al.* (1963) reported that plants deprived of combined nitrogen for six months did not show deficiency symptoms, though changes, if any, in nitrogen content were not measured. It is the author's experience that *Psychotria* plants generally show very slow growth, and it is quite possible that a six months' starvation period is not long enough to induce visible symptoms of nitrogen deficiency.

According to Miehe (1916, 1919), although *Ardisia* plants in a nitrogen-free rooting medium survive for quite a long time, they remain small and bear pale leaves. Hanada (1954) cultivated *Ardisia* plants for two years with and without combined nitrogen, and on the basis of the results concluded that *Ardisia* fixes nitrogen. However, nitrogen deficiency symptoms appeared in the plants denied combined nitrogen and their growth was somewhat inferior to that of plants given combined nitrogen, which were themselves very slow-growing.

Thus in most growth experiments in nitrogen-free rooting media deficiency symptoms have appeared, indicating that for proper growth, plants of these three genera require an external source of combined nitrogen. Failure to observe deficiency symptoms in some experiments has little meaning if the slow growth of these plants is taken into account.

^{15}N *and* C_2H_2 *experiments on nodulated plants*

Tests using ^{15}N were carried out on *Pavetta* seedlings, detached leaves and shoots by Grobbelaar *et al.* (1971), on detached leaves of *Psychotria* by Bond (1959) and by Becking (1971), on leaf homogenates of that genus by Silver *et al.* (1963), and on *Ardisia* plants and detached leaves by Hofstra & Koch-Bosma (1970). Negative results were reported in all tests, except for those with leaf homogenates. In the latter, small but significant enrichment was sometimes observed, but was not readily reproducible.*

* See footnote on p. 549 of this volume.

The acetylene method was used by Grobbelaar *et al.* (1971) on *Pavetta* detached leaves; by Becking (1971, Chapter 38) on *Pavetta* and *Psychotria* twigs still attached to the plant in the field and also on detached leaves of the latter genus; by Silvester & Astridge (1971) on *Psychotria* detached leaves; by Hofstra & Koch-Bosma (1970) on *Ardisia* plants and detached leaves. Apart from a very small acetylene reduction observed for *Pavetta* leaves by Grobbelaar's group, nitrogenase activity was not detected in any of these experiments, carried out under a wide variety of conditions, some of them quite near to natural.

Observations on nodule-free plants

Although direct evidence on the question of the occurrence of fixation can only be obtained from the examination of nodulated plants or, though with less certainty, of the isolated endophyte, indirect evidence has been sought from the study of the bacteria-free plants of the species under discussion which sometimes occur naturally and can also be obtained by heat-treatment of the seed. Many authors (e.g. Becking, 1971) have found that in fertile soil or in cultures supplied with nitrate such plants fail to grow properly and instead show a stunted, distorted shoot and are short-lived. The visual symptoms are not those of nitrogen-deficiency and analysis shows that the tissues contain a normal level of nitrogen (Becking, 1971).

These findings indicate that in the normal nodulated plant, even if fixation of nitrogen occurs, the plant benefits in some other way as well, possibly, as several authors have suggested, through a release by the bacteria of some growth substance essential at least to the growth of the shoot, though perhaps not of the roots, since it has been reported (Miehe, 1919; Němec, 1932; de Jongh, 1938) that the roots of bacteria-free plants are initially healthy. Becking (1971) presented some evidence that cytokinin was the growth substance concerned.

Other arguments

Besides the above major arguments, others of dubious value have been adduced. Thus the local use of *Pavetta* leaves as green manure (Rao, 1923), and the high nitrogen content of *Dioscorea* acumens as compared with the rest of the leaf (Orr, 1923), have been thought to support the notion that the nodules were sites of fixation. On the other hand, the observation that *Psychotria* leaves do not have a higher nitrogen content than leaves from nearby trees (Löhr, 1968), and that leaf margins from non-nodulating *Ardisia* species generally contain more nitrogen than the middle part, as is the case with nodulated *Ardisia* (Němec, 1932; Hofstra & Koch-Bosma, 1970), have been considered to support the view that the nodules are not nitrogen-fixing.

Besides the four genera so far considered, claims have been made for the presence of nitrogen-fixing leaf nodules, or comparable structures, in other genera. Thus Rao (1923) reported that leaf nodules were present on *Chomelia asiatica* (Rubiaceae) and that bacteria isolated from the nodules were able to grow on a nitrogen-free medium; however it should be noted that according to Durand & Jackson (1906) the name *C. asiatica* Kuntze is a synonym for *Tarenna zeylanica* Gaertner, a species reported by Bremekamp (1960) to be not nodule-bearing. Stevenson (1953, 1958, 1959) reported that *Coprosma* (Rubiaceae) bears stipular bacterial nodules, and that growth experiments and ^{15}N tests indicated an occurrence of fixation. On the other hand, Silvester & Astridge (1971) observed a strong reduction in growth of *Coprosma* seedlings in a rooting medium free of combined nitrogen, while excised leaves and glandular stipules did not reduce acetylene. Van Hove & Craig (1973), using the light and electron microscope, were in fact unable to find any bacteria in stipular glands and other *Coprosma* tissues supposed to contain them.

Stevenson (1953) suggested that various members of the families Myoporaceae, Myrtaceae and Verbenaceae, including species of *Myoporum*, *Metrosideros*, *Leptospermum* and *Vitex* also exhibit a nitrogen-fixing bacterial symbiosis, since, according to her observations, small bacteria-filled glands, associated with red pigments, are present on the surface and also within their leaves. These species were re-examined by van Hove & Craig (1973) who failed to find bacteria in any of the structures mentioned by Stevenson.

The author's own acetylene-reduction tests for the presence of nitrogenase activity in species bearing leaf nodules or comparable structures will now be described.

Materials and methods

The species studied comprised *Psychotria calva* Hiern, *Pavetta nitidula* Welw. ex Hiern, *Neorosea andongensis* (Hiern) N. Halle, and *Dioscorea sansibarensis* Pax (Syn.: *D. macroura* Harms). The leaf nodules of *Neorosea* (Rubiaceae) were recently described by van Hove (1972) and shown to be similar to those of *Pavetta* and *Psychotria*. Plants of these species are grown in the garden attached to the author's laboratory, and were transplanted to it a few years ago from various regions of Zaïre. Voucher specimens (*exsicc*, van Hove 75, 78, 7, 28) are kept in the herbaria of the Université nationale du Zaïre and of the Jardin botanique national de Belgique (BR), Brussels.

In laboratory experiments nodulated midribs from *P. calva* leaves were cut out, leaving a strip of blade approx. 3 mm wide on each side. Ten midribs were put in a 25 ml penicillin vial. These operations were carried out in the morning and in the evening. In one set of experiments, vials were filled with distilled water and stoppered; water was then removed and replaced by an appropriate atmosphere (C_2H_2/Ar/O_2: 15/63/22, or C_2H_2/Ar: 15/85). In another series the air was replaced by flushing the vials with 5 volumes of the above gas mixtures. Thus, in the first case, very humid conditions prevailed in the vials, in the second, humidity free conditions prevailed. Newly unfolded young leaves were used in one set of assays and mature leaves from the third node down from the tip in the other. In order to assure good diffusion of gas in and out of the nodules, they were opened by a longitudinal scalpel incision in one series of experiments.

D. sansibarensis excised glandular acumens were tested in the same way, as well as *P. calva* germinating seeds (40 seeds/vial) with root tips just extruding. For seed experiments incision treatment was omitted.

Vials were incubated for 12 hours (24 h for seedlings) in darkness and in light, at 20 °C and 30 °C. All the treatments were carried out at least twice.

In field experiments, *P. calva*, *N. andongensis* and *P. nitidula* branches still attached to the plant were placed in a tightly closed plastic bag, mornings and evenings. Only actively growing twigs with well-nodulated leaves were selected (during some periods of the year, plants only produce non-nodulated leaves). Bags contained between 20 and 30 leaves ($\pm$20 g fresh weight). Five-hundred millilitres of acetylene was injected into approximately 21 bags. After incubation periods of 0, 2, 12, and 24 hours under natural conditions (except that direct insolation was avoided), 100 μl gas was withdrawn, and injected immediately into a gas chromatograph. Again, all the treatments were done at least twice, using different plants for each test.

Unfortunately, it was not possible to use the same method for *D. sansibarensis*, as in these tall climbing plants bacteria-containing tissues represent a negligible part of the plant.

Ethylene production was measured by gas chromatography (Varian aerograph 1840-L, flame ionisation detector), using an activated alumina column 3.2 mm × 185 cm, at 150 °C with a nitrogen flow of 25 ml per minute.

Results

In none of the laboratory experiments with nodules or acumens was any production of ethylene detected, although the method employed allowed a detection of 5 nmole per vial. With germinating seeds a very small quantity (maximum 68 nmole/g fresh weight/24 h) of ethylene was occasionally detected; the erratic occurrence of these positive results was taken to suggest that they were due to contaminant organisms which can develop on the testa or in the seed during germination.

In respect of field experiments, previous tests had shown that the minimum amount of ethylene detectable was approximately 200 nmole per plastic bag. This minimum was never reached in the experiments, although conditions appeared to be very favourable for the activity of any nitrogenase that might have been present. When *Azolla africana* Desv. was added to the system, acetylene reduction proceeded at a linear rate of 730 nmole/g dry weight/h for at least 12 hours, a very similar activity being observed when *Azolla* was incubated in vials in short-term experiments.

Discussion

The author had the advantage of having at his disposal plants growing under conditions which approximated very closely to those prevailing in the natural places of growth of the various species in the wild, where, it may be assumed, any special properties which they possess will be most fully developed. Despite this, the results obtained with leaves and shoots of all four species were uniformly negative for nitrogenase activity. The symbiotic tissue, i.e. the nodules, represents only a small proportion of the enclosed leafy material, but as indicated already substantial samples were inserted into the containers used for the tests, and anything more than negligible activity could scarcely have evaded detection. No extra carbon dioxide was added to the gas mixtures employed, in order to provide for a photosynthetic rate above that at compensation point, but it is rather inconceivable that in photosynthetic organs there should be an immediate dependence of nitrogenase activity on a high rate of photosynthesis.

The tests here reported on *Dioscorea* represent the first direct examination of this particular leaf–bacterial association for nitrogen fixation. The tests on germinating seed of *Psychotria* were included because the seeds are known to be infected with the leaf-nodule endophyte, so that any capacity for fixation should also be detectable at this stage. In fact none was observed.

Research on the possibility of nitrogen fixation in leaf nodules originated from a postulated analogy between these structures and root nodules. While an association between a green plant and a nitrogen-fixing micro-organism appears to be a very rational system, the furnishing of fixed nitrogen does not exhaust the list of benefits which bacteria can convey to a green plant with which they are associated, and there is no reason for expecting *a priori* leaf nodules to be nitrogen fixing.

As indicated in the introductory section of the chapter, reports of the possession of nitrogen-fixing properties by organisms isolated from leaf nodules, but not proved to be the actual endophytes, are of little significance. The more reliable evidence derivable from growth experiments or from ^{15}N or acetylene tests by previous authors on the nodulated plants, or parts of them, in the main supports a conclusion that even if nitrogen fixation occurs at all in leaf nodules it is too small in amount to play any significant role in the nitrogen economy of the plants or their ecosystems; our results support this conclusion. The explanation of the beneficial effect of the nodules on the plant must lie in a different direction.

The author wishes to thank Professor G. Bond, University of Glasgow, and Professor C. Evrard, Université nationale du Zaïre, for their very helpful criticism of the manuscript. Professor Evrard identified the material used and provided some of it. Thanks are also due to Dr E. Petit, Jardin botanique national de Belgique, who confirmed the identification by comparison with authenticated specimens. Excellent facilities were provided for the literature search during the author's sabbatical leave at the Applied Biochemistry Division, DSIR, New Zealand.

References

Becking, J. H. (1971). The physiological significance of the leaf nodules of *Psychotria*. *Plant and Soil*, Special Volume, 361–74.

Bettelheim, K. A., Gordon, J. F. & Taylor, J. (1968). The detection of a strain of *Chromobacterium lividum* in the tissues of certain leaf-nodulated plants by the immunofluorescence technique. *Journal of General Microbiology*, **54**, 177–84.

Bond, G. (1959). The incidence and importance of biological fixation of nitrogen. *Advancement of Science*, **15**, 382–6.

Bremekamp, C. E. B. (1960). Les 'Psychotria' bacteriophiles de Madagascar. *Notulae Systematicae*, **16**, 41–54.

Centifanto, Y. M. & Silver, W. S. (1964). Leaf-nodule symbiosis. I. Endophyte of *Psychotria bacteriophila*. *Journal of Bacteriology*, **88**, 776–81.

de Jongh, Ph. (1938). On the symbiosis of *Ardisia crispa* (Thunb.) A. DC. *Verhandelingen der koninklijke nederlandsche Akademie van Wetenschappen, afdeeling Natuurkunde*, Sect. 2, **37**, 1–74.

Durand, T. & Jackson, B. D. (1906). *Index Kewensis*, supplement 1, p. 95. Oxford University Press, London.

Grobbellaar, N., Strauss, J. M. & Groenewald, E. G. (1971). Non-leguminous seed plants in Southern Africa which fix nitrogen symbiotically. *Plant and Soil*, Special Volume, 325–34.

Hanada, K. (1954). Über die Blattknoten der *Ardisia*-Arten. Isolierung der Bakterien und ihre stickstoffbindende Kraft in Reinkultur. *Japanese Journal of Botany*, **14**, 235–68.

Hofstra, J. J. & Koch-Bosma, T. (1970). Some aspects of the leaf nodule symbiosis in *Ardisia crispa*. *Acta botanica Neerlandica*, **19**, 665–70.

Humm, H. J. (1944). Bacterial leaf nodules. *Journal of the New York Botanical Garden*, **44**, 193–9.

Lieske, R. (1926). *Kurzes Lehrbuch der allgemeinen Bakteriologie*. Berlin.

Löhr, E. (1968). Stickstoffgehalt in Blättern von Psychotria mit N-bindenden endophyten. *Physiologia Plantarum*, **21**, 1156–8.

Miehe, H. (1912). Über Symbiose von Bakterien mit Pflanzen. *Biologisches Zentralblatt*, **32**, 46–50.

Miehe, H. (1914). Weitere Untersuchungen über die Bakteriensymbiose bei *Ardisia crispa*. I. Die Mikroorganismen. *Jahrbuch für wissenschaftliche Botanik*, **53**, 1–54.

Miehe, H. (1916). Über die Knospensymbiose bei *Ardisia crispa*. *Berichte der Deutschen botanischen Gesellschaft*, **34**, 576–80.

Miehe, H. (1919). Weitere Untersuchungen über die Bakteriensymbiose bei *Ardisia crispa*. II. Die Pflanze ohne Bakterien. *Jahrbuch für wissenschaftliche Botanik*, **58**, 29–65.

Němec, B. (1932). Über Bakteriensymbiose bei *Ardisia crispa*. *Mémoires de la Société royale des Sciences de Bohème*, **19**, 1–23.

Orr, M. Y. (1923). The leaf glands of *Dioscorea macroura*, Harms. *Notes from the Royal Botanic Garden, Edinburgh*, **14**, 57–72.

Rao, K. A. (1923). A preliminary account of symbiotic nitrogen fixation in non-leguminous plants with special reference to *Chomelia asiatica*. *Agricultural Journal of India*, **18**, 132–43.

Silver, W. S., Centifanto, Y. M. & Nicholas, D. J. D. (1963). Nitrogen fixation by the leaf-nodule endophyte of *Psychotria bacteriophila*. *Nature, London*, **199**, 396–7.

Silvester, W. B. & Astridge, S. (1971). Reinvestigation of *Coprosma* for ability to fix atmospheric nitrogen. *Plant and Soil*, **35**, 647–50.

Stevenson, G. (1953). Bacterial symbiosis in some New Zealand plants. *Annals of Botany*, **17**, 343–5.

Stevenson, G. (1958). Nitrogen fixation by non-nodulated plants and by nodulated *Coriaria arborea*. *Nature, London*, **182,** 1523–4.

Stevenson, G. (1959). Fixation of nitrogen by non-nodulated seed plants. *Annals of Botany*, **23,** 622–35.

van Hove, C. (1972). Structure and initiation of nodules in the leaves of *Neorosea andongensis* (Hiern) N. Hallé. *Annals of Botany*, **36,** 259–62.

van Hove, C. & Craig, A. S. (1973). A reinvestigation of bacterial symbiosis in *Coprosma*, *Myoporum*, *Metrosideros*, *Leptospermum* and *Vitex*. *Annals of Botany*, **37,** 1013.

von Faber, F. C. (1911). Über das ständige Vorkommen von Bakterien in den Blättern verschiedener Rubiaceen. *Bulletin du département de l'Agriculture aux Indes néérlandaises*, **46,** 1–3.

von Faber, F. C. (1912). Das erbliche Zusammenleben von Bakterien und tropischen Pflanzen. *Jahrbuch für wissenschaftliche Botanik*, **51,** 285–375.

von Faber, F. C. (1914). Die Bakteriensymbiose der Rubiaceen. *Jahrbuch für wissenschaftliche Botanik*, **54,** 243–64.

40. A discussion of the results of cross-inoculation trials between *Alnus glutinosa* and *Myrica gale*

C. RODRIGUEZ-BARRUECO & G. BOND

While working in the laboratory of G.B. during the period 1964–7, C.R.B. examined a number of soil samples from west Scotland for the presence of the *Alnus glutinosa* and *Myrica gale* endophytes, as indicated by the induction of nodulation in test seedlings of the host species planted in pots filled with the soils. The results (Rodriguez-Barrueco, 1968) revealed that every soil sample that caused nodulation in the *A. glutinosa* test plants also induced nodulation in *M. gale*, though plants of the latter were not growing at the sampling sites. Later, in his own laboratory, C. R. B. (unpublished data) similarly tested numerous soil samples from the Salamanca region of Spain, where *A. glutinosa* is a fairly common plant while (unlike the situation in west Scotland) *M. gale* is unrecorded. None the less there was again a good correspondence between the results, since fifteen of the eighteen soils that induced *A. glutinosa* to nodulate had the same effect on *M. gale*.

It occurred to C.R.B. that one of the possible explanations of these findings was that the *A. glutinosa* endophyte is capable of inducing nodulation in *M. gale* plants. He appreciated that this would be contrary to the findings of Fletcher (1953) in trials carried out in the laboratory of G.B. His experience (summarised by Bond, Fletcher & Ferguson, 1954) was that cross-inoculation of the above type, and also in the reverse direction, was inoperative; Becking (1970) reported to the same effect.

C.R.B. decided to re-investigate the matter. In 1966 and again in 1971, during periods of work in Glasgow, he set up *Alnus glutinosa* and *Myrica gale* seedlings, raised from surface-sterilised seed, in water culture in Crone's nitrogen-free culture solution, the beakers and the plastic covers having been initially sterilised. The seedlings were inoculated as required by the application to their roots of a suspension prepared by grinding appropriate nodule material in the proportion of 10 g or 5 g in 100 ml distilled water. About 0.25 ml of a suspension was applied to each plant. The nodules used in the preparation of the inocula were taken from stock plants growing in water culture in the glasshouse; in some instances the nodules were merely thoroughly washed in water before use, while in others they were shaken for 4 minutes in a solution of 0.2 % mercuric chloride in 0.5 % hydrochloric acid followed by repeated washing with water. All trials were on a reasonable scale, usually with twenty plants or more per treatment. Many uninoculated control plants were included in the trials and none became nodulated.

The results obtained with inoculated plants may be summarised as follows:

(1) *M. gale* plants formed nodules freely in response to the application to them of inocula prepared from *A. glutinosa* nodules, irrespective of whether or not the latter had received the mercuric chloride treatment. Practically every plant formed nodules, and they appeared within the normal time for *M. gale*, i.e. 3–4 weeks from inoculation. In due course the occurrence of fixation in the nodules was signified by the greening of the leaves. Very recently, in G.B.s laboratory, Mrs S. Mian obtained similar results in trials in which weaker inocula were used, namely 1.0 and 0.1 g nodule material suspended in 100 ml water.

(2) *A. glutinosa* plants did not nodulate in response to the application to their roots of an inoculum prepared from *M. gale* nodules taken from stock plants.

(3) An inoculum prepared from nodules which had been induced to form on *M. gale* plants by an *A. glutinosa* inoculum, as in (1) above, caused *A. glutinosa* plants to nodulate. All the test plants nodulated when the *M. gale* nodules had not been treated with mercuric chloride, but when that treatment had been applied only about 25 % of the plants nodulated, and the number of nodules formed per plant was small. Again there was evidence of nitrogen fixation proportionate to the number of nodules present.

The results under headings (1) and (3) came as a considerable surprise to both authors. C.R.B. was inclined to see them as confirming his original surmise, while G.B. reserved judgement, partly because to accept the results at their face value, and taking into account the results under (2) above, entailed the improbable conclusion that two types of nodules can occur on *M. gale* plants. At the time of the 1966 trials an examination under the light microscope of the infected cells in nodules formed in trials (1) above failed to reveal the presence of *Alnus*-type vesicles, though this was not felt to be conclusive evidence on the identity of the endophyte since the extent to which the host cell can control the form assumed by an endophyte is unknown. Further information on endophyte structure is given below.

In recent discussion between the authors, G.B. has suggested the following explanation for the inoculation results; C.R.B. subscribes to this, though he feels that further investigation is merited.

In west Scotland at the present time *A. glutinosa* is of common and *M. gale* of frequent occurrence. Both species were undoubtedly more abundant in the recent past, prior to the clearance and draining of land for agriculture. The habitats occupied by the two species are not sharply differentiated, except at lower levels of pH, and it is not unusual to find them growing in fairly close proximity. It is thus conceivable that samples of *A. glutinosa* nodules brought in from sites where *M. gale* plants are not present may, none the less, be superficially contaminated with infective bodies of the *M. gale* endophyte which have survived from earlier times (Rodriguez-Barrueco, 1968), so that when inocula prepared from such nodules are applied to *A. glutinosa* plants in the glasshouse the *M. gale* organism is introduced into the cultures at the same time. Thus the root systems of stock plants of *A. glutinosa* will carry the *M. gale* organism as a contaminant, and inocula prepared from their nodules will also include the latter organism. This is now thought to be the explanation of the results under (1) above. As stated, in certain of those trials *A. glutinosa* nodules which had been immersed for a period in mercuric chloride were used. Here it has to be noted that Quispel (1954) showed that it is impossible to kill all the surface contaminants of *A. glutinosa* nodules by immersion for reasonable periods in mercuric chloride or other sterilising fluids, since after such treatment a substantial proportion of the nodules, when plated out on a nutrient medium, showed evidence of the continuing presence of common saprophytic micro-organisms. It is thus likely that some of the infective bodies of the *M. gale* endophyte thought to be present on the nodules will still be viable, though there is no way of verifying this.

The results reported under (3) above are thought to have been due to the presence on the *M. gale* nodules of the *A. glutinosa* organism as a contaminant, derived from the inoculum which originally induced nodulation in the *M. gale* plants. The greater effectiveness of the mercuric chloride treatment here may be due to the *A. glutinosa* endophyte being more susceptible, or to the surface of the *M. gale* nodule being freer of fissures – certainly no lenticels are present.

The fact that Fletcher and also Becking found that an *A. glutinosa* inoculum failed to induce nodulation in *M. gale* plants appears to have been due to their inocula happening to be free of contamination with the *M. gale* endophyte. It must be presumed that their alder nodules were derived, directly or indirectly, from a field site at which the *M. gale* organism was not present. In respect of Professor Fletcher's trials no record of the site of procurement of nodules now exists.

In considering the results under (2) above it might be anticipated, in the light of remarks already made, that if field nodules of *A. glutinosa* from sites in west Scotland are frequently contaminated with the *M. gale* endophyte, then the opposite would also be true. However, although *M. gale* plants grow over a quite wide range of soil pH (see Boullard's findings on p. 452 of this volume), in west Scotland at the present time the best stands of the species are found on acid soils. Thus Bond (1951) recorded that soil pH at twenty-six sites for *M. gale* ranged from 3.7–4.8, with a mean value of 4.2. Trials by Ferguson & Bond (1953) showed that at pH 4.2 there was little activity on the part of the *A. glutinosa* endophyte in inducing nodulation in its host plant; possibly the endophyte cannot survive this degree of acidity. Rodriguez-Barrueco (1968) found little evidence of its presence at *M. gale* sites where the soil pH was slightly below 4.0. In G.B.s laboratory stock plants of *M. gale* have always been inoculated from nodules brought in from acid sites. Hence they are unlikely to have been contaminated with the *A. glutinosa* endophyte, and inocula prepared from their nodules will not be expected to induce nodulation in *A. glutinosa* test plants.

In 1972 C.R.B. in his own laboratory re-examined – now under the electron microscope – the appearance of the endophyte in nodules induced to form on *M. gale* plants by an alder inoculum. Structures resembling the vesicles of the alder endophyte, except that they were considerably smaller (width 1.5–2.5 μm, instead of 3–5 μm) were observed in quantity. At that time the available information, based entirely on light microscope studies, on the structure of the *M. gale* endophyte indicated that it formed non-compartmented club-shaped enlargements of the tips of the hyphae. However Gardner (Chapter 34) has reported that study under the electron microscope shows that these 'clubs' are compartmented much as in the alder endophyte, so that when cut transversely or obliquely they resemble alder vesicles, though on a smaller scale. It is highly likely that the structures which C.R.B. found in *M. gale* nodules apparently induced to form by an alder inoculum are similar to those described by Gardner, i.e. they are typical *M. gale* endophyte structures.

The explanation of the results of C.R.B.s tests on soil samples drawn from the Salamanca area has still to be considered. As stated, *M. gale* plants do not occur in the area at the present time. Unfortunately there are no records to show whether or not the species occurred there in earlier times. If it did not, it is necessary to conclude that at least the *M. gale* endophyte is not, in its distribution, restricted to areas where the host plant is growing or has grown.

The authors' experiences now recorded serve to exemplify the hazards of cross-inoculation trials between species growing in the same region.

References

Becking, J. H. (1970). Plant–endophyte symbiosis in non-leguminous plants. *Plant and Soil*, **32**, 611–54.

Bond, G. (1951). The fixation of nitrogen associated with the root nodules of *Myrica gale* L., with special reference to its pH relation and ecological significance. *Annals of Botany*, N.S. **15**, 447–59.

Bond, G., Fletcher, W. W. & Ferguson, T. P. (1954). The development and function of the root nodules of *Alnus*, *Myrica* and *Hippophaë*. *Plant and Soil*, **5**, 309–23.

Ferguson, T. P. & Bond, G. (1953). Observations on the formation and function of the root nodules of *Alnus glutinosa* (L.) Gaertn. *Annals of Botany*, N.S. **17**, 175–88.

Fletcher, W. W. (1953). A study of the root nodules of *Myrica gale* L. PhD Thesis, University of Glasgow.

Quispel, A. (1954). Symbiotic nitrogen-fixation in non-leguminous plants. I. Preliminary experiments on the root-nodule symbiosis of *Alnus glutinosa*. *Acta botanica Neerlandica*, **3**, 495–511.

Rodriguez-Barrueco, C. (1968). The occurrence of the root-nodule endophytes of *Alnus glutinosa* and *Myrica gale* in soils. *Journal of General Microbiology*, **52**, 189–94.

41. The possibility of extending the capacity for nitrogen fixation to other plant species

A summary of proceedings at an open session held during the Edinburgh meeting*

Chairman: Professor G. BOND (*United Kingdom*)

Introductory remarks by the chairman

This session was included in the programme because a feeling had been expressed within the organising committee that since the main task allocated to Section PP had been to study the basic processes involved in food production by plants and to suggest how this could be increased, it was desirable that Subsection PP-N should evince an interest in the topic indicated by the title.

In the past it has seemed that a root nodule symbiosis was the most satisfactory way of endowing a crop plant with a 'private' source of combined nitrogen. Bacterial nitrogen-fixing nodules are only encountered at the angiosperm level, and are of two types, tenanted respectively by *Rhizobium* and by an actinomycete. The latter endophyte appears to have been the more adaptable, since its variants live in symbiosis with host plants from seven families of varied taxonomic position, while *Rhizobium* has confined its attention almost exclusively to one family. However, the symbiotic affinities of the existing nodular actinomycetes are limited to their particular hosts, so that although the roots of most wild or crop plants in Scotland are in contact with infective bodies of the *Alnus* endophyte, no nodules form in response to them except in *Alnus*. The actinomycetes of *Alnus*-type nodules cannot at present be isolated into pure culture, and so are presumably unsuited for use in genetic experiments.

As to the reasons why relatively so very few angiosperms participate in nodule symbioses, rather discouragingly the most likely one seems to be that the initiation of such a symbiosis requires some most exceptional concomitance of circumstances and events which rarely happens in the

* Prepared by the chairman from a tape-recording of the proceedings.

natural course of things. The fact that the present nodule-forming species belong to what are probably very old families, and that monocotyledonous plants provide no confirmed instances of nodule symbiosis, can be regarded as favouring that view, though it should not be forgotten that grasses will form nodular structures in response to invasion by other agents, such as fungi or nematodes. It is conceivable that after some more millions of years of evolution, more plants will have entered into nodule symbiosis. The question is, can the process be speeded up?

Recent discoveries concerning rhizosphere fixation prompt the consideration of whether additional plant species can be induced to participate in this apparently looser type of symbiosis. Although it seems obvious that the rhizosphere system cannot be as efficient as a nodule symbiosis, there is now evidence of quite large gains of nitrogen by the higher plant partners.

Rather than attempting to bring further plants into one of the existing types of symbiosis, some workers are considering the possibility of imparting a faculty for fixation to higher plants themselves.

Invited paper by E. C. Cocking (*United Kingdom*)

New approaches which might eventually lead to the achievement of the objective indicated in the title of the session were described. The isolation of protoplasts from the organs of higher plants, e.g. leaves, roots and nodules, which can be secured by the use of cell wall degrading enzymes, now permits the higher plant system to be handled in the same way as micro-organisms. The removal of the barrier presented by the cell wall permits virus particles or bacteria to enter the protoplasts by a process of invagination of the plasma membrane, so that, for example, rhizobia can be taken up by tobacco protoplasts. When cultured on a suitable medium the protoplasts may regenerate a cell wall and undergo division, and in certain instances, e.g. in *Petunia*, they have gone on to form a whole plant again. Occasionally the protoplasts of different species fuse together, and if the fusion cell can be induced to undergo division and produce a whole plant, then hybridisation between sexually-incompatible plants becomes a possibility. In this way the Brookhaven group (**1**) have obtained hybrids between two species of tobacco which are comparable with those obtainable (in this particular instance) by sexual hybridisation, encouraging the belief that the same may eventually be achieved with much more disparate plants, e.g. legumes and cereals.

Although isolated protoplasts are considered by the speaker to have many advantages over tissue cultures in the present context (**2**), significant results have been obtained by Doy's group (**3**) with a particular callus culture which originally could not use galactose or lactose as a carbon source. However after treating the callus with a *gal* or a *lac* carrying transducing phage, the callus acquired the ability to utilise the corresponding sugar. Although the question remains whether these changes were of permanent nature and would persist in any plant that might be regenerated from the callus, the results have led to speculation on the possibility of transferring *nif* genes (**4**) from nitrogen-fixing bacteria to higher plant cells, and thus to confer on them an ability to fix nitrogen. It appears that this would involve the breaking down of a natural barrier whose existence is suggested by the fact that at present no eukaryote organism fixes nitrogen directly.

(**1**) Carlson, P. S., Smith, H. H. & Dearing, R. D. (1972). Parasexual interspecific plant hybridisation. *Proceedings of the National Academy of Sciences, USA*, **69**, 2292–4.

(**2**) Cocking, E. C. (1973). Plant cell modification: problems and perspectives. *Colloques internationaux CNRS* No. 212, 327–41.

(**3**) Doy, C. H., Gresshoff, P. M. & Rolfe, B. G. (1973). Biological and molecular evidence for the transgenosis of genes from bacteria to plant cells. *Proceedings of the National Academy of Sciences, USA*, **70**, 723–6.

(**4**) Streicher, S. L., Gurney, E. G. & Valentine, R. C. (1972). The nitrogen fixation genes. *Nature, London*, **239**, 495–9.

Open discussion

DÖBEREINER (*Brazil*) Though there is much that we do not yet understand about rhizosphere fixation, there is now evidence that under tropical conditions certain forage grasses and cereals benefit substantially by acquiring nitrogen fixed in the rhizosphere. The study of how this association could be extended to other plants and to temperate regions might lead to an increased food supply more quickly than the approaches suggested by Professor Cocking seem likely to do. Obviously it ought to be easier to transfer a character from one grass to another grass than from a legume to a grass.

NUTMAN (*United Kingdom*) The work described by Professor Cocking is of much academic interest, but seems unlikely to make any contribution in the foreseeable future towards solving the problem of the world shortage of protein. This is an urgent problem, and its solution cannot await the development of entirely new technologies, whose prospects of success, though much publicised, are often greatly exaggerated, leading to unwarranted optimism in various quarters. It will be far better to concentrate resources on improving the efficiency with which we utilise the nitrogen-fixing systems which Nature has provided. There is much that can be done to increase the area devoted to leguminous crops and to improve methods of cultivation, so that improved yields will result. The protein shortage would evaporate if a sustained effort were to be made in these directions.

QUISPEL (*The Netherlands*) Although the rhizosphere system–which it certainly seems in order to class as a type of symbiosis – cannot be as efficient as the highly-organised root nodule symbiosis, it does appear to have the merit of involving less specificity. Another way of extending fixation to further crop plants might be to take advantage of the ability of the fungus *Endogone* to set up endotrophic mycorrhiza with a very wide range of plants, and to attempt to introduce a faculty for fixation into this organism.

WERNER (*Germany*) Although whole plants have been in some instances obtained from protoplast or tissue cultures, this has not yet been achieved from important crop plants such as most cereals or soya bean. Also the experience with animal hybrid cells is that they are usually genetically unstable in further culture, so that sooner or later the incorporated chromosomes of one or the other species are eliminated.

COCKING, in replying to the first point, expressed his confidence that regeneration would be achieved in the near future from legume single-cell cultures, while on the second he stated that provided regeneration of the whole plant from hybrid cells could be obtained quickly, the chances of breakdown would be reduced.

GRAHAM (*Colombia*) It appears at the moment that most of the plants participating in nitrogen-fixing rhizosphere associations are of the C4 type. Since some plant genera include both C3 and C4 species, it might be profitable to attempt, by means of the protoplast technique, to convert the C3 species into the C4 type, to permit rhizospheral participation.

In reply, COCKING agreed that this was a theoretical possibility, but pointed out that at present there were difficulties in culturing grass and cereal protoplasts.

GIBSON (*Australia*) The evidence is that a nitrogen-fixing system needs anaerobic conditions in order to function, and it is uncertain whether these would be provided in a cell of some unusual host species into which bacteria with nitrogen-fixing properties had been introduced by suitable manipulation.

COCKING agreed that at present it was unknown whether the necessary localised low level of oxygen could arise in such a cell, as it does presumably in a normal root nodule cell. He and his colleague (Dr Davey) are testing the ability of protoplast systems to survive anaerobiosis, to find whether survival is long enough for the activity of introduced bacteria to be tested.

HARDY (*USA*) Industrial fixation of nitrogen is an expensive process. In ten years time fifty million tons of such nitrogen is likely to be required in order to keep the 'green revolution' going. In view of this prospect it is right that a certain proportion of available research funds should be devoted to exploratory work of the kind described by Professor Cocking, but although the single-cell system is a desirable one, the fact remains that to date nitrogen fixation has not been demonstrated in protoplasts containing introduced rhizobia, whereas it has in callus systems similarly infected.

Replying, COCKING agreed that at present this system could not compete with the callus culture in terms of actual achievement, but he held to the hope that eventually a synchronous infection of protoplast cultures would be secured, contrasting favourably with the proportion of cells that become infected in a callus culture.

VALENTINE (*USA*) Exploratory work on nitrogen fixation is not necessarily expensive. All the work in his own laboratory which led to the coining of the term *nif* genes had been done with the help of just two graduate students.

POSTGATE (*United Kingdom*) The introduction of methods for the culture of single cells of higher plants is a very important advance. Progress is also rapid on the microbiological side, and there is a prospect of preparing a phage carrying the *nif* genes. It would be unimaginative not to try to transfer these genes to plants. The prospect of success is probably remote, but provided that false hopes are not raised by optimistic announcements it is only proper that such exploratory work should go on, in parallel with traditional research seeking to make better use of existing nitrogen-fixing systems.

CHAIRMAN It is certain that many members of the audience, like the chairman himself, will have been glad of this opportunity to acquaint themselves of the progress, or at any rate the prospects, in this field of endeavour, and thanks are expressed to Professor Cocking and to those who took part in the discussion.

Index

Index

Index